AF181961

IUTAM SYMPOSIUM ON GEOMETRY AND STATISTICS OF TURBULENCE

FLUID MECHANICS AND ITS APPLICATIONS
Volume 59

Series Editor: R. MOREAU
MADYLAM
Ecole Nationale Supérieure d'Hydraulique de Grenoble
Boîte Postale 95
38402 Saint Martin d'Hères Cedex, France

Aims and Scope of the Series

The purpose of this series is to focus on subjects in which fluid mechanics plays a fundamental role.

As well as the more traditional applications of aeronautics, hydraulics, heat and mass transfer etc., books will be published dealing with topics which are currently in a state of rapid development, such as turbulence, suspensions and multiphase fluids, super and hypersonic flows and numerical modelling techniques.

It is a widely held view that it is the interdisciplinary subjects that will receive intense scientific attention, bringing them to the forefront of technological advancement. Fluids have the ability to transport matter and its properties as well as transmit force, therefore fluid mechanics is a subject that is particulary open to cross fertilisation with other sciences and disciplines of engineering. The subject of fluid mechanics will be highly relevant in domains such as chemical, metallurgical, biological and ecological engineering. This series is particularly open to such new multidisciplinary domains.

The median level of presentation is the first year graduate student. Some texts are monographs defining the current state of a field; others are accessible to final year undergraduates; but essentially the emphasis is on readability and clarity.

For a list of related mechanics titles, see final pages.

IUTAM Symposium on
Geometry and Statistics of Turbulence

Proceedings of the IUTAM Symposium
held at the Shonan International Village Center,
Hayama (Kanagawa-ken), Japan,
November 1–5, 1999

Edited by

T. KAMBE

University of Tokyo, Japan

T. NAKANO

Chuo University, Japan

and

T. MIYAUCHI

Tokyo Institute of Technology, Japan

SPRINGER-SCIENCE+BUSINESS MEDIA, B.V.

A C.I.P. Catalogue record for this book is available from the Library of Congress.

ISBN 978-90-481-5614-6 **ISBN 978-94-015-9638-1 (eBook)**
DOI 10.1007/978-94-015-9638-1

Printed on acid-free paper

All Rights Reserved
© 2001 Springer Science+Business Media Dordrecht
Originally published by Kluwer Academic Publishers in 2001
Softcover reprint of the hardcover 1st edition 2001

No part of the material protected by this copyright notice may be reproduced or utilized in any form or by any means, electronic or mechanical, including photocopying, recording or by any information storage and retrieval system, without written permission from the copyright owner.

IUTAM Symposium *Hayama, Kanagawa, Japan November 1–5, 1999*

List:

1. H. K. Moffatt
2. T. Tatsumi
3. I. Imai
4. R. Rubinstein
5. J. Jimenez
6. K.R. Sreenivasan
7. T. Kambe
8. C. Doering
9. U. Frisch
10. H. Frisch
11. K.K. Nomura
12. T. Matsumoto
13. S. Katoh
14. Y. Okada
15. B. Shraiman
16. S. Kida
17. V. Yakhot
18. P. Constantin
19. R. Pasmanter
20. T. Miyauchi
21. P.-H.Chavanis
22. J.-P. Laval
23. K. Angele
24. K. Horiuti
25. T. Itano
26. K. Yoshida
27. Y. Tsuji
28. P. Orlandi
29. Y. Miyake
30. J. Peinke
31. R. Piva
32. J.C. Wells
33. A. Noullez
34. M. Sano
35. T. Nakano
36. Y. Kaneda
37. N. Tokugawa
38. G. Kawahara
39. Y. Kitano
40. S. Tsujimura
41. R. Antonia
42. T. Gotoh
43. H. Hanazaki
44. T. Miyazaki
45. O. Sano
46. S. Toh
47. K. Yamamoto
48. M. Takaoka
49. N. Takahashi
50. Y. Fukunishi
51. K. Suzuki
52. Y. Hattori
53. D. Fukayama
54. Y. Fukumoto
55. M. Hino
56. S. Goto
57. N. Hatakeyama
58. M. Umeki
59. A. Ishizawa
60. T. Ishihara
61. K. Ohkitani
62. F. Hamba
63. T. Hasegawa
64. Y. Suzuki

Scientific Committee

T. Kambe (*Chairman*)
V.I. Arnold, *Russia*
R. Benzi, *Italy*
P. Constantin, *USA*
U. Frisch, *France*
J. Jiménez, *Spain*
H.K. Moffatt, *UK*
T. Tatsumi, *Japan*

Advisory (Local) Committee

I. Imai, *Tokyo*
H. Ohashi, *Tokyo*
T. Tatsumi, *Kyoto*

Local Organizing Committee

T. Kambe, *Tokyo* (*Chairman*)
Y. Fukumoto, *Fukuoka*
F. Hamba, *Tokyo*
Y. Kaneda, *Nagoya*
T. Kawamura, *Tokyo*
T. Miyauchi, *Tokyo*
T. Nakano, *Tokyo*
M. Umeki, *Tokyo*
K. Yamamoto, *Tokyo*
T. Gotoh, *Nagoya*
K. Ishii, *Nagoya*
N. Kasasgi, *Tokyo*
S. Kida, *Toki*
T. Miyazaki, *Tokyo*
K. Ohkitani, *Kyoto*
M. Yamada, *Tokyo*
A. Yoshizawa *Tokyo*

Sponsors of Symposium

This Symposium was held under the sponsorship of

International Union of Theoretical and Applied Mechanics (IUTAM)

and the support of

Science Council of Japan.

The financial support has been made from the following foundations and corporations:

Inoue Foundation for Science
The Kajima Foundation
Hewlett-Packard Japan, Ltd.
NEC Corporation

Special registration:

Zexel Corporation
YKK Corporation
Fuji Research Institute Corporation
Sanyo Electric Co., Ltd.
Daikin Industries, Ltd.
Seika Corporation

Contents

PREFACE

This volume contains the papers presented at the IUTAM Symposium on *Geometry and Statistics of Turbulence*, held in November 1999, at the *Shonan* International Village Center, Hayama (Kanagawa-ken), Japan.

The Symposium was proposed in 1996, aiming at organizing concentrated discussions on current understanding of fluid turbulence with emphasis on the statistics and the underlying geometric structures. The decision of the General Assembly of *International Union of Theoretical and Applied Mechanics* (IUTAM) to accept the proposal was greeted with enthusiasm. Turbulence is often characterized as having the properties of mixing, intermittency, non-Gaussian statistics, and so on. Interest is growing recently in how these properties are related to formation and evolution of structures. Note that the *intermittency* is meant for passive scalars as well as for turbulence velocity or rate of dissipation.

There were eighty-eight participants in the Symposium. They came from thirteen countries, and fifty-seven papers were presented. The presentations comprised a wide variety of fundamental subjects of mathematics, statistical analyses, physical models as well as engineering applications. Among the subjects discussed are (*a*) Degree of self-similarity in cascade, (*b*) Fine-scale structures and degree of Markovian property in turbulence, (*c*) Dynamics of vorticity and rates of strain, (*d*) Statistics associated with vortex structures, (*e*) Topology, structures and statistics of passive scalar advection, (*f*) Partial differential equations governing PDFs of velocity increments, (*g*) Thermal turbulences, (*h*) Channel and pipe flow turbulences, and others.

We would like to thank all the contributors for making this Proceedings a landmark of turbulence research open to the coming next century. The papers collected in this volume were reviewed by all members of the Local Organizing Committee.

The assistance by the members of the *Scientific committee* is gratefully acknowledged. Sincere thanks are also extended to the *Advisory committee* and also to all the members of *Local organizing committee* for their efforts making the Symposium very successful. We are particularly grateful to Prof. F. Hamba as a Treasurer for this Symposium.

Financial support to the Symposium was provided by the IUTAM. In addition, generous supports are gratefully acknowledged for the two foundations: *Inoue Foundation for Science* and *The Kajima Foundation* and for the eight corporations: *Hewlett-Packard Japan, NEC Corporation, Zexel Corporation, YKK Corporation, Fuji Research Institute Corporation, Sanyo Electric Corporation, Daikin Industries,* and *Seika Corporation* .

Finally the cooperation of the Kluwer academic publishers for publication of the present Proceedings is very much appreciated as well as the financial assistance.

July 2000
Editors

Tsutomu Kambe
Tohru Nakano
Toshio Miyauchi

A

General and Mathematical

Opening Address

Mathematical Physics of Turbulence

Tomomasa TATSUMI
International Institute for Advanced Studies
9-3 Kizugawadai, Kizu-cho, Kyoto 619-0225, Japan

It is my great honour and pleasure to give an official Opening Address in the "IUTAM Symposium on Geometry and Statistics of Turbulence" on behalf of the President of IUTAM, Professor Werner Schiehlen.

1. IUTAM

The Opening Address is intended to give the audience the scope and activity of IUTAM, in particular on the significance of its Symposia including the present Symposium. In this context, however, I am not talking about the the history and statistics of IUTAM in detail but would like to draw your attension to its very origin.

IUTAM (International Union of Theoretical and Applied Mechanics) was founded in 1947, shortly after the World War II, as an international organization under ICSU (International Council of Scientific Unions), so it has a long history of more than half a century.

1.1. ICAM and ICTAM

One of the main activities of IUTAM is to organize ICTAM (International Congress of Theoretical and Applied Mechanics) every four years at various cities in the world. ICTAM has an eminent prehistory before the War under the name of ICAM (International Congress of Applied Mechanics), which dates back to 1922 when an international conference on Hydro- and Aeromechanics was held at Innsbruck, Austria, under the joint chairmanship of Theodore von Karman, then the director of Aerodynamical Institute of Aachen Polytechnic, and Tullio Levi-Civita, an Italian mathematician in Rome (after Battimelli (1988)).

T. Kambe et al. (eds.), IUTAM Symposium on Geometry and Statistics of Turbulence, 3–12.
© 2001 *Kluwer Academic Publishers.*

Karman's idea of having such an international meeting was motivated from his recognition of a large discrepancy between the rapid developments in the field of hydro- and aeromechanics on one hand, and only a limited space devoted to such problems at scientific meetings on the other hand. He proposed to break the dependence of hydro- and aeromechanics from their mother disciplines like mathematics, physics and engineering.

His proposal was enthusiastically supported by Levi-Civita and they succeeded in organizing the above-mentioned conference. In spite of the wake of the hostility between the nations after the World War I, thirty three scientists gathered to the conference from Germany, Austria, Holland, Scandinavia and Italy, including big names such as Ludwig Prandtl, Karman's former teacher, and Werner Heisenberg, who was sent by Arnold Sommerfeld to report on his research on turbulence.

In this conference Karman found a new partner Jan Burgers, a young Dutch scientist who was at Innsbruck, and they proceeded to organize an international congress covering the wider subjects of Applied Mechanics in 1924 at Delft, the Netherlands. This Congress was also very successful with 200 participants and 76 papers read covering the whole field of applied mechanics.

This is indeed the beginning of the series of ICAM and ICTAM which have been held every four years except for the interruption due to the World War II. The last Congress was the 19th Congress held in 1996 at Kyoto, Japan and the next will be the 20th Congress to be held in 2000 at Chicago, USA.

1.2. IUTAM Symposia

Another main activity of IUTAM is to sponsor about ten IUTAM Symposia per year and also a few Summer and Winter Schools of intructional nature. IUTAM Symposia are selected from many proposals made by the members of IUTAM and the participants are limited to those invited by the Symposia. The subjects of the Symposia are all concerned with the current topics of scientific importance in the fields of IUTAM and their Proceedings are to be published in principle by Kluwer Academic Publishers.

Our Symposium on "Geometry and Statistics of Turbulence" is indeed one of such IUTAM Symposia which has been supported and adopted by the General Assembly of IUTAM.

2. Fluid Mechanics and Turbulence

During about eighty years since the time of Innsbruck Conference, enormous developments have been achieved in the fields of Hydro- and Aeromechanics, or Fluid Mechanics in modern usage.

2.1. Fluid Mechanics

It should be noted that such developments have been built up on the basis of the *real fluid* which is associated with the compressibility and viscosity and governed by the Navier-Stokes equation of motion. Thanks to the ingenious works on the *shock wave* by Sir Geoffrey Taylor (1910) and the *boundary layer* by Ludwig Prandtl (1904), all singularities accompanied with the *perfect fluid* without viscosity have been regularized and given the physical reality.

Now, it is generally believed that all fluid motions, including highly complex flows at large Reynolds numbers, can be described as the solutions of the Navier-Stokes equation, and this belief has been well supported so far by the results of experiments and numerical simulations. In this sense, Karman's idea of dealing with hydro- and aeromechanics as an independent research sector from their mother disciplines seems to be well justified.

2.2. Turbulence

Probably, only exceptional case to such a belief may be *turbulence*. Although our knowledge on turbulent flows has been largely progressed in recent years, there seems to be still lacking the consistent mathematical physics of turbulence. If we interpret "Geometry and Statistics" of turbulence in the title of this Symposium as "Mathematics and Physics", our purpose is nothing but to build up such Mathematical Physics of Turbulence.

It is usually said that a substantial difficulty of the theory of turbulence lies in its lack of *scale-separation* between the large- and small-scale components of turbulence, which makes it impractical to employ the so-called *quasi-particle* description. This is indeed the cause of very limited success of the classical notion of the *mixing length*, which was introduced to the turbulence research taking analogy from the *mean-free-path* of the gas theory.

The genuine *scale-separation* of turbulence has been achieved by Kolmogorov (1941), who assumed the *independence* of small-scale components of turbulence from large-scale varieties of turbulent flows at large Reynolds numbers and dealt with the equilibrium state of the small-scale components. This idea was developed by himself into the theory of *locally isotropic turbulence,* resulting in the celebrated laws of the 2/3 power variance of the velocity difference at two points or equivalently the -5/3 power energy spectrum, both coined by the name of Kolmogorov.

Probably, the *independence* of the small-scale components of turbulence from the large-scale structures can be formulated more adequately in the statistical framework of turbulence. Now we shall attempt to explore this possibility in the velocity distributions of turbulence.

3. Statistics of Turbulence

It is well known that the most complete statistical description of turbulence is provided by the probability distribution functional of its velocity field $\boldsymbol{u}(\boldsymbol{x}, t)$ at all values of the coordinate $\boldsymbol{x}$ and the time t. Actually all statistical informations of turbulence can be derived from this velocity distribution functional.

The equation governing this distribution functional was given by Hopf (1952) in terms of its Fourier transform or the characteristic functional. Thus, the Hopf equation provides us with the fundamental equation of the statistical theory of turbulence. Since, however, there is no mathematical method available for solving this functional equation generally, we have to be satisfied by two particular solutions given in this paper, that is the normal (Gaussian) distribution functionals for infinite and very small values of the Reynolds number R..

3.1. Velocity Distributions

A practical method for dealing with this functional equation is to expand it into an infinite sequence of equations for the distribution functions of the velocities at the discrete points, $\boldsymbol{x}_1$, $\boldsymbol{x}_2$, $\boldsymbol{x}_3$, ... In this way, Lundgren (1967) and Monin (1967) derived independently an infinite system of equations for the n-point velocity distributions, n being positive integers. Thus, this infinite set of equations obtained by Lundgren and Monin also constitute the fundamental equations of turbulence.

There arises, however, a serious diffuculty concerning this set of equations since the equation for the n-th order distribution always involves the $(n+1)$-th order distribution, so that any finite subset of equations is not closed. Such unclosedness of the equations is common to all successive approximations in turbulence and we have to introduce a *closure* assumption for connecting the highest-order distribution with those of lower-orders in order to make the equations solvable.

So far a number of closure assumptions have been proposed in relation with the statistical moment description of turbulence, but no such work seems to have been made for the velocity distribution formalism. Here, a closure assumption for the velocity distributions will be proposed using an idea which is similar to Kolmogorov's scale-separation principle mentioned in § 2.2.

3.2. One and Two-point Velocity Distributions

Denoting the velocities at two points $\boldsymbol{x}_1$ and $\boldsymbol{x}_2$ and time t by $\boldsymbol{u}(\boldsymbol{x}_1, t)$ and $\boldsymbol{u}(\boldsymbol{x}_2, t)$ respectively, we define the one and two-point velocity distributions as follows:

$$f(\boldsymbol{u}_1, \boldsymbol{x}_1, t) = \langle \delta(\boldsymbol{u}(\boldsymbol{x}_1, t) - \boldsymbol{u}_1) \rangle \qquad (1)$$

$$f^{(2)}(\boldsymbol{u}_1, \boldsymbol{u}_2; \boldsymbol{x}_1, \boldsymbol{x}_2; t) = \langle \delta(\boldsymbol{u}(\boldsymbol{x}_1, t) - \boldsymbol{u}_1)\, \delta(\boldsymbol{u}(\boldsymbol{x}_2, t) - \boldsymbol{u}_2) \rangle \qquad (2)$$

where $\boldsymbol{u}_i$ (i = 1, 2) denote the stochastic variables corresponding to $\boldsymbol{u}(\boldsymbol{x}_i, t)$, δ the three-dimensional delta function and $\langle\ \rangle$ the mean value with respect to the initial distribution.

For *homogeneous* turbulence, these distributions are written as,

$$f(\boldsymbol{u}_1, \boldsymbol{x}_1, t) = f(\boldsymbol{u}_1, t) \tag{3}$$

$$f^{(2)}(\boldsymbol{u}_1, \boldsymbol{u}_2; \boldsymbol{x}_1, \boldsymbol{x}_2; t) = f^{(2)}(\boldsymbol{u}_1, \boldsymbol{u}_2; \boldsymbol{r}, t) \tag{4}$$

with $\boldsymbol{r} = \boldsymbol{x}_2 - \boldsymbol{x}_1$ and required to satisfy the condition of zero-mean velocity,

$$\int \boldsymbol{u}_1 f(\boldsymbol{u}_1, t)\mathrm{d}\,\boldsymbol{u}_1 = \int \boldsymbol{u}_2 f(\boldsymbol{u}_2, t)\mathrm{d}\,\boldsymbol{u}_2$$
$$= \int \boldsymbol{u}_1 f^{(2)}(\boldsymbol{u}_1, \boldsymbol{u}_2; \boldsymbol{r}, t)\mathrm{d}\,\boldsymbol{u}_1 = \int \boldsymbol{u}_2 f^{(2)}(\boldsymbol{u}_1, \boldsymbol{u}_2; \boldsymbol{r}, t)\mathrm{d}\,\boldsymbol{u}_2 = 0 \tag{5}$$

By the definition, these distributions must satisfy the reduction conditions,

$$\int f(\boldsymbol{u}_1, t)\mathrm{d}\,\boldsymbol{u}_1 = \int f(\boldsymbol{u}_2, t)\mathrm{d}\,\boldsymbol{u}_2 = 1 \tag{6}$$

$$\int f^{(2)}(\boldsymbol{u}_1, \boldsymbol{u}_2; \boldsymbol{r}, t)\mathrm{d}\,\boldsymbol{u}_1 = f(\boldsymbol{u}_2, t), \quad \int f^{(2)}(\boldsymbol{u}_1, \boldsymbol{u}_2; \boldsymbol{r}, t)\mathrm{d}\,\boldsymbol{u}_2 = f(\boldsymbol{u}_1, t) \tag{7}$$

In general there is no relationship between f and $f^{(2)}$ except for the reduction condition (7), but they are related with each other in the following two limits.

Distant limit: $\lim_{r\to\infty} f^{(2)}(\boldsymbol{u}_1, \boldsymbol{u}_2; \boldsymbol{r}, t) = f(\boldsymbol{u}_1, t)\, f(\boldsymbol{u}_2, t)$ (8)

Coincidence limit: $\lim_{r\to 0} f^{(2)}(\boldsymbol{u}_1, \boldsymbol{u}_2; \boldsymbol{r}, t) = f(\boldsymbol{u}_1, t)\, \delta\,(\boldsymbol{u}_2 - \boldsymbol{u}_1)$ (9)

Eq.(8) shows the independence of $\boldsymbol{u}_1$ and $\boldsymbol{u}_2$ at far distance, while eq.(9) requires the agreement of $\boldsymbol{u}_1$ and $\boldsymbol{u}_2$ for vanishing distance due to the continuity of the fluid.

3.3. Equation for One-Point Velocity Distribution

The equation governing the one-point velocity distribution f is given by Lundgren (1967) and Monin (1967) in the following general form:

$$[\partial/\partial t + (\boldsymbol{u}_1 \cdot \partial/\partial \boldsymbol{x}_1)]\, f(\boldsymbol{u}_1, \boldsymbol{x}_1, t)$$
$$= (4\pi)^{-1}(\partial/\partial \boldsymbol{u}_1 \cdot \partial/\partial \boldsymbol{x}_1) \iint |\boldsymbol{x}_2 - \boldsymbol{x}_1|^{-1} \times$$
$$\times (\boldsymbol{u}_2 \cdot \partial/\partial \boldsymbol{x}_2)^2 f^{(2)}(\boldsymbol{u}_1, \boldsymbol{u}_2; \boldsymbol{x}_1, \boldsymbol{x}_2; t)\, d\boldsymbol{x}_2 d\boldsymbol{u}_2$$
$$- \lim_{\boldsymbol{x}_2 \to \boldsymbol{x}_1} \nu\, (\partial/\partial \boldsymbol{x}_2)^2 \int (\boldsymbol{u}_2 \cdot \partial/\partial \boldsymbol{u}_1) f^{(2)}(\boldsymbol{u}_1, \boldsymbol{u}_2; \boldsymbol{x}_1, \boldsymbol{x}_2; t)\, d\boldsymbol{u}_2 \tag{10}$$

where the presence of $f^{(2)}$ on the right-hand side makes the equation unclosed.

In *homogeneous* turbulence, for which (3) and (4) hold, the transfer term of eq.(10) vanishes, and the pressure term which is written as

$$T_p = (4\pi)^{-1}(\partial/\partial \boldsymbol{u}_1 \cdot \partial/\partial \boldsymbol{x}_1) \iint |\boldsymbol{r}|^{-1} (\boldsymbol{u}_2 \cdot \partial/\partial \boldsymbol{r})^2 f^{(2)}(\boldsymbol{u}_1, \boldsymbol{u}_2; \boldsymbol{r}, t)\, d\boldsymbol{r}\, d\boldsymbol{u}_2$$

also vanishes since the double integral is independent of $\boldsymbol{x}_1$. Thus, eq.(10) becomes

$$\partial f(\boldsymbol{u}_1, t)/\partial t = -\lim_{r\to 0} \nu\, (\partial/\partial \boldsymbol{r})^2 \int (\boldsymbol{u}_2 \cdot \partial/\partial \boldsymbol{u}_1) f^{(2)}(\boldsymbol{u}_1, \boldsymbol{u}_2; \boldsymbol{r}, t)\, d\boldsymbol{u}_2 \tag{11}$$

Although eq.(11) is much simpler than eq.(10), it is still unclosed so that we need a closure assumption for correlating $f^{(2)}$ with f.

3.4. Closure Assumptions

The closure assumptions proposed so far are more or less based upon the distant relation (8) assumed for finite distances $\boldsymbol{r} < \infty$.

$$f^{(2)}(\boldsymbol{u}_1, \boldsymbol{u}_2; \boldsymbol{r}, t) = f(\boldsymbol{u}_1, t)\, f(\boldsymbol{u}_2, t) \tag{12}$$

Since this relation is satisfied by the normal distribution of f, the closure of this kind is usually called the *quasi-normal* approximation. It may be obvious that such an approximations is valid for relatively large values of $\boldsymbol{r}$ but become less satisfactory for smaller $\boldsymbol{r}$ and eventually divergent at the vanishing $\boldsymbol{r}$.

4. Cross-Independence

Now let us consider a new closure assumption starting from the coincidence limit (9), which shows the independence of the velocity difference $\Delta \boldsymbol{u} = \boldsymbol{u}_2 - \boldsymbol{u}_1$ from the velocity $\boldsymbol{u}_1$ in the limit of $\boldsymbol{r} = 0$. If we introduce the *sum* and the *difference* of the velocities $\boldsymbol{u}_1$ and $\boldsymbol{u}_2$, that is,

$$\boldsymbol{u}_+ = \tfrac{1}{2}(\boldsymbol{u}_1 + \boldsymbol{u}_2), \qquad \boldsymbol{u}_- = \tfrac{1}{2}(\boldsymbol{u}_2 - \boldsymbol{u}_1) \tag{13}$$

$\frac{1}{2}$ inserted for convenience, such an assumption amounts to require the independence of the cross-velocities $\boldsymbol{u}_+$ and $\boldsymbol{u}_-$.

4.1. Cross-Velocity Distributions

Just like the velocity distributions of $\boldsymbol{u}_1$ and $\boldsymbol{u}_2$ have been defined by (1)~(4), the distributions of the cross-velocities $\boldsymbol{u}_+$ and $\boldsymbol{u}_-$ are defined as follows:

$$g_+(\boldsymbol{u}_+, \boldsymbol{r}, t) = \langle \delta(\tfrac{1}{2}\{\boldsymbol{u}(\boldsymbol{x}_1, t) + \boldsymbol{u}(\boldsymbol{x}_2, t)\} - \boldsymbol{u}_+)\rangle \tag{14}$$

$$g_-(\boldsymbol{u}_-, \boldsymbol{r}, t) = \langle \delta(\tfrac{1}{2}\{\boldsymbol{u}(\boldsymbol{x}_2, t) - \boldsymbol{u}(\boldsymbol{x}_1, t)\} - \boldsymbol{u}_-)\rangle \tag{15}$$

$$g^{(2)}(\boldsymbol{u}_+, \boldsymbol{u}_-; \boldsymbol{r}, t) = \langle \delta(\tfrac{1}{2}\{\boldsymbol{u}(\boldsymbol{x}_1, t) + \boldsymbol{u}(\boldsymbol{x}_2, t)\} - \boldsymbol{u}_+) \times \times \delta(\tfrac{1}{2}\{\boldsymbol{u}(\boldsymbol{x}_2, t) - \boldsymbol{u}(\boldsymbol{x}_1, t)\} - \boldsymbol{u}_-)\rangle \tag{16}$$

These distributions are also required to satisfy the condition of zero-mean velocity,

$$\int \boldsymbol{u}_\pm g_\pm(\boldsymbol{u}_\pm, \boldsymbol{r}, t)\, d\boldsymbol{u}_\pm = 0 \qquad \text{for} \qquad \boldsymbol{u}_- \perp \boldsymbol{r} \tag{17}$$

and the reduction conditions,

$$\int g_\pm(\boldsymbol{u}_\pm, \boldsymbol{r}, t)\, d\boldsymbol{u}_\pm = 1 \tag{18}$$

$$\int g^{(2)}(\boldsymbol{u}_+, \boldsymbol{u}_-; \boldsymbol{r}, t)\, d\boldsymbol{u}_+ = g_-(\boldsymbol{u}_-, \boldsymbol{r}, t), \quad \int g^{(2)}(\boldsymbol{u}_+, \boldsymbol{u}_-; \boldsymbol{r}, t)\, d\boldsymbol{u}_- = g_+(\boldsymbol{u}_+, \boldsymbol{r}, t) \tag{19}$$

Since $g^{(2)}$ is another expression of $f^{(2)}$, these distributions must be connected with each other by

$$f^{(2)}(\boldsymbol{u}_1, \boldsymbol{u}_2; \boldsymbol{r}, t)\, d\boldsymbol{u}_1 d\boldsymbol{u}_2 = g^{(2)}(\boldsymbol{u}_+, \boldsymbol{u}_-; \boldsymbol{r}, t)\, d\boldsymbol{u}_+ d\boldsymbol{u}_- \tag{20}$$

so that,

$$f^{(2)}(\boldsymbol{u}_1, \boldsymbol{u}_2; \boldsymbol{r}, t) = J^{-3} g^{(2)}(\boldsymbol{u}_+, \boldsymbol{u}_-; \boldsymbol{r}, t) = 2^{-3} g^{(2)}(\boldsymbol{u}_+, \boldsymbol{u}_-; \boldsymbol{r}, t) \tag{21}$$

where $J = \partial((\boldsymbol{u}_1)_\alpha, (\boldsymbol{u}_2)_\alpha)/\partial((\boldsymbol{u}_+)_\alpha, (\boldsymbol{u}_-)_\alpha) = 2$ denotes the second-order Jacobian and α an arbitrary component.

4.2. Cross Independence

The independence of the cross-velocities $\boldsymbol{u}_+$ and $\boldsymbol{u}_-$ considered in the beginning of section can be expressed as

$$g^{(2)}(u_+, u_-, r, t) = g_+(u_+, r, t)\, g_-(u_-, r, t) \qquad (22)$$

which may be called the *cross-independence* in contrast to the ordinary independence (8) of the two-point velocities u_1 and u_2.

It can be confirmed that the relation (22) actually satisfies the coincidence condition (9) in the limit of $r \to 0$. It follows from (14), (15) that

$$\lim_{r\to 0} g_+(u_+, r, t)du_+ = \lim_{r\to 0} \langle \delta(\tfrac{1}{2}\{u(x_1, t) + u(x_2, t)\} - u_+)\rangle\, du_+$$
$$= \langle \delta(u(x_1, t) - u_1)\rangle\, du_1 = f(u_1, t)\, du_1$$
$$\lim_{r\to 0} g_-(u_-, r, t)du_- = \lim_{r\to 0} \langle \delta(\tfrac{1}{2}\{u(x_2, t) - u(x_1, t)\} - \tfrac{1}{2}(u_2 - u_1))\rangle\, du_-$$
$$= \delta(\tfrac{1}{2}(u_2 - u_1))du_- = \delta(\tfrac{1}{2}(u_2 - u_1))d(\tfrac{1}{2}u_2) = \delta(u_2 - u_1)du_2$$

Then, substituting these results to (20) and (22), we find that

$$\lim_{r\to 0} f^{(2)}(u_1, u_2; r, t)du_1 du_2 = \lim_{r\to 0} g^{(2)}(u_+, u_-; r, t)du_+ du_-$$
$$= \lim_{r\to 0} g_+(u_+, r, t)g_-(u_-, r, t)du_+ du_- = f(u_1, t)\,\delta(u_2 - u_1)du_1 du_2$$

so that

$$\lim_{r\to 0} f^{(2)}(u_1, u_2; r, t) = f(u_1, t)\,\delta(u_2 - u_1)$$

which is identical to the condition (9).

This indicates that the cross-independence relation (22) is satisfied with good accuracy for small values of r. Concerning the real extent of its validity, we have to learn from experimental results as shown in the next subsection.

4.2. Kolmogorov's Local Equilibrium Hypothesis

Kolmogorov's theory of turbulence is based on the fundamental assumption of the *independence* of small-scale components of turbulence from its large-scale structures. Such an assumption is essentially similar to the *cross-independence* assumption (22) employed here. The similarity is strengthened by the fact that the small-scale components are represented commonly by the velocity difference $\Delta u = u_2 - u_1 = 2u_-$.

There exists a minor difference in the theories since Kolmogorov's theory assumes temoral steadiness of the local equilibrium while no such restriction is imposed here. Actually we deal with the solutions changing self-similarly in time, but even in this case Kolmogorov's similarity is satisfied between the contemporary values of the statistical variables.

Kolmogorov's scaling laws have been examined experimentally in several aspects by Sreenivasan (1998) and others. Figure 1 shows the conditional mean square of the longitudinal component Δu_r of the velocity difference at the distance $r = |r|$, for the fixed values u_0 of the longitudinal velocity at the middle point. The data was taken from the grid turbulence with the Reynolds number $R_\lambda = 60$, λ being the Taylor microscale.

The experimental points measured at fixed distances r/η, η denoting the Kolmogorov scale, clearly show the statistical independence of the small-scale component Δu_r from the large-scale component u_0. Together with the similar result of the direct numerical simulation at $R_\lambda = 210$, it is concluded that *inertial-range* quantities show no dependence on the large-scale statistics in isotropic turbulence. This conclusion gives an experimental support for Kolmogorov's local equilibrium hypothesis as well as the *cross-independence* assumption (22) so far as the range of r comparable to the inertial range is concerned.

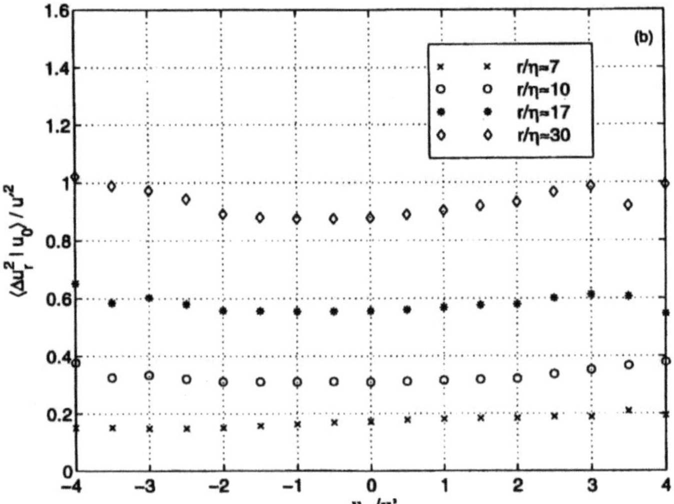

Figure 1. The conditional mean of $(\Delta u_r)^2$ conditioned on the large-scale velocity u_0. Each symbols correspond to a different distance r. u_0 is taken as the velocity at the middle point of the distance r, but other definition such as the average u_+ makes little difference. (Sreenivasan, 1998)

5. One-Point Velocity Distribution

Now, let us proceed to obtain the one-point velocity distribution using the cross-independence relation (22) as the closure assumption.

5.1. Closed Equation for f

First, substituting (21) and (22) into eq.(11), we obtain the following equation for the one-point velocity distribution:

$$\partial f(\boldsymbol{u}_1, t)/\partial t = -\lim_{r\to 0} \nu (\partial/\partial \boldsymbol{r})^2 \int (\boldsymbol{u}_2 \cdot \partial/\partial \boldsymbol{u}_1)\times$$
$$\times 2^{-3} g_+(\boldsymbol{u}_+, \boldsymbol{r}, t)\, g_-(\boldsymbol{u}_-, \boldsymbol{r}, t)\, d\boldsymbol{u}_2 \quad (23)$$

In view of the limit $\boldsymbol{r} \to 0$, we may write

$$\boldsymbol{r} = \boldsymbol{x}_2 - \boldsymbol{x}_1 = \Delta \boldsymbol{x}, \qquad \boldsymbol{u}_- = \tfrac{1}{2}(\boldsymbol{u}_2 - \boldsymbol{u}_1) = \tfrac{1}{2}\Delta \boldsymbol{u}$$
$$\boldsymbol{u}_+ = \tfrac{1}{2}(\boldsymbol{u}_1 + \boldsymbol{u}_2) = \boldsymbol{u}_1 + \tfrac{1}{2}\Delta \boldsymbol{u}, \qquad \boldsymbol{u}_2 = \boldsymbol{u}_1 + \Delta \boldsymbol{u}$$

Then, the right-hand side of eq.(23) is written as

$$T_\nu = -\lim_{r\to 0} \nu\, 2^{-3}(\partial/\partial \Delta \boldsymbol{x})^2 \int((\boldsymbol{u}_1+\Delta \boldsymbol{u})\cdot \partial/\partial \boldsymbol{u}_1)\times$$
$$\times g_+(\boldsymbol{u}_1+\tfrac{1}{2}\Delta \boldsymbol{u},\ \Delta \boldsymbol{x},\ t)\, g_-(\tfrac{1}{2}\Delta \boldsymbol{u},\ \Delta \boldsymbol{x},\ t)\, d\Delta \boldsymbol{u}$$
$$= -\lim_{r\to 0} \nu\, (\partial/\partial \Delta \boldsymbol{x})^2 \int((\boldsymbol{u}_1+2\Delta \boldsymbol{u})\cdot \partial/\partial \boldsymbol{u}_1)\times$$
$$\times g_+(\boldsymbol{u}_1+\Delta \boldsymbol{u},\ \Delta \boldsymbol{x},\ t)\, g_-(\Delta \boldsymbol{u},\ \Delta \boldsymbol{x},\ t)\, d\Delta \boldsymbol{u}$$

Substituting the expansion,

$$\lim_{r\to 0} g_+(\boldsymbol{u}_1+\Delta \boldsymbol{u},\ \Delta \boldsymbol{x},\ t) = f(\boldsymbol{u}_1+\Delta \boldsymbol{u},\ t) = (1+\Delta \boldsymbol{u}\cdot \partial/\partial \boldsymbol{u}_1)\, f(\boldsymbol{u}_1,\ t)$$

into T_ν and taking only the lowest non-zero term, we obtain

$$T_\nu = -(2/3)\,\nu\, C_g\, |\partial/\partial \boldsymbol{u}_1|^2 f(\boldsymbol{u}_1,\ t) \tag{24}$$
$$C_g = \lim_{r\to 0} (\partial/\partial \Delta \boldsymbol{x})^2 \int |\Delta \boldsymbol{u}|^2 g_-(\Delta \boldsymbol{u},\ \Delta \boldsymbol{x},\ t)\, d\Delta \boldsymbol{u} \tag{25}$$

where the isotropic relation, $(\Delta \boldsymbol{u}\cdot \partial/\partial \boldsymbol{u}_1)^2 = (1/3)|\Delta \boldsymbol{u}|^2\, |\partial/\partial \boldsymbol{u}_1|^2$, has been used. C_g is a positive constant representing the curvature of the variance of the distribution g_- at the origin..

Thus, eq.(23) is written in a closed form,

$$\partial f(\boldsymbol{u},\ t)/\partial t = -\alpha^2 \triangle f(\boldsymbol{u},\ t) \tag{26}$$
$$\alpha^2 = (2/3)\,\nu\, C_g \tag{27}$$

where $\triangle = |\partial/\partial \boldsymbol{u}|^2$ and the suffix of $\boldsymbol{u}_1$ has been omitted.

It may clearly be seen from eqs.(25)~(26) that the action of the small-scale components $\Delta \boldsymbol{u}$ against the large-scale components $\boldsymbol{u}$ has solely been confined to the integral (25) and made a constant α^2 in eq.(26). This proves the success of the scale-separation by means of the *cross-independence* assumption.

5.2. Normal Velocity Distribution

It may be seen from eqs.(25) and (27) that C_g as well as α^2 generally change in time. Thus, it is necessary to specify their time-dependence in order to solve the equation.(26). If we asume the change in time of the kinetic energy E of turbulent fluid per unit mass and the energy dissipation rate ε as,

$$E = \tfrac{1}{2}\langle|\boldsymbol{u}|^2\rangle = \tfrac{1}{2}\langle u_i{}^2\rangle \propto t^{-1} \tag{28}$$
$$\varepsilon = -dE/dt = \tfrac{1}{2}\nu\langle(\partial u_i/\partial x_j + \partial u_j/\partial x_i)^2\rangle \propto t^{-2} \tag{29}$$

(i, j =1, 2, 3) respectively, it follows that $\alpha^2 \propto t^{-2}$.

Then, eq.(26) is written as

$$\partial f(\boldsymbol{u},\ t)/\partial t = -\alpha_0^2 t^{-2} \triangle f(\boldsymbol{u},\ t) \tag{30}$$

where $\alpha^2 = \alpha_0^2 t^{-2}$ with a constant α_0^2.

Eq.(30) can be solved generally, but here we refer only to the particular solution,

$$f(\boldsymbol{u},\ t) = (t/4\pi\alpha_0^2)^{3/2} \exp[-|\boldsymbol{u}|^2 t/4\alpha_0^2] \tag{31}$$

which is the isotropic *normal* (*Gaussian*) distribution changing self-similarly in time.

It follows from (31) that the energy and the energy dissipation change in time as

$$E = 3\alpha_0^2 t^{-1}, \qquad \varepsilon = 3\alpha_0^2 t^{-2} \tag{32}$$

in accordance with their assumed decay.

Figure 2 shows the evolution of the distribution (31) in time. At $t = 0$ it represents a uniform distribution with zero probability density. For $t > 0$ it grows up as a *normal* distribution with increasing height and decreasing breadth in time t, and for $t \to \infty$ it tends to the δ distribution at $|\boldsymbol{u}| = 0$ corresponding to the dead-still state.

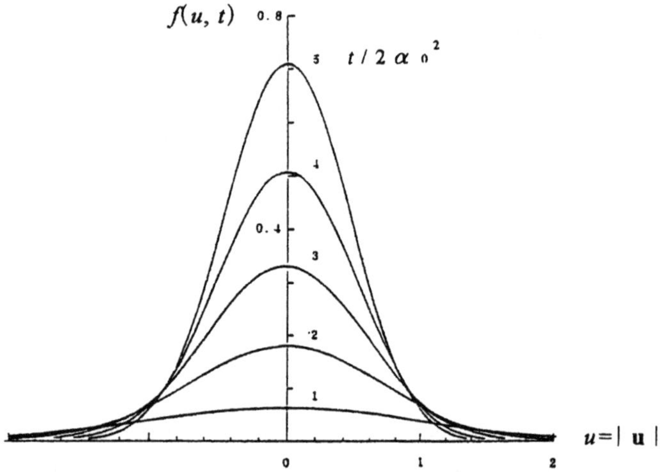

Figure 2. One-point normal velocity distribution $f(\mathbf{u}, t)$ of isotropic turbulence.

Here I shall conclude my Opening Address, in which I took liberty of including my own view and idea on the *statistics* of turbulence. Concerning the *geometry* of turbulence, I entrust to the Opening Lecture by Professor Keith Moffatt.

6. References

Battimelli, G. (1988) The early international congresses of applied mechanics, in S. Juhasz (ed.) *IUTAM -- A Short History*, Springer Verlag. pp. 9-13.

Hopf, E. (1952) Statistical hydromechanics and functional calculus. *J. Rat. Mech. Analysis*, **1**, 87-123.

Kolmogorov, A.N. (1941) The local structure of turbulence in incompressible viscous fluid for very large Reynolds numbers. *Dokl. Akad. Nauk. SSSR*, **30**, 301-305.

Lundgren, T.S. (1967) Distribution functions in the statistical theory of turbulence. *Phys. Fluids* **10**, 969-975.

Monin, A.S. (1967) Equations of turbulent motion. *PMM J. Appl. Math. Mech.* **31**, 1057-1068.

Prandtl, L. (1904) Ueber Fluessigkeitsbewegung bei sehr kleiner Reibung. *Verh. III Intern. Math. Kongr. Heidelberg*. pp. 484-491.

Sreenivasan, K.R. and Dhruva, B. (1998) Is there scaling in high-Reynolds-number turbulence? *Prog. Theo. Phys. Suppl.* No.130. 103-120.

Taylor, G. I. (1910) The conditions necessary for discontinuous motion in gases. *Proc. Roy. Soc. London*, A **84**, 371-377; *Scientific Papers*, Vol. III, Cambridge Univ. Press, 1963, pp. 1-6.

THE TOPOLOGY OF SCALAR FIELDS IN 2D AND 3D TURBULENCE

H.K. Moffatt
Isaac Newton Institute for Mathematical Sciences,
20 Clarkson Road,
Cambridge CB3 0EH, UK

1 Introduction

I am honoured to present the Opening Lecture at this Symposium on the *Geometry and Statistics of Turbulence.* I vividly recall a previous IUTAM Symposium in Tokyo in 1983 on the subject *Turbulence and Chaotic Phenomena in Fluids*; at that Symposium, I contributed a lecture with the title "Simple topological aspects of turbulent vorticity dynamics". During the last two decades, increasing attention has been paid to characteristic structures detected in both experimental work and in direct numerical simulation (DNS); indeed, it is these developments that have provided the main motivation for the present Symposium, and that will be described in many of the lectures on the programme.

In this introductory lecture, it may be appropriate to resume the theme of my 1983 lecture, but at a more basic level: I propose to consider the generic structure of scalar fields in both 2D and 3D turbulence, the topological description of these fields, and the manner in which the topology may change with time. I shall also seek to describe how, at the simplest level, the geometry of these fields may be related to the most basic statistical property – the field spectrum.

While the scalar field presents a tractable problem which it is sensible to consider as a starting point, the problem of providing a similar description of solenoidal vector fields such as velocity $\mathbf{u}$ or vorticity $\boldsymbol{\omega}$ in 3D is very much more difficult. Such fields generically exhibit chaos (as for example in the ABC-flow studied by Dombre et al 1986, or the quadratic STF-flow studied by Bajer & Moffatt 1990); and the problem of classifying such chaotic flows appears prohibitively difficult at present. This will continue to present a major challenge well into the new millennium.

T. Kambe et al. (eds.), IUTAM Symposium on Geometry and Statistics of Turbulence, 13–22.
© 2001 *Kluwer Academic Publishers.*

2 Streamline topology in 2D turbulence

Consider first the problem of 2D turbulence in a periodic domain (topologically a torus). Let $\psi(x, y, t)$ be the streamfunction for the flow. The critical points at any instant of ψ are the points where $\nabla\psi = 0$, i.e. they are the (instantaneous) stagnation points of the flow. Taking origin O at one such critical point, the streamfunction has local Taylor expansion

$$\psi = \psi_0 + c_{ij} x_i x_j + O(|\mathbf{x}|^3)\,, \tag{1}$$

where $(x_1, x_2) \equiv (x, y)$ and where

$$c_{ij} = \tfrac{1}{2}\left(\partial^2\psi/\partial x_i \partial x_j\right)_{\mathbf{x}=0} = c_{ji}\,. \tag{2}$$

Let λ_1, λ_2 be the eigenvalues of c_{ij}. If we choose the axes Ox, Oy to be along the corresponding eigenvectors, then (1.1) takes the form

$$\psi = \psi_0 + \lambda_1 x^2 + \lambda_2 y^2 + O(|\mathbf{x}|^3)\,. \tag{3}$$

Non-degeneracy of the critical point means that $\lambda_1\lambda_2 \neq 0$. The *index* i of the critical point is defined as the number of negative eigenvalues, i.e. $i = 0$, 1 or 2 in this case. If $i = 0$, both λ_1 and λ_2 are positive and ψ is clearly minimal at O; if $i = 2$, λ_1 and λ_2 are negative and ψ is maximal at O. In either case, the streamlines $\psi =$ cst. are elliptic in the neighbourhood of O, the flow being clockwise if $i = 0$, anticlockwise if $i = 2$. If $i = 1$, then one of (λ_1, λ_2) is positive, the other negative, and O is a saddle point of ψ, the streamlines being locally hyperbolic with asymptotes

$$x/y = \pm(-\lambda_2/\lambda_1)^{1/2}\,. \tag{4}$$

Regarding the periodic domain as (topologically) a torus, Euler's index theorem implies that

$$\sum_{i=0}^{2}(-1)^i n_i = n_0 - n_1 + n_2 = 0\,, \tag{5}$$

i.e. the number of elliptic points $n_0 + n_2$ equals the number of hyperbolic points n_1. A uniform flow for which $n_0 + n_2 = n_1 = 0$ provides a trivial example.

The streamlines through the saddle points (or separatrices) play an important role in relation to the field topology. In general, the values of ψ at the critical points within the periodic box will be all different. This means that, in general, 'heteroclinic' separatrices connecting one critical point to another do not occur. All separatrices are homoclinic in the sense that each connects back to a single critical point. They may connect in two distinct ways as indicated in figure 1; the first is a figure-of-eight configuration, while the second may be described as an 'inverted figure-of-eight'.

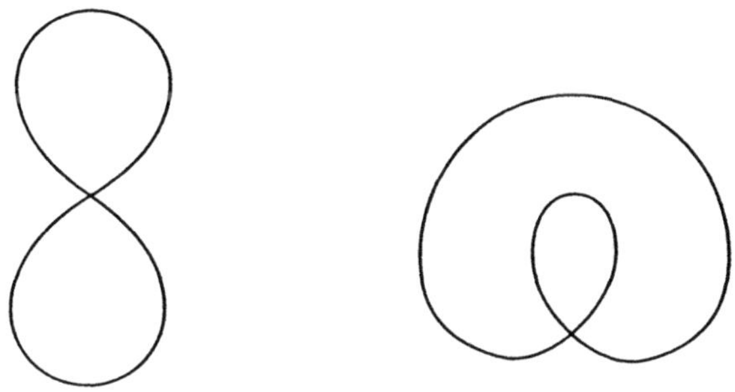
(a) (b)

Figure 1: Homoclinic separatrices associated with a hyperbolic stagnation point. *a)* figure-of-eight structure; *b)* inverted figure-of-eight.

A flow may be topologically simplified through the coalescence of an elliptic point and a hyperbolic point (leaving (5) satisfied). This generic transition may be represented (locally in x, y and t) by the streamfunction

$$\psi_1 = -x^2 + 3yt + y^3 , \tag{6}$$

which, for $t < 0$, has an elliptic critical point at $(0, \sqrt{-t})$ and a hyperbolic critical point at $(0, -\sqrt{-t})$. At $t = 0$, these points coalesce, and for $t > 0$, the flow has no critical points. It is interesting to note that at the instant of coalescence, the streamline $\psi = 0$ has a cusp at $x = y = 0$ (figure 2); at this instant, the critical point is degenerate.

Conversely, a flow may be topologically 'complexified' by time-reversal of this process: at any point within a flow, a local distortion may introduce a hyperbolic-elliptic pair (otherwise known as a saddle-node bifurcation), as illustrated in figure 3. This process indicates that the 'generic' instantaneous separatrix structure of a 2D flow consists of an array of nested figure-of-eights and inverted figure-of-eights, more complex flows containing higher-order nested structures.

The existence of the streamfunction (6) indicates that the transition indicated in figure 2 is kinematically possible; to show that it is dynamically possible, we need to consider the Navier-Stokes equation in dimensionless form

$$\frac{\partial}{\partial t}(\nabla^2\psi) - \frac{\partial(\psi, \nabla^2\psi)}{\partial(x,y)} = \frac{1}{Re}\nabla^4\psi . \tag{7}$$

With ψ_1 given by (6), we have

$$\frac{\partial}{\partial t}(\nabla^2\psi_1) = 0 , \quad \frac{\partial(\psi_1, \nabla^2\psi_1)}{\partial(x,y)} = -12x \tag{8}$$

and (7) is not satisfied. However, if we redefine ψ_1 as

$$\psi_1 = -x^2 + y^3 + 3ty + (Re/10)x^5 \tag{9}$$

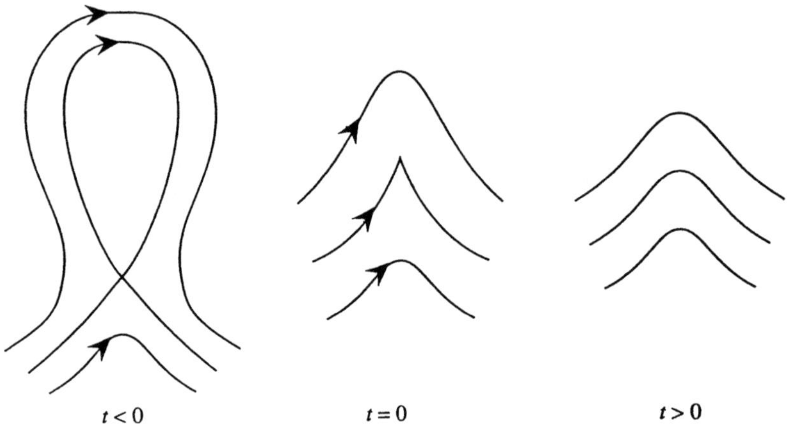

Figure 2: Coalescence of an elliptic point and a hyperbolic point represented by the streamfunction $\psi = -x^2 + yt + y^3$.

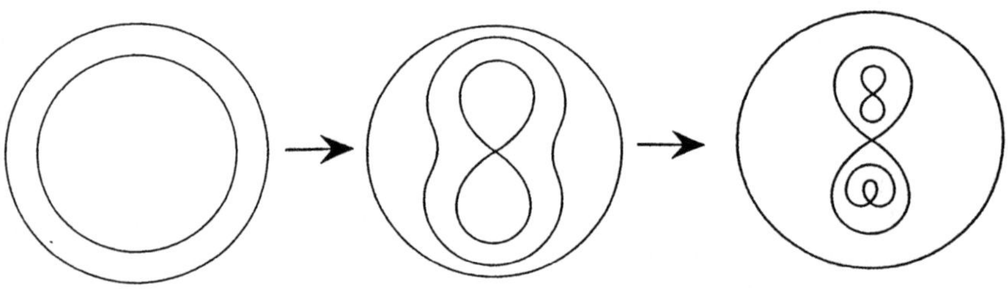

Figure 3: Flow complexification, through successive introduction of hyperbolic-elliptic pairs.

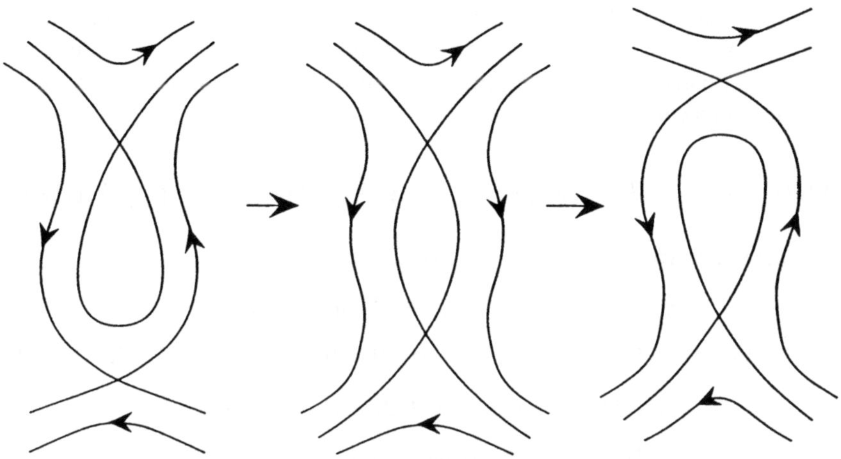

Figure 4: Change of separatrix topology through heteroclinic connexion

then (7) is satisfied at leading order, i.e. at the order of terms linear in x and y. Thus, the transition of figure 2 *is* dynamically possible, and since t may be replaced by $-t$, this transition is possible in either direction. (Alternative modifications of (6) are equally possible.)

As mentioned above, heteroclinic connexions are not generally present; however, they may occur instantaneously in an unsteady flow. This is illustrated by the streamfunction

$$\psi_2(x,y,t) = \tfrac{1}{2}(x^2+y^2) - y^4/4\epsilon^2 - At(y-y_0)^2 \tag{10}$$

where ϵ is small. When $t=0$, there are saddle points at $(0,\pm\epsilon)$ with heteroclinic connexion. Also, (7) is satisfied at leading order provided $A = 3(\epsilon^2 Re)^{-1}$. As t passes through zero, the topology of the flow changes as indicated in figure 4. The whole diagram lies within the region $|\mathbf{x}| = O(\epsilon)$. Thus again, this type of 'heteroclinic transition' appears to be both kinematically and dynamically possible. Note that

$$\frac{\partial(\psi_2, \nabla^2\psi_2)}{\partial(x,y)} = -3xy/\epsilon^2 = O(1) \tag{11}$$

but this term can be compensated by the introduction of higher-order terms in (9).

Similar considerations apply to the vorticity field $\omega = -\nabla^2\psi$ in 2D turbulence. The critical points of ω are now those points where $\nabla\omega = 0$, and these critical points are again elliptic or hyperbolic in equal numbers (for a periodic domain). The elliptic points (maxima or minima) may be identified with the centres of the 'concentrated vortices' that ultimately emerge in freely decaying turbulence at high Reynolds number. In the Euler limit ($Re = \infty$), the iso-vorticity contours $\omega =$ cst. move with the fluid, and so the topology of the ω-field is conserved. Thus topological transitions can now occur *only* through the agency of viscosity. Examples of such transitions can be easily constructed. For example, a transition eliminating an inverted figure-of-eight occurs whenever a strong positive vortex engulfs and eliminates through viscous action a neighbouring weak negative vortex.

3 Eddies and vortices in a 2D field of turbulence

Consider first a simple flow whose streamfunction is

$$\psi = A_1 \sin k_1 x \sin k_1 y + A_2 \sin k_2 x \sin k_2 y \tag{12}$$

where $k_2 \gg k_1$, $A_1 > 0$, $A_2 \geq 0$. The vorticity distribution $\omega = -\nabla^2 \psi$ is then

$$\omega = 2A_1 k_1^2 \sin k_1 x \sin k_1 y + 2A_2 k_2^2 \sin k_2 x \sin k_2 y\,. \tag{13}$$

If $A_2 = 0$, then $\omega = 2k_1^2 \psi$ and the isovorticity curves $\omega =$ cst. coincide with the streamlines $\psi =$ cst.. The critical points are at $(n\pi/k_1,\ m\pi/k_1)$, and the separatrices are all heteroclinic – a degenerate situation. The number of critical points in a square of side πL is $(k_1 L)^2$.

Suppose now that A_2/A_1 is slowly increased from zero. The x-component of velocity on $y = 0$ is

$$u = \partial\psi/\partial y|_{y=0} = A_1 k_1 \sin k_1 x + A_2 k_2 \sin k_2 x\,, \tag{14}$$

and it is easy to see that a first saddle-node bifurcation occurs at $k_2 x = 3\pi/2$ when $A_2/A_1 \doteqdot 3\pi k_1^2/2k_2^2$. As A_2/A_1 increases further, more saddle-node bifurcations occur until $A_2 k_2 = A_1 k_1$ when the number of critical points of the field (12) has increased to $(k_2 L)^2$.

Similar considerations obviously apply to the field ω. Note that, when A_2/A_1 is in the range

$$\frac{k_1^3}{k_2^3} < \frac{A_2}{A_1} \lesssim \frac{3\pi}{2}\frac{k_1^2}{k_2^2}\,, \tag{15}$$

the number N of 'eddies' (i.e. extrema of ψ) per unit area is $(k_1/\pi)^2$, while the number M of 'vortices' (i.e. extrema of ω) per unit area is $(k_2/\pi)^2$.

Similarly, we may seek to estimate these numbers for a field of 2D turbulence with energy spectrum $E(k)$ having an inertial range $\sim k^{-\lambda}$ between a maximum at k_1 and a viscous cut-off at $k_2 (\gg k_1)$. The corresponding velocity v_k at wavenumber k scales like $(kE(k))^{\frac{1}{2}} \sim k^{\frac{1}{2}(1-\lambda)}$, so the number of eddies per unit area will be of order k_2^2 or k_1^2 according as $\lambda < 1$ or $\lambda > 1$. Similarly the number of vortices per unit area is controlled by $k^2 v_k \sim k^{\frac{1}{2}(5-\lambda)}$, and is of order k_2^2 or k_1^2 according as $\lambda < 5$ or $\lambda > 5$. Note that, for λ in the interesting range $1 < \lambda < 5$, the number of *eddies* per unit area is $O(k_1^2)$ while the number of *vortices* per unit area is $O(k_2^2)$; this is of course consistent with the fact that far more structure is normally seen in the vorticity field than in the ψ-field in DNS of 2D turbulence.

4 Scalar field structure in 3D turbulence

Consider now the generic structure of a scalar field $s(\mathbf{x})$ in 3D. The critical points are again those points where $\nabla s = 0$; and in the neighbourhood of a non-degenerate critical point O, the field admits an expansion of the form

$$s = s_0 + \lambda_1 x^2 + \lambda_2 y^2 + \lambda_3 z^2 + O(|\mathbf{x}|^3)\,, \tag{16}$$

where $\lambda_1\lambda_2\lambda_3 \neq 0$. The index i is again the number of negative eigenvalues, so that $i = 0, 1, 2$ or 3; the index theorem is now

$$\Sigma_{i=0}^{3}(-1)^i n_i = n_0 - n_1 + n_2 - n_3 = 0 \tag{17}$$

for the case of a field periodic in all three directions (topologically, we are then in T^3). If $i = 0$, we have a minimum of s (denoted M_0); if $i = 3$, we have a maximum (denoted M_3); and if $i = 1$ or 2, we have a saddle point of 'type 1' or 'type 2' (denoted S_1, S_2). The separatrix surface $\Sigma : s = s_0$ through a saddle point O is locally a cone with elliptic cross-section,

$$\lambda_1 x^2 + \lambda_2 y^2 + \lambda_2 z^2 = 0\,, \tag{18}$$

and generically, this cone must 'connect' to itself in homoclinic manner. There are four ways (figure 5) in which it can do this, in which Σ takes the following forms:

1. a 'constricted' sphere Σ_1 (to visualise this, tie a loop of string round the equator of an inflated balloon and pull it tight);
2. an inverted constricted sphere Σ_2 (one 'bulb' of the closure now contains the other);
3. a 'pinched' sphere Σ_3 (push the poles of a balloon inwards till they make contact);
4. a constricted torus Σ_4 (which may be knotted) (tie a string round the small circumference of a toroidal balloon and tighten it).

(We may also pinch a torus, giving a 'pretzel' of two holes; and so on.)

Consider now how we may complexify an s-field starting from a region inside a surface $s =$ cst. of spherical topology within which there is a single extremum, a maximum M_3 say. We may perturb the field in three ways (see figure 6):

1. by introducing another maximum M_3 and a saddle point S_2 of type 2; this topological transition is represented by

$$T_1 : (n_0, n_1, n_2, n_3) \to (n_0, n_1, n_2 + 1, n_3 + 1) \tag{19}$$

and generates a separatrix surface of type Σ_1;

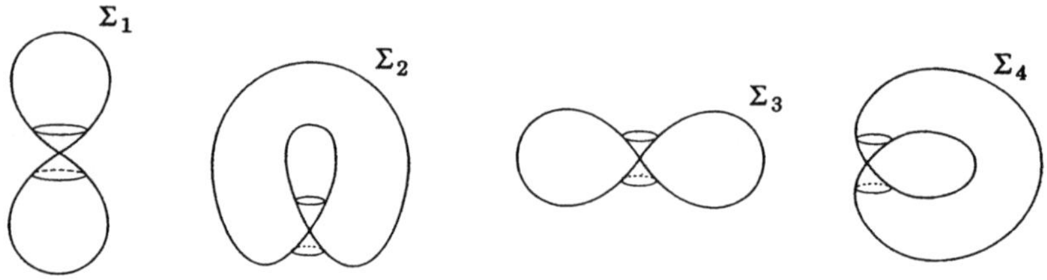

Figure 5: Four possible homoclinic surface connections

2. by introducing a minimum M_0 and a saddle S_1; this corresponds to the transition
$$T_2 : (n_0, n_1, n_2, n_3) \to (n_0 + 1, n_1 + 1, n_2, n_3) \tag{20}$$
and generates a separatrix surface of type Σ_2;

3. by introducing two saddle points S_1 and S_2; here M_3 lies inside a torus Σ_4 constricted at S_2, which in turn lies inside a sphere Σ_3 pinched at S_1; this corresponds to the transition
$$T_3 : (n_0, n_1, n_2, n_3) \to (n_0, n_1 + 1, n_2 + 1, n_3)\,. \tag{21}$$

The transitions T_1, T_2, T_3 are all evidently compatible with (18).

Successive complexifications will lead to a field s whose separatrix surfaces form a complex of nested surfaces of types $\Sigma_1, \Sigma_2, \Sigma_3, \Sigma_4$, with (Σ_3, Σ_4) always occurring as a pair with Σ_4 inside Σ_3. It would appear that this type of construction should provide the generic structure of any scalar field $s(\mathbf{x})$. Note that, because of the possibility of the transition T_3, the number $n_1 + n_2$ of saddles may be much greater than the number $n_0 + n_3$ of extrema.

5 The passive scalar field $\theta(\mathbf{x}, t)$ in turbulent flow

Suppose now that $s = \theta(\mathbf{x}, t)$ is a passive scalar field satisfying the advection-diffusion equation
$$\frac{D\theta}{Dt} \equiv \frac{\partial\theta}{\partial t} + \mathbf{u}\cdot\nabla\theta = \kappa\nabla^2\theta \tag{22}$$
in a turbulent flow $\mathbf{u}(\mathbf{x}, t)$. Note first that if $\kappa = 0$, then all surfaces (including separatrix surfaces) $\theta =$ cst. are frozen in the fluid; hence their topology is invariant, and the set of numbers $\mathbf{n} = (n_0, n_1, n_2, n_3)$ is constant. If $\mathbf{n} = 0$ at some initial instant (e.g. if $\theta = \mathbf{c}\cdot\mathbf{x}$ at $t = 0$, a field of uniform gradient $\mathbf{c}$), then (when $\kappa = 0$) $\mathbf{n} = 0$ for all $t > 0$: turbulence alone cannot generate maxima or minima or saddle points of θ; it needs the cooperation of diffusion ($\kappa > 0$) to achieve this. Some aspects of this problem have been explored by Gibson (1968).

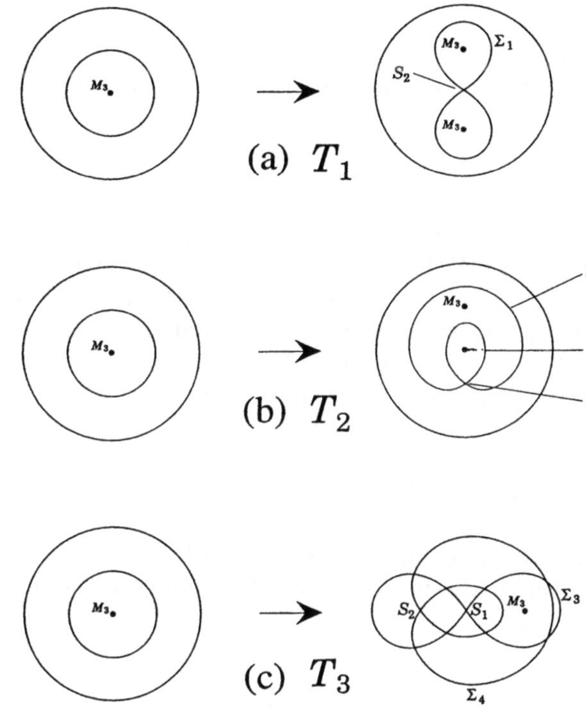

Figure 6: Three topological transitions by which a scalar field may be successively complexified.

If κ is small (equivalently, if the Peclet number is large) it may take a long time for $\mathbf{n}(t)$ to attain statistical equilibrium, starting from a situation in which $\mathbf{n}(0) = 0$. The behaviour of $\mathbf{n}(t)$ as a function of time under the action of a 'prescribed' field of turbulence, is a problem that should be amenable to DNS.

Here, we simply speculate on the (statistical) equilibrium established after a long time, assuming that the turbulence is statistically steady and the Peclet number large. Suppose first that the Prandtl number $Pr = \nu/\kappa$ is $\lesssim O(1)$. Then (Batchelor, Howells & Townsend 1959), the spectrum $\Gamma(\kappa)$ of θ^2 has a $k^{-5/3}$-range ($k_0 \lesssim k \lesssim k_c = (\epsilon/\kappa^3)^{1/4}$) and a relatively steep ($k^{-17/3}$) decrease for $k \gtrsim k_c$. The spectrum of $\nabla\theta$ is $k^2\Gamma(k)$ and

$$(k \cdot k^2\Gamma(k))^{1/2} \sim k^{2/3} \quad (k_0 \lesssim k \lesssim k_c)\,. \tag{23}$$

By analogy with the 2D situation already considered, the increase with k in (23) indicates that the topology of θ is controlled at the upper end of the range, and that the number $N = |\mathbf{n}(t)|$ of critical points of θ is of order $k_c^3 = (\epsilon/\kappa^3)^{3/4}$ per unit volume.

Suppose now that $\nu/\kappa \gg 1$. Then (Batchelor 1959) $\Gamma(k) \sim k^{-5/3}$ for $k_0 \lesssim k \lesssim k_v = (\epsilon/\nu^3)^{1/4}$, and there is a further conduction subrange

$$\Gamma(k) \sim k^{-1} \qquad (k_v \lesssim k \lesssim k_c = (\epsilon/\nu k^2)^{1/4})\,. \tag{24}$$

In this subrange,

$$(k \cdot k^2\Gamma(k))^{1/2} \sim k \tag{25}$$

and this suggests that the topology of θ is again controlled at the conduction cut-off k_c, implying

$$N \sim k_c^3 \sim (\epsilon/\nu k^2)^{3/4} \quad (\kappa \ll \nu) \,. \tag{26}$$

There is reason however to doubt the validity of this result, which would imply a large number (of order $(\nu/\kappa)^{3/2}$) of critical points of θ within every 'Kolmogorov sphere' of radius $(\nu^3/\epsilon)^{1/4}$; within any such sphere, the velocity gradient is (as assumed by Batchelor) approximately uniform, and it is difficult to see how transitions of types T_1, T_2 and T_3 can be induced by such a flow on such a scale.

For this reason, it seems more likely that in all circumstances, the number of critical points per unit volume is given by

$$N \sim \min\left((\epsilon/\kappa^3)^{3/4}, (\epsilon/\nu^3)^{3/4}\right) \,. \tag{27}$$

This estimate should again be amenable to testing by DNS.

I am grateful to Boris Khesin and Paul Glendinning for helpful discussions on the topic of this paper.

References

[] Bajer, K. and Moffatt, H.K. (1990) On a class of steady confined Stokes flows with chaotic streamlines, *J. Fluid Mech.*, **212**, 337–363.

[] Batchelor, G.K. (1959) Small-scale variation of convected quantities like temperature in turbulent fluid, I. General discussion and the case of small conductivity, *J. Fluid Mech.*, **5**, 113–133.

[] Batchelor, G.K., Howells, I.D. & Townsend, A.A. (1959) Small-scale variation of convected quantities like temperature in turbulent fluid, II, *J. Fluid Mech.*, **5**, 134.

[] Dombre, T., Frisch, U., Greene, J.M., Hénon, M., Mehr, A. and Soward, A.M. (1986) Chaotic streamlines in the ABC flows. *J. Fluid Mech.*, **167**, 353–391.

[] Gibson, C.H. (1968) Fine structure of scalar fields mixed by turbulence, I.Zero gradient points and minimal gradient surfaces. *Phys.Fluids*, **11**, 2305–2315.

Bounds for Turbulent Transport

Peter Constantin
Department of Mathematics
The University of Chicago

November 23, 1999

We would like to pursue a mathematical approach to turbulence that is able to predict bulk average quantities of experimental and practical significance. Among such bulk average quantities, the Nusselt number N, the enhanced bulk average heat transfer due to convection, is perhaps the simplest physical objective of rigorous mathematical study of turbulence.

The question we wish to address here is: what are the optimal bounds on N as a function of Rayleigh number R? What is the effect of the Prandtl number? Theoretical results ([1], [2], [3]) predict a maximal ultimate behavior of $N \sim \sqrt{R}$. The experimental ([4]) findings indicate however that

$$N \sim R^q$$

where the reported values for q belong approximately to the interval $[\frac{2}{7}, \frac{1}{3}]$ for large R. The exponents 2/7 and 1/3 have been discussed by several authors, ([5], [6]). The recent results of ([7]) favor an exponent of 3/10 with logarithmic corrections. We will describe below some very simple rigorous results.

The mathematical formulation of the problem is based on the three dimensional Boussinesq equations for Rayleigh-Bénard convection ([8]). These are a system of equations coupling the three dimensional Navier-Stokes equations

$$\frac{\partial \boldsymbol{u}}{\partial t} + \boldsymbol{u} \cdot \nabla \boldsymbol{u} + \nabla p = \sigma \Delta \boldsymbol{u} + \sigma R T \boldsymbol{e}_z, \quad \nabla \cdot \boldsymbol{u} = 0 \tag{1}$$

T. Kambe et al. (eds.), IUTAM Symposium on Geometry and Statistics of Turbulence, 23–31.
© 2001 *Kluwer Academic Publishers.*

to a heat advection-diffusion equation

$$\frac{\partial T}{\partial t} + \boldsymbol{u} \cdot \nabla T = \Delta T. \tag{2}$$

The five unknowns, incompressible velocity, $\boldsymbol{u} = (u, v, w)$, pressure p, and temperature T are functions of position $\boldsymbol{x} = (x, y, z)$ and time t. R is the Rayleigh number and σ is the Prandtl number. $\boldsymbol{e}_z = (0, 0, 1)$ is the unit vector in the vertical direction. For simplicity of exposition the equations have been non-dimensionalized: the vertical variable z is scaled so that it belongs to the interval $[0, 1]$, the horizontal independent variables (x, y) belong to a square $Q \subset \boldsymbol{R}^2$ of side L. The boundary conditions are as follows: all functions $((u, v, w), p, T)$ are periodic in x and y with period L; u, v, and w vanish for $z = 0, 1$, and the temperature obeys $T = 0$ at $z = 1$, $T = 1$ at $z = 0$. We write

$$\|f\|^2 = \frac{1}{L^2} \int_0^L \int_0^L \int_0^1 |f(x, y, z)|^2 \, dx \, dy \, dz$$

for functions and vectors f. We use $< \cdots >$ for long time average:

$$\langle f \rangle = \limsup_{t \to \infty} \frac{1}{t} \int_0^t f(s) ds.$$

We write $\overline{f}$ the horizontal average

$$\overline{f} = \frac{1}{L^2} \int_0^L \int_0^L f(x, y) \, dx \, dy.$$

When a function depends on additional variables we write only the remaining variables after integration, so for instance

$$\|\nabla \boldsymbol{u}(\cdot, t)\|^2 = \frac{1}{L^2} \int_0^L \int_0^L \int_0^1 |\nabla \boldsymbol{u}(x, y, z, t)|^2 \, dx \, dy \, dz.$$

The Nusselt number is given in terms of the long time average of the vertical heat flux:

$$N = 1 + \left\langle \int_0^1 b(z) \, dz \right\rangle \tag{3}$$

with

$$b(z, t) = \frac{1}{L^2} \int_0^L \int_0^L w(x, y, z, t) T(x, y, z, t) \, dx \, dy = \overline{wT}(z, t). \tag{4}$$

A consequence of the equations of motion are two additional formulas for the Nusselt number:

$$\left\langle \|\nabla T\|^2 \right\rangle = N \tag{5}$$

and

$$\left\langle \|\nabla \boldsymbol{u}\|^2 \right\rangle = R(N-1). \tag{6}$$

The classical result of Howard, conditioned on on assumptions about statistical averages, is that N is bounded at very large Rayleigh numbers by a multiple of $R^{\frac{1}{2}}$. The same kind of bound can be derived without any conditions ([9]). This bound is valid for all solutions, aspect ratios L and Prandtl numbers σ and is also valid in a rotating frame, at arbitrary rotation speed.

The system is not isotropic: the direction of gravity is singled out. Consider a function $\tau(z)$ that satisfies $\tau(0) = 1$, $\tau(1) = 0$, and express the temperature as

$$T(x,y,z,t) = \tau(z) + \theta(x,y,z,t). \tag{7}$$

The role of τ is that of a convenient background ([10], [11], [12]) that carries the inhomogeneous boundary conditions; thus θ obeys the same homogeneous boundary conditions as the velocity. Note that, because τ does not depend on x, y one has

$$T(x,y,z,t) - \overline{T}(z,t) = \theta(x,y,z,t) - \overline{\theta}(z,t). \tag{8}$$

The equation obeyed by θ is

$$(\partial_t + u \cdot \nabla - \Delta)\,\theta = \tau'' - w\tau' \tag{9}$$

where we used $\tau' = \frac{d\tau}{dz}$. The horizontal average of the vertical velocity vanishes identically because of incompressibility

$$\overline{w}(z,t) = 0.$$

Therefore the quantity $b(z,t)$ can be written as

$$b(z,t) = \frac{1}{L^2}\int_0^L\int_0^L w(x,y,z,t)\left(\theta(x,y,z,t) - \overline{\theta}(z,t)\right)\,dx\,dy = \overline{w(T-\overline{T})}(z,t) \tag{10}$$

Multiplying the equation (9) by θ and integrating we obtain

$$N + \left\langle \|\nabla\theta\|^2 \right\rangle = 2\left\langle -\int_0^1 \tau'(z)b(z)dz \right\rangle + \int_0^1 (\tau'(z))^2 \, dz. \tag{11}$$

We will choose the background profile τ for simplicity to be a smooth profile concentrated in a boundary layer of width δ,

$$\tau(z) = P\left(\frac{z}{\delta}\right)$$

with $P(0) = 1$ and $P(s) = 0$ for $s \geq 1$. Using only elementary facts (fundamental theorem of calculus, the boundary conditions and the Schwartz inequality) it is easy to see from (10) that

$$|b(z,t)| \leq z\|\nabla \boldsymbol{u}(\cdot,t)\| \cdot \|\nabla(T - \overline{T})(\cdot,t)\| \tag{12}$$

holds for any z. Let us define

$$n = \left\langle \|\nabla(T - \overline{T})\|^2 \right\rangle \tag{13}$$

Note that, from the definition of n and (5) it follows that

$$n \leq N.$$

From (12), (11) and (6) we obtain

$$N \leq \frac{C}{\delta} + 2D\delta\sqrt{R(N-1)}\sqrt{n}$$

with

$$C = \int_0^1 \left(\frac{dP(s)}{ds}\right)^2 ds \tag{14}$$

and

$$D = \int_0^1 s\left|\frac{dP(s)}{ds}\right| ds. \tag{15}$$

Optimizing in $\delta < 1$ we get

$$\delta^{-1} = \sqrt{\frac{2D}{C}}\{R(N-1)n\}^{\frac{1}{4}} + 1$$

and letting $P \to s$ we obtain

$$N \leq 2(Rn)^{\frac{1}{4}}(N-1)^{\frac{1}{4}} + 1 \tag{16}$$

Thus we have proved

Theorem 1 *Let*

$$n = \left\langle \|\nabla(T - \overline{T})\|^2 \right\rangle$$

Then the Nusselt number (3, 5, 6) for three dimensional Rayleigh-Bénard convection satisfies

$$N \leq 2^{\frac{4}{3}} (Rn)^{\frac{1}{3}} + 1.$$

If one has no additional information then, using $n \leq N$ in the inequality above we obtain the square-root bound

$$N \leq 4\sqrt{R} + 1$$

(The prefactor is not optimal. The search for optimal prefactors is better motivated for other systems, where the power law obtained rigorously coincides with the one observed in experiments. When that is the case then the rigorous results can match experiments with remarkable accuracy ([13])). The exponent 1/3 (or anything less than 1/2 for that matter) have not been proven rigorously for the general system. The theorem above brings in the exponent 1/3 in the general Boussinesq system conditionally, for slowly varying n. Another way by which the Nusselt number dependence on the Rayleigh number can be lower is if the Prandtl number is very high or in rotating ([14]) convection. If the Prandtl number is infinity then the upper bound is closer to 1/3. The equations of motion for infinite Prandtl number Rayleigh-Bénard convection ([15]) are

$$-\Delta \boldsymbol{u} + \nabla p = RTe_z, \quad \nabla \cdot \boldsymbol{u} = 0 \tag{17}$$

coupled with the advection-diffusion equation (2). Because (17) is time independent we say that T obeys an active scalar equation.

An important observation, true even for the general case (1) is that, in view of the boundary conditions and incompressibility, not only the vertical component of velocity w but also its normal derivative $\frac{\partial w}{\partial z}$ vanish at the vertical boundaries. Therefore we can write

$$b(z,t) = \overline{\int_0^z dz_1 \int_0^{z_1} \frac{\partial^2 w}{\partial z^2}(\cdot, s, t) ds \int_0^z \left(\frac{\partial (T(\cdot, \sigma, t) - \overline{T}(\sigma, t))}{\partial z} \right) d\sigma}. \tag{18}$$

Consequently

$$|b(z,t)| \leq z^{\frac{5}{2}} \left\| \frac{\partial^2 w(\cdot, t)}{\partial z^2} \right\|_{L^\infty} \|\nabla (T - \overline{T})(\cdot, t)\| \tag{19}$$

where $\|f\|_{L^\infty}$ is the sup-norm. One can express $\frac{\partial^2 w}{\partial z^2}$, in terms of $T - \overline{T}$. Indeed, eliminating the pressure from (17) one has

$$\Delta^2 w = -R\Delta_h(T - \overline{T}) \tag{20}$$

where Δ_h is the Laplacian in the horizontal directions x and y. Using the boundary conditions we may write this as

$$w = -R(\Delta_{DN}^2)^{-1}\Delta_h(T - \overline{T}) \tag{21}$$

where $(\Delta_{DN}^2)^{-1}$ is the inverse bilaplacian with homogeneous Dirichlet and Neumann boundary conditions. Taking two z derivatives then yields

$$\frac{\partial^2 w}{\partial z^2} = -RB(T - \overline{T}) \tag{22}$$

where the linear operator B is given by

$$B = \frac{\partial^2}{\partial z^2}(\Delta_{DN}^2)^{-1}\Delta_h \tag{23}$$

The temperature equation obeys a maximum principle so that

$$0 \le T \le 1$$

holds pointwise in space and time. Therefore

$$0 \le |T - \overline{T}| \le 1$$

holds. The operator B is not bounded in L^∞ but obeys a logarithmic extrapolation estimate. This means that higher derivatives enter logarithmically in the bound; the estimate

$$\left\langle \|B(T - \overline{T})\|_{L^\infty}^2 \right\rangle \le C_1^2 \left\{1 + \log_+ R\right\}^4 \tag{24}$$

follows from the bounds in ([16]). The constant C_1 can be computed explicitly. Therefore, using the same kind of background as above in (11) we obtain

$$N \le \int_0^1 (\tau'(z))^2\, dz + 2\int_0^1 z^{\frac{5}{2}}|\tau'(z)| \left\langle \left\| \frac{\partial^2 w(\cdot, t)}{\partial z^2} \right\|_{L^\infty} \|\nabla(T - \overline{T})\| \right\rangle dz \tag{25}$$

and consequently

$$N \leq \frac{C}{\delta} + 2E\delta^{\frac{5}{2}} C_1 R \left\{1 + \log_+ R\right\}^2 \sqrt{n} \tag{26}$$

with C defined in (14), C_1 coming from (24) and

$$E = \int_0^1 s^{\frac{5}{2}} P(s) ds. \tag{27}$$

optimizing in $\delta < 1$ we find

$$\delta^{-1} = \left[\frac{5E}{C} C_1 R \left\{1 + \log_+ R\right\}^2 \sqrt{n}\right]^{\frac{2}{7}} + 1$$

and then letting $P \to s$ we deduce

Theorem 2 *There exists a constant C_2 such that the Nusselt number for the infinite Prandtl number equation is bounded by*

$$N \leq 1 + C_2 R^{\frac{2}{7}} \left\{1 + \log_+ R\right\}^{\frac{4}{7}} n^{\frac{1}{7}}.$$

If no additional information is given then, using $n \leq N$ in the theorem above we recover the result

$$N \leq 1 + C_2^{\frac{7}{6}} R^{\frac{1}{3}} \left\{1 + \log_+ R\right\}^{\frac{2}{3}}$$

of ([16]). The theorem brings in the exponent 2/7 as long as n is not varying too much with R.

Discussion

There is no proof that the Nusselt number cannot ever scale like $R^{\frac{1}{2}}$ for exceedingly large R. The experimental data do not seem however to encounter this behavior. Mathematically, the Nusselt number represents the maximum (among all possible invariant measures) of the expected value of the diameter of the global attractor in the energy dissipation norm. The functions on the attractor are not arbitrary, and may have certain properties that explain the experimentally observed bounds ([11]). In this paper we showed that if the ratio

$$\frac{n}{N} = \frac{\left\langle \|\nabla(T - \overline{T})\|^2 \right\rangle}{\langle \|\nabla T\|^2 \rangle}$$

is small then the Nusselt number dependence on Rayleigh is depleted.

Acknowledgment This work was partially supported by NSF-DMS9802611 and by the ASCI Flash Center at the University of Chicago under DOE contract B341495.

References

[1] L.N. Howard, (1964), Heat transport in turbulent convection, J. Fluid Mechanics **17** 405-432.

[2] L.N. Howard, (1964) Convection at high Rayleigh number, Applied Mechanics, Proc. 11th Cong. Applied Mech. (Ed. H. Grtler), pp. 1109-1115.

[3] R.H. Kraichnan, (1962), Turbulent thermal convection at arbitrary Prandtl number, Phys. Fluids 5, 1374-1389.

[4] F. Heslot, B. Castaing, and A. Libchaber, (1987), Transitions to turbulence in helium gas, Phys. Rev. A 36, 5870-5873 .

[5] B. Castaing, G. Gunaratne, F. Heslot, L. Kadanoff, A. Libchaber, S. Thomae, X.-Z. Wu, S. Zaleski and G. Zanetti, (1989), Scaling of hard thermal turbulence in Rayleigh-Benard convection, JFM **204**, 1-30.

[6] B.I. Shraiman and E.D. Siggia (1990), , Heat transport in high-Rayleigh-number convection, , Phys. Rev. A 42, 3650-3653.

[7] J.J. Niemela, L. Skrbek, K.R. Sreenivasan, and R.J. Donnelly, (1999) "Turbulent convection at very high Rayleigh numbers", submitted to Nature.

[8] S. Chandrasekhar, (1961) Hydrodynamic and hydromagnetic stability, Oxford University Press, Oxford.

[9] C. R. Doering, P. Constantin, (1996), Variational bounds on energy dissipation in incompressible flows III. Convection, Phys. Rev E, **53** 5957-5981.

[10] C. R. Doering, P. Constantin, (1992), Energy dissipation in shear driven turbulence, Phys.Rev.Lett. **69**, 1648-1651.

[11] P. Constantin, C. R. Doering, (1996), Heat transfer in convective turbulence, Nonlinearity **9**, 1049-1060.

[12] P. Constantin, C. R. Doering, (1995), Variational bounds in dissipative systems, Physica D **82**, 221-228.

[13] C. Doering, P. Constantin, (1998), Bounds for heat transport in a porous layer, J. Fluid Mechanics **376**, 263-296.

[14] P. Constantin, C. Hallstrom, V. Putkaradze, (1999), Heat transport in rotating convection, Physica **D 125**, 275-284.

[15] S.-K. Chan, (1971), Infinite Prandtl Number Turbulent Convection, Stud. Appl. Math 50, 13-49.

[16] P. Constantin, C. R. Doering, (1999), Infinite Prandtl number convection, J. Stat. Phys., **94**, 159-172.

BULK DISSIPATION IN SHEAR LAYERS WITH SUCTION

CHARLES R. DOERING AND RODNEY A. WORTHING
Department of Mathematics, University of Michigan
Ann Arbor, Michigan 48109-1109

AND

EDWARD A. SPIEGEL
Department of Astronomy, Columbia University
New York, New York 10027

Introduction

Although turbulence presents some of the most challenging unsolved problems in theoretical physics and applied mathematics, there are quantities we may hope to be able to bound even when we cannot compute them exactly. In this report we describe some of the ways of doing this within the context of a flow in a shear layer with injection and suction at the boundaries. We obtain bounds on the rate of viscous energy dissipation, valid for turbulent flows as well as for any laminar (steady or unsteady) flows, that may be directly compared to that in an exact solution for a flow with the same boundary conditions.

The quantitative study of upper bounds for turbulence was largely initiated by Howard in the early 1960s with his formulation of a variational principle for bounds on the ensemble averaged rate of energy dissipation ε for statistically stationary flows [1]. That theory was developed by Busse and collaborators [2] and applied to a variety of shear flow and convection problems. In this work we utilized the "background field" method, a mathematical device introduced in 1941 by Hopf [3] which has been further developed and extended during the 1990s [4, 5, 6, 7, 8, 9, 10, 11, 12, 13, 14, 15, 16, 17]. This approach produces estimates on long time averages of bulk dissipation without any statistical hypotheses. Applied to problems with sufficient symmetry so that Howard's statistical stationarity hypotheses may be invoked, the two approaches have been found to share a related mathematical structure [18] and have in many cases produced similar estimates [19].

T. Kambe et al. (eds.), IUTAM Symposium on Geometry and Statistics of Turbulence, 33–38.
© 2001 *Kluwer Academic Publishers.*

Wang [20] recently used Hopf's approach for general flow geometries to show that the dissipation rate ε is typically bounded independent of the viscosity (ν) when $\nu \to 0$ as long as there is no flux at the boundaries. If there is flux (a normal component) at the boundary, then Hopf's original estimates, where the *a priori* bound varies exponentially with the Reynolds number (Re), are generally the only known limits. For the flows considered in this work we find that the upper bound for the specific problem under consideration *with* mass flux at the rigid boundaries does not exhibit such an unphysical exponential dependence on the viscosity or the Reynolds number. On the other hand, in many cases of turbulent shear flows along highly symmetric smooth boundaries, the empirical logarithmic friction law [21] suggests that ε scales $\sim \nu^0$ with corrections proportional to $(\log Re)^{-2}$ as $Re \to \infty$. In the presence of suction, we find that the lower estimate of any upper bound on ε provided by the exact solution is itself *above* that predicted by the logarithmic friction law, in accord with the upper bound. Hence this model and the analysis described here is an important example for upper bound theory: the upper bound scaling in the vanishing viscosity limit is sharp and only the prefactor could possibly be improved in the high Reynolds number limit. The details of the methods of analysis and the calculations, both analytic and computational, can be found in our complete presentation of this project [22].

Statement of the problem

We consider a layer of incompressible (unit density) Newtonian fluid with constant kinematic viscosity ν confined between parallel rigid planes separated by distance h. The bottom plate, at $y = 0$, is stationary and the top one at $y = h$ moves with speed U^* in the x-direction. There is also uniform injection of fluid into the layer with speed (flux) V^* on the top plane and fluid is removed uniformly at the same speed on the bottom plane. The velocity field is $\mathbf{u} = \mathbf{i}u_x + \mathbf{j}u_y + \mathbf{k}u_z$ and the conditions at the (rigid) boundaries are thus

$$\mathbf{u} = -\mathbf{j}V^* \quad \text{at} \quad y = 0, \tag{1}$$

$$\mathbf{u} = \mathbf{i}U^* - \mathbf{j}V^* \quad \text{at} \quad y = h. \tag{2}$$

In the interior, the velocity field and the pressure field $p(\mathbf{x}, t)$ are governed by the Navier-Stokes equations

$$\frac{\partial \mathbf{u}}{\partial t} + \mathbf{u} \cdot \nabla \mathbf{u} + \nabla p = \nu \Delta \mathbf{u} \tag{3}$$

$$\nabla \cdot \mathbf{u} = 0. \tag{4}$$

We consider periodic boundary conditions on all dependent variables in the horizontal directions with periods L_x and L_z.

The two nondimensional control parameters of this problem are the Reynolds number $Re = \frac{hU^*}{\nu}$ and the entry angle, θ, defined by $\tan\theta = \frac{V^*}{U^*}$.

We use the notation

$$\|f\|_2 = \left(\int_0^{L_x} dx \int_0^h dy \int_0^{L_z} dz |f(x,y,z)|^2 \right)^{1/2}, \tag{5}$$

for the L_2 norm and

$$< f > = \limsup_{T\to\infty} \frac{1}{T} \int_0^T f(t)dt. \tag{6}$$

for the largest possible long time average of functions of time. We identify the largest possible long time averaged bulk energy dissipation rate per unit mass in a solution as

$$\varepsilon = \frac{\nu}{hL_xL_z} \left\langle \|\nabla \mathbf{u}\|_2^2 \right\rangle. \tag{7}$$

A simple exact steady solution of the problem is the laminar flow

$$\begin{aligned} u_x(y) &= U^* \frac{1 - e^{-V^*y/\nu}}{1 - e^{-Re\tan\theta}} \\ u_y &= -V^* \\ u_z &= 0. \end{aligned} \tag{8}$$

For $\theta = 0$ this solution reduces to plane Couette flow $\mathbf{u}_{Couette} = \mathbf{i}U^*y/h$. For $0 < \theta < 90°$ and $Re\tan\theta > 1$ the flow has a classical boundary layer structure near the suction plane at $y = 0$. That is, the velocity deviates from a nearly constant bulk flow only in a layer of thickness

$$\delta_{laminar} = \frac{\nu}{V^*} = \frac{h}{Re\tan\theta} \tag{9}$$

above the suction boundary.

The energy dissipation rate in the steady solution is

$$\varepsilon_{laminar} = \frac{v}{h} \int_0^h \left(\frac{\partial u_x}{\partial y} \right)^2 dy = \frac{U^{*3}}{h} \frac{\tan\theta}{2\tanh(\frac{1}{2}Re\tan\theta)} \tag{10}$$

Not unexpectedly, this expression reduces to the energy dissipation rate in planar Couette flow as $Re \to 0$ or $\theta \to 0$:

$$\lim_{Re\tan\theta\to 0} \varepsilon_{laminar} = \varepsilon_{Couette} = \nu \frac{U^{*2}}{h^2}. \tag{11}$$

Large values of the Reynolds number may be achieved in various ways for the flow geometry considered here, but we concentrate on two just particular extreme limits. First note that for fixed velocities and geometry, the Reynolds number increases as the viscosity decreases: $Re \to \infty$ as $\nu \to 0$. The laminar flow has two distinct simple limits for vanishing viscosity. For $\theta = 0$, $\mathbf{u}_{Couette}$ retains its structure as $\nu \to 0$. But for $\theta \neq 0$, the vanishing viscosity limit is the constant flow field

$$\lim_{\nu \to 0} \mathbf{u}_{laminar} = \mathbf{i}U^* - \mathbf{j}V^* \quad (\theta \neq 0). \tag{12}$$

That is, the limiting velocity vector field is constant parallel flow in the bulk, continuous at the injection boundary, and *discontinuous* at the suction boundary; $\delta_{laminar} \to 0$ as $Re \to \infty$ at fixed $\theta \neq 0$ and h.

For $\theta = 0$, $\varepsilon_{Couette}$ vanishes as $\nu \to 0$, but for $\theta \neq 0$, the emerging discontinuity at the suction boundary results in a residual dissipation:

$$\lim_{\nu \to 0} \varepsilon_{laminar} = \frac{\tan\theta}{2} \frac{U^{*3}}{h} \qquad (\theta \neq 0). \tag{13}$$

This residual dissipation is a manifestation of ν^0 scaling in the vanishing viscosity limit, i.e., the energy dissipation rate is nonvanishing and independent of the viscosity as $Re \to \infty$. Dimensional analysis insists that ε be composed of the cube of a velocity scale divided by a length scale, exactly the content of equation (13) with a geometry-dependent prefactor, i.e., the angle θ. Such scaling is often associated with high Reynolds number energy dissipation in the presence of a turbulent energy cascade, but this flow is an example of a steady laminar flow in which the dissipation takes place on an ever smaller length scale ($\delta_{laminar}$) which disappears in the zero viscosity limit. One of the major points of this study was to compare this laminar dissipation rate to an upper limit valid for *any* solution of the Navier-Stokes equations, even solutions corresponding to turbulent flows.

The Reynolds number also becomes large for fixed velocities and viscosity as the layer thickness increases: $Re \to \infty$ as $h \to \infty$. The semi-infinite layer is a trivial limit for $\theta = 0$ at fixed U^*, but for $\theta \neq 0$, the laminar solution with suction on the boundary has the nontrivial limit

$$\lim_{h \to \infty} \mathbf{u}_{laminar} = \mathbf{i}U^*(1 - e^{-y/\delta_{laminar}}) - \mathbf{j}V^* \quad (\theta \neq 0). \tag{14}$$

Then the energy dissipation rate per unit horizontal area is

$$\varepsilon'_{laminar} = \lim_{h \to \infty} \nu \int_0^h \left(\frac{\partial u_x}{\partial y}\right)^2 dy = \frac{\tan\theta}{2} U^{*3} \tag{15}$$

independent of the viscosity.

Summary of results

We explored the stability characteristics of the steady laminar solution in the $Re - \theta$ plane. Energy stability theory [23] was used to search for sufficient conditions for absolute stability, establishing that for sufficiently large values of the injection angle ($\theta > 3°$) or sufficiently small values of the Reynolds number ($Re < 82$), the steady laminar flow is indeed absolutely stable. We also used linear stability theory [24] to establish sufficient conditions for instability, finding that for sufficiently small but nonzero angles ($0 < \theta < .001°$) the laminar flow is linearly unstable at high Reynolds numbers. This is consistent with a previous stability analysis for the semi-infinite layer [25]. Instability of the steady flow does not prove that turbulence necessarily follows, but it is highly suggestive that turbulent flows may appear at high Reynolds numbers with sufficiently small injection angles.

Using the background field method we proved [22] that for any flow with Reynolds numbers $Re \geq 2\sqrt{2}$,

$$\varepsilon \leq \varepsilon_B \equiv \frac{1}{2\sqrt{2}} \left(1 + \frac{8}{3}\tan^2\theta \left(1 - \frac{3\sqrt{2}}{2}\frac{1}{Re}\right)\right) \frac{U^{*3}}{h}. \tag{16}$$

This is the explicit upper bound valid for any solution, steady, unsteady or turbulent, of the Navier-Stokes equations with these boundary conditions. This estimate displays the ν^0 scaling at high Reynolds numbers:

$$\varepsilon_B \sim \frac{1}{2\sqrt{2}} \left(1 + \frac{8}{3}\tan^2\theta\right) \frac{U^{*3}}{h} \quad \text{as} \quad Re \to \infty. \tag{17}$$

Note also that the upper bound on the energy dissipation rate per unit horizontal area of the suction boundary is finite in the limit of a semi-infinite fluid layer:

$$\varepsilon' \leq \varepsilon'_B \equiv \lim_{h\to\infty} h\varepsilon_B = \frac{1}{2\sqrt{2}} \left(1 + \frac{8}{3}\tan^2\theta\right) U^{*3}. \tag{18}$$

The upper bound on the energy dissipation rate—valid even for turbulent solutions of the Navier-Stokes equations—scales precisely the same as that in the steady laminar solution with regard to the viscosity as $\nu \to 0$. The laminar dissipation and the upper bound on turbulent dissipation exhibit a residual dissipation in the vanishing viscosity limit when $\theta \neq 0$. That is, both $\varepsilon_{laminar}$ and ε_B obey

$$\varepsilon \sim \frac{U^{*3}}{h}\mathcal{F}(\theta) \quad \text{as} \quad Re \to \infty. \tag{19}$$

Hence the turbulent bound and high Reynolds number laminar energy dissipation for flows in this geometry differ only by just a prefactor that depends only on the injection angle. This establishes the sharpness of the upper bound's scaling for these boundary conditions. This system provides a mathematically accessible example of the delicacy of corrections to high Reynolds number scaling—such as logarithmic terms as appearing in the law of the wall—to perturbations in the boundary conditions. This observation is consistent with the sensitivity of logarithmic corrections to wall roughness or other disturbances in turbulent shear flows [26, 27].

References

1. Howard, L. N., *J. Fluid Mech.* **17**, 405 (1963); *Ann. Rev. Fluid Mech.* **4**, 473 (1972).
2. Busse, F. *Adv. Applied Mech.* **18**, 77 (1978).
3. Hopf, E., *Math. Annalen* **117**, 764 (1941).
4. Doering, C. R., and Constantin, P., *Phys. Rev. Lett.* **69**, 1648 (1992).
5. Marchioro, C., *Physica D* **74**, 395 (1994).
6. Doering, C. R., and Constantin, P., *Phys. Rev. E* **49**, 4087 (1994).
7. Constantin, P. and Doering, C. R., *Physica D* **82**, 221 (1995).
8. Gebhardt, T., Grossmann, S., Holthaus, M. and Löhden, M., *Phys. Rev. E* **51**, 360 (1995).
9. Constantin, P. and Doering, C. R., *Phys. Rev. E* **51**, 3192 (1995).
10. Nicodemus, R., Grossmann, S. and Holthaus, M., *Physica D* **101**, 178 (1997).
11. Nicodemus, R., Grossmann, S. and Holthaus, M., *J. Fluid Mech.* **363**, 281 (1998).
12. Doering, C. R., and Constantin, P., *Phys. Rev. E* **53**, 5957 (1996).
13. Constantin, P. and Doering, C. R., *Nonlinearity* **9**, 1049 (1996).
14. Doering, C. R., and Hyman, J. H., *Phys. Rev. E* **55**, 7775 (1997).
15. Constantin, P., Hallstrom, C., and Putkaradze, V., *Physica D* **125** 275 (1999).
16. Constantin, P. and Doering, C. R., *J. Stat. Phys.* **94**, 159 (1999).
17. Doering, C. R., and Constantin, P., *J. Fluid Mech.* **376**, 263 (1998).
18. Kerswell, R., *Physica D* **121**, 175 (1998).
19. Kerswell, R., *Physica D* **100**, 355 (1997).
20. Wang, X., *Physica D* **99**, 555 (1997).
21. Prandtl, L., *Angew. Math. Mech.* **5**, 136 (1925). For a recent experimental realization, see Lathrop, D. P., Feinberg, J. and Swinney, H. L. *Phys. Rev. A* **46** 6390 (1992).
22. Doering, C. R., Spiegel, E. A., and Worthing, R. A., preprint, 1999.
23. Joseph, D. D., *Stability of Fluid Motions*, New York: Springer, 1976.
24. Drazin, P. and Reid, W., *Hydrodynamic Stability*, Cambridge, England: Cambridge University Press, 1981.
25. Hocking, L., *Quart. J. Mech. & Appl. Math.* **28**, 341 (1974).
26. See for example, Vennard, J., and Street, R., *Elementary Fluid Mechanics*, New York: John Wiley & Sons, 1980, ch. 9, pp. 372-380.
27. Cadot, O., Couder, Y., Daerr, A., Douady, S. and Tsinober, A., *Phys. Rev. E.* **56**, 427 (1997).

Trying a metric on atmospheric flows

Rubén A. Pasmanter[1] and Xue-Li Wang
KNMI, P.O.Box 201, 3730 AE De Bilt, Netherlands

1 Introduction

It is not obvious how to introduce, in a natural and intrinsic way, a metric structure in the configuration space of a fluid, i.e., in a space with $(\rho, T, \vec{u}) \equiv$ (density, temperature, velocity) as coordinates. Such a lack of geometric structures imposes many restrictions on the kind of computations that one can perform: it is impossible to talk of "the distance" between two states of the fluid, i.e., between two positions with coordinates $(\rho_1, T_1, \vec{u}_1)$ and $(\rho_2, T_2, \vec{u}_2)$ respectively; neither is it possible to talk of "the norm" of the vector formed by the rate of change of the dynamical variables $(d\rho/dt, dT/dt, d\vec{u}/dt)$; it is not possible to consider the angle between two such vectors; there is no volume element defined in configuration space, therefore, it does not make sense to talk about "the density" of a distribution of points in that space; etc. These limitations are too restrictive since, to name but a few examples: 1) A norm is needed when studying the (in)stability of flows, especially when the linear growth of the perturbations is only transient [[6]]; 2) The angle between two directions is required in order to determine how strongly a given perturbation projects along an optimal perturbation [[7]]; 3) A volume element is needed in order to determine the density distribution of ensemble simulations of flows [[8]]. Therefore, it is often opted to circumvent these limitations by introducing an acceptable, if somewhat arbitrary, metric tensor [[9, Hanifi A., Schmid P. and Henningson D.S., Transient growth in compressible boundary layer; Phys. Fluids 8 (1996) 826–837.]]. In the case of incompressible, isentropic flows, the kinetic-energy metric is justifiably employed [[6, 9]].

In [[5]] we constructed a natural metric tensor in the configuration space of a compressible fluid in local thermal equilibrium and in uniform motion. In the present work, we try this metric on atmospheric flows, i.e., on strongly non-uniform flows, and show that even in this case the metric leads to interesting and useful results. In order to avoid possible misunderstandings: we are now working on the derivation of a metric that includes the gradients of the extensive variables from the outset; this metric will be much better suited to deal with non-uniform flows than the one considered in the present work.

[1]Corresponding author; e-mail: *pasmante@knmi.nl*

T. Kambe et al. (eds.), IUTAM Symposium on Geometry and Statistics of Turbulence, 39–46.
© 2001 *Kluwer Academic Publishers.*

2 Metric of a gas in thermal equilibrium

In [[5]] we constructed a natural metric tensor in the configuration space of a compressible fluid in local thermal equilibrium and in uniform motion. In this Section, we briefly review its derivation. The thermodynamical variables $(\rho, T, \vec{u})$ can be seen as the parameters defining the probability distribution of the microscopic, thermal fluctuations. The distance between two probability distributions [[2]], e.g., parametrized by $(\rho, T, \vec{u})$ and by $(\rho + d\rho, T + dT, \vec{u} + d\vec{u})$ respectively, is given by the so-called Fisher metric [[3]]. It is computed as follows: 1) For a system in local thermal equilibrium and occupying a volume V, the probability density of finding N identical particles of mass m possessing a total linear momentum $\vec{M}(\xi_N)$ and a total energy $E(\xi_N)$ is

$$p_N(\xi_N; \gamma, \beta, \vec{\kappa}) = \exp[\gamma N - \beta E(\xi_N) + \vec{\kappa} \cdot \vec{M}(\xi_N)]/\hbar^N N! \mathcal{Z}(\gamma, \beta, \vec{\kappa}), \tag{1}$$

where ξ_N stands for the positions and the momenta of the particles and the grand-canonical partition function $\mathcal{Z}(\gamma, \beta, \vec{\kappa})$ is

$$\mathcal{Z}(\gamma, \beta, \vec{\kappa}) := \sum_N^{\infty} \int d\xi_N \exp\left[\gamma N - \beta E(\xi_N) + \vec{\kappa} \cdot \vec{M}(\xi_N)\right]/\hbar^N N!. \tag{2}$$

The intensive parameters characterizing the thermal bath are: the inverse temperature β, the chemical potential $\mu = \gamma/\beta$ and the velocity $\vec{u} = \vec{\kappa}/\beta$. In accordance with Galilean invariance, the chemical potential of the fluid moving with velocity $\vec{u}$ is shifted by an amount $mu^2/2$ with respect to the chemical potential of the fluid at rest. One identifies $\ln \mathcal{Z}(\gamma, \beta, \vec{\kappa})$ as the (dimensionless) thermodynamic grand potential, i.e., $\ln \mathcal{Z}(\gamma, \beta, \vec{\kappa}) = \beta PV$, where $P = P(\gamma, \beta, \vec{\kappa})$ is the pressure. 2) In the intensive-variables coordinate system $\theta := (\gamma, \beta, \vec{\kappa})$, the components of the Fisher metric tensor, are given by $g_{ij}(\theta) := \langle (\partial \ln p_N/\partial \theta^i) (\partial \ln p_N/\partial \theta^j) \rangle$, where the pointed brackets indicate an average taken over the distribution defined in (1). These components turn out to equal,

$$g_{ij}(\theta) = \frac{\partial^2 \ln \mathcal{Z}(\theta)}{\partial \theta^i \partial \theta^j} = V \frac{\partial^2 \beta P(\theta)}{\partial \theta^i \partial \theta^j}. \tag{3}$$

In the case of an ideal gas $\ln \mathcal{Z}_{ideal}(\gamma, \beta, \vec{\kappa}) = V \beta^{-3/2} \exp(\gamma + (m\kappa^2/2\beta))$. In this case, the metric is diagonal when expressed in the coordinate system $(\rho, T, \vec{u})$; the contribution of a volume element V to the squared distance between two states $(\rho, T, \vec{u})$ and $(\rho + d\rho, T + dT, \vec{u} + d\vec{u})$ turns out to be

$$(dL)^2_{ideal} = \frac{V\rho}{m} \left[\left(\frac{d\rho}{\rho}\right)^2 + \frac{3}{2}\left(\frac{dT}{T}\right)^2 + m\beta\, d\vec{u} \cdot d\vec{u} \right]. \tag{4}$$

This metric is not flat, however, there are two, conjugate flat connections. For more details, see [[5]].

3 Atmospheric states

The metric described in the previous Section is the most natural one when dealing with fluids in thermal equilibrium and in uniform motion. In practice, however, one is more interested in non-uniform or even turbulent motion. Work is now under way [[11]] in order to generalize the results in the previous Section to flows with gradients in their configuration variables, e.g., in ρ, T and in $\vec{u}$. While waiting for the results of this generalization, we decided to check the performance of (4) when it is applied to mesoscopic-scale motions in the atmosphere. One of the first things one wants to check is whether the three terms in eq. (5) are, statistically speaking, of the same order of magnitude. Suppose that at two different times t_1 and t_2, we are given two atmospheric configurations which are characterized by velocity, temperature and density fields, i.e., by $\vec{u}_i(\vec{R}) \equiv \vec{u}(\vec{R}, t_i)$, $T_i(\vec{R}) \equiv T(\vec{R}, t_i)$ and by $\rho_i(\vec{R}) \equiv \rho(\vec{R}, t_i)$ with $i = 1$ and 2, respectively. Define $\Delta\rho(\vec{R}) \equiv \rho_2 - \rho_1, \Delta T(\vec{R}) \equiv T_2 - T_1$, etc. From the available atmospheric data it follows that, on the average, $|\Delta\rho| \ll \rho$, $|\Delta T| \ll T$ and $m\Delta\vec{u} \cdot \Delta\vec{u} \ll kT$. Therefore, in equation (4) we can replace the infinitesimals $d\vec{u}, dT$ and $d\rho$ by the finite differences $\Delta\vec{u}, \Delta T$ and $\Delta\rho$; subsequently we sum the contributions from each volume element $V \leftrightarrow d^3R$ so that the total distance between these two atmospheric states is

$$(\Delta L)^2 = \int d^3R\, \rho \left[\left(\frac{\Delta\rho}{\rho}\right)^2 + \frac{3}{2}\left(\frac{\Delta T}{T}\right)^2 + \frac{m}{kT}\Delta\vec{u}\cdot\Delta\vec{u} \right] \tag{5}$$

where the quantities ρ, $\Delta\rho$, T, ΔT and $\Delta\vec{u}$ are all functions of the position $\vec{R}$. The constants in (5) are: Boltzmann's constant k (1.38066×10^{-23} Joule/Kelvin) and m, the mass of an 'air molecule' ($28.9644 \times 1.67265 \times 10^{-27}$ kgr). Notice that $(\Delta L)^2$ is a dimensionless quantity; its value is independent of the spatial coordinate system **and** of the coordinates system used to describe the dynamical variables.

Most data are available on isobaric surfaces; this turns out to be convenient for our calculations because, when passing from the space coordinates $\vec{R} \equiv (x, y, z)$ to pressure coordinates (x, y, p), the following simplifications take place: 1) Due to the hydrostatic balance, the volume element times the density, $dx\, dy\, dz\, \rho$, appearing in the integral (5) becomes $g^{-1} dx\, dy\, dp$ (g being the gravity constant); 2) At fixed pressure, density and temperature are inversely proportional to each other so that the first term in the integral (5) can be approximated as

$$\left(\frac{\Delta\rho}{\rho}\right)_P^2 \cong \left(\frac{\Delta T}{T}\right)_P^2,$$

as long as $\Delta\rho \ll \rho$, $\Delta T \ll T$. Then the first two terms in the integral (5) can be combined into one so that (5) becomes

$$(\Delta L)^2 \cong g^{-1} \int dxdydp \left[\frac{5}{2}\left(\frac{\Delta T}{T}\right)^2 + \frac{m}{kT}\Delta\vec{u}\cdot\Delta\vec{u} \right]. \tag{6}$$

For the sake of simplicity, we shall call the first term "the temperature contribution". It can be shown [[10]] that, on the average, by using isobaric data with $p_1 = p_2$ one underestimates the contribution of the first two terms in (5).
For later use, we introduce also the definition of a "distance" in terms of the geopotential height $Z(x, y, p, t)$:

$$(\Delta L)^2_H := g^{-1} \int dxdydp\,[\Delta Z]^2, \tag{7}$$

where $\Delta Z \equiv Z(x, y, p, t_2) - Z(x, y, p, t_1)$.

All the results presented below were computed on fixed pressure levels, i.e., the integral over p in the integrals (6) and (7) were *not* performed. Moreover, the constant of gravity g has been ignored and the results of integrating over the horizontal coordinates (x, y) have been divided by the total area of integration. Consequently, in the results presented below, $(\Delta L)^2$ is dimensionless while $(\Delta L)^2_H$ has dimensions length^2.

We can express $(\Delta L)^2$ in terms of the gradients of the geopotential $\Phi(x, y, p, t) = gZ(x, y, p, t)$ as follows: 1) Assuming hydrostatic balance, the temperature on the p-surface is related to the vertical gradient of the geopotential by $T = -R^{-1}p(\partial\Phi/\partial p)$, 2) At midlatitudes, assuming geostrophic balance, the (horizontal) velocity is related to the horizontal gradient of the geopotential by $\vec{u} = f^{-1}\vec{z} \wedge \vec{\nabla}\Phi$ where f is the Coriolis parameter; therefore

$$\left.\frac{\Delta T}{T}\right|_P = \frac{\Delta(\partial\Phi/\partial p)}{\partial\Phi/\partial p} \quad \text{and} \quad \frac{m}{kT}\Delta\vec{u}\cdot\Delta\vec{u} = -\frac{1}{f^2 p}\frac{\partial p}{\partial\Phi}\left|\Delta\vec{\nabla}\Phi\right|^2.$$

Consequently, when these assumptions are valid, a very small value of (6) with the volume integration restricted to the extra-tropical regions means that the corresponding geopotentials have very similar values over the 3D, extra-tropical atmosphere. Similarly, a very small (7) restricted to the extra-tropical areas implies a very small value of (6) restricted to the same extra-tropical area. However, horizontal and vertical gradients of $\Phi(x, y, p, t)$ contribute with different weights to (6); moreover, in tropical regions and whenever deviations from geostrophy are appreciable, it is not possible to relate $(\Delta L)^2$ to the geopotential gradients.

4 Data description and results

We used ECMWF analysis data on the temperature, zonal wind velocity, meridional wind velocity[2] and geopotential height fields on the 850, 500, 250, 100 hPa isobars of 14 Northern hemisphere winters (December through February) corresponding to the years 1982 through 1995. These data sets contain 4 time steps per day at 00 GMT, 06 GMT, 12 GMT and 18 GMT each. We wanted to check whether the three terms in eq. (5) are, statistically speaking, of the same order of magnitude. We took the fields of the first 15 days in December 1995 at 00 GMT

[2]The vertical velocity gives a negligible contribution to the integral (1).

and computed, separately, the two terms in eq. (6) for fields ranging from 6hs up to 360hs after the chosen day for each of these 15 days and for each isobar; finally, we averaged the results over the 15 initial days.
The results are presented in Figs. 1 and 2. The general picture one gets agrees with the theoretical expectations: at short time differences, less than one day, the growth is dominated by the deterministic part of the dynamics ("ballistic" regime) and $(\Delta L)^2 \propto (t_2 - t_1)^2$; at later time differences, approximately between one and five days, the time-correlation ("memory") is lost and the motion acquires a random-walk or diffusive character so that $d(\Delta L)^2/dt \propto 2D$ where D is a diffusion-like constant; finally, $|\Delta L|$ reaches the average distance between two arbitrary points on the attractor and saturation sets in around a time difference of approximately 15 days. However, a systematic, weak increase is observed also at these relatively large time differences. Since the data has not been detrended this trend is probably due to systematic seasonal effects; in fact, a couple of tests that we have performed validate this conjecture.

The "temperature contribution" has approximately the same value on all pressure surfaces. On the lowest and highest pressure surfaces, i.e., on the 850 hPa and 100 hPa pressure surfaces, the square of the "temperature term" $\frac{5}{2}\left(\frac{\Delta T}{T}\right)^2$ is comparable to the kinetic energy term $\frac{m}{kT}\Delta\vec{u}\cdot\Delta\vec{u}$. On the other pressure surfaces, the kinetic energy term is appreciably larger than the square of the "temperature term", the largest difference occurs on the 250 hPa pressure surface where the kinetic-energy term is one order of magnitude larger than the temperature term. We interpret this as a manifestation of large velocity fluctuations in the atmospheric jets. In order to check this and in order to get an idea about any possible spatial dependence of the results, we looked at the Northern (see Fig. 1) and Southern (not shown) extratropical and at the tropical (see Fig. 2) regions separately. As one can see, both contributions to $(\Delta L)^2$ are a factor 1/8-th to 1/6-th smaller in the tropics than in the extratropics, the only exception being the 100 hPa pressure surface where they are approximately 1/2 (kinetic-energy term) to 1/4-th (temperature term) smaller than in the extra-tropics. All the differences between tropical and extratropical regions, as well as those between Northern and Southern extratropical regions, are compatible with associating stronger velocity fluctuations with the jet areas.

A similar study was performed using $(\Delta L)_H^2$, the geopotential-height definition of the "distance" as in equation (7). For this computation we used ECMWF analysis of geopotential height on 850, 500, 250 hPa pressure levels for the same period, i.e., December 1995. In this case, tropical squared "distances" are, approximately, twenty times smaller than those in the extratropics; the 250 hPa pressure surface, which strongly overlaps with the jets, gives a squared "distance" five times larger than the 850hPa surface. The most striking differences observed between the geopotential-height results and those obtained with (6) are: 1) when using the geopotential-height distance $(\Delta L)_H^2$, the lack of saturation even after 17days[3] is particularly evident on the 500 hPa and 250 hPa pressure surfaces in

[3]We extended the maximum time-lag to 20 days and still did not find signs of saturation on

the extratropics, 2) in the tropical area, the tidal oscillation is more clearly visible with the geopotential-height definition and 3) the difference between tropical and extratropical regions is stronger with the geopotential-height distance $(\Delta L)^2_H$.

5 Conclusions and comments

The definition of a "distance" as given by eq. (5) was obtained by considerations about fluids in local thermal equilibrium and in uniform motion. In spite of this, we have found that it gives reasonable results also in the atmosphere. More specifically: 1) Even in the jet regions, the kinetic-energy term is, at most (remember that some density fluctuations have been neglected by taking isobaric coordinates), 10 times larger than the other contributions. 2) The magnitude of the wind and temperature terms are of the same order on the 850 hPa level over the whole globe. The level on which the wind contribution is more dominant is the 250 hPa surface. 3) Extra-tropical regions contribute, per unit area, approximately 5 times the contribution of tropical ones. If one uses the geopotential-height measure then this difference grows to become a factor of approximately 25, i.e., the geopotential measure gives an overwhelming weight to the extra-tropical regions.

We have extended the distance definition by including humidity. It turned out that this variable gives a contribution comparable to the temperature terms only in the lower tropical atmosphere; this agrees with one's expectations. We performed other checks, like searching best analogues in the record available to us. For more details, please refer to [[10]]. As already mentioned, we are developing the metric generated by non-uniform flows, i.e., with gradients in the density, temperature and velocity.

References

[1] Amari S., *Differential-Geometrical Methods in Statistics,* Lecture Notes in Statistics nr. 28, Springer Verlag, 1985.

[2] Eguchi S., *A differential geometric approach to statistical inference on the basis of contrast functionals*, Hiroshima Math. J. **15** (1985) 341–391.

[3] Fisher R.A., *Theory of statistical estimation*, Proc. Cambridge Philos. Soc. **22** (1925), 700–725.

[4] Nomizu K. and Simon U., *Notes on cojugate connections* in *Geometry and topology of submanifolds IV*, eds.: Dillen F. and Verstraelen L., World Scientific, pp. 152–173, 1992.

[5] Pasmanter R.A., *Metric structures of laminar flows*, Physica A **258** (1998) 311-328.

the 250 hPa level. This is probably due to a sistematic seasonal trend.

[6] Trefethen L.N., Trefethen A.E., Reddy S.C. and Driscole T.A., *Hydrodynamic stability without eigenvalues*, Science **261** (1993) 578; Reddy S. C. and Henningson D. S., *Energy growth in viscous channel flows*, J. Fluid Mech. **252** (1993) 209–238.

[7] Butler K. M. and Farrell B. F., *Three-dimensional optimal perturbations in viscous shear flow*, Phys. Fluids **A 4** (1993) 1637–1650.

[8] Palmer T., Molteni F., Mureau R., Chapelet P. and Tribbia J., *Ensemble predictions*, pp. 21–66 in *Validation of models over Europe, Vol. I*, ECMWF-Seminar Proceedings September 1992, Reading.

[9] Hanifi A., Schmid P. and Henningson D.S., *Transient growth in compressible boundary layer*, Phys. Fluids **8** (1996) 826–837.

[10] Pasmanter R. A. and Wang, X. L., *Experimenting with a similarity measure of atmospheric flows*, KNMI-report 9803. Available also as `http://www.knmi.nl/~wang/ruben/ruben.html`.

[11] Pasmanter R. A., *The metric structure generated by non-uniform, compressible flows*, in preparation.

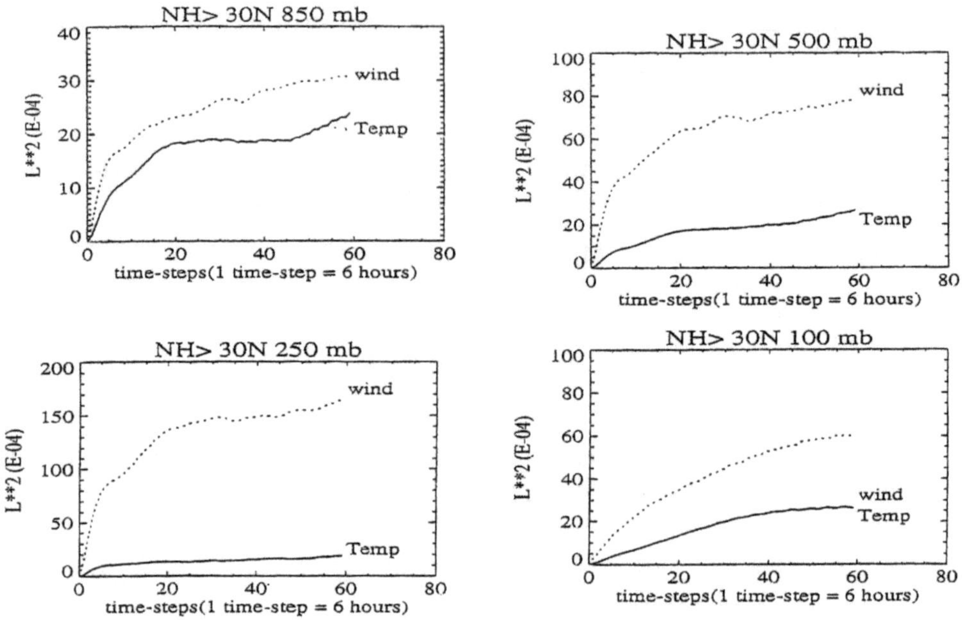

Fig. 1. Average growth with time of the wind and temperature contributions to the distance in the Northern extra-tropics (>30°N) on the 850mb, 500mb, 250mb and 100mb isobar levels. Notice the difference in scales.

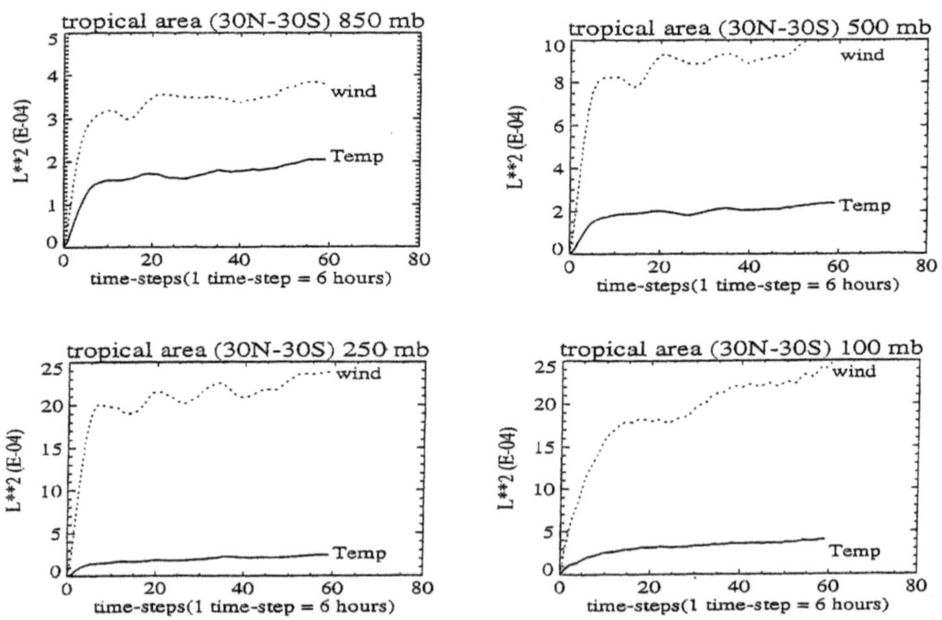

Fig. 2. As in Fig. 1 but now in the tropical region, 30° N–30°S. Notice the difference in scales, also with respect to Fig. 1.

ON THE ANALOGY BETWEEN TWO-DIMENSIONAL VORTICESAND STELLAR SYSTEMS

P.H. CHAVANIS
Université Paul Sabatier
31062 Toulouse, France

1. Introduction

Two-dimensional flows with high Reynolds numbers have the striking property of organizing spontaneously into coherent structures. The robustness of Jupiter's Great Red Spot, a huge vortex persisting for more than three centuries in a turbulent shear between two zonal jets, is probably related to this general phenomenon. Similarly, it is striking to observe that galaxies themselves follow a kind of organization revealed in the Hubble classification or in de Vaucouleur's $R^{1/4}$ law for the surface brightness of ellipticals (Binney & Tremaine, 1987). We shall discuss some analogies between stellar systems and 2D vortices and show that their structure and organization can be understood from relatively similar statistical mechanics [for a short review on this subject see Chavanis (1998a)].

2. Equilibrium states of stellar systems and 2D vortices

2.1. THERMODYNAMICS OF STELLAR SYSTEMS

Basically, a stellar system can be considered as a collection of N point masses in gravitational interaction described by the Hamilton equations:

$$m\frac{d\mathbf{r}_i}{dt} = \frac{\partial H}{\partial \mathbf{v}_i}, \qquad m\frac{d\mathbf{v}_i}{dt} = -\frac{\partial H}{\partial \mathbf{r}_i} \tag{1}$$

$$H = \sum_{i=1}^{N} \frac{1}{2} m v_i^2 - \sum_{i<j} \frac{Gm^2}{r_{ij}}. \tag{2}$$

When N is large, it is neither feasable nor useful to describe the motion of each individual star in detail and one is forced to recourse to statistical methods. We shall be particularly interested in the distribution function

T. Kambe et al. (eds.), IUTAM Symposium on Geometry and Statistics of Turbulence, 47–53.
© 2001 *Kluwer Academic Publishers.*

$f(\mathbf{r}, \mathbf{v}, t)$ which gives the average density of stars with position $\mathbf{r}$ and velocity $\mathbf{v}$ at time t. For short times, the evolution of the system is collisionless and the distribution function is solution of the Vlasov-Poisson equations

$$\frac{\partial f}{\partial t} + \mathbf{v}\frac{\partial f}{\partial \mathbf{r}} + \mathbf{F}\frac{\partial f}{\partial \mathbf{v}} = 0 \tag{3}$$

$$\mathbf{F} = -\nabla\Phi, \qquad \Delta\Phi = 4\pi G \int f d^3\mathbf{v} \tag{4}$$

where Φ is the gravitational potential. At later times, the collisions will substantially deviate the stars from their unperturbed mean field trajectories. By analogy to what happens in a gas as a result of "molecular chaos", one expects that the system will ultimately relax towards an isothermal distribution

$$f = Ae^{-\beta m(\frac{v^2}{2}+\Phi)} \tag{5}$$

which maximizes the Boltzmann entropy at fixed mass and energy. Simple arguments of kinetic theory show that the relaxation time is given by $t_{relax} \sim \frac{N}{\ln N} t_D$ where t_D is the dynamical time. This relaxation time is consistent with the age of globular clusters so most methods of statistical mechanics were initially developed for such systems.

However, the statistical mechanics of self-gravitating systems makes problems. This is due to the unshielded long-range nature of the potential and its singularity at short distances. The problems are first apparent when we realize that the solutions of the Boltzmann-Poisson system (4) (5) have an *infinite mass* since the star density decays only as r^{-2} at large distances. This means that there is no equilibrium state in an infinite domain: one can always increase entropy by spreading the density profile. One is forced therefore to invoke *incomplete relaxation* and use truncated models instead of (5). The Michie-King model takes into account the escape of high energy stars and gives relatively good fits with globular clusters. Another possibility, used by theorists, is to work inside a box against which the stars bounce ellastically. However, even when the system is confined to a box, statistical equilibrium does not exist in general: for low energies, the system takes a "core-halo" structure and can always create entropy by making the core denser and denser (and hotter and hotter) up to a black hole singularity. This instability, known as *gravothermal catastrophe*, can be related to the negative specific heat of self-gravitating systems: by losing heat, they grow hotter and evolve away from equilibrium!

For elliptical galaxies, the relaxation time is much larger than the age of the universe and their evolution is collisionless. Yet, these systems seem to have reached a universal equilibrium state as evidenced by de Vaucouleur's $R^{1/4}$ law. In 1967, Lynden-Bell showed that the solutions of the Vlasov-Poisson equations are not smooth but involve intermingled filaments at smaller and smaller scales. This is the result of a *mixing process* in phase space associated for example with the damped oscillations of a protogalaxy initially far from mechanical equilibrium. During this process, the strong fluctuations of the gravitational field redistribute energy between stars and provide a mechanism analogous to collisions in a gas. On account of these fluctuations, the system is expected to achieve, on a coarse-grained scale, an equilibrium state on a very short time scale $\sim t_D$. This is called *violent relaxation.* Due to the collisionless nature of the evolution, the coarse-grained distribution function can only decrease by internal mixing and must satisfy everywhere $\overline{f} \leq \eta_0$, where η_0 is the maximum phase-space density of the initial state. This results in an effective "exclusion principle", like in quantum mechanics, and leads to the Fermi-Dirac distribution at equilibrium

$$\overline{f} = \frac{\eta_0}{1 + \lambda e^{\beta\eta_0(\frac{v^2}{2}+\Phi)}}. \tag{6}$$

This distribution function also suffers the infinite mass problem and specific truncated models must be introduced. However, when the system is confined to a box there exists now a *global* entropy maximum for any accessible energy. For low energies, it is made of a degenerate core surrounded by a halo of stars as calculated by Chavanis & Sommeria (1998a). Like in quantum mechanics, the exclusion principle has a stabilizing role and removes the problems associated with the singularity of the gravitational potential at short distances.

2.2. CLASSIFICATION OF 2D VORTICES

Two-dimensional incompressible and inviscid flows are described by the Euler equations

$$\frac{\partial \omega}{\partial t} + \mathbf{u}\nabla\omega = 0 \tag{7}$$

$$\mathbf{u} = -\mathbf{z} \wedge \nabla\psi, \qquad \Delta\psi = -\omega \tag{8}$$

where ψ is the stream function and ω the vorticity. It is often convenient to approximate the vorticity field by a cloud of point vortices with individual circulation γ. This system has a Hamiltonian structure

$$\gamma\frac{dx_i}{dt} = \frac{\partial H}{\partial y_i}, \qquad \gamma\frac{dy_i}{dt} = -\frac{\partial H}{\partial x_i} \tag{9}$$

$$H = -\frac{1}{2\pi}\sum_{i<j}\gamma^2 \ln r_{ij} \tag{10}$$

where the coordinates (x, y) are canonically conjugate. Point vortices behave therefore like particles in interaction (like stars or electric charges) except that they produce directly a velocity, not an acceleration. This is therefore a very peculiar Hamiltonian system.

The statistical mechanics of point vortices was first considered by Onsager (1949) who showed that *negative temperatures* were possible. He predicted that, at negative temperatures, point vortices would cluster into "supervortices". His theory was further developed by Joyce & Montgomery (1973) in a mean field approximation. They found that the equilibrium state is given by

$$\omega = Ae^{-\beta\gamma\psi} \tag{11}$$

where β is the inverse temperature (the Lagrange multiplier associated with the energy constraint). At negative temperatures, the mean field equation obtained by substituting (11) into (8) is formally similar to the system (4)(5) and the interaction is "attractive" like gravity: the vortices cluster into large-scale structures. At positive temperatures, the interaction is "repulsive" like for electric charges, and the vortices are pushed against the boundaries. Consequently, the analogy between 2D vortices and stellar systems is intimately related to the existence of negative temperatures in 2D turbulence! However, unlike gravity, there is no gravothermal catastrophe because the Hamiltonian is purely "potential": point vortices cannot collapse otherwise their energy would not be conserved.

The analogy between stellar systems and 2D vortices is not limited to the point vortex approximation. The Euler equations for continuous vorticity fields are similar to the Vlasov-Poisson system and can generate a mixing process associated for example with the nonlinear developement of the Kelvin-Helmholtz instability or with vortex merging. As a result, a large-scale coherent structure is formed which can be considered as a maximum entropy (i.e most mixed) state with distribution:

$$\overline{\omega} = \frac{\sigma_0}{1 + \lambda e^{\beta\sigma_0\psi}} \tag{12}$$

similar to Lynden-Bell's statistics (Miller 1990, Robert & Sommeria 1991). In general, the relaxation is incomplete because the fluctuations necessary to provide ergodicity are effective only in a finite region of space and for a finite period of time. Therefore, isolated vortices can be considered as restricted maximum entropy states or "maximum entropy bubbles".

Unlike the stellar case, vorticity can be positive and negative which implies a wider variety of structures like monopoles, rotating or translating dipoles or tripoles. The interesting problem in that context is to obtain a classification of the "zoology" of vortices met in 2D flows. A step in that direction was made by Chavanis & Sommeria (1996,1998b) in a particular limit of the statistical theory (the strong mixing limit) where the maximization of entropy becomes equivalent to a kind of enstrophy minimization.

3. Relaxation towards equilibrium

3.1. DYNAMICAL FRICTION OF STARS

The relaxation of stars towards the isothermal distribution (5) was treated by Chandrasekhar (1943) by analogy with Brownian motion. Since the deflections due to two-body encounters are weak but numerous, he wrote a Fokker-Planck equation for the evolution of the distribution function

$$\frac{\partial f}{\partial t} = \frac{1}{2}\frac{\partial^2}{\partial v^\mu \partial v^\nu}\left(f\frac{\langle \Delta v^\mu \Delta v^\nu \rangle}{\Delta t}\right) - \frac{\partial}{\partial v^\mu}\left(f\frac{\langle \Delta v^\mu \rangle}{\Delta t}\right) \tag{13}$$

and calculated the moments of the velocity increments from a heuristic Langevin equation

$$\Delta \mathbf{v} = -\xi \mathbf{v}\Delta t + \mathbf{B}(\Delta t). \tag{14}$$

The term $\mathbf{B}(\Delta t)$ is a stochastic force and a *dynamical friction* $-\xi\mathbf{v}$ must be introduced in order to recover the Maxwell-Boltzmann distribution at equilibrium. When substituted in the Fokker-Planck equation, this yields

$$\frac{\partial f}{\partial t} = \frac{\partial}{\partial \mathbf{v}}\left(D\frac{\partial f}{\partial \mathbf{v}} + \xi f \mathbf{v}\right) \tag{15}$$

where D is the diffusion coefficient and ξ the coefficient of dynamical friction. The condition that the Maxwell-Boltzmann distribution (5) is a stationary solution of (15) requires that ξ and D be related according to the Einstein formula $\xi = D\beta m$ which expresses the "fluctuation-dissipation" theorem. Equation (15) can also be obtained from a general Maximum Entropy Production Principle (Chavanis, Sommeria & Robert 1996). In this formalism, β is the Lagrange multiplier associated with the conservation of energy and the Einstein relation is automatically satisfied. The same variational principle can be advocated for the coarse-grained relaxation of collisionless stellar systems and leads to the equation (Chavanis 1998b):

$$\frac{\partial \overline{f}}{\partial t} + \mathbf{v}\frac{\partial \overline{f}}{\partial \mathbf{r}} + \overline{\mathbf{F}}\frac{\partial \overline{f}}{\partial \mathbf{v}} = \frac{\partial}{\partial \mathbf{v}}\left\{D\left(\frac{\partial \overline{f}}{\partial \mathbf{v}} + \beta \overline{f}(\eta_0 - \overline{f})\mathbf{v}\right)\right\}. \tag{16}$$

The nonlinearity $\overline{f}(\eta_0 - \overline{f})$ of the effective friction accounts for the degeneracy discovered by Lynden-Bell at equilibrium. From equations (15) and (16) one can derive stationary distributions with a *finite mass* known as Michie-King models (possibly degenerate, Chavanis 1998b):

$$f = A(e^{-\beta m\epsilon} - e^{-\beta m\epsilon_m}), \qquad \overline{f} = \eta_0\frac{e^{-\beta\eta_0\epsilon} - e^{-\beta\eta_0\epsilon_m}}{\lambda + e^{-\beta\eta_0\epsilon}}. \tag{17}$$

These models take into account the escape of stars with energy $\epsilon \geq \epsilon_m$.

3.2. SYSTEMATIC DRIFT OF POINT VORTICES

Physically, the dynamical friction experienced by a body moving through a stellar system arises from the attraction that the body experiences for the region of enhanced density that tails behind it as a wake behind a ship. It is therefore the result of a polarization process: the star attracts the surrounding mass and in response the system exerts a "back-reaction" which is responsible for its decceleration. We can use a similar approach to show that a point vortex in 2D turbulence experiences a *systematic drift* normal to its mean field velocity. This drift can be calculated precisely with a linear response theory which yields (Chavanis 1998c):

$$\langle \mathbf{V} \rangle_{drift} = -D\beta\gamma\nabla\psi. \tag{18}$$

The drift coefficient $\xi = D\beta\gamma$ is an amuzing generalization of the Einstein relation to the case of point vortices. The direction of the drift is consistent with Onsager theory: when $\beta < 0$, the drift is "attractive" and the vortices cluster into supervortices and when $\beta > 0$ the drift is "repulsive" and the vortices are ejected against the boundaries. This can be understood as follows. Consider a collection of N point vortices at equilibrium. When $\beta < 0$, the density of these "field" vortices decreases from the center to the periphery of the domain. A "test" vortex moving through this medium modifies locally the vorticity field and produces a polarization cloud which amounts typically to a rotation of the surrounding vortices. This creates more density behind it than in front of it. Therefore, the retroaction of the field vortices leads to a drift of the test vortex directed inward.

If we apply this reasoning to all point vortices in the system, we can derive the following Fokker-Planck equation (Chavanis 1998c):

$$\frac{\partial \omega}{\partial t} + \langle \mathbf{V} \rangle \nabla \omega = \nabla(D(\nabla\omega + \beta\gamma\omega\nabla\psi)) \tag{19}$$

for the relaxation of the average vorticity. At equilibrium the drift is precisely balanced by random scattering and the distribution (11) is settled. Equation (19) can also be obtained from the MEPP. Its generalization to continuous vorticity fields yields (Robert & Sommeria 1992):

$$\frac{\partial \overline{\omega}}{\partial t} + \overline{\mathbf{u}}\nabla\overline{\omega} = \nabla\left\{ D\left(\nabla\overline{\omega} + \beta\overline{\omega}(\sigma_0 - \overline{\omega})\nabla\psi \right) \right\}. \tag{20}$$

Equations (19)(20) conserve energy, increase entropy and converge towards the equilibrium distributions (11) (12). The diffusion coefficient is space dependant (related to the local fluctuations of the vorticity) and can "freeze" the system in a subdomain (Robert & Rosier 1997), giving further support to the concept of incomplete relaxation and "maximum entropy bubbles".

References

Binney, J. and Tremaine, S. (1987) Galactic Dynamics (Princeton Series in Astrophysics)

Chandrasekhar, S. (1943) Principles of stellar dynamics (Dover)

Chavanis, P.H. (1998a) From Jupiter's Great Red Spot to the structure of galaxies: statistical mechanics of two-dimensional vortices and stellar systems, *Annals of the New York Academy of Sciences*, **Vol. 867**, pp. 120-141

Chavanis, P.H. (1998b) On the coarse-grained evolution of collisionless stellar systems, *Mon. Not. R. Astr. Soc.*, **Vol. 300**, pp. 981-991

Chavanis, P.H. (1998c) Systematic drift experienced by a point vortex in two-dimensional turbulence, *Phys. Rev. E*, **Vol. 58**, pp. R1199-R1202

Chavanis, P.H. and Sommeria, J. (1996) Classification of self-organized vortices in two-dimensional turbulence: the case of a bounded domain, *J. Fluid Mech.*, **Vol. 314**, pp. 267-297

Chavanis, P.H. and Sommeria, J. (1998a) Degenerate equilibrium states of collisionless stellar systems, *Mon. Not. R. Astr. Soc.*, **Vol. 296**, pp. 569-578

Chavanis, P.H. and Sommeria, J. (1998b) Classification of robust isolated vortices in two-dimensional hydrodynamics, *J. Fluid Mech.*, **Vol. 356**, pp. 259-296

Chavanis, P.H., Sommeria, J. and Robert, R. (1996) Statistical mechanics of two-dimensional vortices and collisionless stellar systems, *Astrophys. J.*, **Vol. 471**, pp. 385-399

Joyce, G. and Montgomery, D. (1973) Negative temperature states for the two-dimensional guiding-center plasma, *J. Plasma Physics*, **Vol. 10**, pp. 107-121

Lynden-Bell, D. (1967) Statistical mechanics of violent relaxation in stellar systems, *Mon. Not. R. Astr. Soc.*, **Vol. 136**, pp. 101-121

Miller, J. (1990) Statistical mechanics of the Euler equation in two dimensions, *Phys. Rev. Lett.*, **Vol. 65**, pp. 2137-2140

Onsager, L. (1949) Statistical hydrodynamics, *Nuovo Cimento Suppl.*, **Vol. 6**, pp. 279-287

Robert, R. and Rosier, C. (1997) The modelling of small scales in two-dimensional turbulent flows: a statistical mechanics approach, *J. Stat. Phys.*, **Vol. 86**, pp. 481-515

Robert, R. and Sommeria, J. (1991) Statistical equilibrium states for two-dimensional flows, *J. Fluid Mech.*, **Vol. 229**, pp. 291-310

Robert, R. and Sommeria, J. (1992) Relaxation towards a statistical equilibrium state in two-dimensional perfect fluid dynamics, *Phys. Rev. Lett.*, **Vol. 69**, pp. 2776-2779

References

B

Coherent Structures, Intermittency, and Cascade

SELF-SIMILARITY AND COHERENCE IN THE TURBULENT CASCADE

JAVIER JIMÉNEZ
School of Aeronautics, U. Politécnica, 28040 Madrid, Spain
and
Centre for Turbulence Research, Stanford University

1. Introduction

It has been known for some time that intermittency is an inevitable consequence of multiplicative cascades, which arise naturally from two assumptions [8]:

1. Causal locality, which implies that a variable v_n, associated with the cascade step n, depends *only* on the value of a single 'parent' in the preceding cascade step, so that its probability distribution p_n is

$$p_n(v_n) = \int W(v_n|v_{n-1}; n)p_{n-1}(v_{n-1})\, \mathrm{d}v_{n-1}. \tag{1}$$

   This is in contrast to more complicated functional dependences, such as on the values of v_{n-1} in some extended spatial neighbourhood, or on several previous cascade stages.
2. Scale similarity, which implies that the transition probability distribution W is independent of the cascade step n, and depends only on the ratio v_n/v_{n-1}.

The model assumes that each step generates eddies with smaller length scales Δx, usually $\Delta_0 2^{-n}$. After the initial effects are forgotten, the probability distribution of v_n is completely determined by the distribution W of the 'breakdown coefficients' v_n/v_{n-1} [6]. These assumptions are known to describe well some of the statistical properties of isotropic turbulence, and in particular the scaling exponents of the velocity structure functions [11],

$$S_p = \langle|\Delta u_{\Delta x}|^p\rangle \sim \Delta x^{\zeta_p}, \tag{2}$$

where $\langle\rangle$ stands for global averaging, and

$$\Delta u_{\Delta x} = u(x + \Delta x/2) - u(x - \Delta x/2). \tag{3}$$

T. Kambe et al. (eds.), IUTAM Symposium on Geometry and Statistics of Turbulence, 57–66.
© 2001 *Kluwer Academic Publishers.*

This is so even if there is no precise implied dynamical model of how the multiplicative process is related to the Navier-Stokes equations [7].

The converse of the previous discussion is not true, and intermittency does not necessarily imply a multiplicative cascade. Structural observations of numerical and experimental turbulent flows show for example that the vorticity is partially organized into coherent filaments with large aspect ratios, whose lifetimes are long compared to those which would correspond to their smallest dimensions [10]. The concept of 'scale' is difficult to apply to these anisotropic structures, and several models have been proposed in which the flow is described in terms of self-similar and coherent components.

It was argued in [8, 10] that such differentiation into components is a natural consequence of intermittency itself, and that any intermittent *field*, as opposed to a set with no spatial topology, is likely to develop a coherent component which is essentially different from the background. This is because the first of the two assumptions above is violated, and the evolution of the field variable generically depends on global, besides local, information. The nonlocality of the interactions introduces the average intensity of the fluctuations as an extra scale for the cascade, self-similarity is lost, and strong structures behave differently from weaker ones. Thus, while self-similarity is natural for variables, it is not a generic behaviour for fields in which several elements at the same stage of the cascade are coupled to each other.

In such cases, such as in three-dimensional turbulence, a full description of the strong structures should include their geometry, which controls how they interact with the background, but this has only been possible at the moderate Reynolds numbers accessible by numerical simulations. Such flows have very short inertial ranges, and it is difficult in them to study structures whose dimensions are neither in the dissipative nor in the integral range of scales. This was nevertheless the way in which vortex filaments with diameters of the order of the Kolmogorov scale were first identified. Only later could similar signals be educed from experimental flows at higher Reynolds numbers. For a summary, see [7].

It was argued theoretically in [10] that these dissipation-scale structures should only be the most obvious manifestations of coherence, and that a continuum of both weaker and stronger ones should be expected. The former would have larger diameters and weaker vorticities, and could perhaps be related to the low-pressure filaments observed in some experiments [3]. The latter would have diameters below the Kolmogorov scale and velocity differences comparable to those of the presently observed vorticity filaments. Reliable numerical and experimental data are lacking for the sub-Kolmogorov scales, but the purpose of this note is to discuss whether evidence can be found in the available experiments for, or against, organized structures in the inertial range of scales.

Experimental data are analyzed in the next section, of which a more extended version can be found in [9]. The structure of filtered direct numerical simulation fields is briefly described in §3, and conclusions are offered in §4.

2. Analysis of experimental data

The scaling properties of the inertial range have traditionally been characterized by the probability density functions (p.d.f.s) of the velocity differences at different separations, and in particular by their structure functions. It has been repeatedly noted, however, that statistical moments are poor discriminants of the differences between probability distributions [4, 9, 12], so that, even if is found that the structure functions scale approximately as powers, as in equation (2), it is difficult to conclude from that observation that the p.d.f.s of the breakdown coefficients are truly independent of the length scale.

Here we study directly the probability distributions, using three time series of the longitudinal velocity component in approximately isotropic turbulence in low-temperature helium gas [2]. The Reynolds numbers are $Re_\lambda = 155$, 760 and 1600. The range between the integral length L_ε and the Kolmogorov scale η is $8,400$ at the highest Reynolds number, and each set contains $10^4 - 10^5$ integral scales. For more details, see [9]. Only the two highest Reynolds numbers have clear power-law ranges in their spectra, and they are the ones used below. The remaining set, whose Reynolds number is comparable to those of numerical simulations, does not collapse well with them.

We first discuss the p.d.f.s of the breakdown coefficients

$$W(q_{2\Delta x}) = W(\varepsilon_{\Delta x}/2\varepsilon_{2\Delta x}). \tag{4}$$

for the averaged dissipation

$$\varepsilon_{\Delta x} = \frac{1}{\Delta x} \int_{x-\Delta x/2}^{x+\Delta x/2} (\partial_x u)^2 \, \mathrm{d}x. \tag{5}$$

They are roughly symmetrical in (0, 1), and approximately bell-shaped. It was suggested in [14] that they can be characterized by their mid-point values, $W(0.5)$, which is where their maxima approximately occur. This value varies with the separation [14] and with the dissipation in the parent interval, which can be expressed as a velocity scale $\delta u = (\varepsilon_{\Delta x} \Delta x)^{1/3}$, arguing against strict self-similarity of the cascade. It is given in figure 1, where it is seen that the fluctuations with the lowest parent dissipations at each length scale have narrower distributions, whose taller central peaks

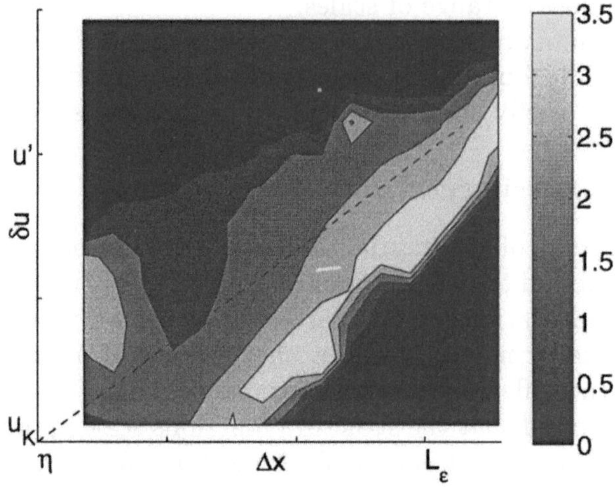

Figure 1. Midpoint value of the conditional p.d.f.s of the breakdown coefficients for the surrogate averaged dissipation, as a function of the averaging length and of the 'parent' velocity increment. $Re_\lambda = 1600$. The diagonal dashed line represents the standard Kolmogorov cascade. $u_k = \nu/\eta$ is the global Kolmogorov dissipative velocity scale, and u' is the one-component r.m.s. velocity. Areas in which there are not enough points to collect statistics are marked in black.

correspond to the light region in the lower diagonal strip of the figure. This implies that eddies with weak fluctuations break down into sub-segments whose dissipations tend to be half of that of their parents, as would be expected in an incoherent situation in which the parents are composed of several uncorrelated smaller pieces. More intense fluctuations have more spread distributions, with lower central peaks, suggesting coherence. This effect is most pronounced at the smallest scales.

Another way of characterizing the cascade is to study directly the conditional probability distributions of the velocity increments as a function of the velocity increments of their parent intervals. This was done in [9] for the data sets discussed here. If the cascade were completely self-similar the conditional p.d.f.s would be universal, and their means and standard deviations would be proportional to the velocity increment of the parent interval,

$$\langle \Delta u_{\Delta x} | \Delta u_{2\Delta x} \rangle = \Delta u_{2\Delta x}/2, \qquad \langle \Delta u^2_{\Delta x} | \Delta u_{2\Delta x} \rangle = \Delta u^2_{2\Delta x}/2^{2/3}. \tag{6}$$

The second relation assumes Kolmogorov's inertial scaling for the energy spectrum. The first relation in (6) is approximately satisfied by the data, but the second one is not, and the form of the conditional distributions depends on Δx. The conditional standard deviations are given in figure 2 for several separations in the inertial range. The dotted lines are equation

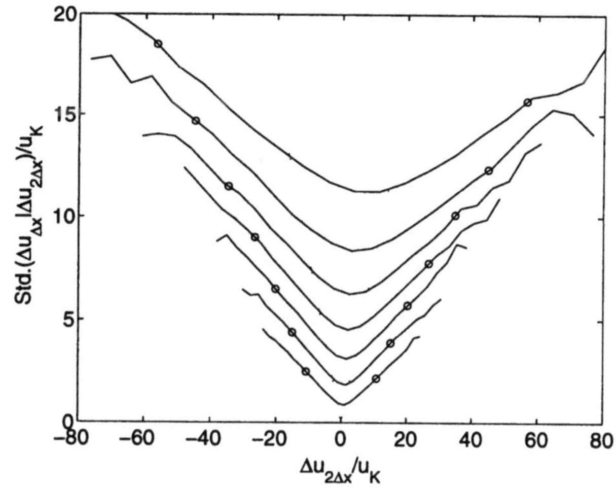

Figure 2. Standard deviation of the conditional distributions of the velocity increments, as a function of the separation and velocity increments of their 'parent' intervals. Each line represents a logarithmically spaced separation in the range $2\Delta x/\eta = 25(\times 2)1600$, and is normalized with the global Kolmogorov velocity scale u_k. The dotted lines would correspond to complete self similarity of the cascade, and the open circles are the abscissae of the velocity thresholds used to compute figure 3. $Re_\lambda = 1600$.

(6) and do not describe the experiments well. The measured deviations are bounded below by a global additive 'noise' which is of the order of the Kolmogorov velocity scale at that particular separation, $(\varepsilon \Delta x)^{1/3}$. This is consistent with the idea introduced above that the breakdown of the weak fluctuations is controlled by the background. The situation is different for the stronger fluctuations, whose intensities depend more linearly on the velocity differences of their parent intervals. It was shown in [9] that it is possible to approximately describe figure 2 as the superposition of a random additive process, corresponding to the minimum conditional standard deviation at each separation, and a multiplicative one that generates the linear tails. A similar conclusion was reached in [5].

Several refinements of this analysis are possible. It was for example noted in [9] that velocity increments are used in this context as band-pass filters to isolate a range of length scales, and that sharper filters could give different results. It was also remarked in [13] that this and other flows are not isotropic, and that the scaling improves if this is taken into account by analyzing increments associated with a given mean velocity. Both corrections were tested here, the former by using a five-point band-pass filter instead of (3), and the latter by repeating the analysis only for those segments for which the mean velocity is in a narrow range around the global mean. Figure 2 is essentially unchanged by both corrections.

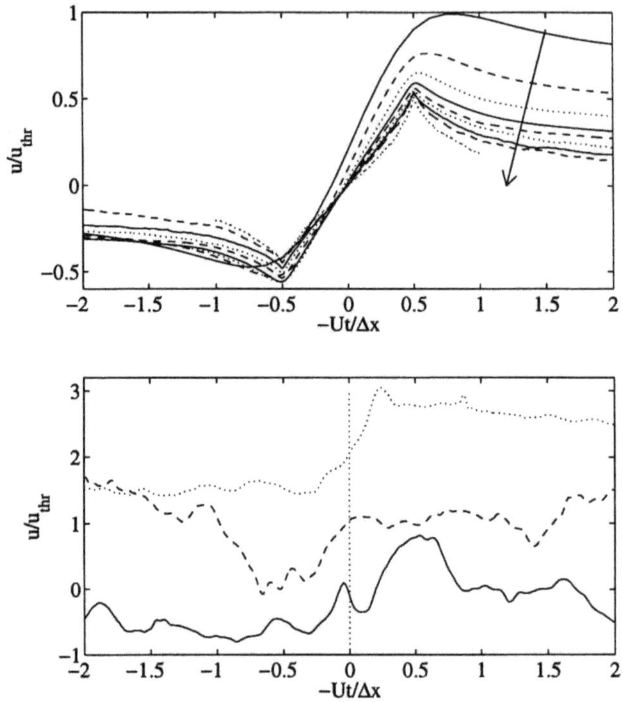

Figure 3. Top: Mean velocity, conditioned on $\Delta u_{\Delta x} > u_{thr}$, where the threshold is given in figure 2. Velocities and abscissae are normalized with the conditioning values, and $\Delta x/\eta = 12(\times 2)3000$, increasing in the direction of the arrow. $Re_\lambda = 1600$. Bottom: Three of the velocity traces used for the averages in the top figure. $\Delta x/\eta = 100$, $u_{thr} \approx u'$. The traces are arbitrarily offset vertically for clarity. They are consistent with: ——— , a positive vortex advected along the x axis; ---- , a negative vortex; ········ , a vortex layer or an stagnation flow.

2.1. EDUCED VELOCITY TRACES

To gain some understanding of the approximately self-similar intense structures implied by the linear tails of figure 2, we compiled averaged velocity traces conditioned on strong velocity increments across a given distance. Some precautions had to be taken to avoid the smearing that would result from counting each structure more than once. Disjoint 'active' segments were defined, each of which was the largest connected union of segments of width Δx for which $\Delta u_{\Delta x} > u_{thr}$, where u_{thr} was fixed to three times the global r.m.s. value of $\Delta u_{\Delta x}$. The velocities around the mid-points of all the active segments were then extracted and averaged together. The thresholds u_{thr} are plotted in figure 2, and fall in the approximately linear part of the conditional standard deviation curves. It was checked that the results are independent of the precise threshold as long as it falls in the linear part of the standard deviation curves.

The resulting conditionally averaged velocities are shown in figure 3 for separations ranging from the dissipative to the integral range of length scales. They are roughly antisymmetric, with widths and intensities which are consistent with those of the conditioning algorithm. Somewhat similar traces were found in [1] for the dissipative range, and were interpreted there as the effect of single vortices advected at an angle to the mean stream. In the present analysis the averaged structures in figure 3 are not representative of individual traces, some of which are shown in the bottom part of the figure. Different traces can be interpreted as caused by different kinds of structures. A hump of width δ and height u_δ is consistent with a positive vortex of circulation $2\pi\delta u_\delta$ passing at a distance $O(\delta)$ below the probe (or with a negative one above it). A negative hump is also consistent with a vortex, but a step which does not return to the mean velocity within a distance comparable to its width can only be explained by a vortex sheet or by a plane stagnation strain. The three kinds of traces are found in our sample, as well as others which are harder to interpret, and neither type is clearly dominant. Because of the relative alignment of each of them with respect to the detection criterion, which is visible in the samples in the bottom part of figure 3, their average results in the antisymmetric traces found in the top of the figure. The interpretation of figure 3, and especially of the part of the traces outside the detection interval $(-0.5, 0.5)$, should be that there is some coherence at all the scales tested. This is for example not true for weaker conditioning increments, for which the averaged traces fall to the overall mean velocity just outside the detection interval.

The traces for the smaller separations are more asymmetric than those for the larger ones, which could be interpreted as an indication that sheets or stagnation structures become more common at the smaller scales.

3. Direct numerical simulations

Given the difficulties in interpreting the conditional velocity traces obtained from one-dimensional signals, it is tempting to compare them with the results of similar conditioning in three-dimensional direct numerical simulations, even if their Reynolds numbers are lower. In this section we present two figures from an isotropic simulation at $Re_\lambda = 168$ [10], for which $L_\varepsilon/\eta = 290$. A plane section of the vorticity magnitude, without filtering, is shown at the top of figure 4, and shows a complicated pattern of vortex sheets whose width is a few Kolmogorov scales. Detailed inspection shows that the most intense vorticity is in the form of circular vortices, some of which are associated to sheets, presumably corresponding to the filaments observed in three-dimensional visualizations [10].

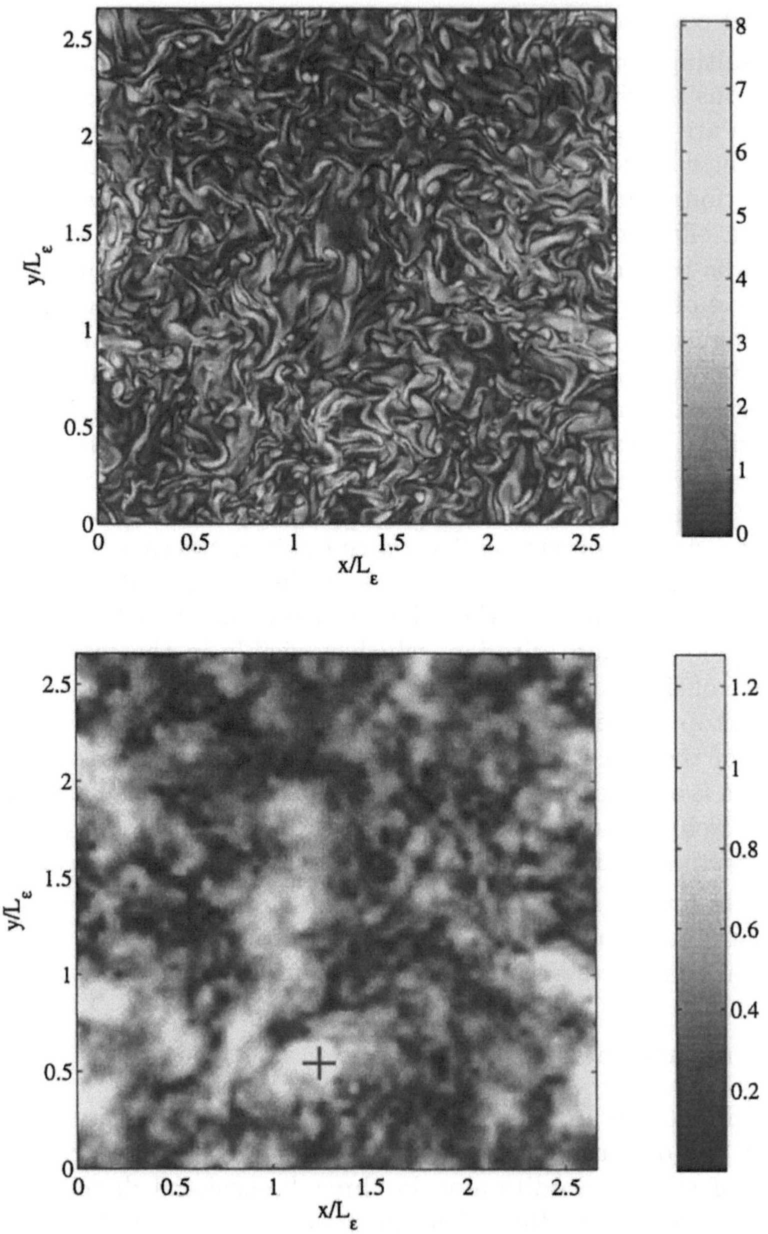

Figure 4. Top: Vorticity magnitude in a plane section across a triply periodic computational box of isotropic turbulence at $Re_\lambda = 170$ [10]. The smallest visible features are approximately 2–3 Kolmogorov lengths across. Bottom: 'Discrete vorticity' magnitude in the same plane. The velocity increments are computed over $\Delta x/\eta = 50$, which is the size of black cross. The scale bars to the right are normalized with the global r.m.s. vorticity magnitude.

To obtain an equivalent representation for the velocity differences, we define a 'discrete vorticity',

$$\Omega_{\Delta x,i} = \Delta x^{-1}\varepsilon_{ijk}\Delta u_{k,\Delta x_j}, \tag{7}$$

where ε_{ijk} is the fully antisymmetric unit tensor, and the velocity increments replace the usual derivatives. That quantity, although not a true vector, is an approximation to the result of filtering the vorticity over a box of size Δx^3, and gives a sense of how the velocity increments separate into rotational and potential components. The magnitude of $\Omega_{\Delta x}$ is showed at the bottom of figure 4, and is dominated by a strong 'vortex' at the location of the cross. It can be shown from three-dimensional representations that the most intense values of the discrete vorticity magnitude are organized into roughly tubular objects, although with smaller aspect ratios than those found in the dissipation range, and that 'vortex' lines of $\Omega_{\Delta x}$ run along their axes and connect neighbouring objects as would be expected of true vortices. It can be tested independently that a probe moving in the neighbourhood of each of these objects, more or less perpendicularly to their axes, would see velocity increments at scale Δx which are consistent with the passage of a vortex, but a two-dimensional section of $\Delta u_{\Delta x}$ contains other structures which are difficult to explain in this way.

The comparison of the two parts of figure 4 shows that the discrete 'vortex' is not simply a smoothed version of a single dissipation-scale structure, but the collective effect of a particularly strong sheet and a cluster of smaller vortex cores.

4. Conclusions

We have shown that available experimental evidence is inconsistent with a strictly self-similar multiplicative cascade in the inertial range. A better description is that the velocity increments at scale Δx are related to those at their 'parent' interval by the superposition of a local multiplicative process and of a 'global' additive noise whose intensity is proportional to the standard deviation of $\Delta u_{2\Delta x}$. This is consistent with the theoretical arguments outlined in §1, which suggest that this superposition is a generic property of cascades in *fields*, and results in the separation of the field into weak and strong components with different behaviours. This separation is illustrated here by the conditional distributions of the breakdown coefficients of the dissipation, which are shown to depend on the relative intensity of their parent intervals with respect to their Kolmogorov-predicted average.

The multiplicative cascade is recovered for fluctuations which are much stronger than the background, and such fluctuations were studied both through conditional averaging and, to a lesser extend, through visualizations of numerical fields at lower Reynolds numbers. It was noted that the symmetry of the conditionally obtained averages is spurious, and can be interpreted as the superposition of different structures that include both quasi-circular vortices and vortex or stagnation sheets. The latter seem to become more prevalent at smaller separations.

It is impossible with the present data to distinguish between the different structural models, but the comparison with the results of DNS suggest that inertial-scale structures, including vortices, exist, and that at least some of them consist of clusters of smaller structures. The dynamics of the inertial range of scales, in isotropic turbulence or otherwise, is one of the greatest unknowns in our present understanding of turbulence. The ambiguities of the present study underline the need for high Reynolds number experiments which are more complete than time traces of one velocity component.

This work was supported in part by the Spanish CICYT under contract PB95-0159, and by the Training and Mobility programme of the EC under grant CT98-0175.

References

1. Belin, F., Maurer, J., Tabeling, P. and Willaime, H. (1996) Observation of intense filaments in fully developed turbulence, *J. Physique II*, **6**, 573-583.
2. Belin, F., Maurer, J., Tabeling, P. and Willaime, H. (1997) Velocity gradient distributions in fully developed turbulence: experimental study, *Phys. Fluids*, **9**, 3843–3850.
3. Cadot, O., Douady, S., and Couder, Y. (1995) Characterization of the low-pressure filaments in the three-dimensional turbulent shear flow, *Phys. Fluids*, **7**, 630–646.
4. Chhabra, A.B. and Sreenivasan, K.R. (1992) Scale invariant multipliers in turbulence, *Phys. Rev. Lett.* **68**, 2762–2765.
5. Friedrich, R. and Peinke, J. (1997) Description of a turbulent cascade by a Fokker-Plank equation, *Phys. Rev. Lett.* **78**, 863–866.
6. Frisch, U. (1995) *Turbulence. The legacy of A.N. Kolmogorov.* Cambridge U. Press.
7. Jiménez, J. (1998) Small scale intermittency in turbulence, *Eur. J. Mech. B/Fluids*, **17**, 405–419.
8. Jiménez, J. (1999) Intermittency and cascades, *J. Fluid Mech.*, in press.
9. Jiménez, J., Moissy, F., Tabeling, P. and Willaime, H. (1999) Scaling and structure in isotropic turbulence, *Proc. of the I. Newton Inst. Symp. on Intermittency*, June 1999. (C. Vassilicos ed.)
10. Jiménez, J. and Wray, A.A. (1998) On the characteristics of vortex filaments in isotropic turbulence, *J. Fluid Mech.*, **373**, 255–285.
11. Meneveau, C. and Sreenivasan, K.R. (1991) The multifractal nature of the energy dissipation, *J. Fluid Mech.*, **224**, 429–484.
12. Nelkin, M. (1995) Inertial range scaling of intense events in turbulence, *Phys. Rev.* E, **52**, R4610–4611.
13. Sreenivasan, K.R. and Dhruva, B. (1998) Self-similarity and coherence in the turbulent cascade, *Prog. Theor. Physics Suppl.*, **130**, 103–120.
14. Van Atta, C.W. and Yeh, T.T. (1975) Evidence for scale similarity of internal intermittency in turbulent flows at large Reynolds numbers, *J. Fluid Mech.*, **71**, 417–440.

COHERENT FINE SCALE STRUCTURE IN TURBULENCE

T. MIYAUCHI AND M. TANAHASHI
Department of Mechano-Aerospace Engineering,
Tokyo Institute of Technology
2-12-1 Ohokayama, Meguro-ku, Tokyo 152-8552, Japan

1. Introduction

The understanding of fine scale structure in turbulence is very important in the development of turbulence theory. In the theory of fine scale motion in turbulence, tube-like vortices are frequently assumed to represent the phenomenon of intermittency (Townsend, 1951; Tennekes, 1968; Lundgren, 1982; Pullin and Saffman, 1993; Saffman and Pullin, 1994; Hatakeyama and Kambe, 1997). Recently, direct numerical simulations of the turbulent flows have shown that similar tube-like structures can be observed by the visualization of intense vorticity regions in turbulence (Kerr, 1985; She *et al.*, 1990; Vincent and Meneguzzi, 1991; Ruetsch and Maxey, 1991; Jimenez *et al.*, 1993). In this study, the universality of the fine scale structure is discussed by analyzing DNS database of various turbulent flows. Fine scale eddies in homogeneous isotropic turbulence, turbulent mixing layer, turbulent channel flows and MHD homogeneous turbulence are compared to show the scaling law of the tube-like structures in turbulence. The relationships between coherent fine scale eddy and the characteristic length of turbulence are also discussed in homogeneous isotropic turbulence.

2. Universal Scaling Law of Coherent Fine Scale Eddies

2.1. INVESTIGATED FLOW FIELDS

In this study, DNS database of homogeneous isotropic turbulence, turbulent mixing layer, turbulent channel flows and MHD homogeneous turbulence are used. In Table 1, numerical parameters of DNS are listed. In Fig.1, contour surfaces of the second invariant of the velocity gradient tensor ($Q = (W_{ij}W_{ij} - S_{ij}S_{ij})/2$) are shown for four turbulent flows with positive threshold. Figure 1 suggests that all of the investigated turbulent flows

T. Kambe et al. (eds.), IUTAM Symposium on Geometry and Statistics of Turbulence, 67–76.
© 2001 *Kluwer Academic Publishers.*

TABLE 1. DNS database of turbulence. HIT: homogeneous isotropic turbulence, TTML: temporally developing turbulent mixing layer, TCF: turbulent channel flow and MHT: MHD homogeneous turbulence (magnetic Prandtl number = 0.1). Re_λ: Reynolds number based on u'_{rms} and λ, Re_δ: Reynolds number based on the velocity difference and momentum thickness of a mixing layer, Λ: the most unstable wave length in a mixing layer, Re_τ: Reynolds number based on the friction velocity, h: channel half width, $|B_0|$: magnitude of imposed magnetic field.

ID	Re	$N_x \times N_y \times N_z$	$L_x \times L_y \times L_z$
HIT1	$Re_\lambda = 37.1$	$256 \times 256 \times 256$	$2\pi \times 2\pi \times 2\pi$
HIT2	$Re_\lambda = 66.1$	$216 \times 216 \times 216$	$2\pi \times 2\pi \times 2\pi$
HIT3	$Re_\lambda = 87.9$	$216 \times 216 \times 216$	$2\pi \times 2\pi \times 2\pi$
TTML	$Re_\delta = 1034$	$216 \times 325 \times 144$	$4\Lambda \times 6\Lambda \times 8\Lambda/3$
TCF1	$Re_\tau = 100$	$128 \times 129 \times 128$	$4\pi h \times 2h \times 2\pi h$
TCF2	$Re_\tau = 180$	$192 \times 193 \times 160$	$4\pi h \times 2h \times 2\pi h$
MHT1	$Re_\lambda = 38.3(\lvert B_0\rvert = 0.5)$	$256 \times 256 \times 256$	$2\pi \times 2\pi \times 2\pi$
MHT2	$Re_\lambda = 38.3(\lvert B_0\rvert = 1.0)$	$256 \times 256 \times 256$	$2\pi \times 2\pi \times 2\pi$

consist of similar tube-like fine scale eddies. The characteristics of fine scale eddies have been investigated by Tanahashi *et al.* (1997a; 1997b) in homogeneous isotropic turbulence, Tanahashi *et al.* (1997c; 1998) in turbulent mixing layer, Tanahashi *et al.* (1999a; 1999b) in turbulent channel flows and Tanahashi *et al.* (1999c) in MHD homogeneous turbulence. Detailed descriptions of DNS and characteristics of fine scale eddies can be found in the corresponding references.

2.2. DIAMETER AND AZIMUTHAL VELOCITY OF COHERENT FINE SCALE EDDIES

Figure 2 shows mean azimuthal velocity profiles of the tube-like fine scale eddies in various turbulent flows. The identification scheme of the center of the tube-like eddies is described in Tanahashi *et al.* (1997b). By this identification scheme, center of the swirling motion can be determined directly from the local flow pattern on the cross-sections with the local maximum of the second invariant along the tube-like structure. The ensemble average is conducted over the velocity distributions on these detected cross-sections. The mean azimuthal velocity profiles in Fig. 2 are normalized by Kolmogorov micro scale (η) and r. m. s. of velocity fluctuations (u'_{rms}). The volume averaged values of η and u'_{rms} are used in homogeneous isotropic turbulence and MHD homogeneous turbulence. In the turbulent mixing layer, η and u'_{rms} at the center of the shear layer are used . In the turbu-

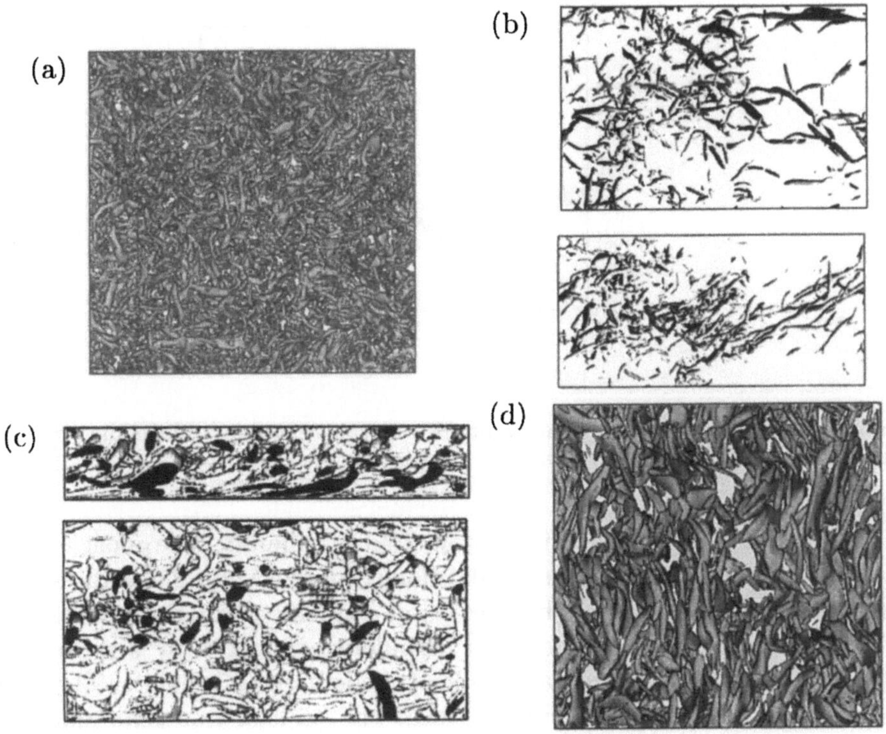

Figure 1. Contour surfaces of the second invariant in homogeneous isotropic turbulence with $Re_\lambda = 37.1$ (a), turbulent mixing layer with $Re_\delta = 1048$ (b), turbulent channel flow with $Re_\tau = 180$ (c) and MHD homogeneous turbulence with $|B_0| = 1.0$ (d). The level of threshold is $Q^* = 0.03$ for (a), (c) and (d) and $Q^* = 0.056$ for (b) where Q^* is normalized by η and u'_{rms}.

lent channel flows, η and u'_{rms} which are defined as a function of y^+ are used. Hereafter, the symbol * denotes a variable normalized by η and u'_{rms}. The mean azimuthal velocity profile of the fine scale eddies in all investigated flows can be approximated by the Burgers' vortex. Mean diameter and maximum azimuthal velocity is always about 10η and about 0.6 u'_{rms}.

Under the normalization by η and u'_{rms}, statistical properties of the fine scale eddy such as pdfs of diameter, azimuthal velocity, circulation etc. in all kinds of turbulent flows coincide very well. In the case of the turbulent mixing layer, these fine scale eddies compose so-called large scale structure (Brown and Roshko, 1974). Near the wall, well-known streamwise vortices and hairpin like vortices are also identified as fine scale eddies and they obey the same scaling law. In the case of MHD homogeneous turbulence, imposed magnetic field causes strong anisotropy. This anisotropic feature can be explained by the orientation of coherent fine scale eddies.

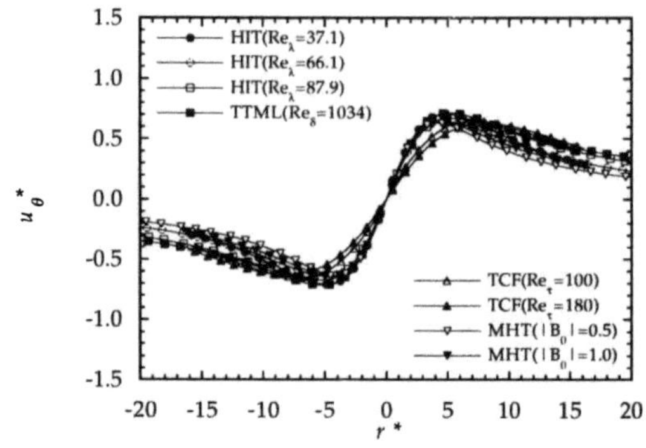

Figure 2. Mean azimuthal velocity profiles of coherent fine scale eddies in various turbulent flows normalized by η and u'_{rms}.

Although the geometry of the flow field is different, the most expected diameter and maximum azimuthal velocity of the tube-like eddies are 8η and $0.5u'_{rms}$, respectively. These features coincide with the results of the vorticity filaments in homogeneous turbulence reported by Jimenez *et al.* (1993; 1998) except for the magnitude of the azimuthal velocity. The scaling law of the fine scale eddies implies that the educed fine scale eddies are universal. We can call them as 'coherent fine scale structure' in turbulence because they show distinct coherency and universality.

2.3. THREE-DIMENSIONAL FEATURES OF COHERENT FINE SCALE EDDIES AND CHARACTERISTIC LENGTH SCALES OF TURBULENCE

In this section, three-dimensional features of coherent fine scale eddies are discussed using the results obtained in homogeneous isotropic turbulence (Tanahashi *et al.*, 1999d). Figure 3 shows the distribution of the axes of coherent fine scale eddies in homogeneous isotropic turbulence with $Re_\lambda = 87.9$. Visualized region is selected to be $14\lambda \times 14\lambda \times 14\lambda$ and visualized diameters of the axes are drawn to be proportional to the square root of the second invariant of the velocity gradient tensor on the axis. Larger diameter corresponds to larger second invariant. The parts with large second invariant tend to be nearly straight. In homogeneous isotropic turbulence, length of the axes of coherent fine scale eddies shows relatively large variance ranging from the order of the integral length scale (l_E) to near 10 times l_E as discussed later.

Each fine scale eddy has several sharp kinks and can be divided into several segments that are defined by the parts between local minimums of the second invariant on the axis. Most of the segments correspond to the straight parts of the axis in Fig. 3 and the second invariant on the

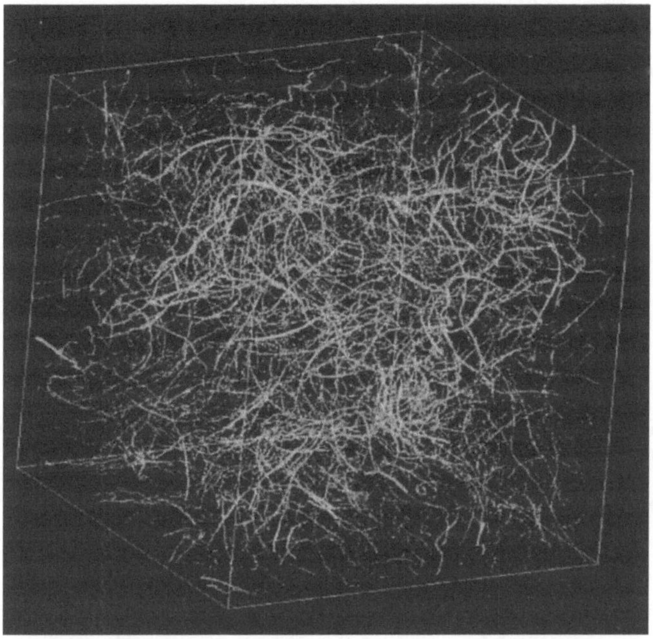

Figure 3. Axes of coherent fine scale eddies in a decaying homogeneous isotropic turbulence with $Re_\lambda = 87.9$. Visualized region is $14\lambda \times 14\lambda \times 14\lambda$ and visualized diameters of the axes are drawn to be proportional to $\sqrt{Q_c^*}$.

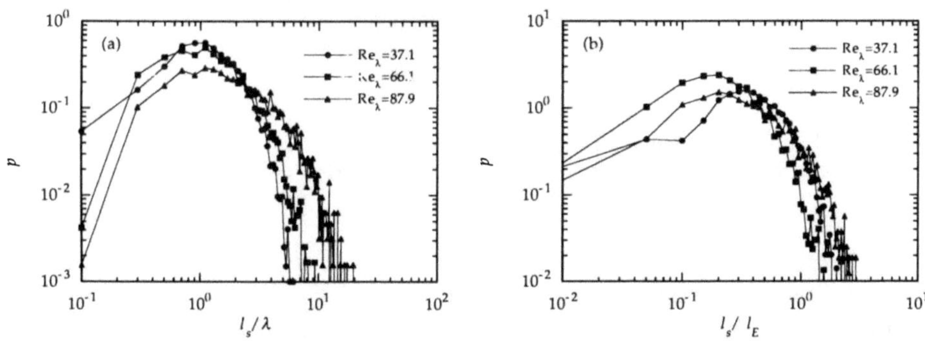

Figure 4. Pdfs of the segment length of coherent fine scale eddies in homogeneous isotropic turbulence. (a): segment length is normalized by λ, (b): by l_E.

axis show local minimums at the kinks (Tanahashi *et al.*, 1999d). Figure 4(a) shows the probability density function (pdf) of the segment length (l_s) of coherent fine scale eddies in homogeneous isotropic turbulence. The segment length is normalized by the Taylor micro scale (λ). The pdfs of the

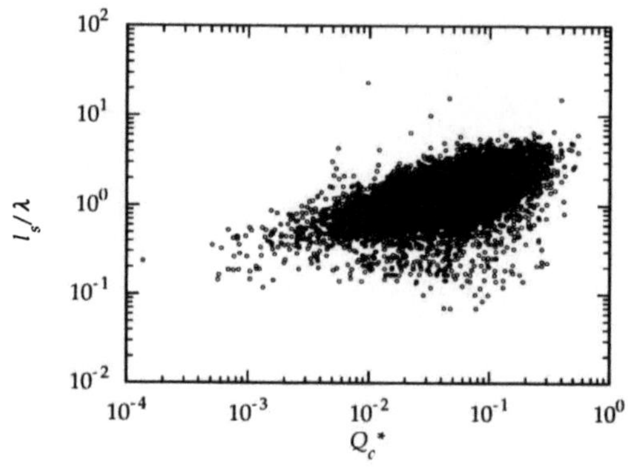

Figure 5. Scatter plots of the segment length and the maximum second invariant in the segment in homogeneous isotropic turbulence with $Re_\lambda = 37.1$.

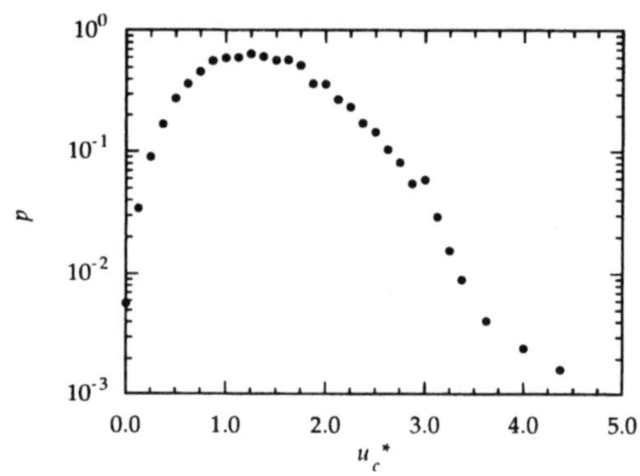

Figure 6. Probability density function of advection velocity in the radial direction of the segment in homogeneous isotropic turbulence with $Re_\lambda = 37.1$.

segment length show peaks at λ for all cases. The results imply that the coherent fine scale structure is directly related to the Taylor micro scale. The tails of pdf for large l_s/λ seem to depend on Re_λ, as shown in Fig. 4(a). The probability of long segments becomes high with the increase of Re_λ. In Fig. 4(b), the segment length is re-normalized by l_E. The probability decreases significantly at the length of l_E, which indicates that segment length is between the Kolmogorov micro scale and the integral length scale: $\eta \leq l_s(\approx \lambda) \leq l_E$.

The length of each segment shows strong correlation with the maximum second invariant in the segment ($Q^*_{c,max}$). The relation between l_s and $Q^*_{c,max}$ for $Re_\lambda = 37.1$ is shown in Fig. 5. The segments having large second invariant tend to be long, which coincides with the observation in

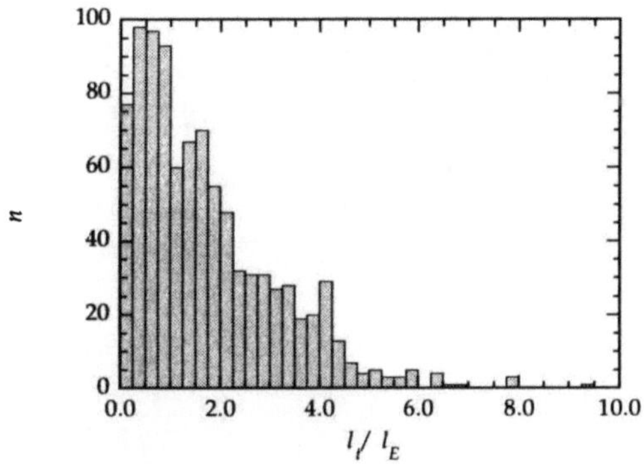

Figure 7. Histograms of the length of the axis in homogeneous isotropic turbulence with $Re_\lambda = 87.9$.

Fig. 3, and the length increases with the second invariant by a power of 1/2 ($l_s/\lambda \propto (Q^*_{c,max})^{1/2}$). Similar correlation can be observed for the cases with different Re_λ.

Furthermore, segments of the coherent fine scale eddies have relatively large advection velocity in the radial direction. Figure 6 shows the pdf of the radial advection velocity in the cross-sections with $Q^*_{c,max}$ for the case of $Re_\lambda = 37.1$. The radial advection velocities are normalized by u'_{rms}. As shown by Tanahashi *et al.* (1999d), the magnitudes of the axial flows along the axis are of the order of u'_{rms}. Although the total advection velocity of the fluid elements near the central axis is represented by the sum of the axial velocity and radial advection velocity, only the radial advection velocity is discussed because the axial velocity does not correspond to the advection of the eddy. In the case of the stretched Burgers' vortex, the axial velocity is induced by the outer strain field, even though the vortex has no advection velocity. Tanahashi *et al.* (1999d) also showed that the radial advection velocity scarcely changes its direction and magnitude in the segment. The pdfs of the advection velocity show a peak at $1.25u'_{rms}$ for all cases and are independent of Re_λ. The largest advection velocity exceeds $4u'_{rms}$. This result suggests that the movement of the axis of the coherent fine scale eddy is very active in turbulence.

Jimenez *et al.*(1993; 1998) have shown that the mean length of the intense vorticity filaments in homogeneous turbulence is about $2 \sim 3l_E$ and tends to decrease with the increase of Re_λ. The histogram of length of the axis are shown in Fig. 7 for $Re_\lambda = 87.9$. The length of the axis is normalized by l_E. For all cases, a peak of the histogram can be observed near l_E. Therefore, the most expected length is of the order of the integral length scale. The longest one reaches to $8l_E$. In the present study, the mean

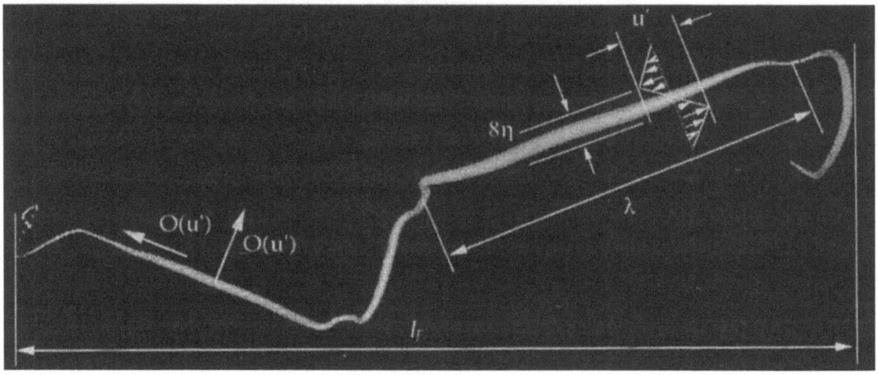

Figure 8. Characteristics of the coherent fine scale eddy in homogeneous isotropic turbulence. The presented axis is educed from the results with $Re_\lambda = 37.1$

length is $2.07l_E$ for $Re_\lambda = 66.1$ and $1.74l_E$ for $Re_\lambda = 87.9$. Although the definitions of the intense vorticity filaments by Jimenez *et al.* (1993; 1998) and the coherent fine scale eddies in the present study are quite different, the mean length is almost the same. However, the most expected length in this study is shorter than that by Jimenez *et al.* (1993; 1998).

3. Summary

In this study, characteristics of coherent fine scale eddies in homogeneous isotropic turbulence, turbulent mixing layer, turbulent channel flows and MHD homogeneous turbulence are compared to show the universal scaling law of tube-like fine scale eddies observed in fully-developed turbulence. It is shown that tube-like eddies in all investigated turbulent flows have the most expected diameter of 8η and the most expected maximum azimuthal velocity of $0.5u'_{rms}$.

Characteristics of the coherent fine scale eddies are presented schematically in Fig. 8 using a typical axis educed in homogeneous isotropic turbulence with $Re_\lambda = 37.1$. On the axis, the coherent fine scale eddies show axial velocity of the order of u'_{rms}. The coherent fine scale eddies can be divided into several segments, length of which is of the order of the Taylor micro scale, and each segment shows large advection velocity in the radial direction which is also of the order of u'_{rms}. The length of these eddies is of the order of the integral length scale. Therefore, the coherent fine scale structure includes three important characteristic length scales of turbulence: η, λ and l_E, and all characteristic velocities of the coherent fine scale eddies: azimuthal velocity, axial velocity and advection velocity, are of the order of u'_{rms}.

Acknowledgments

This work is partially supported by the Ministry of Education, Culture and Sports under Grant No. 09450088.

References

Brown, G.L. and Roshko, A. (1974) On density effects and large structure in turbulent mixing layers, *J. Fluid Mech.*, Vol. 64, pp.775–816.

Hatakeyama, N. and Kambe, T. (1997) Statistical laws of random strained vortices in turbulence, *Phys. Rev. Let.*, Vol. 79, pp. 1257–1260.

Jimenez, J., Wray, A.A., Saffman, P.G. and Rogallo, R.S. (1993) The structure of intense vorticity in isotropic turbulence, *J. Fluid Mech.*, Vol. 255, pp. 65–90.

Jimenez, J. and Wray, A.A. (1998) The structure of intense vorticity in isotropic turbulence, *J. Fluid Mech.*, Vol. 373, pp. 255–90.

Kerr, R.M.V. (1985) Higher-order derivative correlations and the alignment of small-scale structures in isotropic numerical turbulence, *J. Fluid Mech.*, Vol. 153, pp. 31–58.

Lundgren, T.S. (1982) Strained spiral vortex model for turbulent fine structure, *Phys. Fluids*, Vol. 25, pp.2193–2203.

Pullin, D.I. and Saffman, P.G. (1993) On the Lundgren-Townsend model of turbulent fine scales, *Phys. Fluids*, Vol.A5, pp. 126–145.

Ruetsch, G.R. and Maxey, M.R. (1991) Small-scale features of vorticity and passive scalar fields in homogeneous isotropic turbulence, *Phys. Fluids*, Vol. A3, pp. 1587–1597.

Ruetsch, G.R. and Maxey, M.R. (1992) The evolution of small-scale structures in homogeneous isotropic turbulence, *Phys. Fluids*, Vol. A4, pp. 2747–2760.

Saffman, P.G. and Pullin, D.I. (1994) Anisotropy of the Lundgren-Townsend model of fine-scale turbulence, *Phys. Fluids*, Vol. A6, pp.802–807.

She, Z.-S., Jackson, E. and Orszag, S.A. (1990) Intermittent vortex structures in homogeneous isotropic turbulence, *Nature*, Vol. 344, pp.226–228.

Tanahashi, M., Miyauchi, T. and Yoshida, T. (1996) Characteristics of small scale vortices related to turbulent energy dissipation, *Transport Phenomena in Thermal-Fluids Engineering (Proc. 9th Int. Symp. Transport Phenomena)*, Vol. 2, pp.1256–1261, Pacific Centre of Thermal-Fluids Engineering.

Tanahashi, M., Miyauchi, T. and Ikeda, J. (1997a) Identification of coherent fine scale structure in turbulence, *Simulation and Identification of Organized Structures in Flows (Proc. the IUTAM Symp. Simulation and Identification of Organized Structure in Flows, 1997)*, pp.131–140, Kluwer Academic Publishers, 1999.

Tanahashi, M., Miyauchi, T. and Ikeda, J. (1997b) Scaling law of coherent fine scale structure in homogeneous isotropic turbulence, *Proc. 11th Symp. Turbulent Shear Flows*, Vol. 1, pp.4-17–4-22.

Tanahashi, M., Miyauchi, T. and Matsuoka, K. (1997c) Coherent fine scale structures in turbulent mixing layers, *Turbulence, Heat and Mass Transfer(Proc. 2th Int. Symp. Turbulence, Heat and Mass Transfer, 1997)*, Vol. 2, pp. 461–470, Delft University Press.

Tanahashi, M., Miyauchi, T. and Matsuoka, K. (1998) Statistics of coherent fine scale structure in turbulent mixing layer, *Proc. IUTAM/IUCG Symp. Developments in Geophysical Turbulence, 1998* to be published.

Tanahashi, M., Shoji, K., Das, S. K. and Miyauchi, T. (1999a) Coherent fine scale structure in turbulent channel flow (in Japanese), *Trans. Jpn. Soc. Mech. Eng.*, Vol. 65, No. 638, pp.3244-3251.

Tanahashi, M., Shiokawa, S., Das, S. K. and Miyauchi, T. (1999b) Scaling of fine scale eddies in near-wall turbulence (in Japanese), *J. Jpn. Soc. Fluid. Mech.*, Vol. 18, No. 4, pp.256-261.

Tanahashi, M., Tsujimoto, T., Karim, Md. F., Fujimura, D. and Miyauchi, T. (1999c)

Anisotropy of MHD homogeneous isotropic turbulence (2nd report: Coherent fine scale eddies and Lorentz force) (in Japanese), *Trans. Jpn. Soc. Mech. Eng.*, Vol. 65, No. 640, pp.3884-3890.

Tanahashi, M., Iwase, S., Uddin, Md. A. and Miyauchi, T. (1999d) Three-dimensional features of coherent fine scale eddies in turbulence, *Turbulence and Shear Flow Phenomena - 1 (Proc. 1st Int. Symp. Turbulence and Shear Flow Phenomena, 1999)*, Eds. S. Banaerjee & J. K. Eaton, pp. 79–84, Begell House Inc.

Tennekes, H. (1968) Simple model for the small-scale structure of turbulence, *Phys. Fluids*, Vol. 11, pp.669–761.

Townsend, A.A. (1951) On the fine-scale structure of turbulence, *Proc. R. Soc. Lond.*, Vol. A208, pp.534–542.

Vincent, A. and Meneguzzi, M. (1991) The spatial structure and statistical properties of homogeneous turbulence, *J. Fluid Mech.*, Vol. 225, pp.1–20.

COHERENT FINE SCALE EDDIES IN TURBULENT SHEAR FLOWS AND IT'S RELATION TO ANISOTROPY

M. TANAHASHI AND T. MIYAUCHI
Department of Mechano-Aerospace Engineering,
Tokyo Institute of Technology
2-12-1 Ohokayama, Meguro-ku, Tokyo 152-8552, Japan

1. Introduction

The relation between coherent fine scale structure and anisotropy of turbulence is investigated by using DNS data of turbulent mixing layers and turbulent channel flows. These turbulent flows include tube-like fine scale eddies which are commonly observed in homogeneous isotropic turbulence (Kerr, 1985; She *et al.*, 1990; Vincent and Meneguzzi, 1991; Jimenez *et al.*, 1993; Tanahashi *et al.*, 1996; Tanahashi *et al.*, 1997a). In turbulent mixing layers, large scale structures which were found out by Brown and Roshko (1974) are composed of coherent fine scale eddies (Tanahashi *et al.*, 1997c; Tanahashi *et al.*, 1998). In turbulent channel flows, well-known streamwise vortices possess the same feature as the coherent fine scale eddies (Tanahashi *et al.*, 1999a; Tanahashi *et al.*, 1999b). Characteristics of fine scale eddies in turbulent mixing layers and turbulent channel flows coincide with those in homogeneous isotropic turbulence: the most expected diameter and maximum azimuthal velocity are about 10 times of Kolmogorov micro scale (η) and 0.5 times of r. m. s. velocity fluctuation (u'_{rms}). However, global characteristics of turbulence statistics are quite different in these flow fields. In this study, the origin of these difference is discussed from a viewpoint of the coherent fine scale eddies in turbulence.

2. DNS Database and Identification of Fine Scale Eddies

2.1. DNS DATABASE

DNS data of a temporally developing turbulent mixing layer and turbulent channel flows are analyzed. The temporally-developing turbulent mixing layer has been simulated by using spectral methods in all directions and

T. Kambe et al. (eds.), IUTAM Symposium on Geometry and Statistics of Turbulence, 77–82.
© 2001 *Kluwer Academic Publishers.*

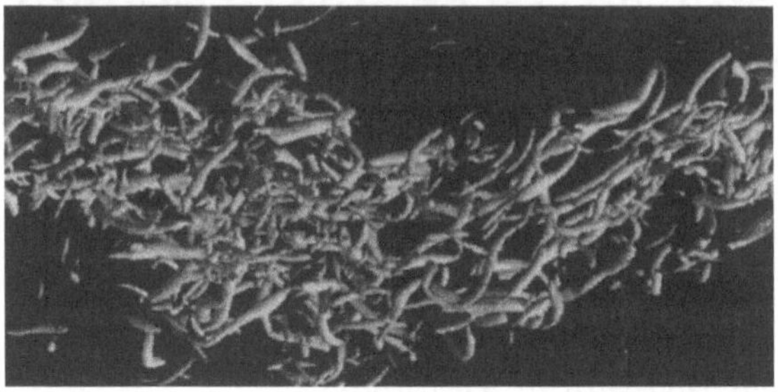

Figure 1. Contour surfaces of second invariant in a fully-developed turbulent mixing layer ($Re_\delta = 1034$)

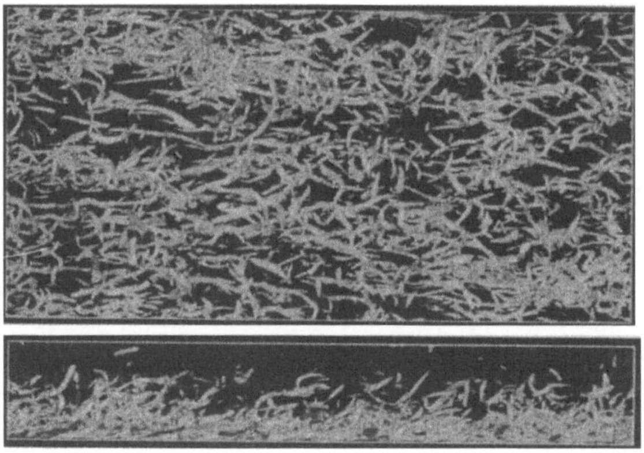

Figure 2. Contour surfaces of second invariant in a fully-developed turbulent channel flow ($Re_\tau = 400$).

Reynolds number based on initial vorticity thickness and a velocity difference across the mixing layer is 500. At a fully developed state, Reynolds number based on momentum thickness (Re_δ) reaches to 1034. Since the length of the calculation domain in the streamwise direction is selected to be 4Λ (Λ: most unstable wavelength), twice parings of large-scale structures (Brown and Roshko, 1974) occur. DNS of turbulent channel flows was conducted by using spectral methods in the streamwise and spanwise direction and 4th-order finite difference method in the transverse direction. Reynolds number based on the friction velocity (Re_τ) is 180 and 400. Detailed descriptions of numerical schemes and calculation conditions can be found in Tanahashi and Miyauchi (1995) and Tanahashi *et al.* (1997c) for the turbulent mixing layer and Tanahashi *et al.* (1999a; 1999b) for turbulent channel flow.

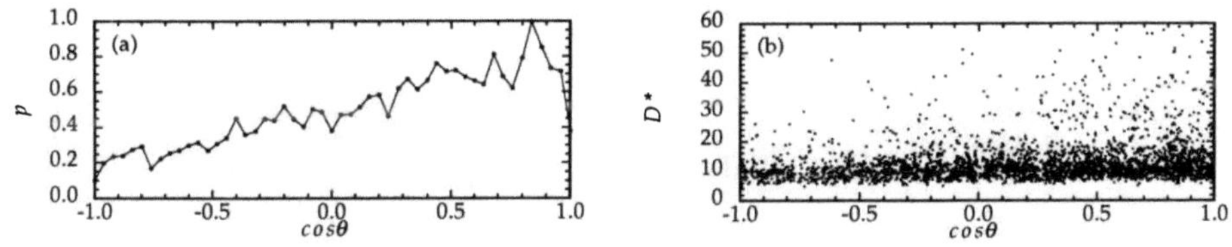

Figure 3. The pdf of the inclination angle (a) and the diameter (b) of coherent fine scale eddies in a fully-developed turbulent mixing layer ($Re_\delta = 1034$)

2.2. IDENTIFICATION SCHEME OF COHERENT FINE SCALE EDDIES

Figures 1 and 2 show the contour surfaces of the second invariant of the velocity gradient tensor defined by $Q = (S_{ij}S_{ij} - W_{ij}W_{ij})/2)$, where S_{ij} and W_{ij} is symmetric and asymmetric part of the velocity gradient tensor $A_{ij} = \partial u_i/\partial u_j$. As reported by previous works (Moser and Rogers, 1991; Moser and Rogers, 1993; Tanahashi and Miyauchi, 1995), the transition to turbulence of mixing layers occurs after the first pairing and before the second pairing of Kelvin–Helmholtz rollers. Kelvin–Helmholtz rollers become large by successive parings, and they consist of many fine scale eddies as shown in Fig.1. In the case of the turbulent channel flow, streamwise vortices near the wall and hairpin like vortices can be visualized by the positive Q region, which coincides with the results by the λ_2–definition (Jeong *et al.*, 1997) and by the Δ–definition (Blackburn *et al.*, 1996).

From these flow fields, coherent fine scale eddies are educed by using the identification scheme based on the local flow pattern (Tanahashi *et al.*, 1997b). In this identification scheme, the cross-sections which include the local maximum Q on the axis of fine scale eddies are determined.

3. Coherent Fine Scale Eddies in Turbulent Shear Flows

Characteristics of fine scale eddies in turbulent mixing layers and turbulent channel flows show similar features: the most expected diameter and maximum azimuthal velocity are about 10η and $0.5u'_{rms}$. However, the coherent fine scale eddies in two turbulent shear flows show different spatial distribution, which is related to anisotropic nature of each turbulent flow field.

Figure 3 shows the pdf of the inclination angle of the rotating axis of coherent fine scale eddies and the relationships between the diameter and the inclination angle in a turbulent mixing layer. The diameter in Fig. 3(b) is normalized by η at the center of the shear layer. The inclination angle of the rotating axis of a coherent fine scale eddy with respect to the direction of mean vorticity vector is defined by $\cos\theta = (\omega_c \cdot e_\omega)/(|\ \omega_c\ ||\ e_\omega\ |)$, where

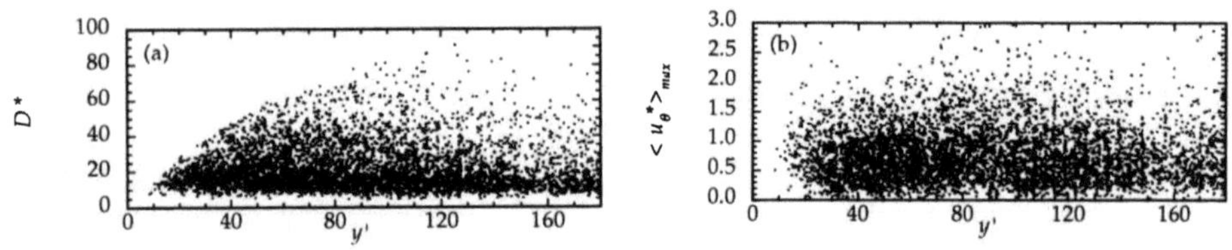

Figure 4. The diameter (a) and the maximum azimuthal velocity (b) of the coherent fine scale eddies as a function of the distance from the wall in a fully-developed turbulent channel flow ($Re_\tau = 180$).

ω_c and e_ω represent the vorticity vector at the center of a coherent fine scale eddy and the direction of mean vorticity vector ($= (0, 0, -1)$) in the mixing layer, respectively. The value of $\cos\theta$ becomes 1 for fine scale eddies co-rotating with large-scale structure, and becomes -1 for counter-rotating case. If the turbulence is isotropic, the pdf does not depend upon $\cos\theta$.

The probability density in Fig. 3(a) increases linearly with $\cos\theta$, which means that the number of co-rotating eddies is larger than that of counter-rotating ones. Figure 3(b) shows that the diameter of the coherent fine scale eddies depends on the direction of the axis. Large eddies tend to be co-rotating. With the decrease of the diameter, the directional dependencies become weaker. The coherent fine scale eddies with 10η diameter are distributed nearly uniformly and do not depend on $cos\theta$. These results suggest that randomness of distribution of coherent fine scale eddy increases with the decrease of the diameter, which seems to coincide with the assumption of local isotropy. Tanahashi *et al.* (1998) have shown that the most expected value of maximum azimuthal velocity ($< u_\theta^* >_{max}$) is almost independent on $cos\theta$ and is about $0.5u'_{rms}$, while the variances in $< u_\theta^* >_{max}$ are large for the co-rotating coherent fine scale eddies. Therefore, most of fine scale eddies with large D^* and $< u_\theta^* >_{max}$ are co-rotating with the large scale structure.

Figure 4 shows the diameter and the maximum azimuthal velocity of the coherent fine scale eddies in the turbulent channel flow with $Re_\tau = 180$ as a function of the distance from the wall. The diameter and the maximum azimuthal velocity are normalized by η and u'_{rms} at y^+ where the coherent fine scale eddy exists. In the turbulent channel flow, η and u'_{rms} are defined as a function of y^+. In the wall unit, the maximum azimuthal velocity increases and the diameter decreases with the decrease of the distance from the wall, whereas the most expected diameter and the maximum azimuthal velocity become independent on y^+ under the normalization by η and u'_{rms} as shown in Fig. 4. They are 10η and $0.5u'_{rms}$ respectively.

In the case of turbulent channel flows, the spatial distribution of the axes

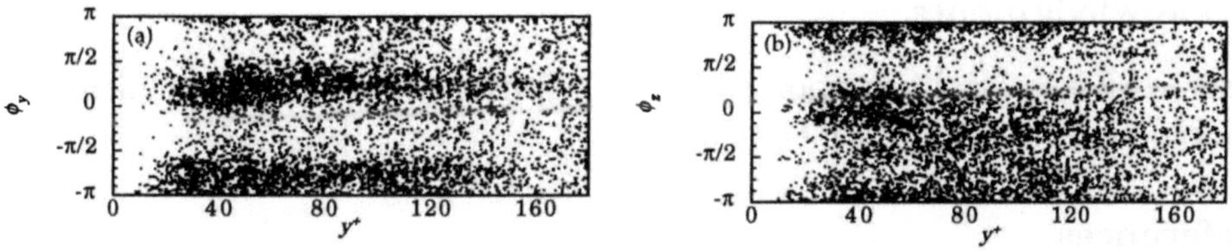

Figure 5. The inclination angle (a) and the tilting angle (b) of the coherent fine scale eddies as a function of the distance from the wall in a fully-developed turbulent channel flow ($Re_\tau = 180$).

shows strong directional dependence with the decrease of the distance from the wall. Figure 5 shows the inclination angles and the tilting angles of the coherent fine scale eddy. The inclination angle and the tilting angle are defined by the similar method of Jeong *et al.* (1997). The distributions of both angles show strong dependence on direction near the wall. These structures correspond to hairpin like eddies and streamwise vortices observed near the wall (Tanahashi *et al.*, 1999a; Tanahashi *et al.*, 1999b). The variance of the diameters becomes relatively small near the wall as shown in Fig. 4(a) and probability of the diameters shows a very sharp peak near 10η (Tanahashi *et al.*, 1999b), which means that diameter of almost all coherent fine scale eddies near the wall is about 10η. Therefore, the anisotropy near the wall occurs in the smallest coherent fine scale eddies. These results suggest that the anisotropic feature in near-wall turbulence is significantly different from that in free-shear turbulence from the view point of the fine scale structure.

4. Summary

In this study, spatial distribution of the coherent fine scale eddies in turbulent mixing layer and turbulent channel flow are compared to understand the relation between the anisotropy of turbulence and the coherent fine scale eddies. In turbulent mixing layer, coherent fine scale eddies with small diameter ($\approx 10\eta$) tend to be distributed randomly in the space, while those with relatively large diameter are co-rotating with large scale structure(Kelvin-Helmholtz rollers). In turbulent channel flows, directional dependence or alignment of the rotating axis appears near the wall. The aligned coherent fine scale eddies correspond to the streamwise vortices, which are one type of the smallest coherent fine scale eddies which has about 10η as diameter. These results suggest that anisotropy is caused by coherent fine scale eddies with relatively large diameter in the free-shear flows and by those with smallest diameter in wall-shear turbulence.

Acknowledgments

This work is partially supported by the Ministry of Education, Culture and Sports under Grant No. 09450088.

References

Blackburn, H. M., Mansour, N. N. & Cantwell, B.J. (1996) Topology of fine-scale motions in turbulent channel flow, *J. Fluid Mech.*, Vol. 310, pp.269–292.

Brown, G.L. and Roshko, A. (1974) On density effects and large structure in turbulent mixing layers, *J. Fluid Mech.*, Vol. 64, pp.775–816.

Jimenez, J., Wray, A.A., Saffman, P.G. and Rogallo, R.S. (1993) The structure of intense vorticity in isotropic turbulence, *J. Fluid Mech.*, Vol. 255, pp. 65–90.

Jeong, J., Hussain, F., Schoppa, W. and Kim, J. (1997) The structure of intense vorticity in isotropic turbulence, *J. Fluid Mech.*, Vol. 332, pp. 185–214.

Kerr, R.M.V. (1985) Higher-order derivative correlations and the alignment of small-scale structures in isotropic numerical turbulence, *J. Fluid Mech.*, Vol. 153, pp. 31–58.

Moser, R.D. and Rogers, M.M. (1991) Mixing transition and the cascade to small scales in a plane mixing layer, *Phys. Fluids*, Vol. A3, pp. 1128–1134.

Moser, R.D. and Rogers, M.M. (1993) The three-dimensional evolution of a plane mixing layer: pairing and transition to turbulence, *J. Fluid Mech.*, Vol. 247, pp.275–320.

She, Z.-S., Jackson, E. and Orszag, S.A. (1990) Intermittent vortex structures in homogeneous isotropic turbulence, *Nature*, Vol344, pp.226–228.

Tanahashi, M. and Miyauchi, T. (1995) Small scale eddies in turbulent mixing layer, *Proc. 10th Symp. Turbulent Shear Flows*, Vol. 1, pp. P-79–P-84.

Tanahashi, M., Miyauchi, T. and Yoshida, T. (1996) Characteristics of small scale vortices related to turbulent energy dissipation, *Transport Phenomena in Thermal-Fluids Engineering (Proc. 9th Int. Symp. Transport Phenomena, 1996)*, Vol. 2, pp.1256–1261, Pacific Centre of Thermal-Fluids Engineering.

Tanahashi, M., Miyauchi, T. and Ikeda, J. (1997a) Identification of coherent fine scale structure in turbulence, *Simulation and Identification of Organized Structures in Flows (Proc. the IUTAM Symp. Simulation and Identification of Organized Structure in Flows, 1997)*, pp.131–140, Kluwer Academic Publishers, 1999.

Tanahashi, M., Miyauchi, T. and Ikeda, J. (1997b) Scaling law of coherent fine scale structure in homogeneous isotropic turbulence, *Proc. 11th Symp. Turbulent Shear Flows*, Vol. 1, pp.4-17–4-22.

Tanahashi, M., Miyauchi, T. and Matsuoka, K. (1997c) Coherent fine scale structures in turbulent mixing layers, *Turbulence, Heat and Mass Transfer(Proc. 2th Int. Symp. Turbulence, Heat and Mass Transfer, 1997)*, Vol. 2, pp. 461–470, Delft University Press.

Tanahashi, M., Miyauchi, T. and Matsuoka, K. (1998) Statistics of coherent fine scale structure in turbulent mixing layer, *Proc. IUTAM/IUCG Symp. Developments in Geophysical Turbulence, 1998* to be published.

Tanahashi, M., Shoji, K., Das, S. K. and Miyauchi, T. (1999a) Coherent fine scale structure in turbulent channel flow (in Japanese), *Trans. Jpn. Soc. Mech. Eng.*, Vol. 65, No. 638 pp.3244–3251.

Tanahashi, M., Shiokawa, S., Das, S. K. and Miyauchi, T. (1999b) Scaling of fine scale eddies in near-wall turbulence (in Japanese), *J. Jpn. Soc. Fluid. Mech.*, Vol. 18, No. 4, pp.256–261.

Vincent, A. and Meneguzzi, M. (1991) The spatial structure and statistical properties of homogeneous turbulence, *J. Fluid Mech.*, Vol. 225, pp.1–20.

SHEAR STRESS AND DISSIPATION INTERMITTENCY IN HIGH-REYNOLDS-NUMBER TURBULENCE

YOSHIYUKI TSUJI
Department of Energy Engineering and Science, Nagoya University Furo-cho, Chikusa-ku, Nagoya city, 464-8603, JAPAN

1. Introduction

Energy dissipation and instantaneous shear stress fluctuations are considered for high-Reynolds-number (up to $R_\lambda \approx 10^4$) turbulent flow field using the concept of maximum norm.

The instantaneous fluctuation of the Reynolds shear stress, defined as $\gamma_{ij} = -u_i u_j, i \neq j$, has previously been explored in several ways. To compliment these classical analyses, we apply a new idea to characterize intermittency in shear stress fluctuations. Attention is paid to the rapid change in γ_{ij}, with positive and negative signs, and the maximum norm is introduced to quantify their behavior in the inertial range. With the help of multifractal analysis, the intermittency exponent, μ_S, is obtained for different flow fields (boundary layer, grid turbulence, round jet) and Reynolds numbers ($90 \leq R_\lambda \leq 10000$). It is found to be $0.1 \leq \mu_S \leq 0.15$ independent of the mean strain rate [6]. In addition to this, the maximum norm approach is applied to the energy dissipation field at a very high Reynolds number to obtain its scaling behavior, including the corresponding intermittency exponent [7].

2. Experimental Conditions

The high-Reynolds number atmospheric turbulent boundary layers are analyzed, whose typical features are summarized in Table 1. We used the x-probe for the measurements, with a sensitive length of 0.7mm and diameter of 5μm. The data were measured at the meteorological tower at Brookhaven National Laboratory. The measurement station is located at about 35 m above ground. The velocity signals are sampled by a 12-bit A/D converter at a frequency of 5 kHz $\sim$ 10 kHz with a low-pass filter no

T. Kambe et al. (eds.), IUTAM Symposium on Geometry and Statistics of Turbulence, 83–88.
© 2001 *Kluwer Academic Publishers.*

TABLE 1. U is a local mean velocity, u' and v' are r.m.s. values of streamwise and vertical component, respectively. $\langle \varepsilon \rangle$ is the energy dissipation obtained by the assumption of Taylor's frozen flow hypothesis, λ and η are Taylor micro scale and Kolmogorov scale. The turbulence Reynolds number is defined as $R_\lambda \equiv u'\lambda/\nu$.

U (m/s)	u', v'(m/s)	$-\langle uv \rangle / u'v'$	$\langle \varepsilon \rangle$(m^{2}s^{-3})	λ(cm)	η(mm)	R_λ
2.82	0.77, 0.49	0.39	1.06×10^{-2}	10.9	0.72	5,940
5.16	1.87, 1.40	0.42	8.40×10^{-2}	9.6	0.44	12,240
5.67	1.87, 1.40	0.42	8.40×10^{-2}	11.3	0.46	15,630

greater than half the sampling frequency; the measurements are made for one hour for each data set. The probe is positioned about 3 m from the tower. As the tower is made of angle irons, the wind passes through them. We assume that there is little effect of the tower on the measurements. Another important problem is the wind direction. We equipped the tower with a wind vane and observed it from the ground. The probe direction was corrected in terms of the flow condition in case there were any dramatic changes during the measurements. A more detailed analysis of the wind direction is found in [2]. Since only the γ_{12} component is measured, we use the character γ instead of γ_{12} for convenience hereafter.

3. Maximum Norm and Scaling Exponents

For statistical analysis, the time axis is divided into small boxes with length r. The largest positive and smallest negative values of γ within the box are expressed by $max(r)$ and $min(r)$, respectively (see Fig. 1). Therefore, the difference between $max(r)$ and $min(r)$ should contain the typical features of γ oscillation. We introduce the maximum norm;

$$M_r \equiv [max(r) - min(r)] \,, \tag{1}$$

which is defined in each box with size r. It is, however, necessary to add the following condition:

$$max(r) > \delta_B \quad \text{and} \quad min(r) < -\delta_B \,. \tag{2}$$

This condition is to disregard fluctuations which are contained in the narrow band $[-\delta_B, \delta_B]$. Note also that $max(r)$ and $min(r)$ have a positive and negative sign, respectively. It is well known that a large part of γ fluctuates around the zero value. We assume that the electrical noise may affect these fluctuations around the zero-crossing. Although the band width δ_B is set as a function of the Reynolds number [6], in a high Reynolds

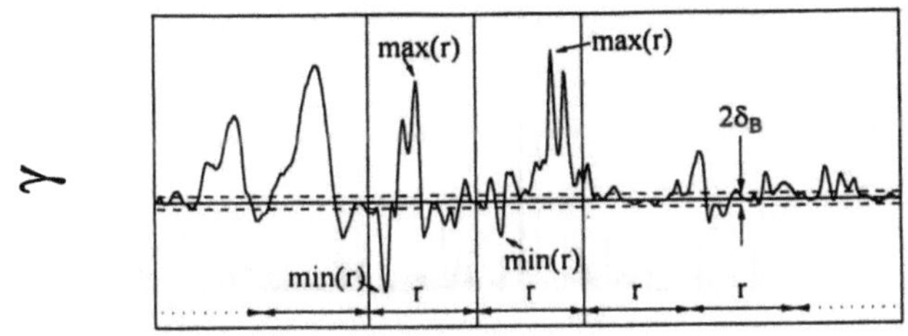

Figure 1. The horizontal axis is divided into small boxes whose size is r. The largest and the smallest values of γ fluctuation are defined as $max(r)$ and $min(r)$, respectively. The broken lines indicate the narrow band with $2\delta_B$ width.

number flow it is not a problem to set it as $\delta_B = 0$. Then the condition of Eq. (2) is that $max(r)$ is positive and $min(r)$ is negative in a finite box size r, which is expressed in the units of sampling intervals.

3.1. SCALING EXPONENTS IN DISSIPATION FIELD

For analyzing the dissipation fluctuation, there is no additional complications on the definition of M_r. Since the dissipation rate is never negative, the definition becomes $M_r \equiv max(r)$. The maximum norm is a positive random quantity which may be used to characterize intermittent fluctuations.

Figure 2 shows typical examples of M_r distribution in shear stress and the dissipation field. The horizontal axis is a unit of box size r, and the vertical one is normalized by its average, $\langle M_r \rangle$. Although the distributions depend on the size r, they look intermittent for various box sizes in both the dissipation and inertial ranges. In this figure, the box size is set at $r = 155$ which is in the inertial range, and it is apparent that the dissipation field is more intermittent.

The maximum norm exponents, ν_q, are introduced as follows, in analogy to the local energy dissipation rate, ε_r,

$$\langle M_r^q \rangle \, / \, \langle M_r \rangle^q = C_q' \, (r/L)^{-\nu_q} \; . \tag{3}$$

For convenience, the characters ν_q^S and ν_q^D are used as the exponents for the maximum norm of shear stress and dissipation, respectively. It should be noted that, there is a significant difference between ε_r and M_r. When $q = 1$, the local averaged energy dissipation equals the total average, that is $\langle \varepsilon_r \rangle = \langle \varepsilon \rangle = constant$. However, for the maximum norm this condition is not satisfied [6].

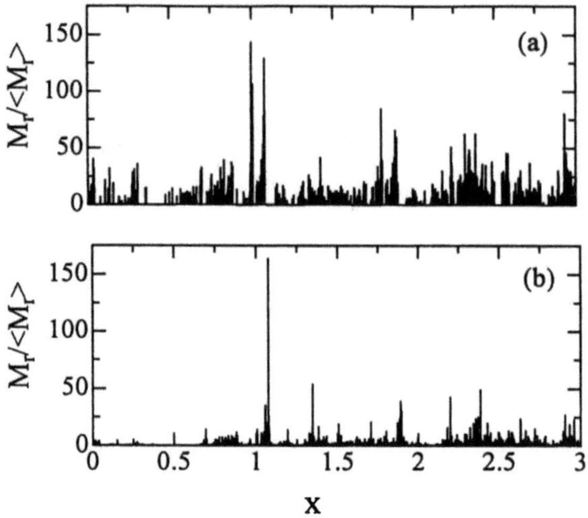

Figure 2. An example of maximum norm distribution for (a) shear stress fluctuation for the box size $r = 155$, (b) dissipation fluctuation for the same box size. This box size is located within the inertial range. r has a units of sampling intervals.

We have computed the moments, $\langle M_r^q \rangle$, and confirmed that the maximum norm, for this case, has firmly a power law distribution in the inertial range. Thus, the slope is equal to the exponent ν_q^D.

3.2. SCALING EXPONENTS IN SHEAR STRESS FIELD

As the shear stress, γ, fluctuates with positive and negative signs, the maximum norm is introduced to characterize them. We have found that the maximum norm has also a power law distribution for this case in the inertial range. Experimental accuracy limited the analysis up to the sixth order, but we have analyzed the lower-order moments which contain valuable information about the central region of the pdfs.

These scaling exponents are summarized in Fig. 3. The dissipation exponents, ν_q^D (solid circles), are larger than those for the shear stress for $q > 1$. Hence, the dissipation field is more intermittent, which is consistent with the qualitative observation of γ indicating the intermittent phenomena. For lower-order moments ($0.2 \leq q \leq 0.8$) the scaling exponents are negative, and the shear stress has larger exponents.

If the locally averaged dissipation, ε_r, is approximated by the log-normal distribution (this condition is expressed as **K62**) [4], the scaling exponents are given as

$$\tau_q = -\mu q(1-q)/2 = -\tau_2 q(1-q)/2\,, \tag{4}$$

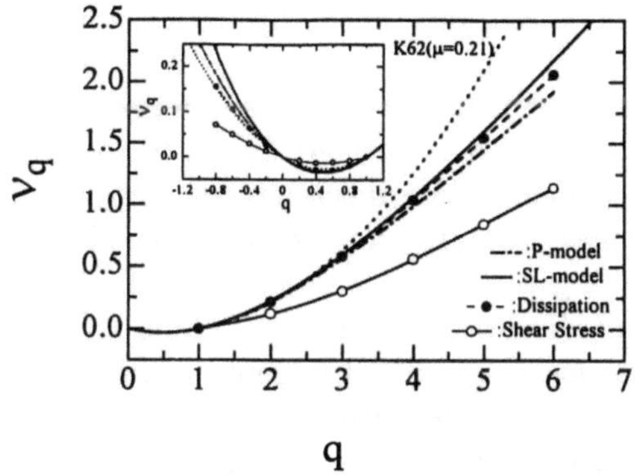

Figure 3. Scaling exponents of the maximum norm for shear stress (ν_q^S:circle) and the dissipation fluctuation (ν_q^D:solid circle). They are compared with the exponents τ_q predicted by **K62** assumption, Eq. (4), which is expressed by the dotted line. The other model (Eq. (5)) evaluating τ_q is plotted by the solid line. p-model (the parameter p is set at the recommended value, 0.7) is plotted by dashed line.

where μ ($= \tau_2$) is the intermittency exponent. In Fig. 3 we compare the predicted exponents, τ_q, from **K62** and those obtained from the present measurements, ν_q^D and ν_q^S. We find there is good agreement between the predicted and measured values for $-0.8 \leq q \leq 3.0$. However, there is a noticeable difference in high-order moments. She et al., [5] proposed a phenomenology involving a hierarchy of fluctuation structures associated with the vortex filaments. From this they derived a relation with no adjustable parameters, which agrees very well with ν_q^D when q is positive, but negative-moment values differ slightly from these. Another excellent model, called P-model [1], was plotted in the figure. The open parameter was set at the recommended value, 0.7, and they agree well with the experimental data for positive and negative q values.

It is remarkable that the maximum norm exponents in energy dissipation match those of locally averaged dissipation, that is, $\nu_q^D \simeq \tau_q$. We believe this is a strong reason to believe that the maximum norm is a valuable quantity in characterizing the small-scale intermittency in turbulence.

3.3. INTERMITTENCY EXPONENT

There is much interest in quantifying the intermittency exponents. In this section, the maximum norm concept is used to extract the intermittency exponents for high Reynolds number turbulence data. The exponent using the conventional approach yields a value of $\mu = 0.22$ for the average rate of energy dissipation. We can get the intermittency exponent for the maximum norm by Eq. (3), in which ν_2 is the exponent in analogy to τ_2. It is found that in the dissipation field $\nu_2^D = 0.21$ and in the shear stress $\nu_2^S = 0.12$; here we use the characters μ_D and μ_S to indicate the intermittency exponent calculated using the maximum-norm approach for dissipation and shear stress, respectively. That is, we find $\mu_D = 0.21$, $\mu_S = 0.12$. We wish to make two important remarks. First, the maximum norm exponent, μ_D, agrees remarkably well with μ derived by ε_r. Second, $\mu_S < \mu_D$; indicating that the shear stress field is less intermittent than the dissipation field.

To explain the first remark, notice that the largest value of the dissipation rate within a given box size, M_r, may be a significant portion of the locally averaged dissipation field, ε_r. Thus, the qualitative properties of these two quantities may be equivalent. A more detailed discussion of the second remark on the difference in the scaling exponent between shear stress and dissipation is given in the reference. We should also note that the vorticity and scalar dissipation field are more intermittent than the energy dissipation field [3]. Despite much numerical and experimental research along this theme, a clear explanation has not yet been presented.

4. Conclusion

Shear stress and dissipation fluctuations are considered in high Reynolds number turbulence. A new concept of the maximum norm is introduced to characterize their spike-like oscillations. The scaling exponents, ν_q, are naturally introduced, and the intermittency exponent is obtained. The classical scaling exponents, τ_q defined by local averaged energy dissipation ε_r, are very close to ν_q at least for $-0.8 \leq q \leq 4.0$. Therefore, the intermittency exponents are $\mu = \tau_2 = 0.22$ and $\mu_D = \nu_2 = 0.21$, respectively. For the shear stress fluctuation, we obtained the exponent $\mu_S = 0.12$ by the maximum norm, thus confirming that the shear stress field is less intermittent than the energy dissipation field.

References

1. C. Meneveau and K. R. Sreenivasan (1987) Simple Multifractal Model for Fully Developed Turbulence, *Phys. Rev. Lett*, **Vol. 59**, p. 1424
2. E. van Doorn, B. Dhruva, K. R. Sreenivasan and V. Cassella (1999) The Statistics of Wind Direction and Its Change: The More Things Chnage the More They Stay the Same, *Phys Fluids A*, submitting
3. K. R. Sreenivasan and R. A. Antonia (1997) The Phenomenology of Small-Scale Turbulence, *Annu. Rev. Fluid Mech.*, **Vol. 29**, p. 435
4. A. N. Kolmogorov (1962) A Refinement of Previous Hypotheses Concerning the Local Structure of Turbulence in a Viscous Incompressible fluid at High Reynolds Number, *J. Fluid Mech.*, **Vol. 13**, p. 82
5. Z-S. She and E. Leveque (1994) Universal Scaling Laws in Fully Developed Turbulence, *Phys. Rev. Lett.*, **Vol. 72**, p. 336
6. Y. Tsuji and B. Dhruva (1998) On the Intermittency Feature of Shear Stress Fluctuation in Turbulence, *J. Phys. Soc. Japan*, **Vol. 67**, p. 1234
7. Y. Tsuji and B. Dhruva (1999) Intermittency Feature of Shear Stress Fluctuation in High-Reynolds-Number Turbulence, *Physics of Fluids A*, **Vol. 11**, p. 3017

CHARACTERISTIC TIME SCALES AND ENERGY TRANSFER IN MHD TURBULENCE

YUJI HATTORI
Faculty of Engineering, Kyushu Institute of Technology, Tobata-ku, Kitakyushu 804-8550, Japan

AND

AKIHIRO ISHIZAWA
Naka Fusion Research Establishment, Japan Atomic Energy Research Institute, Ibaraki 311-0193, Japan

Abstract. Statistical properties of MHD turbulence are studied by GOY-type shell models and direct numerical simulation. We introduce the Alfvén effect with randomized coherent magnetic field into the shell model. It leads to $k^{-3/2}$ spectrum and is consistent with Iroshnikov-Kraichnan picture of MHD turbulence. The PDF's of shell models are compared with those of wavelet coefficients obtained from DNS data.

1. Introduction

Much attention has been payed to MHD turbulence not only in the context of plasma physics but also in the context of turbulence theory. However, the statistical properties of MHD turbulence are not fully understood. There are two plausible theories for the scaling laws of homogeneous and isotropic MHD turbulence. One is the well-known theory by Kolmogorov proposed for hydrodynamic turbulence; it predicts $k^{-5/3}$ spectrum (hereafter referred to as K41 spectrum). The other is proposed by Iroshnikov and Kraichnan independently[1]; it predicts $k^{-3/2}$ spectrum (hereafter referred to as IK spectrum). In the latter, the Alfvén effect plays a key role. When coherent magnetic structures exist in MHD turbulence, the Alfvén waves introduce the characteristic time scale $\tau_A = (B_C k)^{-1}$, which is called Alfvén time, to the energy transfer. Since the Alfvén time is much smaller than the hydrodynamic local time scale $(v_k k)^{-1}$, the Alfvén effect weakens the energy

T. Kambe et al. (eds.), IUTAM Symposium on Geometry and Statistics of Turbulence, 89–94.
© 2001 *Kluwer Academic Publishers.*

transfer so that the energy dissipation rate is proportional to the Alfvén time: $\epsilon \propto (B_C k)^{-1}$. Then dimensional analysis yields $k^{-3/2}$ spectrum. The data from solar wind seem to support $k^{-5/3}$ spectrum[2]. On the other hand, the spectrum is close to $k^{-3/2}$ in two-dimensional simulations[3].

In this paper, we study the statistical properties of MHD turbulence by direct numerical simulation (DNS) and GOY-type shell models. We pay special attention to the characteristic time scales in MHD turbulence and their effect on the energy transfer.

2. GOY-type Shell model of MHD turbulence

The shell models of MHD turbulence are studied by several authors[4-7]. The extension of the original GOY-type shell model to MHD turbulence was first studied by Biskamp[6]. Here we consider an MHD version of Sabra model for hydrodynamic turbulence[8]. It is written as

$$\begin{aligned}\frac{\mathrm{d}Z_n^\pm}{\mathrm{d}t} &= i\alpha k_{n+1} Z_{n+1}^{\pm *} Z_{n+2}^{\mp} + i\beta k_{n+1} Z_{n+1}^{\mp *} Z_{n+2}^{\pm} - i\alpha k_n Z_{n-1}^{\pm *} Z_{n+1}^{\mp} \\ &\quad + i\gamma k_n Z_{n-1}^{\mp *} Z_{n+1}^{\pm} + i\beta k_{n-1} Z_{n-2}^{\pm} Z_{n-1}^{\mp} + i\gamma k_{n-1} Z_{n-2}^{\mp} Z_{n-1}^{\pm} \\ &\quad - \nu k_n^2 Z_n^\pm \pm i k_n B_C Z_n^\pm + f_n^\pm, \qquad (1)\end{aligned}$$

where $Z_n^\pm$ is a representative mode of Elsässer variable $\boldsymbol{Z}^\pm = \boldsymbol{u} \pm \boldsymbol{B}$ corresponding to the wavenumber $k_n = k_0\lambda^n$, α, β and γ are constants, ν corresponds to diffusivity (unit magnetic Prandtl number is assumed) and $f_n^\pm$ is an external force. As in the case of normal GOY-type models[6], the total energy $E_T = \frac{1}{4}\sum_n(|Z_n^+|^2 + |Z_n^-|^2)$ and the cross helicity $H_C = \frac{1}{4}\sum_n(|Z_n^+|^2 - |Z_n^-|^2)$ are conserved in the inviscid limit $\nu = 0$. Furthermore, the magnetic quantity $H_M = \sum_n k_n^\delta |B_n|^2$, which can be regarded as the squared magnetic potential in two-dimensional case or the magnetic helicity in three-dimensional case, becomes an inviscid invariant by imposing

$$\frac{\beta}{\alpha} = -\frac{\lambda^{2\delta} - \lambda^{2\delta} + 1}{\lambda^{2\delta} + \lambda^{2\delta} - 1}, \quad \frac{\gamma}{\alpha} = -\frac{\lambda^{2\delta} - \lambda^{2\delta} - 1}{\lambda^{2\delta} + \lambda^{2\delta} - 1}.$$

The term $ik_n B_C Z_n^\pm$, first discussed by Gloaguen *et al.*[4], is introduced to take account of the Alfvén effect in the presence of large-scale coherent magnetic field. Biskamp[6] found that when B_C is a non-zero constant the energy spectrum has two ranges of different scaling exponents; $k^{-5/3}$ spectrum is observed in the lower wavenumber range and the spectrum is flatter than $k^{-3/2}$ spectrum in the higher wavenumber range (as seen in figure 1). Although the above choice seems to be consistent with Iroshnikov-Kraichnan picture, it does not lead to $k^{-3/2}$ spectrum. The reason for this unexpected result is that the Alfvén term $ik_n B_C Z_n^\pm$ was too coherent for constant B_C so that the average of the nonlinear terms in eq. (1) almost vanish for the inertial time scale.

We apply two methods for incorporating the Alfvén effect correctly: in the first one we take small λ instead of the usually chosen value $\lambda = 2$; in the second one the coherent magnetic field B_c is randomized as

$$B_{C,n} = \overline{B_C} + B'_{C,n}, \quad \frac{\mathrm{d}B'_{C,n}}{\mathrm{d}t} = -\frac{B'_{C,n}}{\tau_n} + g(t), \quad \tau_n = \frac{1}{k_n \overline{B_C}},$$

where $g(t)$ is Gaussian white noise. Both methods are designed to relax the coherency of the Alfvén term. In addition, forcing is introduced by fixing $Z_{n_0}^\pm$ since both constant forcing and random forcing, which are often applied to shell models, can lead to "dynamic alignment"; dynamic alignment itself occurs in MHD turbulence, but it is undesirable for the present purpose since we need turbulent behavior for a long time to obtain statistical properties from shell models.

3. Results

3.1. NUMERICAL METHODS

The shell model (1) is numerically solved by Runge-Kutta method. The constants are chosen as $\alpha = -0.5, \beta = 1.5, \gamma = -1$. The diffusivity is fixed to $\nu = 2 \times 10^{-8}$. The first modes are fixed: $Z_1^\pm = 0.5 \pm 0.5i$. For constant B_C case, three sets of the wavenumber ratio λ and the number of modes N are used: $(\lambda, N) = (2, 26), (2^{1/2}, 45), (2^{1/4}, 85)$. For random B_C case, both the average and the fluctuation amplitude are set to be equal to $B_1 = (Z_1^+ - Z_1^-)/2i$: $\overline{B_{C,n}} = (\overline{B'^2_{C,n}})^{1/2} = B_1$; the wavenumber ratio λ is 2 and the number of modes N is 26.

Direct numerical simulation of two-dimensional MHD turbulence is performed by pseudo-spectral method. The total number of modes is 2048^2. The velocity field and magnetic field are analyzed by wavelet decomposition. See Ishizawa and Hattori[9] for the details of DNS.

3.2. ENERGY SPECTRUM

Figure 1 shows the energy spectrum for constant B_C case. The spectrum is multiplied by $k^{3/2}$ to show the difference from IK spectrum clearly. Figure 1(a) shows the spectra for three values of B_C: $B_C/B_1 = 0.1, 0.3, 1$. The spectrum is close to K41 spectrum for small k and it is flatter than IK spectrum for large k. The dividing wavenumber becomes smaller as B_C becomes larger. These results are observed by Biskamp[6]. Figure 1(b) shows the spectra for different λ with $B_C/B_1 = 1$. The spectrum for $\lambda = 2^{1/4}$

seems to close to IK spectrum, but it still has two scaling regions. The scaling exponent in the large wavenumber range is seen to depend on λ.

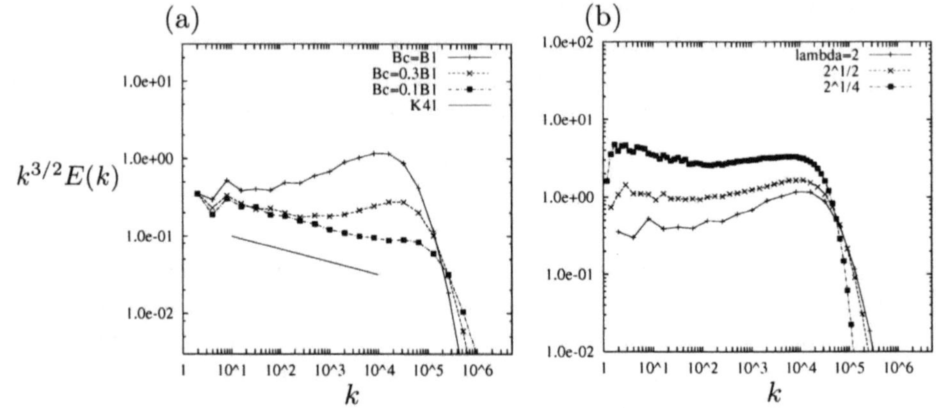

Figure 1. Energy spectrum multiplied by $k^{3/2}$. Constant B_C.

On the other hand, the energy spectrum for random B_C case shown in figure 2 is close to IK spectrum.

In order to clarify why IK spectrum is observed for random B_C case and not observed for constant B_C case, the following correlation between neighboring cells is calculated

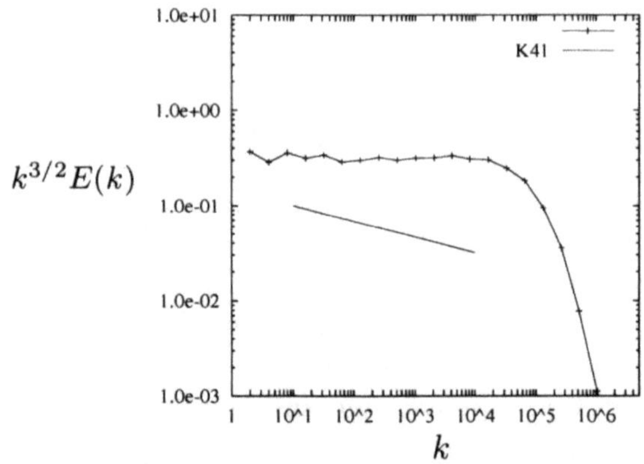

Figure 2. Energy spectrum multiplied by $k^{3/2}$. Random B_C.

$$Cor(n, n+1; \tau) = \frac{\overline{E'_n(t)E'_{n+1}(t+\tau)}}{\sqrt{\overline{E'^2_n}}\sqrt{\overline{E'^2_{n+1}}}}, \quad E_n = \frac{1}{4}(|Z^+_n|^2 + |Z^-_n|^2), \quad E_n = \overline{E_n} + E'_n.$$

Figure 3 shows this correlation. For constant B_C case, the correlation is small for larger n, showing that the energy is not transferred effectively. For random B_C case, the correlation has a maximum of $0.35 \sim 0.5$ for all n shown in the figure. It shows that the energy is transferred effectively.

3.3. PDF

Probability distribution function (PDF) of $Re(u_n)$ for the shell models and wavelet coefficients $\tilde{u_x}^{jr}_{pq}$ obtained from DNS are shown in figures 4 and 5,

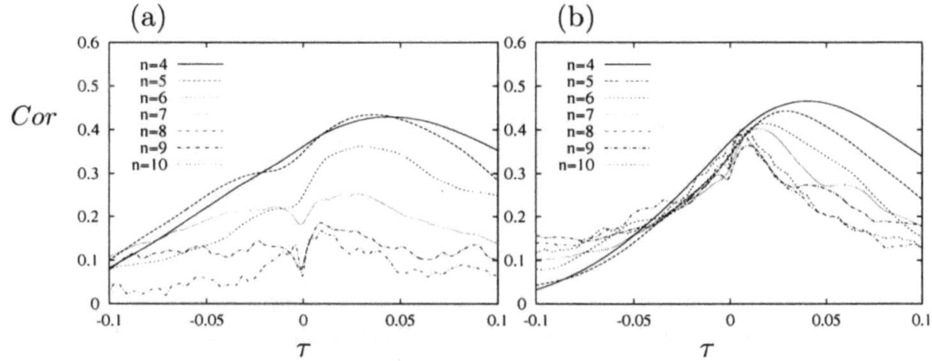

Figure 3. Correlation between neighboring shells: (a) Constant B_C, (b) random B_C.

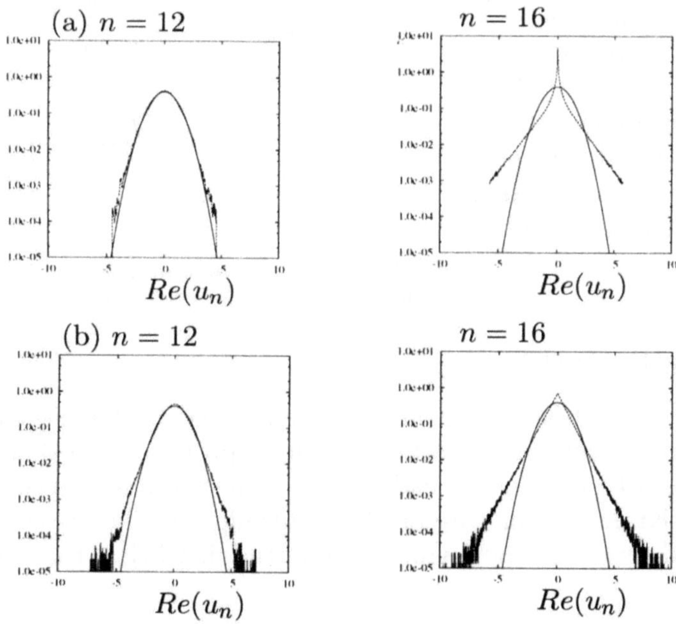

Figure 4. Normalized PDF of velocity modes: (a) Constant B_C, (b) random B_C.

respectively. Two scales are chosen: one is in the inertial range ($n = 12$ for shell models and $j = 5$ for DNS); the other is in the dissipation range ($n = 16$ for shell models and $j = 7$ for DNS). In the inertial range, PDF of $Re(u_n)$ is nearly Gaussian for constant B_C case; it has stretched tail for random B_C case. In the dissipation range, PDF shows extremely intermittent nature for constant B_C case; it is almost exponential for random B_C case. PDF of $\tilde{u_x}_{pq}^{jr}$ obtained from DNS data appears to be close to random B_C case in the inertial range and to constant B_C case in the dissipation range. Figure 6 shows PDF of the corresponding magnitudes, $|u_n|$ for shell models and $[\sum_j \{(\tilde{u_x}_{pq}^{jr})^2 + (\tilde{u_y}_{pq}^{jr})^2\}]^{1/2}$ for DNS, in the dissipation range. It shows that

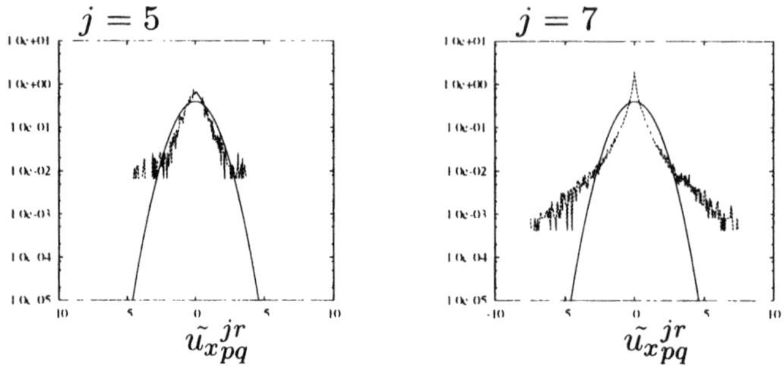

Figure 5. Normalized PDF of wavelet coefficients obtained from DNS data.

PDF of wavelet coefficients is similar to that of $|u_n|$ for random B_C case. Yamada *et al.*[10] found a universal form of PDF in the dissipation range for hydrodynamic case. At present, however, we have not found a form like that for magnetohydrodynamic case.

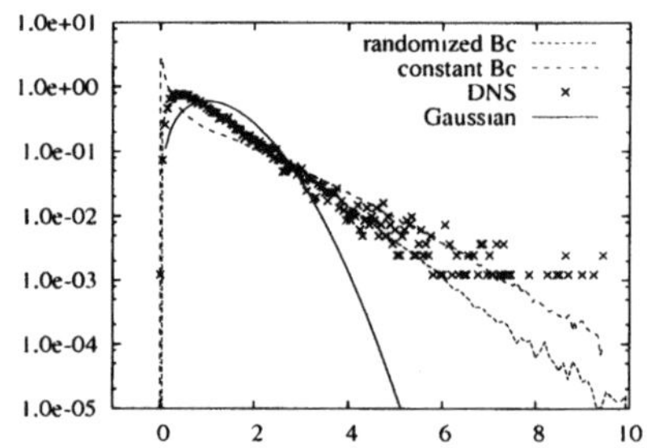

Figure 6. PDF's of magnitude of velocity modes and wavelet coefficients.

4. Concluding Remarks

Statistical properties of MHD turbulence are studied by GOY-type shell models and direct numerical simulation. We have proposed the shell model with randomized Alfvén effect. It leads to $k^{-3/2}$ spectrum and is consistent with Iroshnikov-Kraichnan picture of MHD turbulence. The present mod' fied model would serve as a powerful tool for studying MHD turbulence.

References

1. P. S. Iroshnikov: Astron. Zh. **40** (1963) 742.; R. H. Kraichnan: Phys. Fluids **8** (1965) 1385.
2. D. A. Roberts, M. L. Goldstein, W. H. Matthaeus and L. W. Klein: in *Proceedings of the Workshop on Turbulence and Nonlinear Dynamics in MHD Flows* (eds. M. Meneguzzi, A. Pouquet and P. L. Sulem), (1988) 87.
3. D. Biskamp and H. Welter: Phys. Fluids B **1** (1989) 1964.
4. C. Gloaguen, J. Léorat, A. Pouquet and R. Grappin: Physica D **17** (1985) 154.
5. V. Carbone: Phys. Rev. E **50** (1994) R671.
6. D. Biskamp: Phys. Rev. E **50** (1994) 2702.
7. P. Frick and D. Sokoloff: Phys. Rev. E **57** (1998) 4155.
8. V. S. L'vov, E. Podivilov, A. Pomyalov, I. Procaccia and D. Vandebroucq: Phys. Rev. E **58** (1998) 1811.
9. A. Ishizawa and Y. Hattori: J. Phys. Soc. Jpn. **67** (1998) 441.
10. M. Yamada, S. Kida and K. Ohkitani: in *Unstable and Turbulent Motion of Fluid* (ed. S.Kida), (1993) 188.

C

Probability Density Functions and Structure Functions

STATISTICS OF TRANSVERSE VELOCITY DIFFERENCES IN TURBULENCE

VICTOR YAKHOT[1]
Department of Aerospace and Mechanical Engineering
Boston University
Boston, MA 02215

1. Introduction

Beginning with classic Kolmogorov papers, the ability to theoretically describe statistical properties of velocity differences became, justly or not, a measuring stick of our understanding of turbulence. Development of turbulence theory proceeded in two directions. First, the equations of motion, resulting from the relatively simple and physically appealing second-order closures were so successful that they evolved into a design tool widely used in engineering community [1]. Since all such models can be formulated in terms of perturbation expansions in powers of an $O(1)$ parameter [2], their success in description of the large-scale features of the flow somewhat resembles remarkable precision achieved by the small-parameter-lacking Hartree-Foc approximations in prediction of the properties of atoms and molecules in quantum mechanics. Frequent complains of mathematicians and physisists about "total" lack of understanding of turbulence are typically related to our inability even to formulate a theoretical (dynamic) framework yielding the the experimentally observed small-scale structure of the flow [3].
Early attempts to attack the problem were based on various expansions in powers of the renormalized Reynolds number (RN) $Re^* = u_{rms}L/\nu$ where u_{rms} and L stand for the rms velocity and integral scale of turbulence, respectively. The parameter ν denotes the renormalized, "dressed", viscosity. Since $\nu \approx u_{rms}L$, the coupling constant, which is a measure of the strength of nonlinear interaction, $Re^* \approx O(1)$. Thus, the RN as a basis for the perturbative treatment, must be disqualified. In fact, the expansion in powers of Re^* is equivalent to the one in powers of non-linearity.
The fatal blow to perturbative treatments of the small-scale statistics of turbulence was discovery of intermittency or multi-scaling of turbulence

T. Kambe et al. (eds.), IUTAM Symposium on Geometry and Statistics of Turbulence, 97–106.
© 2001 *Kluwer Academic Publishers.*

which is formulated as follows: consider the two points $\mathbf{x}$ and $\mathbf{x}'$ and define $\mathbf{r} = \mathbf{x} - \mathbf{x}'$. Assuming that the x-axis is parallel to the displacement vector $\mathbf{r}$ we can measure the moments of velocity differences:

$$S_n = \overline{(u(x+r) - u(x))^n} \equiv \overline{(\Delta u)^n} \propto r^{\xi_n} \propto (Pr)^{\frac{n}{3}-\delta_n} \qquad (1)$$

The parameter P in (1) stands for the rate of turbulence production which in the steady state is equal to the mean dissipation rate $\mathcal{E}$. Numerous experiments, conducted in a wide variety of flows, revealed that instead of expected "normal" $\xi_n = n/3$, the exponents ξ_n were "strange" (anomalous) and could not be derived on dimensional grounds. This means that if a perturbative treatment is possible at all, the number of closures (dynamic constraints) needed to obtain first n exponents is at least equal to n. This fact was the main reason for the failure of Kraichnan's "decimation" schemes, proposed in the end of 80's [4].
Realization of futility of the closure-based renormalized perturbation expansions to describe small-scale intermittency led, as it often happens in life, to the opposite extreme: the closures became a taboo, not taken seriously by many researchers in the field. Recently, this taboo was somewhat shaken by a counterexample, produced by Kraichnan for the case of a passive scalar advected by a random velocity field [5]. It was shown later [6] that Kraichnan's anzatz corresponded to an expression for the dissipation contribution to the equation for the probability density of the scalar differences. Kraichnan's suggestion, though incorrect for the scalar problem, demonstrated the ability of a single closure model, applied directly to the equation for the probability density, generate an infinite set of anomalous exponents. Recent Polyakov's operator product expansion was the first theoretical approach to yield a closed equation for the pdf of velocity differences in a non-linear problem of Burgers equation driven by a random force [7]. The expression for the probability density (pdf) of velocity differences was derived from this equation together with an explanation of the observed bi-fractality of velocity field.
In this paper we derive an exact equation for the generating function of velocity difference in turbulence. It is shown how combining this equation with plausable and simple models for the pressure and dissipation terms leads to the results which are in a good agreemnt with expremental data on both two- and three-dimensional turbulence.

2. Equation for the Generating Function

The equations of motion are (density $\rho \equiv 1$):

$$\partial_t v_i + v_j \partial_j v_i = -\partial_i p + \nu \nabla^2 v_i + f_i; \qquad \partial_i v_i = 0 \qquad (2)$$

where $\mathbf{f}$ is a forcing function mimicking the large-scale turbulence production mechanism and in a statistically steady state the mean pumping rate $P = \overline{\mathbf{f} \cdot \mathbf{v}}$. In what follows we will be mainly interested in the probability density function of two-point velocity difference $\mathbf{U} = \mathbf{u}(\mathbf{x}') - \mathbf{u}(\mathbf{x}) \equiv \Delta\mathbf{u}$. The generating function is: $Z =< exp(\lambda \cdot \mathbf{U}) >$. It is easy to see that in the incompressible case the equation for the generating function of velocity differences is [8]-[9]:

$$\frac{\partial Z}{\partial t} + \frac{\partial^2 Z}{\partial \lambda_\mu \partial r_\mu} = I_f + I_p + D \tag{3}$$

with

$$I_f =< \lambda \cdot \Delta\mathbf{f} e^{\lambda \cdot \Delta\mathbf{u}} >; \; I_p = -\lambda \cdot < e^{\lambda \cdot \Delta\mathbf{u}} (\nabla_{\mathbf{2}} p(x_2) - \nabla_{\mathbf{1}} p(x_1)) >;$$

$$\mathrm{D} = \nu\lambda \cdot < (\nabla_{\mathbf{2}}^2 \mathbf{v}(\mathbf{x_2}) - \nabla_{\mathbf{1}}^2 \mathbf{v}(\mathbf{x_1})) e^{\lambda \cdot \mathbf{U}} > \approx \lambda^2 < (\mathcal{E}(x+r) + \mathcal{E}(x)) e^{\lambda \cdot \mathbf{U}} > + O(\lambda^4) \tag{4}$$

The most interesting and surprising feature of these equations is the fact that, unlike in the problem of Burgers turbulence, the advectice contributions are represented in (3) in a closed form. To completely close the problem the expressions for I_p and D are needed. The equations (3)-(4) formulate the turbulence theory in terms of "only" two uknowns I_p and D. The Kolmogorov refined similarity hypothesis stating that $(\Delta u)^3 = \phi \mathcal{E}_r r$ where ϕ is a scale-independent random process and $\mathcal{E}_r$ is a dissipation rate averaged over a ball of radius r around point x, can be a promising starting point to a closure for the dissipation term D. This will be done below. The pressure term in (3)-(4) is also of a very specific and rather limited nature: all we have to know is the correlation functions $< U_i U_j \cdots U_m \Delta \nabla p >$. Thus, the definite targets needed for derivation of the closed equation for Z-functions are well-defined.

The generating function can depend only on three variables: $\eta_1 = r$; $\eta_2 = \frac{\lambda \cdot \mathbf{r}}{r} \equiv \lambda cos(\theta)$; $\eta_3 = \sqrt{\lambda^2 - \eta_2^2}$; and

$$Z_t + [\partial_{\eta_1} \partial_{\eta_2} + \frac{d-1}{r} \partial_{\eta_2} + \frac{\eta_3}{r} \partial_{\eta_2} \partial_{\eta_3} + \frac{(2-d)\eta_2}{r\eta_3} \partial_{\eta_3} - \frac{\eta_2}{r} \partial_{\eta_3}^2] Z = I_f + I_p + D \tag{5}$$

where, to simplify notation we set $\partial_{i,\alpha} \equiv \frac{\partial}{\partial x.\alpha}$ and $v(i) \equiv v(\mathbf{x_i})$. The functions I_p, I_f and D are easily extracted from the above definitions.

3. Two-Dimensional Flow in the Inverse Cascade Range

Now we will be interested in the case of the two-dimensioanl turbulence in the inverse cascade range. If a two-dimensional (2d) fluid is stirred by a random (or non-random) forcing, acting on a scale $l_f = 1/k_f$, the produced energy is spent on creation of the large-scale ($l > l_f$) flow which cannot be dissipated in the limit of large Reynolds number as $\nu \to 0$. This is a direct and most important consequence of an additional, enstrophy conservation, law, characteristic of two dimensional hydrodynamcs [12]. As a result, the dissipation terms are irrelevant in the inverse cascade range and we set $D = 0$ in (5) and hope that in two dimensions the situation is greatly simplified. This hope is supported by recent numerical and physical experiments [11]-[13] showing that as long as the integral scale $L_i \propto t^{\frac{3}{2}}$ is much smaller than the size of the system, the velocity field at the scales $L_i >> l >> l_f$ is a stationary close-to-gaussian process characterized by the structure functions (1) with the Kolmogorov exponents $\xi_n = n/3$. In a recent paper Bofetta, Celatti and Vegrassola [13] reported the results of very accurate numerical simulations of two-dimensional turbulence generated by a random force. No deviations from gaussian statistics of transverse velocity differences as well as from the Kolmogorov scaling $\xi_n = n/3$ were detected. Moreover, the measured pdf of longitudinal differences could be represented as a sum: $P(\Delta u) = P_s(\Delta u) + P_a(\Delta u)$ were $P_s(x) = P_s(-x)$ and $P_a(x) = -P_a(-x)$ with $P_s(x)$ indistinguishable from the gaussian. The generating function in the new variables can be written as: $Z =< exp(\eta_2 \Delta u + \sqrt{d-1}\eta_3 \Delta v)$ so that the pdf of transverse velocity differences is obtained in the limit $\eta_2 \to 0$. With $D = 0$, the solution of the problem of $2d$- turbulence is reduced to evaluation of the pressure term I_p which, when $\eta_2 \to 0$, which can be rewritten as:

$$I_p \approx \eta_3 < (\partial_y p(0) - \partial_y p(r)) exp(\sqrt{d-1}\eta_3 \Delta v + \eta_2 \Delta u) > \qquad (6)$$

The Navier-Stokes equations are invariant under transformation: $v \to -v$ and $y \to -y$. That is why: $< (\partial_y p(0) - \partial_y p(r))(\Delta v)^m > \neq 0$ if $m = 2n+1$ with $n > 1$ and is equal to zero if $m = 2n$. The pressure gradient $\partial_y p = \partial_y \partial_i \partial_j \partial^{-2} \Delta v_i \Delta v_j$ and the difficulty in calculation of I_p is in the integral over the entire space defined by the inverse Laplacian ∂^{-2}. The huge simplification, valid in 2d only, comes from the fact that all contributions to the left side of equation (5) as well as I_f are independent on time. This means that the integrals involved in the pressure terms cannot be infra-red divergent since in a two-dimensional flow $L = L(t) \propto t^{\frac{3}{2}}$. The expression (6) can formally be represented as: $I_p =< i_p(U, V, r) e^{\lambda \cdot \mathbf{V}} >$, where $i_p = \lambda \cdot < (\nabla_2 \mathbf{p}(2) - \nabla_1 \mathbf{p}(1)) | U, V, r >$ is the conditional expectation value of $\Delta \nabla p$ for the fixed $U = \Delta u$; $V = \Delta v$; r. We also have that $i_p \to \alpha^2 i_p$ when U, $V \to \alpha U$; αV. Based on this and taking into account that $< (\Delta v)^{2n+1}(u_x^2 + v_y^2 + u_y v_x) >= 0$, we have

$$\partial_y < \Delta\nabla p(\Delta v)^{2n+1} >\approx \frac{1}{r}\overline{(\Delta v)^{2n+1}((a\Delta u)^2 + (c\Delta v)^2 + b\Delta u\Delta v)} \quad (7)$$

where a, b, c are as yet undetermined constants. This expression is exact, provided the coefficients $a; b; c$ are finite in the limit $l_f \to 0$. This is the only assumption of the theory. All, but last, terms in the right side of (7) are equal to zero. Thus:

$$\begin{aligned} I_p = b\frac{\eta_3}{r} < \Delta u \Delta v e^{\eta_2\Delta u + \eta_3 \Delta v} > &= b\frac{\eta_3}{r}\frac{\partial}{\partial\eta_2} < \Delta v e^{\eta_2\Delta u + \eta_3\Delta v} > \\ &= b\frac{\eta_3}{r}\frac{\partial^2}{\partial\eta_2\partial\eta_3} Z(\eta_2 = 0, \eta_3, r) \quad (8) \end{aligned}$$

Let us consider a strirring force $\mathbf{f} = (f_x, 0)$, acting on the same scale l_f. It has been tested numerically that at the scales $l > l_f$ the flow rapidly recovers isotropy [14]. In this case in the limit $\eta_2 \to 0$ the information about the forcing function disappears from the equation:

$$Z_t + [\partial_r\partial_{\eta_2} + \frac{d-1}{r}\partial_{\eta_2} + \frac{\eta_3}{r}\partial_{\eta_2}\partial_{\eta_3} - I_p]Z = 0 \quad (9)$$

Taking I_p given by the second in (8) equation and assuming that the field Δv obeys gaussian statistics (see below) we can write:

$$I_p \approx b\eta_3^2 \frac{\overline{(\Delta v)^2}}{r}\frac{\partial Z_3}{\partial \eta_2} \quad (10)$$

Substituting (10) into (9) we discover the most remarkable and unusual feature of the resulting equation: it can be represented as: $\partial_{\eta_2}(\mathcal{L}Z)$ where $\mathcal{L}$ is a differential operator which in the limit $\eta_2 \to 0$ does not include the variable η_2. Integrating this equation over η_2 and choosing an arbitrary function $\Psi(\eta_3, r)$ in such a way that the remaining equation satisfy the normalizability constraint $Z(\eta_2 = \eta_3 = 0, r) = 1$, gives:

$$\frac{\partial Z_3}{\partial r} + \frac{\eta_3}{r}\frac{\partial Z_3}{\partial \eta_3} - \frac{\gamma P}{(Pr)^{\frac{1}{3}}}\eta_3^2 Z_3 = 0 \quad (11)$$

and

$$Z = Exp(\frac{3\gamma}{8}\eta_3^2(Pr)^{\frac{2}{3}}) \quad (12)$$

with parameter γ defining the width of the gaussian. The first-order differential equation (11) tells us that the dynamics of transverse velocity differences is goverened by a linear Langevin equation, driven by a solution-dependent forcing. This equation is non-local in the physical space but local in the wave-number space.

Now let us consider the third equation for I_p (8). Repeating the above procedure gives an identity from which no information about generating function can be extracted. This is explained very simply: consider the forced diffusion equation: $\partial_t C = \kappa\nabla^2 C + f$ with the same as above forcing function. The steady state Fokker-Plank equation in this case is: $\lambda^2 PZ - \lambda^2 < \mathcal{E}e^{\lambda C} >= 0$, telling us that $< (\mathcal{E} - \overline{\mathcal{E}})e^{\lambda C} >= 0$ or in other words, that the dissipation rate and scalar fluctuations are statistically independent. All this is consistent with the derived linearity of dynamic equation for transverse velocity differences. It is clear that, to obtain the above solution in this case, we have to modify the function $\Psi(\eta_3, r)$ accordingly.

4. Three-Dimensional Flow

The difficulty of the three-dimensional case comes from the dissipation contributions. This is easily understood on the basis of the Navier-Stokes equations giving $\mathcal{E} = O(\partial_x u^3)$. The dissipation term can be written as:

$$D =< d_p(\Delta u, \Delta v, r)e^{\lambda \mathbf{\Delta v}} > \tag{13}$$

where $d_p = \lambda\cdot < (\nabla_2^2 \mathbf{v}(2) - \nabla_1^2 \mathbf{v}(1))|\Delta u; \Delta v; r >$ is a conditional expectation value of $\Delta\nabla p$ for fixed $\Delta U;\ \Delta v:\ r$. The dissipation term is local in physical space and, since the displacement $\mathbf{r} = (r, 0, 0)$, d_p is invariant as $v \to -v$. In addition, d_p is invariant as $u \to -u$: and $x \to -x$. The combination satisfying this set of constraints in the limit $\eta_2 \to 0$ is:

$$\begin{aligned} D &\approx c\eta_3 \partial_{\eta_2}\partial_{\eta_3}\partial_r Z \\ &\approx c\eta_3^2 < \Delta u \Delta v \frac{\partial \Delta v}{\partial r} e^{\eta_2 \Delta u + \sqrt{d-1}\eta_3 \Delta v} > + O(\partial_r \partial_{\eta_2} Z) \end{aligned} \tag{14}$$

The locality of this expression is clear since $\frac{\partial \Delta v}{\partial r} = \frac{\partial v(2)}{\partial x_2} + \frac{\partial v(1)}{\partial x_1}$. The last term in the right side of (14), simply modifies the coefficient in front of the first term in the left side the equation (5) and does not generate anything new. The expression (14) resembles Kolmogorov's refined similarity hypothesis, connecting the dissipation rate, averaged over a region of radius r, with $(\Delta u)^3$. At the present time we cannot prove (14). Thus, in the limit $\eta_2 \to 0$:

$$[\partial_{\eta_1}\partial_{\eta_2}+\frac{2}{r}\partial_{\eta_2}+\frac{\eta_3}{r}\frac{\partial^2}{\partial_{\eta_2}\partial\eta_3}-\frac{\eta_2}{r}\frac{\partial^2}{\partial\eta_3{}^2}+b\frac{\eta_3}{r}\frac{\partial^2}{\partial\eta_2\partial\eta_3}+c\eta_3\partial_{\eta_2}\partial_{\eta_3}\partial_r]Z(\eta_2=0,\eta_3,r)=0 \tag{15}$$

where unknown coefficients b and c will be determined below. Integrating (15) over η_2 gives

$$[\partial_{\eta_1}+\frac{2}{r}+\frac{\eta_3}{r}\frac{\partial}{\partial\eta_3}+b\frac{\eta_3}{r}\frac{\partial}{\partial\eta_3}+c\eta_3\partial_{\eta_3}\partial_r]Z(\eta_2=0,\eta_3,r)=\Psi(\eta_3,r) \tag{16}$$

We must choose $\Psi(\eta_3)$ in such a way that the generating function $Z(0,0,r)=1$. Inverse Laplace transform gives the equation for the pdf $P(\Delta v,r)$

$$\frac{\partial P}{\partial r}+\frac{1+3\beta}{3r}\frac{\partial}{\partial V}VP-\beta\frac{\partial}{\partial V}V\frac{\partial P}{\partial r}=0 \tag{17}$$

Since $S_3^t=0$, the coefficients in (17) are chosen to give $s_3^t=\overline{|\Delta v|^3}=a_3Pr$ with an undetermined amplitude a_3. This is an assumption of the present theory, not based on rigorous theoretical considerations. Seeking the solution in a form $S_n^t=<(\Delta v)^n>\propto r^{\xi_n}$ we obtain

$$\xi_n=\frac{1+3\beta}{3(1+\beta n)}n\approx\frac{1.15}{3(1+0.05n)}n \tag{18}$$

which was derived in [15] together with $\beta\approx 0.05$. It follows from (17) that: $P(0,r)\propto r^{-\kappa}$ where $\kappa=\frac{1+3\beta}{3(1-\beta)}\approx 0.4$ for $\beta=0.05$. Very often the experimental data are presented as $P(X,r)$ where $X=V/r^{\mu}$ with $2\mu=\xi_2\approx 0.696$ for $\beta=0.05$. This gives $P(X=0,r)\propto r^{-\kappa+\mu}\approx r^{-0.052}$ compared with the experimental data by Sreenivassan [16]: $-\kappa+\mu\approx -0.06$. Introducing $P(V,r)=r^{-\kappa}F(\frac{V}{r^{\kappa}},r)=r^{-\kappa}F(Y,r)$ and $-\infty<y=Ln(Y)<\infty$, substituting this into (17) and evaluating the Fourier transform of the resulting equation gives:

$$(1-\beta)r\frac{\partial F}{\partial r}+\beta\kappa(ik-k^2)F-ik\beta r\frac{\partial F}{\partial r}=0 \tag{19}$$

with the result: $F\propto r^{\gamma(k)}$, where $\gamma(k)=\beta\kappa\frac{-ik+k^2}{1-\beta-i\beta k}Ln(r/L)$ with $r/L<<1$. We have to evaluate the inverse Fourier transform:

$$F=\int_{-\infty}^{\infty}dke^{-iky}e^{\gamma(k)} \tag{20}$$

in the limit $y=O(1)$ and $r\to 0$ so that $Ln(r/L)\to-\infty$. The integral can be calculated exactly. However, the resulting expression is very involved. Expanding the denominator of $\gamma(k)$ gives :

$$F = \int_{-\infty}^{\infty} dk e^{-ik(y+\frac{\beta\kappa(Ln(r))}{1-\beta})} e^{-\frac{\beta\kappa(1+\beta)|Ln(r)|}{(1-\beta)}k^2} \tag{21}$$

and

$$F \propto \frac{1}{\sqrt{\Omega(r)}} exp(-\frac{(Ln(\xi))^2}{4\Omega}) \tag{22}$$

with $\xi = V/r^{\frac{\kappa}{1-\beta}}$ and $\Omega(r) = 4\beta\kappa\frac{1+\beta}{1-\beta}|Ln(r/L)|$.
To understand the range of validity of this expression, let us evaluate $< V^n >$ using the expression (22) for the pdf. Simple integration, neglecting $O(\beta^2)$ contributions, gives: $< V^n > \propto r^{\alpha_n}$ with $\alpha_n = (1+3\beta)(n - \beta(n^2+2))/3$. Comparing this relation with the exact result (18) we conclude that the expression for the pdf, calculated above, is valid in the range $n >> 1$ and $\beta n << 1$. The properties of the pdf in the range $3 \le \xi \le 15$ are demonstrated on Fig. 1 for $r/L = 0.1$; 0.01; 0.001. The log-normal distribution (22), is valid in a certain (wide but limited) range of V- variation. It is clear from (17) that neglecting the dissipation terms ($c \propto \beta = 0$) leads to $\xi_n = n/3$, i. e. disappearence of anomalous scaling of moments of velocity differences. This result agrees with the well-developed phenomenology, attributing intermittency to the dissipation rate fluctuations: the stronger the fluctuations, the smaller the fraction of the total space they occupy [1], [3]. To the best of our knowledge, this is the first work leading to multifractal distribution of velocity differences as a result of approximations made directly on the Navier-Stokes equations. The expression (22) is similar to the one obtained in a groundbreaking paper by Polyakov on the scale-invariance of strong interactions, where the multifractal scaling and the pdf were analytically derived for the first time [17]-[18]. In the review paper [18] Polyakov noticed that the exact result can be simply reproduced considering a cascade process with a heavy stream (particle) transformed into lighter streams at each step of the cascade (fission). Due to relativistic effects the higher the energy of the particle, the smaller the angle of a cone, accessible to the fragments formed as a result of fission. Thus, the larger the number of a cascade step, the smaller is the fraction of space occupied by the particles [18].

5. Conclusions

The importance of exact equation (5) is in the fact that it provides a mathematic framework for discussion and understanding of small-scale structure of turbulence. The approximations and models, developed in this paper, show how both close-to-gaussian statistics and Kolmogorov scaling in two dimensions and anomalous scaling and intermittency in 3d appear directly

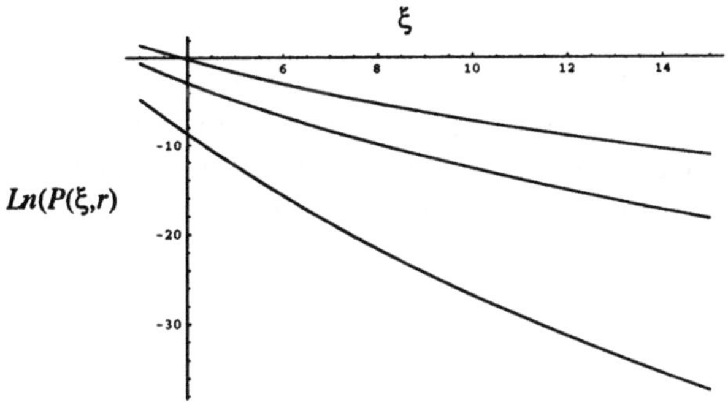

Figure 1. $Ln(P(\xi, r))$. From bottom to top: r/L=0.1; 0.01; 0.001, respectively.

from dynamic equations. All previous attempts were unable not only to produce the results but even to point to feasible directions of research. The presented above theory of two-dimensional turbulence involves only one assumption: locality of the pressure gradient contributions to the expression for I_p which is based on the rigorous conclusion that the time-dependent large- scale features of the flow cannot enter the equation. (5). Still, some corrections to the above results may exist. Elucidation of the role of these corrections is an extremely important task. Numerical test of (7), is straitforward. It has been shown recently that in accord with (7) $< \Delta\partial_y p|\Delta u >=< \Delta\partial_y p|\Delta v >= 0$ even in a three-dimensional turbulent flow [19].

The power of equation (5) has been demonstrated above on an example of a theory of three-dimensional turbulence. It has been shown that (5), combined with the two plausable models, can describe both anomalous scaling and the entire pdf of transverse velocity differences. The assumptions of the theory are: locality of the pressure gradient, as in 2d, and a model expression, similar to Kolmogorov's refined similarity hypothesis for the dissipation terms. The equation (5) and the theory presented above, provide a promising intellectual and mathematical framework for discussion of both semi-empirical and theoretical models aimed at description of experimental data. Moreover, they directly point to the neccessary experiments which can be crucial for development of "the final theory".

The unusual symmetry of the equation for transverse velocity differences is a new feature, never seen before. That is why, in our opinion, state-of-the-art simulations of two-dimensional turbulence aiming at verification of the gaussian statistics of Δv is the most important and urgent task.

6. Acknowledgement

I am grateful to A. Polyakov for most interesting and stimulating discussions. He participated in development of the argument, leading to (8), which is simpler and clearer than the previous one [8]. I benefited a lot from discussions with B. Shraiman, P. Constantin, C. Doering, T. Gotoh, T. Kambe and I. Procaccia.

7. References

1. A.S.Monin and A.M.Yaglom, "Statistical Fluid Mechanics" vol. 1, MIT Press, Cambridge, MA (1971)
2. V. Yakhot and S.A.Orszag, Phys.Rev.Lett. **57**, 1722 (1986)
3. U. Frisch, "Turbulence", Cambridge University Press, 1995
4. R.H. Kraichnan, preprint
5. R.H. Kraichnan, Phys.Rev.Lett., **72**, 1016 (1994)
6. V. Yakhot, Phyds. Rev. E., **55**, 329 (1997)
7. A.M. Polyakov, Phys.Rev. E, **52**, 6183 (1995)
8. V. Yakhot, Phys. Rev.E, in press (1999)
9. V.Yakhot, Phys.Rev.Lett, submitted (1999)
10. R.H.Kraichnan, Phys.Fluids. **10**, 1417 (1967)
11. L.Smith and V. Yakhot, J. Fluid. Mech. **274**, 115 (1994). L.Smith and V. Yakhot, J. Fluid. Mech. **274**, 115 (1994)
12. P. Tabeling and J. Paret, Phys. Fluids, **12**, 3126 (1998)
13. G. Bofetta, A. Cellini, M. Vergassola, Chao-Dyn/9906016
14. V. Borue and V. Yakhot, unpublished
15. V. Yakhot, Phys. Rev.E, Phys. Rev. E, **57**, 1737 (1998)
16. K.R. Sreenivasan, private communication
17. A.M. Polyakov, Sov. Phys. JETP **34**, 1177 (1972)
18. A.M. Polyakov, "Scale Invariance of Strong Interactions and Its Application to Lepton-Hadron Reactions", preprint, Landau Institute for Theoretical Physics, 1971
19. T. Gotoh, private communication (1999)

ON A COMPLETE STATISTICAL CHARACTERIZATION OF TURBULENCE

R. FRIEDRICH[1], ST. LUECK[2], J. PEINKE[2], AND CH. RENNER[2]
[1] *Inst. Theo. Physik, Universität Stuttgart, D-70550 Stuttgart*
[2] *FB 8 Physik, Universität Oldenburg, D-26111 Oldenburg*

1. Introduction

The common picture of fully developed local isotropic turbulence is that the velocity field is driven by external fields on large scales. By this, driving energy is fed into the system at scales larger than the integral length L_0. A cascading process will transport this energy to smaller and smaller scales until at the viscous length scale η the injected energy is finally dissipated by viscous effects [1, 2]. It is commonly believed that this picture of a cascade leads to the universality of the statistical laws of small scale turbulence.

The standard quantity to study the statistics of turbulent fields is the longitudinal velocity increment of the length scale r defined as $v(r) = u(x+r) - u(x)$, where u denotes the velocity component in the direction of the separation r. x is a selected reference point.

To analyze the statistical content of the velocity increments two common ways have been established. On the one side, the structure functions have been evaluated by supposing a scaling behavior

$$< (v(r))^q >= \int dv \; (v)^q \; P(v,r) \propto r^{\zeta_q}, \tag{1}$$

where $P(v,r)$ denotes the probability density function (pdf) for v at the length scale r. Using the recently proposed so-called extended selfsimilarity [3], it has become possible to evaluate the characterizing scaling exponents ζ_q quite accurately c.f. [4, 5]. On the other side, it is interesting to find out how to parameterize directly the evolution of the probability density functions (pdf) $P(v,r)$ c.f. [6, 7].

One major challenge of the research on turbulence is to understand small scale intermittency, which is manifested in a changing form of $P(v,r)$

T. Kambe et al. (eds.), IUTAM Symposium on Geometry and Statistics of Turbulence, 107–116.
© 2001 *Kluwer Academic Publishers.*

or equivalently in a nonlinear q-dependence of the scaling exponents ζ_q, provided that the scaling assumption is valid.

For a long time, the main effort has been put into the understanding of the q-dependence of ζ_q (for actual reviews see [2, 5]), although it is well known that due to the statistics of a finite number of data points (let's say 10^7 data) the determination of scaling exponents ζ_q for $q > 6$ becomes inaccurate [8]. In addition there are different experimental indications that no good scaling behavior is present [9, 10].

There have been less attempts to analyze directly the pdfs. This might be based on the fact that up to now the scaling exponents are regarded as the simplest reduction of the statistical content and that this analysis does not depend on model assumptions. In contrast to this, proposed parameterizations of the form of the pdfs [6, 7], although they are quite accurate, are still based on some additional assumptions on the underlying statistics. Based on the recent finding that the turbulent cascade obeys a Markov process in the variable r and that intermittency is due to multiplicative noise [10, 11, 12], we show in Section III that it is possible to estimate from the experimental data a Fokker-Planck equation, which describes the evolution of the pdfs with r. We show that this Fokker-Planck equation reproduces accurately the experimental probability densities $P(v, r)$ within the inertial range (see Fig. 1). Thus, an analysis of experimental data which quantifies the statistical process of the turbulent cascade and which neither depends on scaling hypotheses nor on some fitting functions for pdfs is possible [13]. Having determined the correct Fokker-Planck equation for an experimental situation of local isotropic turbulence, we show in section VI that with this approach also the statistics of inhomogeneous turbulence can be analyzed. Here, we present results for velocity data in a turbulent wake behind a cylinder. In two recent works it has been shown that the analysis of a Markov process in r also holds for the energy dissipation averaged over distances r [14, 15].

2. Experimental data

The results presented here are based on 10^7 velocity data points measured in two different flows, namely, a free jet and a turbulent wake flow. Local velocity fluctuations were measured with a hot wire anemometer (Dantec Streamline 90N10). A single hot wire probe (55P01) with a spatial resolution of about 1 mm as well as an x-wire probe (55P71) with a spatial resolution of about 1.5 mm were used. The measurements were performed with sampling frequencies up to several kHz.

For the free jet, the stability was verified by measurements of the self-similar profiles of the mean velocity according to [16]. The turbulence measurements were performed by placing the probe on the axis of a free jet of dry air developing downwards in a closed chamber of the size of 2 m x 1 m x 1 m . To prevent a disturbing counterflow of the outflowing air, an outlet was placed at the bottom of the chamber. The distance of the hot wire to the nozzle was 125 nozzle diameters D. As a nozzle we used a convex inner profile [17] with an opening section of $D = 8$ mm and an area contraction ratio of 40. Together with a laminarizing prechamber we achieved a highly laminar flow coming out of the nozzle. At a distance of 0.25 D from the nozzle, no deviation from a rectangular velocity profile was found within the detector resolution. Based on a 12bit A/D converter resolution, no fluctuations of the velocity could be found. For the experimental data used here, the velocity at the nozzle was 45.5 m/s corresponding to a Reynolds number of 2.7 10^4. At the distance of 125 D, we measured a mean velocity of 2.25 m/s, a degree of turbulence of 0.17, an integral length $L_0 = 67$ mm, a Taylor length $\theta = 6,6$ mm(determined according to [18]), a Kolmogorov length $\eta = 0.3$ mm, and a Taylor Reynolds number $R_\theta = 190$.

As a second experimental system, a wake flow was generated behind a circular cylinder inserted in a wind tunnel. Cylinders with two diameters d of 2 cm and 5 cm were used. The wind tunnel [19] we used has the following parameters: cross section 1.6 m x 1.8 m; length of the measuring section 2 m; velocity 25 m/s; residual turbulence level below 0.1 %. To measure longitudinal and transversal components of the local velocity, the x-wire probes were placed at several distances between 8 and 100 diameters of the cylinder. Depending on the used cylinder, the Reynolds numbers based on the cylinder diameter were of 5 10^4 and 1.2 10^5. In the far field, the integral length varied between 10 cm and 30 cm, the Taylor lengths were around 2 mm, the Taylor Reynolds numbers were $R_\theta = 280$ and 650.

The space dependence of the velocity increments v was obtained by the Taylor hypotheses of frozen turbulence. For the structure functions, we found a tendency to scaling behavior for $L_0 \geq r \geq \theta$. Intermittency clearly emerged as $r \rightarrow \theta$, as shown in Fig.1 by the different form of the pdfs for different scales r. Note that the velocity increments given in this paper are normalized to the saturation value of $\sigma_\infty = \sqrt{< v_\infty^2 >}$ for length scales larger than the integral length.

3. Measurement of Kramers-Moyal coefficients

Next, we show how to determine from experimental data appropriate statistical equations to characterize the turbulent cascade. The basic quantity for this procedure is to evaluate the cascade by the statistical dependence of velocity increments of different length scales at the same location x [20]. Either two-increment probabilities $p(v_2, r_2; v_1, r_1)$ or corresponding condi-

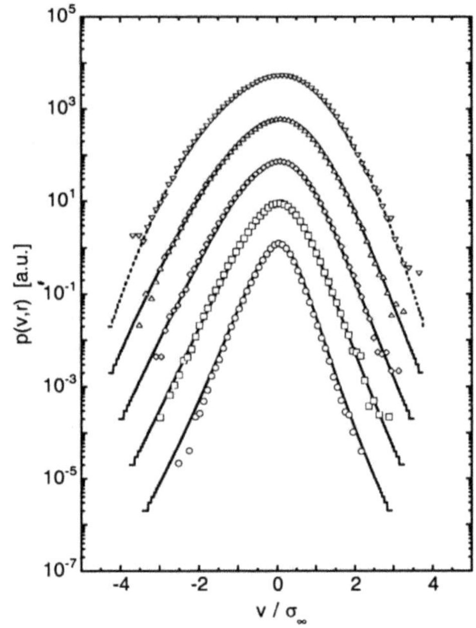

Figure 1. Probability density functions $P(v, r)$ determined from the experimental data of the free jet (bold symbols) and calculated pdfs (lines) by the numerical iteration with the experimentally determined Fokker-Planck equation (5) . The length scales $r = L_0, 0.6L_0, 0.35L_0, 0.2L_0$ and $0.1L_0$ going from up to down. For the numerical iteration as initial condition, an empirical fitting function for the large scale pdf was used which is shown by a broken line. For clarity of presentation the pdfs were shifted in y-direction.

tional probabilities $p(v_2, r_2|v_1, r_1) := p(v_2, r_2; v_1, r_1)/p(v_1, r_1)$ are evaluated from the whole data set. (Here, we use the convention that $r_{i+1} < r_i$, furthermore $v(r_i) = v_i = u(x + r_i) - u(x)$.) Investigating the corresponding three-increment statistics, we could provide evidence that the evolution of these statistics with different r fulfills the Chapman-Kolmogorov equation [10, 11, 21]. Furthermore, evidence of the validity of the Markov process is obtained as long as the step size between different r is larger than a Markov-length L_{mar}, which is in the order of the Taylor length [12]. Thus we know [21] that the evolution of the increments v is given by a master equation without the involvement of memory functions and that the evolution of the pdfs $P(v, r)$ as well as $p(v, r|v_1, r_1)$ is described by a Kramers-Moyal expansion:

$$-r\frac{d}{dr}P(v,r) = \sum_{k=1}^{\infty}[-\frac{\partial}{\partial v}]^k D^{(k)}(v,r)\; P(v,r). \qquad (2)$$

Here, the Kramers-Moyal coefficients $D^{(k)}(v,r)$ are defined by the following conditional moments:

$$\begin{aligned} M^{(k)}(v_1,r_1,\Delta r) &= \frac{r}{n!}\frac{1}{\Delta r}\int dv_2\,(v_2-v_1)^k\,p(v_2,r_2|v_1,r_1) \\ D^{(k)}(v_1,r_1) &= \lim_{\Delta r\to 0} M^{(k)}(v_1,r_1,\Delta r), \end{aligned} \tag{3}$$

where $\Delta r = r_2 - r_1$. The presentation of the Kramers-Moyal expansion and the definition of the Kramers-Moyal coefficients were chosen in a way adapted to assumed scaling behavior, i.e., all was multiplied by r. Knowing the Kramers-Moyal coefficients and thus the evolution of $P(v,r)$, the differential equations for the structure functions are obtained easily [10]

$$-r\frac{d}{dr} < (v(r))^q >= \sum_{n=1}^{q-1} \frac{q!}{(q-n)!} < D^{(n)} v^{q-n} > . \tag{4}$$

If the Kramers-Moyal coefficients have the form $D^{(q)} = d_q v^q$ (where d_q are constants), scaling behavior of (1) is guaranteed with $\zeta_q = -\sum_{n=1}^{q-1}\frac{q!}{(q-n)!}d_n$. For a discussion of this finding in terms of multifractality and multiaffinity see [22].

An important simplification is achieved if the fourth Kramers-Moyal coefficient is zero. In this case the infinite sum of the Kramers-Moyal expansion (2) reduces to the Fokker-Planck equation of only two terms [21]:

$$-\frac{d}{dr}P(v,r) = \left[-\frac{\partial}{\partial v}D^{(1)}(v,r) + \frac{\partial^2}{\partial v^2}D^{(2)}(v,r)\right]P(v,r). \tag{5}$$

Now, only the drift term $D^{(1)}$ and the diffusion term $D^{(2)}$ are determining the statistics completely. From the corresponding Langevin equation characterizing the evolution of a spatially fixed increment, we know that $D^{(1)}$ describes the deterministic evolution, whereas $D^{(2)}$ describes the noise acting on the cascade. If $D^{(2)}$ shows a v-dependency one speaks of multiplicative noise.

In Figs. 2 and 3 we show the evaluated conditional moments $M^{(1)}$ and $M^{(2)}$ at scale $r_1 = L/2$ for $\Delta r = \theta$. For the higher Kramers-Moyal coefficients we found that $M^{(4)} < 10^{-2}M^{(2)}$. Thus, we take the Fokker-Planck equation as the adequate description.

The moment $M^{(1)}$ shows a linear dependence on the velocity increment with small second- and third-order term corrections, while $M^{(2)}$ can be approximated by a polynomial of degree two in v: This behavior is found for all scales r and for all values of Δr. It is therefore reasonable to assume that the Kramers-Moyal coefficients show the same dependence on v as

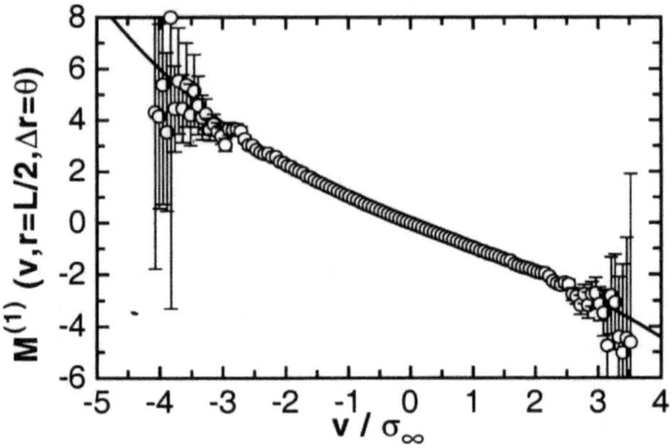

Figure 2. The moments $M^{(1)}(v, r, \Delta r)$ as a function of the velocity increment v for $r = L/2$ and $\Delta r = \theta$ (circles). The data are well fitted by a polynomial of degree three (solid line), where the second and third order term are small.

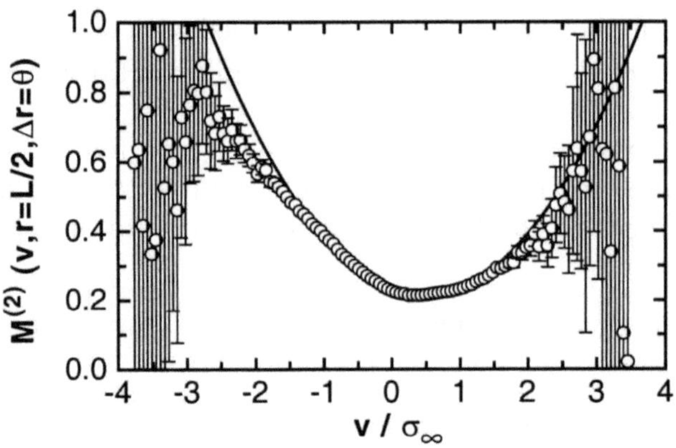

Figure 3. The moment $M^{(2)}(v, r, \Delta r)$ as a function of the velocity increment v for $r = L/2$, $\Delta r = \theta$ (circles) and the fitting polynomial of degree two (solid line). the error bars indicate the increasing uncertainty for large v.

the moments $M^{(k)}$. With this functional dependence we could perform the limit $\Delta r \to 0$. The complete results for the coefficients $D^{(1)}$ and $D^{(2)}$ are ($\rho = \frac{r}{\theta}$):

$$
\begin{aligned}
D^{(1)}(v, r) &= -o(r) - \gamma(r)v + \kappa(r)v^2 - \epsilon(r)v^3 \\
o(r) &= 0.0015\,\rho^{1.6} \\
\gamma(r) &= 0.61\,\rho^{1/3}
\end{aligned}
$$

$$\begin{aligned}
\kappa(r) &= 0.0096\,\rho \\
\epsilon(r) &= 0.0024\,\rho^{1.3} \\
D^{(2)}(v,r) &= \alpha(r) - \delta(r)v + \beta(r)v^2 \\
\alpha(r) &= 0.034\,\rho^{1.2} \\
\delta(r) &= 0.017\,\rho \\
\beta(r) &= 0.062\,\rho^{0.23}.
\end{aligned} \tag{6}$$

Let us briefly comment this result. For the estimation of the limiting values, we considered only Δr values larger than the Markov length. This yields to γ values larger than 1/3, in contrast to the expected value in [10]. A puzzle (see also [25]) we have not solved yet is the finding that the approximation for smaller δ-value yields a γ value of 1/3. Having estimations for $D^{(1)}$ and $D^{(2)}$, it becomes possible to verify this finding by the numerical iteration of the Fokker-Planck equation (5). Here, we start with an approximation for the large scale pdf, as initial condition for the iteration. In Figure 1, the result of such a numerical iteration is shown by solid curves. We find an almost perfect accordance between the numerical solutions and the experimental data. On the basis of this result, it is not possible to decide the correct value of γ. Within the same functional dependencies of $D^{(1)}$ and $D^{(2)}$ it is possible to reproduce the pdf correctly with different combinations of the prefactors (6). If not only the pdf of single increments but also the conditional pdfs are taken as a testing ground for the correctness of the extracted $D^{(1)}$ and $D^{(2)}$ coefficient, we see that only the values given in (6) work well [26].

4. Anisotropic turbulence

Up to now the gained insight into details of local isotropic turbulence has not improved essentially the understanding of real flow situations which are commonly not local isotropic. Thus we investigated with our method, which is mentioned above, also the disorder in the velocity field close to a cylinder. Here, even in the case of high Reynolds numbers, pronounced large scale structures are present. More precisely, we investigated the longitudinal and transversal velocity measured in the near field of the turbulent wake already mentioned.

First, we evaluated the pdf $P(v,r)$ and found out that close to the cylinder at large scales r the pdfs clearly deviate from a Gaussian form. Sometimes even a double hump structure appears. The question whether this situation can also be described by a Markov-process was verified by the evaluation of the Chapman-Kolmogorov equation. We found again that this equation is nicely fulfilled and thus it makes sense to evaluate the Kramers-Moyal coefficients. The result for the first moments $M^{(1)}$ is shown in Fig. 4.

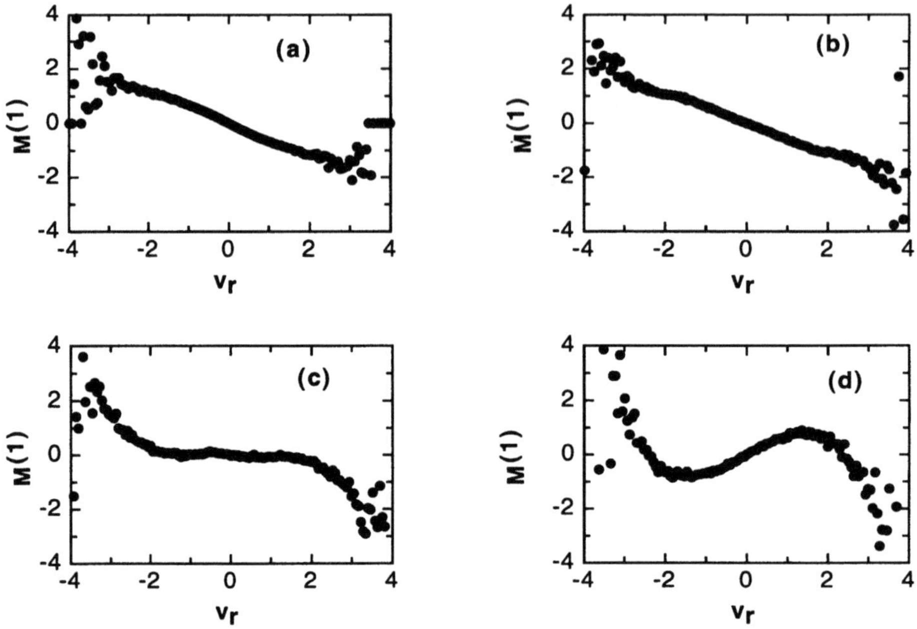

Figure 4. The moment $M^{(1)}(v, r, \Delta r)$ as a function of the transversal velocity increment v for r =8.5 mm (a), 26.6 mm (b), 44.3 mm (c), and 62 mm (d); cylinder diameter 5 cm; Δr =0, 177 mm.

For small scales, $M^{(1)}$ has a similar form like the one found for local isotropic turbulence. This may be taken as an indication of universal small scale turbulence. For large scales r, we find pronounced new structures in $M^{(1)}$. The formerly linear behaviour changes to an n-shaped form. An important effect is that the number of zeros ($M^{(1)}$=0) changes from 1 to three. These zeros correspond to fixed points of the deterministic part of the stochastic process. Fixed points with negative slope are accumulation points. The random part of the stochastic process causes fluctuations around these fixed points. This leads to a building up of maxima in the corresponding pdfs for large scales, which vanish as one comes to smaller scales r (see also [27]). Thus, the symmetric arrangement of the fixed points for large scales corresponds to higher probabilities for finite positive and negative velocity increments. In our interpretation this is a signature of turbulent Karman vortices passing over the detector. These vortices are rotating alternately in opposite senses. In this way, the changing from of $M^{(1)}$ can be taken as a signature of large scale coherent structures within the turbulent near field of the wake flow.

5. Conclusion

We have presented a method to analyze the disorder of the velocity in a turbulent field. We showed how stochastic equations can be extracted directly from the measured data. The presented results from isotropic turbulence and from anisotropic turbulence in the near field of a turbulent wake flow show that this method seems to work in many turbulent situations. The characterization of turbulence by Markov processes, or even by Fokker-Planck equations, comprises that any n-increment distribution $p(v_n, r_n; v_{n-1}, r_{n-1}; \dots v_1, r_1)$ is given, too. More precisely, these n-increment distributions can be expressed as a product of conditioned probabilities, which are given by the Markov process. The knowledge of the probability of the occurrence of n increments on different length scales can be regarded as a complete statistical description of increments.

At last we want to set our finding into the context of other contributions to this issue. Differential equations in r, which describe the evolution of the pdfs, are discussed by Yakhot [28] and by Nakano [29]. The estimations of joint probabilities as an important quantity to characterize the cascade are presented by Jimenez [30]. Our finding of a finite length L_{mar} seems to be connected to the finding of Kambe [31]. The Burger vortex represents the coherent small scale structure ($< 10\eta$). They seem to correspond to the finding that below the length L_{mar} the Markov properties is not valid any more. The random arrangement of these vortices seem to be linked to our finding of a stochastic Fokker-Planck process.

6. Acknowledgments

The financial support by the Deutsche Forschungsgmeinschaft is acknowledged.

References

1. A.S. Monin, A.M. Yaglom, Statistical Fluid Mechanics, (MIT Press, Cambridge 1975).
2. U. Frisch, Turbulence (Cambridge 1995).
3. R. Benzi, S. Ciliberto, C. Baudet, G.R. Chavarria, Physica **D 80**, 385 (1995).
4. A. Arneodo, et. al. Europhys. Lett. **34**, 411 (1996).
5. K.R. Sreenivasan, R.A. Antonia, Annu. Rev. Fluid Mech. **29**, 435 (1997).
6. R. Benzi, L. Biferale, G. Paladin, A. Vulpiani, M. Vergassola, Phys. Rev. **67**, 2299 (1991); P. Kailasnath, K.R. Sreenivasan, G. Stolovitzky, Phys. Rev. **68**, 2767 (1992).
7. B. Castaing, Y. Gagne, E. Hopfinger, Physica **D 46**, 177 (1990).
8. H. Tennekes, J.C. Wyngaard, J. Fluid Mech. **55**, 93 (1972); F. Anselmat, Y. Gagne, E.J. Hopfinger, R.A. Antonia, J. Fluid. Mech **149**,63 (1984); J. Peinke, B. Castaing, B. Chabaud, F. Chilla, B. Hebral, A. Naert, in Fractals in the Natural and Applied Sciences, edt.: M.M. Novak (North Holland, Amsterdam 1994) p.295.
9. R.A. Antonia, B.R. Satyaprakash, A.K.M.F. Hussain, J. Fluid Mech. **119**, 55 (1982);

B. Castaing, Y. Gagne, E.J. Hopfinger, A new View of Developed Turbulence, in New Approaches and Concepts in Turbulence, edts. Th. Dracos and A. Tsinober (Birkhäuser, Basel 1993) see also discussion p. 47 - 60 therein; B. Castaing, Y. Gagne, M. Marchand, Physica **D 68**, 387 (1993); B. Chabaud, A. Naert, J. Peinke, F. Chilla, B. Castaing, B. Hebral, Phys. Rev. Lett. **73** (1994) 3227.

10. R. Friedrich, J. Peinke, Phys. Rev. Lett. **78**, 863 (1997).
11. R. Friedrich, J. Peinke, Physica **D 102**, 147 (1997).
12. R. Friedrich, J. Zeller, J. Peinke, Europhys. Lett. **41**, 143 (1998).
13. It has recently been shown that in an analogous way it is possible to analyze dynamical systems, and to extract the Langevin equation directly from a given data set; S. Siegert, R. Friedrich, J. Peinke, Phys. Lett. **A 243**, 275 (1998).
14. A. Naert, R. Friedrich, J. Peinke, Phys. Rev. E **56**, 6719 (1997).
15. Ph. Marcq, A. Naert, Physica **D 124**, 368 (1998).
16. N. Rajartnan, Turbulent Jets (Elsevier, Amsterdam 1976); Ch. Renner, Diplomarbeit (Bayreuth 1997).
17. B. Bümel and H.E. Fiedler (Berlin) priv. communication.
18. D. Aronson, L. Lofdahl, Phys. Fluids **A5**, 1433 (1993).
19. Wind tunnel of the Lehrstuhl für Strömungsmechanik, University of Erlangen, Germany, was used.
20. J. Peinke, R. Friedrich, F. Chilla, B. Chabaud, and A. Naert, Z. Phys. **B 101**, 157 (1996).
21. H. Risken, The Fokker-Planck Equation, (Springer-Verlag Berlin, 1984); P. Hänggi and H. Thomas, Physics Reports **88**, 207 (1982).
22. J. Peinke, R. Friedrich, A. Naert, Z. Naturforsch. **52a**, 588 (1997).
23. The Fokker-Planck equations for P_{even} and P_{odd} have the same form as (5), only if the coefficient $\delta = 0$, this may clear up the question of the sense of evaluating moments of the absolute values of v, see for example footnote on page 446 in [5].
24. A.M. Oboukhov, J. Fluid Mech. **13**, 77 (1962); A.N. Kolmogorov, J. Fluid Mech. **13**, 82 (1962).
25. R. Badii and P. Talkner, Europhys. Lett. **43**, 284 (1998).
26. Ch. Renner et.al to be published.
27. St. Lueck, J. Peinke, R. Friedrich, Phys. Rev. Lett. **83**, 5495 (1999).
28. V. Yakhot, "Statistics of Transverse Velocity Differences in Turbulence", within this issue.
29. T. Nakano et. al.,"Probability Density Function of Longitudinal Velocity Fields", within this issue; N. Takahashi, T. Kambe, T. Nakano, T. Gotoh, and K. Yamamoto, J. Phys. Soc. Jpn **68**, 86 (1999).
30. J. Jimenez, "Self-similarity and Coherence in the Turbulent Cascade", within this issue.
31. T. Kambe, "Statistical Laws governed by Vortex Structures in Fully Developed Turbulence", within this issue.

STATISTICAL LAWS GOVERNED BY VORTEX STRUCTURES IN FULLY DEVELOPED TURBULENCE

TSUTOMU KAMBE
Dept. of Physics, University of Tokyo,
Hongo, Bunkyo-ku, Tokyo 113-0033, Japan

Abstract. Statistics of fully developed turbulence is modeled by an ensemble of strained vortices (*i.e.* Burgers vortices) distributing randomly in space, and probability density functions (pdfs) for longitudinal and transversal components of velocity difference are estimated by taking statistical averages of isotropy and homogeneity. It is found [15] that the pdfs tend to close-to-exponential forms at small scales, and that there exist two scaling ranges in the structure function of every order, which are identified as the viscous range and inertial range respectively. The pdfs deviate increasingly away from the Gaussian as the separation distance decreases. For the *inertial* range (second scaling range of larger scales), scaling exponents are obtained and found to be close to those known in the experiments and DNSs. It is remarkable that the Kolmogorov's four-fifths law is observed to be valid in a small-scale range. The scaling exponents of higher order structure functions are numerically estimated up to the 25th order. It is found that asymptotic scaling exponents as the order increases are in good agreement with the behavior of recent experiment of Praskovsky & Oncley [4]. The above model analysis is considered to represent successfully the statistical behaviors at small scales (possibly less than the Taylor microscale) and higher orders. The present statistical analysis leads to scale-dependent probability density functions.

1. Introduction

Fully developed turbulence is structured with a number of intense elongated vortices, observed in many laboratory experiments and computer simulations [1]~[13]. Turbulence is often characterized by *intermittency*,

T. Kambe et al. (eds.), IUTAM Symposium on Geometry and Statistics of Turbulence, 117–126.
© 2001 *Kluwer Academic Publishers.*

non-Gaussian statistics with *negative* skewness, *inertial* range, *multi-scaling* and so on. A natural question would be as follows: how the statistical laws in turbulence are influenced by such structures. It is reasonable that random distribution of discrete vortices accounts for the intermittency. The present paper considers whether the non-Gaussian statistics with negative skewness and the inertial range with multi-scaling are predicted by an ensemble of strained vortices distributing randomly in space.

In the limit of high Reynolds numbers, it is considered that there is an invariant measure in the turbulence governed by Navier-Stokes equation: $\partial_t \boldsymbol{v} + (\boldsymbol{v}\cdot\nabla)\boldsymbol{v} = -\nabla p + \nu\Delta\boldsymbol{v}$, $\nabla\cdot\boldsymbol{v} = 0$, where ν is the kinematic viscosity. A probability desnsity function (pdf) $P(v,s)$ is defined for the turbulence, where v is a velocity difference at two points separated by the distance s in the field, v being either the longitudinal $v_l(s)$ or transversal $v_t(s)$. Non-Gaussian turbulence is often investigated in terms of the structure function defined with

$$S_n(s) = \langle v^n \rangle \equiv \int v^n P(v,s)\, dv\,, \tag{1}$$

for the nth-order moment, where $\langle\,\cdot\,\rangle$ is an ensemble average for a fixed s. Behaviors of $S_n(s)$ with respect to the order n and the distance s give us information about the above characteristics of turbulence.

In the homogeneous isotropic turbulence, the second- and third-order structure functions satisfy the *Kolmogorov equation* [16, 17]:

$$S_3(s) - 6\nu\frac{\mathrm{d}}{\mathrm{d}s}S_2(s) = -\frac{4}{5}\,\varepsilon s\,, \tag{2}$$

where ε is the rate of energy dissipation. This exact relation holds at the scales including the viscous range as well as the inertial range. The equation provides the testing ground of the present model analysis, which is not usually done in many phenomenological models of turbulence. In the limit of vanishing viscosity, the third-order structure function in the inertial-range is given by the *four-fifths* law [18]: $S_3(s) = -(4/5)\varepsilon\, s$. In this context, the parameter ε would be termed more appropriately as the constant rate of energy transfer across each wave number shell in the inertial range. The Kolmogorov 1941 theory [16] implies the scaling properties of the structure functions given by dimensional arguments as $S_n(s) \sim (\varepsilon s)^{n/3}$, so the scaling exponent is given by $\zeta_n = n/3$ (referred to as K41 below). It is known that real turbulence is a dynamical system characterized with anomalous exponents different from the K41 phenomenology of *single* scaling.

We will examine the pdfs $P(v,s)$ and the structure functions derived from ensembles of Burgers vortices in detail. Although the model analysis is already reported [12, 19, 20], much improvement has been done in the recent paper [15], described below, and calculations are carried out with a renewed scheme.

2. Scaling exponents

In the homogeneous isotropic turbulence, the structure functions are considered to follow power laws in the *inertial* range as,

$$S_n(s) \equiv \langle [v(\mathbf{x}, \mathbf{s})]^n \rangle \sim s^{\zeta_n}, \tag{3}$$

where $v(\mathbf{x}, \mathbf{s})$ is velocity difference at a distance $\mathbf{s}$ either longitudinal v_l or transversal v_t, and ζ_n is the scaling exponent to be sought.

The first model predicting the scaling exponents based on pdfs would be the lognormal theory for the rate of dissipation which however predicts unfavorable behaviors of scaling exponent ζ_n decreasing when the order n becomes larger [18]. Recent models [21, 22] predicting the scaling exponents of intermittent energy dissipation is based on the notion of filamentary structure in turbulence. It is also proposed by [21] that the scaling exponents are represented in the parametric form: $\zeta_n = hn + C(1 - \beta^n)$, where the following scaling law with a parameter h is assumed:

$$v(s) = v_L \, (s/L)^h \, . \tag{4}$$

Experimental pdfs for v_l are determined by Praskovsky and Oncley [4] at very high Reynolds number turbulence of $R_\lambda = 10^3 \sim 10^4$. This suggests that asymptotic behavior of the scaling exponents is represented as $\zeta_n^{(\mathrm{PO})} \approx 0.16\, n + const$ as $n \to \infty$.[1] In the present analysis below, it will be shown as n becomes large that the exponent ζ_n asymptotes to the lines represented as $\zeta_n \approx 0.158\, n + const$, with different values of the constant for the even n and odd n. It is remarkable that the proportional coefficients are almost coincident. The experimental results of van de Water and Herweijer [23] also favor the asymptotic behavior $\zeta_n/n \to 0.16$, which is interpreted as the minimum of the exponent h in (4) from the multifractal point of view.

In the theory of multifractal model, the scaling behavior (4) of the exponent h is assumed to have a fractal distribution of dimension $D(h)$ in space with a probability measure $\mathrm{d}\mu(h)$ for the parameter h. Then the structure function is given by $S_n(s)/v_L^n \sim \int (s/L)^{nh+3-D(h)} \mathrm{d}\mu(h)$, and in the small scale limit $S_n(s)$ is dominated with the smallest exponent and the exponent ζ_n is given by $\zeta_n = \inf_h \, [nh + 3 - D(h)]$. The function $D(h)$ is supposed convex with a maximum at $h \approx 1/3$. The minimum value $h_{\min}$ of h, if any, corresponds to the most singular scaling. As the order n gets larger, the moment $S_n(s)$ is dominated by the strongest singularity of $h_{\min}$. Therefore in the limit of large n, the scaling exponent would be given asymptotically as

$$\zeta_n \sim n h_{\min} + 3 - D(h) \, . \tag{5}$$

[1] If $\zeta_2 = 0.69$ is used instead of the value $\zeta_2 = 2/3$ of [4], $\zeta_n^{(\mathrm{PO})} \approx 0.17\, n + const$.

3. Burgers vortex

The velocity field of Burgers vortex is represented, at the point $\mathbf{r} = (\sigma, \phi, z)$ in the cylindrical system called *vortex* frame V, as $\mathbf{v}_\mathrm{B}(\mathbf{r}) = (-a\sigma,\ v_\phi(\sigma),\ 2az)$, where a is assumed a positive constant. The total velocity field is a superposition of an azimuthal circulating flow $(0,\ v_\phi(\sigma),\ 0)$ and an irrotational axisymmetric straining field $(-a\sigma,\ 0,\ 2az)$. The vorticity has only axial component: $\omega_\mathrm{B}(\mathbf{r}) = (0,\ 0,\ \omega(\sigma))$. The circulating velocity and the axial vorticity of the exact solution are given as,

$$v_\phi(\sigma) = \frac{K}{2\pi\sigma}\left(1 - \exp\left[-\left(\frac{\sigma}{r_\mathrm{B}}\right)^2\right]\right), \quad \omega(\sigma) = \frac{K}{\pi r_\mathrm{B}^2}\exp\left[-\left(\frac{\sigma}{r_\mathrm{B}}\right)^2\right]$$

respectively, where $r_\mathrm{B} = (2\nu/a)^{1/2}$ and K is the vortex strength. Local rate of energy dissipation is given explicitly as $\varepsilon_\mathrm{loc} = 4\nu(3a^2 + e_{\sigma\phi}^2)$, with $e_{\sigma\phi} = (1/2)(v_\phi'(\sigma) - v_\phi/\sigma)$.

It is to be noted that the Burgers vortex is an exact steady solution of the incompressible Navier-Stokes equation [24] and also a solution in the asymptotic limit of large time [25], and that the combined field $\mathbf{v}_\mathrm{B}(\mathbf{r})$ is characterized with *negative* skewness at points satisfying $a < |e_{\sigma\phi}|$ [1, 19, 26], while neither circulating component ($a = 0$) nor pure straining component ($v_\phi = 0$) is so.

4. Statistical Ensembles

In a *laboratory* frame L, the vortex structures in turbulence are assumed to distribute randomly, isotropically and homogeneously. To begin with, the velocity difference v at two points O and P (separated with a distance s) in the frame L of the origin at O is to be determined, where the velocity difference v is either $v_l(s)$ or $v_t(s)$.

It is assumed that vortices in the laboratory frame are statistically independent to each other, and that only the nearest vortex dominates the velocity difference. In the leading order, the latter would be approximated with the state that there exists only a single Burgers vortex in space. The vortex position is characterized by the nearest point A on the vortex axis to the origin O. Provided that the point A is located at a distance r from O, the difference v can be calculated by using the velocity field $\mathbf{v}_\mathrm{B}(\mathbf{r})$ for various values of separation distance s. In this way one obtains a function $v(s)$. Further, the statistical averages $S_n(s) = \langle v^n \rangle(s)$ $(n = 1, 2, \cdots)$ are also to be determined from the $v(s)$ with respect to an ensemble of vortices of all possible positions and orientations. If the function $v(s)$ had a scaling form like (4), then we would be led to the *viewpoint* similar to the multifractal model.

The statistical average $\langle \cdot \rangle$ is taken triply so as to take account of the isotropy, homogeneity and ensemble of various strengths and sizes of strained vortices. First, the nearest point A of the vortex axis to the origin O is assumed to distribute according to the *Poisson* distribution in 3D space [27]. Then the probability of the point A located at a distance r from O is given by $P_V(V) = (1/\langle V \rangle)\exp(-V/\langle V \rangle)$, whrere V denotes the space volume of no vortex including O and $\langle V \rangle$ is its expectation. Using $V = (4\pi/3)r^3$ and the relation $P_V(V)dV = P_r(r)dr$, one obtains

$$P_r(r) = 3br^2 \exp(-br^3), \tag{6}$$

where $b = (\Gamma(4/3)/r_0)^3$ and $r_0 \equiv \langle r \rangle$ (the expectation of r). The probability $P_r(r)$ corresponds to the event that there is no vortex within the distance r from O, namely the single vortex is located at A (of distance r). An experimental evidence of the Poisson statistics is found in [5].

Further, an ensemble of vortices of various strengths (and various sizes) is considered. Its probability density function $F(K)$ is assumed to be given in the form of a gamma distribution [11, 28]: $F(R_K) = 0.5C^3R_K^2 \exp(-CR_K)$, where $R_K = K/\nu$ is the vortex Reynolds number and C is a normalizing constant. (For the probability function of vortex size r_B, see [15].) Thus, various vortex strengths are taken into account.

The statistical analysis gives the nth-order structure function $S_n(s)$ dominated by a single vortex in turbulence, hence considered to predict statistics for large n and small s. This model would describe the statistics very well at scales less than the Taylor microscale, above which scale the study of Friedrich et al. [29] would be more appropriate. Their important finding is an experimental verification of existense of a *Markov* length L_{mar} of the order of Taylor scale, above which the process is Markovian.

Suppose that there exists a Burgers vortex (of strength K) tangent at the point A to the sphere of radius r centered at the origin O in the laboratory frame L. Correspondingly, we consider the vortex frame V in which the Burgers vortex is fixed with its axis being along the z-axis of the system (σ, ϕ, z) with the velocity field $\mathbf{v}_B(\mathbf{r})$. The origin of the vortex system V is taken at A in the system L. Transformation of the velocity field between the two frames V and L is expressed in terms of the Euler angles denoted by $\boldsymbol{\Omega}$ collectively. The vortex position and orientation are determined by the parameters $\boldsymbol{\Omega}$ and r.

5. Statistical Averages

Choosing a point P denoted by the vector $\mathbf{s}$ in L with $|\mathbf{s}| = s$, the velocity increment between the points P and O, $\Delta \mathbf{v} = \mathbf{v}(P) - \mathbf{v}(O)$, can be calculated as a function of $\boldsymbol{\Omega}$, r, s and K. Denoting a stochastic variable like one of $(\Delta \mathbf{v})^n$ as $X(\boldsymbol{\Omega}, r, s, K)$, an average of X is taken first over $\boldsymbol{\Omega}$.

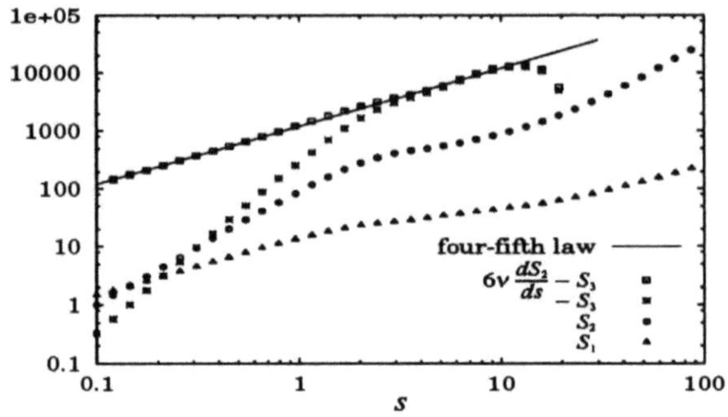

Figure 1. Three lowest order structure functions, $S_n(s)$ $(n = 1, 2, 3)$.

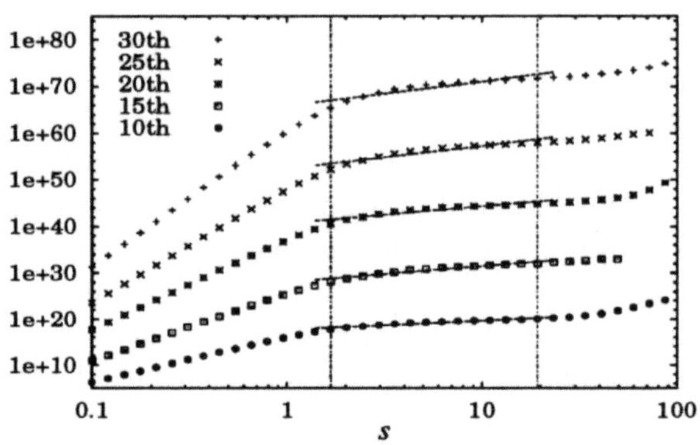

Figure 2. Structure functions with fitting lines in the inertial range.

The resulting value $\langle X \rangle_{\rm sph}$ corresponds to the average over the spherical surface S_r of the sphere V and considered as a function of r, s and K. Next is an average with respect to the distance r with the probability $P_r(r)$: $\langle X \rangle(s, K) = \int \langle X \rangle_{\rm sph}(r, s, K)\, P_r(r)\, dr$. Thus we obtain the statistical properties of isotropy and homogeneity influenced by a single strained vortex of strength K in turbulence. The n-th order structure function is given by

$$S_n(s) = \int \langle v^n \rangle(s, K)\ F(R_K)\ dR_K\,. \tag{8}$$

6. Computed Results

Figure 1 shows three lowest-order structure functions ($v = v_l$, $R_\lambda = 2000$ and $r_0 = 2.5 r_{\rm B}$). It is observed that the Kolmogorov equation, $6\nu S_2'(s) - S_3(s) = (4/5)\,\varepsilon s$, is satisfied reasonably well below the scale $s \approx 15 r_{\rm B}$, where the right hand side $(4/5)\varepsilon s$ is denoted by the solid straight line.

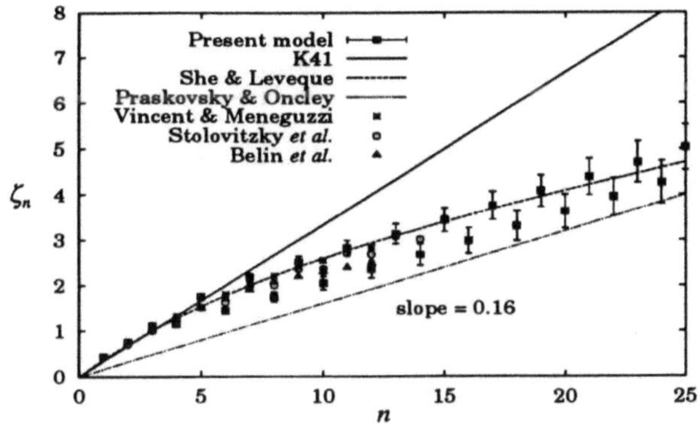

Figure 3. Scaling exponents ζ_n of longitudinal structure functions *v.s.* the order n.

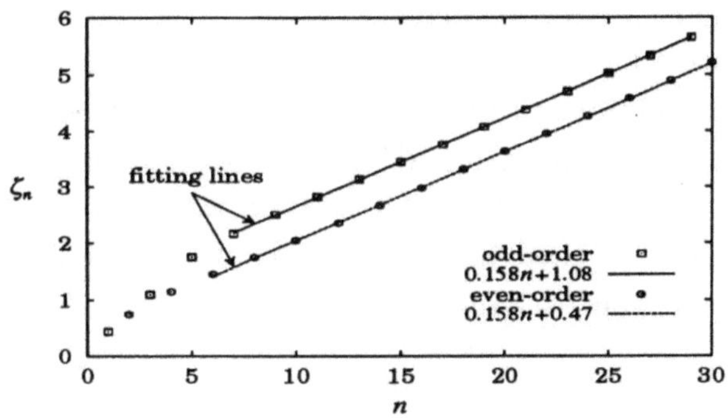

Figure 4. Least square fitting to the linings of present ζ_n for large n.

Only the function $S_1(s)$ is calculated for $|v_l|$, otherwise $S_1(s) = 0$ by the solenoidality. In Fig. 2, the structure functions ($v = v_l$) are illustrated for the higher orders of $n = 10 \sim 30$. It is observed that there exist two scaling ranges in each structure function.

The scaling exponents in the second inertial range are plotted in Fig. 3 and compared with those obtained from other models, DNSs and experiments. In the figure the dotted straight line shows the slope 0.16 predicted by Praskovsky and Oncley [4]. In Fig. 4, two lines of least square fitting to the present reslts for large n are shown. The slopes are estimated separately for the even- and odd-orders, both being found to be remarkably close to the slope 0.16.

Figure 5 shows the pdfs of (a) longitudibal velocity difference v_l and (b) transversal one v_t, normalized by $v_{\rm rms}$ for various values of the separation s. It is observed that the pdfs of v_l are asymmetric, resulting in the negative skewness, while the pdfs of v_t are symmetric. The properties are consistent

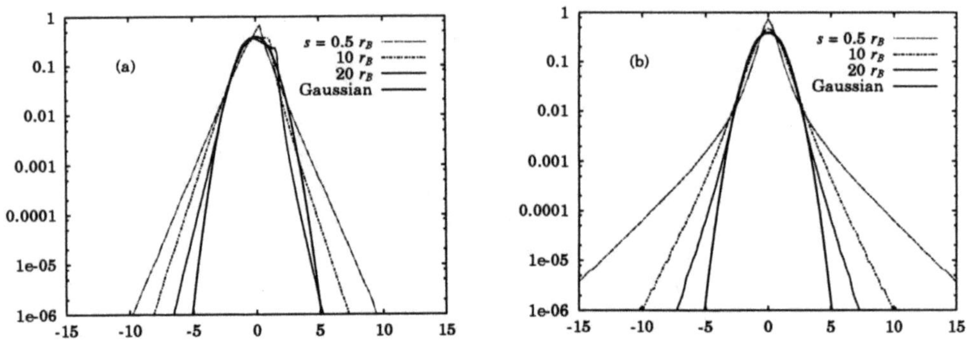

Figure 5. **Probability density functions for (a) $v_l/v_{\rm rms}$, and (b) $v_t/v_{\rm rms}$.**

with those obseerved in DNSs and experiments. The pdfs tend to close-to-exponential forms as s gets smaller.

Taking into account the probability distribution $P(r_{\rm B})$ of vortex radius $r_{\rm B}$, the structure functions of both longitudinal velocity difference $S_n^{(L)}(s)$ and transversal velocity difference $S_n^{(T)}(s)$ are estimated, assuming mutual independence between R_Γ and $r_{\rm B}$. The results are shown in Fig. 6. In the isotropic turbulence, the statistical theory [17] predicts that both of $S_2^{(L)}(s)$ and $S_2^{(T)}(s)$ have the same scaling exponent, and that $S_2^{(T)}(s) = 2S_2^{(L)}(s) = (2/15)(\varepsilon/\nu)s^2$ in the viscous range. It is found in the figure that these properties are approximately satisfied.

7. Scale-Dependent PDFs

An exact relation, $HP + (\partial/\partial U)(GP) = 0$, is known to be valid in the homogeneous turbulence (Pope & Ching, *Phys. Fl.* **5** (1993), 1529), where $H(U,s) = -\langle \nabla_{\mathbf{x}}^2 U(\mathbf{x}) \,|\, U \rangle$, and $G(U,s) = \langle |\nabla_{\mathbf{x}} U(\mathbf{x})|^2 | \, U \rangle$. The functions $H(U,s)$ and $G(U,s)$ are the conditional averages with the value of $U \equiv v_l(\mathbf{x},s)$ fixed. This equation is integrated immediately: $P(U,s) = (N/G(U,s)) \exp[-\int_0^U (H(x,s)/G(x,s))\,dx\,]$. Recent DNS analysis [30] implies that the functions $H(U,s)$ and $G(U,s)$ are represented as quadratic polynomials of U. For large values of the normalized variable $u = U/\overline{V}$ ($\overline{V} = \sqrt{S_2(s)}$), the probability function of u is given asymptotically as,

$$p(u,s) \approx C(s)u^{-2} \exp[-c_\pm(s)\,|u|\,]\,, \quad C(s) = D(s)/a_2(s)\,, \tag{8}$$

($c_\pm$ are numerical constants depending on the sign of u), where $D(s) = N\overline{V}/J_2(s)$, $J_2(s) = \langle\,|\nabla_{\mathbf{x}} U(\mathbf{x})|^2\,\rangle$ and $g(u,s) = G(U,s)/J_2(s) \sim a_2(s)\,u^2$ for large u. Thus the function $u^2P(u)$ tends to an exponential form as $|u|$ becomes large, which is confirmed in Takahashi *et al.* [30]. It can be shown that the present pdfs of Fig. 5 also have such asymptotic property.

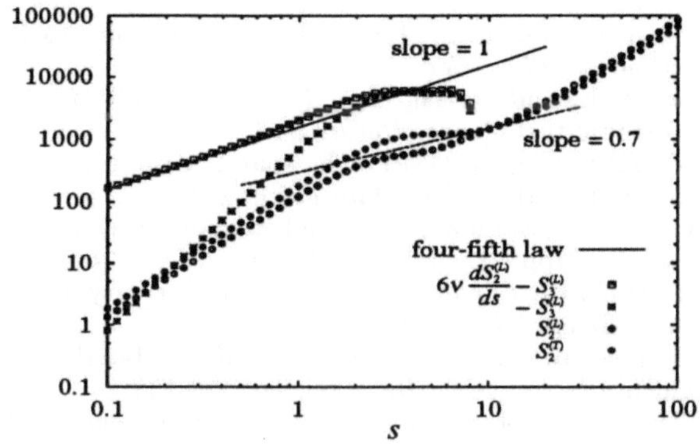

Figure 6. Structure functions $S_2^{(L)}(s)$, $S_3^{(L)}(s)$ and $S_2^{(T)}(s)$, with the probability $P_B(r_B)$.

Praskovsky and Oncley [4] determined the coeffcient $c(s)$ of Eq. (9) experimentally at $O(R_\lambda) = 10^3{\sim}10^4$, which is represented as $b(s) = b_1(s/\eta_K)^\beta$ with $\beta = 0.17$ and $b_1 = 0.58$ $(u > 0)$, 0.51 $(u < 0)$. Then the pdf is represented as $p(u, s) = (D(s)/g(u, s)) \exp[-b(s)\,|u|\,]$. Thus,

$$S_n(s) \;=\; \overline{V}^n D(s) \int \frac{u^n}{g(u,s)} \exp[-b(s)\,|u|\,]\,du \propto s^{\zeta_n} \qquad \text{(for large } n\text{)},$$

where $g(u, s)$ is replaced for large n by $a_2(s)u^2$. Assuming $N/(J_2 a_2) \propto s^\alpha$, the exponent ζ_n for large n is expressed as $\zeta_n \approx (0.5\zeta_2 - \beta)\,n + \alpha + \beta + 0.5\zeta_2 \approx 0.17\,n + const$, where $\beta = 0.17$ and $\zeta_2 \approx 0.69$ are used. As is shown in Fig. 4, the exponents in the present model tend to the asymptotic slope 0.16 (with separate linings between the even and odd orders), which is remarkably close to the above value. This value is also reported to favor the experimental results of van de Water and Herweijer [23].

References

1. A. A. Townsend, "On the fine scale structure of turbulence," Proc. R. Soc. Lond. A **208**, 534 (1951).
2. Anselmet, F., Gagne, Y., Hopfinger, E.J. and Antonia, R.A. (1984) High-order velocity structure functions in turbulent shear flow, *J. Fluid Mech.* **140**, 63-89.
3. Douady, S., Couder, Y., and Brachet, M. E. (1991) Direct observation of the intermittency of intense vorticity filaments in turbulence, *Phys. Rev. Lett.*, **67**, 983.
4. Praskovsky A. and Oncley, S. (1994) Probability density distrinution of velocity differences at very high Reynolds numbers, *Phys. Rev. Lett.*, **73**, 3399.
5. Cadot, O., Douady, S. and Couder, Y. (1995) Characterization of the low-pressure filaments in a three-dimensional turbulent shear flow, *Phys. Fluids*, **7**, 630.
6. Belin,F., Maurer, J., Tabeling, P. and Willaime, H. (1996) Obsevation of intense filaments in fully developes turbulence, *J. Phys. II France*, **6**, 573.
7. Kerr, R.M. (1985) Higher-order derivative correlations and the alignment of small-scale structures in isotropic numerical turbulence, *J. Fluid Mech.* **153**, 31.

8. Yamamoto, K. and Hosokawa, I. (1988) A decaying isotropic turbulence pursued by the spectral method, *J. Phys. Soc. Jpn.* **57**, 1532.
9. She, Z.-S., Jackson, E. and Orszag, S.A. (1990) Intermittent vortex structures in homogeneous isotropic turbulence, *Nature* **344**, 226.
10. Vincent, A. and Meneguzzi, M. (1991) The spatial structure and statistical properties of homogeneous turbulence, *J. Fluid Mech.* **225**, 1.
11. Jimenéz, J., Wray, A. A., Saffman, P. G. and Rogallo, R. S. (1993) The structure of intense vorticity in isotropic turbulence, *J. Fluid Mech.* **255**, 65; Jimenéz, J. and Wray, A. A. (1994) On the dynamics of small-scale vorticity in isotropic turbulence, in *CTR Ann. Res. Briefs - 1994*, 287–312.
12. Kambe, T. and Hosokawa, I. (1995) Statistics of small-scale structures and a dynamical mechanism of cascade, in Ref. [31], pp. 123-130.
13. Saffman, P. G. and Pullin, D. I. (1996) Calculation of velocity structure functions for vortex models of isotropic turbulence, Phys. Fluids **8**, 3072; (1998) Vortex dynamics in turbulence, *Ann. Rev. Fluid Mech.* **30**, 31.
14. Boratav, O.N. and Pelz, R.B. (1997) Structures and structure functions in the inertial range of turbulence, *Phys. Fluids*, **9**, 1400-1415.
15. Kambe, T. and Hatakeyama, N. (1999) Statistical laws and vortex structures in fully developed turbulence, *Preprint.*
16. Kolmogorov, A. N. (1941) The local structure of turbulence in incompressible viscous fluid for very large Reynolds numbers, *C. R. Acad. Sci. URSS* **30**, 301.
17. Landau, L. D. and Lifshitz, E. M. (1987) *Fluid Mechanics* (Pergamon, Oxford).
18. Frisch, U. (1995) *Turbulence* (Cambridge University Press, Cambridge).
19. Hatakeyama, N. and Kambe, T. (1997) Statistical laws of random strained vortices in turbulence, *Phys. Rev. Lett.* **79**, 1257.
20. Hatakeyama, N. and Kambe, T. (1998) Statistical properties of random distribution of strained vortices in turbulence, in *Advances in Turbulence VII*, edited by U. Frisch (Kluwer Academic Publishers, Dordrecht), pp. 339-342.
21. She, Z.-S. and Leveque, E. (1994) Universal scaling laws in fully developed turbulence, *Phys. Rev. Lett.* **72**, 336; She, Z.-S. and Waymire, E. C. (1995) Quantized energy cascade and log-Poisson statistics in fully developed turbulence, *Phys. Rev. Lett.* **74**, 262; She, Z.-S. Universal law of cascade of turbulent fluctuations, *Prog. Theo. Phys.* **130**, 87 (1998).
22. Dubrulle, B. (1994) Intermittency in fully developed turbulence: log-Poisson statistics and generalized scale covariance, *Phys. Rev. Lett.* **73**, 959.
23. Water, W. van de, and Herweijer, J.A. (1999) High-order structure functions of turbulence, *J. Fluid Mech.* **387**, 3.
24. Burgers, J. M. (1948) A mathematical model illustrating the theory of turbulence, *Adv. Appl. Mech.* **1**, 171.
25. Kambe, T. (1984) Axisymmetric vortex solution of Navier-Stokes equation, *J. Phys. Soc. Jpn.* **53**, 13.
26. Pullin, D. I. and Saffman, P. G. (1993) On the Lundgren–Townsend model of turbulent fine scales, *Phys. Fluids A* **5**, 126.
27. W. Feller, *An Introduction to Probability Theory and Its Applications*, volume II (John Wiley & Sons, 1971), 2nd ed.
28. Tanahashi, M., Miyauchi, T, and Ikeda, J. (1997) Scaling law of coherent fine scale structure in homogeneous isotropic turbulence, in *Proceedings of the 11th Symposium on Turbulent Shear Flows* Vol.1, 4-17.
29. Friedrich, R., Lück, St., Peinke, J. and Renner, Ch. (2000) On a complete statistical characterization of turbulence, *in this issue.*
30. Takahashi, N., Kambe, T., Nakano, T., Gotoh, T. and Yamamoto, K. (1999) Probability density function of longitudinal velocity increment in homogeneous turbulence, *J. Phys. Soc. Jpn.* **68**, 86.
31. Meneguzzi, M., Pouquet, A. and Sulem, P. L. (1995) *Small-Scale Structures in 3D Hydrodynamic and Magnetohydrodynamic Turbulence* (Springer).

PROBABILITY DENSITY FUNCTION OF LONGITUDINAL VELOCITY INCREMENT

T. NAKANO AND D. FUKAYAMA
Department of Physics, Chuo University,
Tokyo 112-8551, Japan

T. GOTOH
Department of Systems Engineering, Nagoya Institute of Technology,
Nagoya 466-8555, Japan

AND

K. YAMAMOTO
National Aerospace Laboratory,
Chofu, Tokyo 182-8522, Japan

1. Introduction

In this paper we present an analysis of the PDF for the longitudinal velocity increment with the aid of the conditional average method (CAM). The functional forms of the necessary averages are derived and compared with the results of simulation. The derived PDF is categorically shown; the Gaussian for small amplitudes, the stretched exponential for intermediate ones and the exponential for large ones. The comparison with the PDF obtained in high Reynolds number flows is given.

2. Conditional Averages and PDF

In a statistically homogeneous system, it is rigorously shown that the PDF for the longitudinal velocity increment $w_r(\boldsymbol{x}) \equiv u(\boldsymbol{x} + r\boldsymbol{e}_x) - u(\boldsymbol{x})$, where $u(\boldsymbol{x})$ is the x component of the velocity field at $\boldsymbol{x}$ and $\boldsymbol{e}_x$ is a unit vector along x axis, can be expressed in terms of the conditional averages for viscous term and dissipation rate with a given value of w_r. Although the CAM can be applied to the PDF for the increments of a passive scalar[1] and transversal velocity, we stress that the application to w_r[2] is more

T. Kambe et al. (eds.), IUTAM Symposium on Geometry and Statistics of Turbulence, 127–132.
© 2001 *Kluwer Academic Publishers.*

interesting and important for the following reasons. (1) Statistics of w_r are the most essential to characterize the turbulence and experimentally the most easily accessible. (2) Two conditional averages related to w_r can be interpreted in the dynamical context of the Navier-Stokes turbulence more directly than the ones for other fields.

The conditional averages we examine are

$$H(w_r) = \nu\left\langle \nabla_{\boldsymbol{x}}^2 u(\boldsymbol{x}+r\boldsymbol{e}_x) - \nabla_{\boldsymbol{x}}^2 u(\boldsymbol{x})|w_r \right\rangle \equiv \nu\left\langle \nabla_{\boldsymbol{x}}^2 w_r(\boldsymbol{x})|w_r \right\rangle, \tag{1}$$

$$G(w_r) = \nu\left\langle |\nabla_{\boldsymbol{x}} u(\boldsymbol{x}+r\boldsymbol{e}_x) - \nabla_{\boldsymbol{x}} u(\boldsymbol{x})|^2|w_r \right\rangle \equiv \nu\left\langle |\nabla_{\boldsymbol{x}}\, w_r(\boldsymbol{x})|^2|w_r \right\rangle, \tag{2}$$

where the average is taken with the value of w_r fixed, and ν the viscosity. Combining the both averages, we are led to the PDF $P(w_r)$ in homogeneous turbulence as

$$P(w_r)G(w_r) \propto \exp\left[\int_0^{w_r} \frac{H(x)}{G(x)}\mathrm{d}x\right]. \tag{3}$$

If the functional forms of H and G are figured out, we are able to know the PDF. The derivation of (3) is seen in Ref.2.

Before going to show the behavior of H and G, it is convenient to express w_r, H and G in the following dimensionless forms;

$$u_r = \frac{w_r}{\sqrt{S_2(r)}}, \quad h(u_r) = -\frac{\sqrt{S_2(r)}}{J_2(r)}H(w_r), \quad g(u_r) = \frac{1}{J_2(r)}G(w_r), \tag{4}$$

where $J_2(r) = \nu\langle|\nabla_{\boldsymbol{x}} w_r(\boldsymbol{x})|^2\rangle$ and $S_2(r)$ is the second order structure function. Then the PDF becomes

$$P(u_r)g(u_r) \sim \exp\left[-\int_0^{u_r} \frac{h(x)}{g(x)}\mathrm{d}x\right]. \tag{5}$$

3. Functional Forms of Conditional Averages

In the present section we show the behavior of h and g against u_r for simulated turbulence and obtain their fitted functional forms, and then, attempt to justify those forms from the view point of the turbulence dynamics.

The averages h and g are calculated for decaying turbulence simulated on 512^3 mesh points. Figures 1 and 2 are the curves of h and g against u_r for various values of r/η, η being the Kolmogorov length. It is interesting to notice that $h(u_r)$ is almost independent of separation r, while $g(u_r)$ depends on r. We find that h and g are satisfactorily fitted by the formulae

$$h(u_r) = c_1 u_r + c_2 u_r|u_r|, \qquad g(u_r) = a_0 + a_1|u_r| + a_2 u_r^2. \tag{6}$$

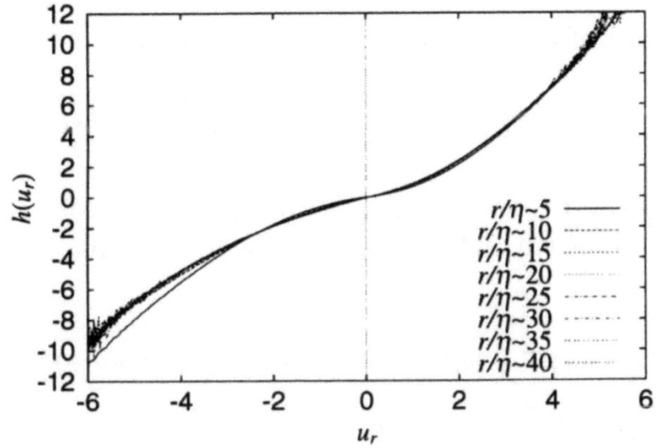

Figure 1. $h(u_r)$ against u_r for various values of r/η.

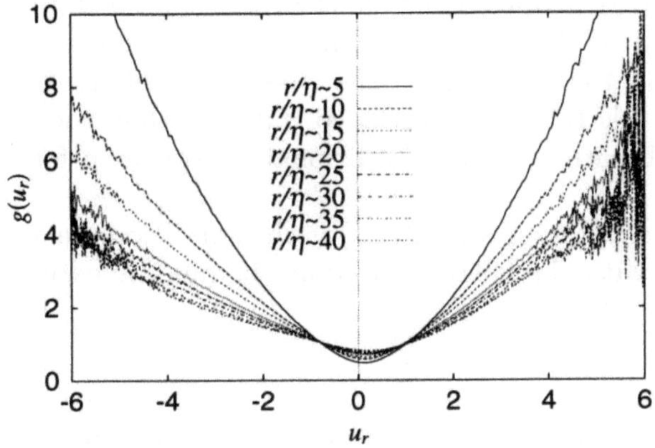

Figure 2. $g(u_r)$ against u_r for various values of r/η.

The parameters c_i are almost independent of r in the inertial range. On the other hand, a_0 increases with r, while a_1 and a_2 decrease.

Next, we discuss what are H and G. To this end we investigate the difference of velocities at $\boldsymbol{x}_2$ and $\boldsymbol{x}_1$:

$$w_i(\boldsymbol{X}, \boldsymbol{r}, t) = u_i(\boldsymbol{x}_2, t) - u_i(\boldsymbol{x}_1, t), \quad \boldsymbol{x}_1 = \boldsymbol{X} - \frac{\boldsymbol{r}}{2}, \; \boldsymbol{x}_2 = \boldsymbol{X} + \frac{\boldsymbol{r}}{2}. \tag{7}$$

The equation for $w_i(\boldsymbol{X}, \boldsymbol{r}, t) \equiv w_i$ is derived from the Navier-Stokes equation of incompressible fluids of unit density as

$$\left(\frac{\partial}{\partial t} + \boldsymbol{V} \cdot \nabla_{\boldsymbol{X}}\right) w_i = -w_k \frac{\partial}{\partial r_k} w_i - \frac{\partial}{\partial X_i} \delta p + \nu \nabla^2_{\boldsymbol{X}} w_i, \tag{8}$$

where δp is the difference of pressures at $\boldsymbol{x}_1$ and $\boldsymbol{x}_2$, and $\boldsymbol{V}$ is the average velocity at the two points. Taking the conditional average of (8) for the fixed value of w_i with the aid of the stationary condition, we are led to

$$w_i(r)H(w_i(r)) = \left\langle w_i \left(\boldsymbol{V} \cdot \nabla_{\boldsymbol{X}} w_i + w_k \frac{\partial}{\partial r_k} w_i + \frac{\partial}{\partial X_i} \delta p \right) \Big| w_i(r) \right\rangle. \quad (9)$$

Hence $w_i(r)H(w_i(r))$ is the conditional average of the energy transfer rate. (Note that $H(w_r)$ denotes $H(w_x(r))$.) In order to turn to G, consider the following identity;

$$\begin{aligned} w_i(r)H(w_i(r)) &= \nu\langle \nabla_{\boldsymbol{X}} \cdot (w_i \nabla_{\boldsymbol{X}} w_i) | w_i(r)\rangle - \nu\left\langle |\nabla_{\boldsymbol{X}} w_i|^2 | w_i(r) \right\rangle \\ &= \nu\langle \nabla_{\boldsymbol{X}} \cdot (w_i \nabla_{\boldsymbol{X}} w_i) | w_i(r)\rangle - G(w_i(r)). \end{aligned} \quad (10)$$

Thus G is the conditional averaged dissipation rate. The difference between $-w_i(r)H(w_i(r))$ and $G(w_i(r))$ is due to the effect of the spatial diffusion, which increases with $|w_i(r)|$. If the diffusion term were neglected, i.e. $G \sim w_r H$, (3) predicts that the PDF would be the power law of w_r to the contrary to observation[3, 4]. It is of interest to indicate that the PDF is determined by the combination of the conditional energy transfer and energy dissipation.

Now let us model H and G. Since $-w_r H(w_r)$ is the conditional average of the energy transfer rate with the fixed value of w_r, it must be expressed as

$$-w_r H(w_r) \sim \frac{w_r^2}{\tau_r}. \quad (11)$$

What is τ_r? When w_r is weak, Kolmogorov's idea should be valid; $\tau_r \sim \bar{\varepsilon}^{-1/3} r^{2/3} \equiv \tau_K(r)$. When w_r is strong, the relaxation time is regulated by itself, i.e. the local turnover time $\tau_r \sim r/|w_r|$. Both relaxational processes must coexist, so that

$$H(w_r) \sim -c_1 \frac{w_r}{\tau_K(r)} - c_2 \frac{w_r |w_r|}{r}, \quad (12)$$

where c_i are independent of r. If the above relation is non-dimensionalized according to (4), (12) reduces to (6). Hence, the comparison of (12) with the simulation is remarkable in the following respects; (1) the combination of linear and quadratic in w_r, and (2) the r-independence of c_i. Of importance is that the above derivation is not based on a perturbative way, implying that the highest order of H is w_r^2.

Finally turn to G. In the inertial region $G(w_r) = 2\nu\langle |\nabla_{\boldsymbol{x}} u(\boldsymbol{x})|^2 | w_r \rangle$. For small amplitude of w_r, $G(w_r)$ must be constant, which is of order of the average dissipation rate. For large amplitude of w_r, $G(w_r)$ cannot be of order of w_r^3, because the PDF would become power-law like. For large amplitude of w_r the first term on the right hand side of (10) is dominant.

to think that $G(w_r)$ is proportional to the energy density w_r^2, although we need more persuasive argument.

4. Classification of PDF

Substituting the fitted forms (6) for $h(u_r)$ and $g(u_r)$ into (5), we can compute the PDF. For small amplitudes of u_r, $h \sim u_r$ and $g \sim$ const, yielding the Gaussian PDF. For extremely large amplitudes, where the deterministic picture is valid, $h \sim u_r|u_r|$ and $g \sim u_r^2$, providing the exponential PDF

$$P(u_r)g(u_r) \sim \exp\left[-\frac{c_2}{a_2}|u_r|\right], \tag{13}$$

where $c_2/a_2 \sim 1/a_2$ is found to increase with r. The above tendency is clearly seen in Fig.3, where the product form $P(u_r)g(u_r)$ is given.

Since the Reynolds number of the present simulation is not large, we had better refer the experimental results of high Reynolds flows[3, 4]. According to Praskovsky and Oncley[3], $P(u_r)$ takes the form $\exp[-b(r)|u_r|]$, where $b(r) \propto r^\beta$ with $\beta = 0.17 \pm 0.01$ for flows with $R_\lambda \approx (2.0 - 12.7) \times 10^3$. Hence, the present result (13) is consistent with the experiments of high Reynolds numbers.

For the intermediate amplitudes the factor $1/g(u_r)$ before the exponential in (3) has a certain effect to yield the PDF flatter than the exponential, i.e. the stretched exponential, as seen from $P(u_r)$ in Ref.2.

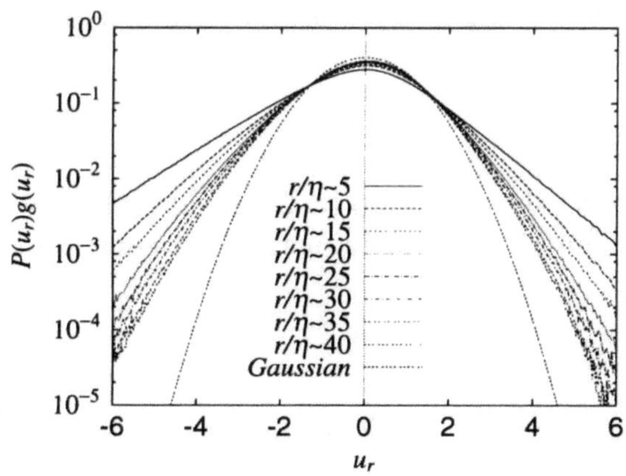

Figure 3. $P(u_r)g(u_r)$ against u_r for various values of r/η.

5. Connection to Work by Yakhot

Recently how the PDF propagates in scale space was investigated experimentally[5] and theoretically[6]. We want to show the connection of the present result with Yakhot's model equation for $P(r,w)$ [6];

$$(B-1)P_r - w\frac{\partial}{\partial w}P_r = -\frac{A}{r}P - \frac{A}{r}w\frac{\partial}{\partial w}P, \tag{14}$$

where $P_r = \partial P/\partial r$. Here w is used for the velocity increment w_r in order to avoid the misunderstanding that w_r is the derivative of velocity with respect to r. The parameters A and B were adjusted in accordance with the observation; $B = 20$ and $A = (3+B)/3 = 23/3$. The interesting results obtained from (14) are seen in Ref.6.

In order to make the connection with (14) let us take derivative of the equation $HP = \partial GP/\partial w$, which is obtained from (3), with respect to r:

$$\left(H - \frac{\partial G}{\partial w}\right)P_r - G\frac{\partial P_r}{\partial w} = -\left(H_r - \frac{\partial G_r}{\partial w}\right)P + G_r\frac{\partial P}{\partial w}, \tag{15}$$

where H_r and G_r are derivatives of H and G with respect to r. Notice that every term in (15) is contained in (14). The comparison of (15) to (14) yields the following relations:

$$\begin{aligned} B-1 &= \frac{w}{G}\left(H - \frac{\partial G}{\partial w}\right), \\ \frac{A}{r} &= \frac{w}{G}\left(H_r - \frac{\partial G_r}{\partial w}\right), \\ \frac{A}{r}w &= -\frac{w}{G}G_r. \end{aligned} \tag{16}$$

With the use of the formulae (6), the parameters A and B can be estimated. The agreement with Yakhot's values is obtained partly. The detailed comparison is referred to Ref.2.

Acknowledgements

T. N and T. G's work was partially supported by Grant-in-Aid for Scientific Research (C) 09640260 by The Ministry of Education, Science, Sports and Culture of Japan.

References

1. Takahashi, N., Kambe, T., Nakano, T., Gotoh, T., and Yamamoto, K. : J. Phys. Soc. Japan **67** (1998) 833-841.
2. Takahashi, N., Kambe, T., Nakano, T., Gotoh, T., and Yamamoto, K. : J. Phys. Soc. Japan **68** (1999) 86-96.
3. Praskovsky, A. and Oncley, S. : Phys. Rev. Lett. **73** (1994) 3399-3402.
4. Gagne,Y., Marchand, M. and Castaing, B. : J. Phys. II **4** (1994) 1-8.
5. Friedrich, R. and Peinke, J. : Phys. Rev. Lett. **78** (1997) 863-866. See also Friedrich et. al. "On a Complete Statistical Characterization of Turbulence" within this issue.
6. Yakhot, V. : Phys. Rev. E **57** (1998) 1737-1751.

STATISTICS OF SMALL-SCALE STRUCTURE OF HOMOGENEOUS ISOTROPIC TURBULENCE

Data-base analysis of direct numerical simulation

T. ISHIHARA, Y. YAMAZAKI, AND Y. KANEDA
Department of Computational Science and Engineering,
School of Engineering, Nagoya University,
Chikusa-ku, Nagoya 464-8603, Japan

1. Introduction

In order to study the properties of the inertial subrange, we have recently generated a data-base by direct numerical simulation (DNS) with 512^3 grid points of forced homogeneous isotropic turbulence obeying the incompressible Navier-Stokes (NS) equations. In the DNS, the enstrophy grows monotonically in time and then attains its peak value at about $t = 12.0$ (in eddy turnover time units), after which the turbulence is in a quasi-steady state. The Taylor scale Reynolds number R_λ at $t = 12$ is 126, and the energy spectrum $E(k)$ is close to $K_o\epsilon^{2/3}k^{-5/3}, (K_o = 2.28)$ in the wavenumber range $8 < k < 18$. Here we study the small-scale structure of turbulence on the basis of the data-base. Jiménez *et al.* [1] analyzed the structure of intense vorticity by using a data-base with 512^3 grid points of a homogeneous isotropic turbulent field of $R_\lambda = 169$. We analyze our data-base by paying attention mainly to the structure at length scales in the inertial subrange. First, we investigate the probability distribution functions (pdf's) of the Eulerian and Lagrangian time-derivative fields and fourth-order velocity moments, to study the intermittent and non-Gaussian nature of turbulence. Next, in order to get some idea on the mechanism by which the turbulence exhibits the intermittent and non-Gaussian nature, we analyze the turbulent field from the viewpoint of vortex dynamics by visualizing the regions with high vorticity magnitude, high dissipation, and high enstrophy generation, as well as the regions where the vorticity and the eigenvector of rate-of-strain tensor align well. Finally, in order to study the flow structures at the scales of the inertial subrange, we analyze geometric relations between the vorticity and rate-of-strain tensor at several coarse-grained levels.

T. Kambe et al. (eds.), IUTAM Symposium on Geometry and Statistics of Turbulence, 133–138.
© 2001 *Kluwer Academic Publishers.*

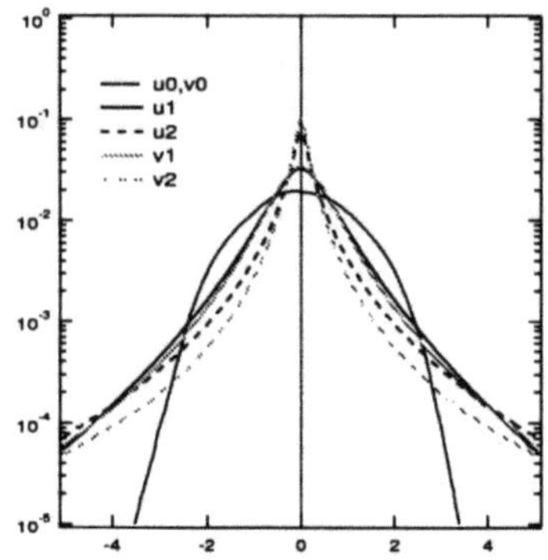

Figure 1. Pdf's of the Eulerian and Lagrangian time-derivative fields; pdf's of the Eulerian 0th-, 1st-, and 2nd-order time-derivative fields are indicated as $u0$, $u1$, and $u2$, respectively, while those of the Lagrangian ones as $v0$, $v1$, and $v2$, respectively.

2. Pdf's of the Eulerian and Lagrangian time-derivative fields

It is well known that the pdf's of turbulent velocity fields are nearly Gaussian, while those of spatial derivative fields deviate from Gaussian, showing an intermittent nature.[2] As compared to the spatial derivatives, little is known about the time derivatives. Since turbulence is a phenomena in space and time, it may be of interest to know not only the spatial derivatives but also time derivatives. Moreover, the latter derivatives, in contrast to the former, have a direct dynamical significance, since they appear in the equations governing the fluid motion. Recently, Kaneda, *et al.*[3, 4] developed a systematic method to generate the Taylor expansions in powers of time difference of the Eulerian and Lagrangian two-time velocity correlations, subject to given dynamics as well as initial and boundary conditions. The expansions are obtained by computing the time-derivatives of the Eulerian and Lagrangian velocity fields. Figure 1 shows the pdf's of the Eulerian and Lagrangian time derivative fields. It can be observed that the Eulerian, as well as Lagrangian, time derivatives of the velocity field are more intermittent for the higher order time derivative, and also that for given order of the time derivative, the Lagrangian derivative is more intermittent than the Eulerian one.

3. Fourth-order velocity moments

Statistical moments like $\langle(\epsilon - \langle\epsilon\rangle)^2\rangle$, $\langle|\nabla p|^2\rangle$, or $\langle|\mathbf{u} \times \boldsymbol{\omega}|^2\rangle$ are among the simplest measures which may show the non-Gaussian nature of turbulence, where $\mathbf{u}$, $\boldsymbol{\omega}$, p and ϵ are, respectively, velocity, vorticity, pressure, and energy

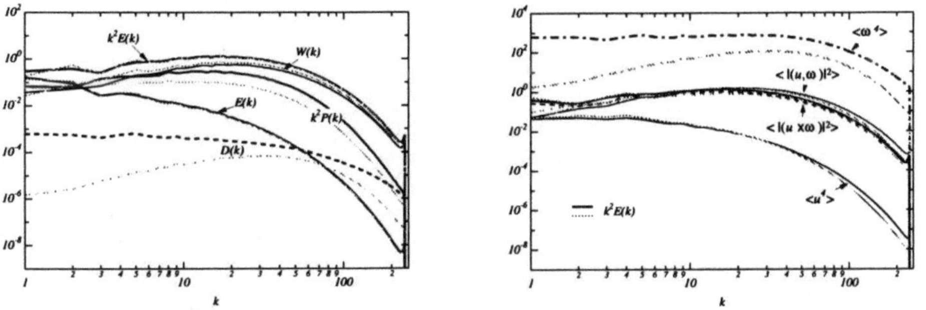

Figure 2. Spectra of fourth-order velocity moments obtained from the DNS data at $t = 12T$ (black lines) and those generated by an artificial Gaussian velocity fields with the same energy spectrum (gray lines).

dissipation rate per unit mass.[5] Here we investigate the spectra $P(k)$, $W(k)$, $D(k)$, and $F(k)$ defined by

$$\langle|\nabla p|^2\rangle = \int_0^\infty k^2 P(k)dk, \quad \left\langle\left|\frac{\partial \mathbf{u}}{\partial t}\right|^2\right\rangle = \int_0^\infty W(k)dk,$$

$$\left\langle(\epsilon - \langle\epsilon\rangle)^2\right\rangle = \int_0^\infty D(k)dk, \quad \text{and } \langle f\rangle = \int_0^\infty F(k)dk,$$

where f is a fourth-order velocity moment such as $|\mathbf{u}|^4$, $|\mathbf{u}\cdot\boldsymbol{\omega}|^2$, $|\mathbf{u}\times\boldsymbol{\omega}|^2$, and $|\boldsymbol{\omega}|^4$. In Fig. 2, we compare the spectra of various fourth-order velocity moments obtained from the data-base against those generated by an artificial Gaussian velocity field with the same energy spectrum. The comparison suggests that $\langle|\mathbf{u}|^4\rangle$, $\langle|\partial\mathbf{u}/\partial t|^2\rangle$, $\langle|\mathbf{u}\times\boldsymbol{\omega}|^2\rangle$, and $\langle\mathbf{u}\cdot\boldsymbol{\omega}\rangle$ are less sensitive to the non-Gaussianity of the real turbulent field than $\langle(\epsilon - \langle\epsilon\rangle)^2\rangle$, $\langle|\nabla p|^2\rangle$, and $\langle|\boldsymbol{\omega}|^2\rangle$.

4. Vortex dynamics

It is clear that the intermittent and non-Gaussian nature of the turbulence is closely related to the NS dynamics. However, the detailed mechanism which generates such flow structures is not well understood. Here we study the flow structures from the viewpoint of vortex dynamics.

The vorticity $\boldsymbol{\omega}$ and its amplitude $\omega = |\boldsymbol{\omega}|$ obey, respectively,

$$\frac{D}{Dt}\omega_i = S_{ij}\omega_j + \nu\Delta\omega_i, \quad \text{and} \quad \frac{D}{Dt}\omega = \alpha\omega + \frac{\omega_i(\nu\Delta\omega_i)}{\omega^2},$$

where $D/Dt \equiv \partial/\partial t + (\mathbf{u}\cdot\nabla)$, ω_i is the ith component of $\boldsymbol{\omega}$, S_{ij} is the rate-of-strain tensor, ν is kinematic viscosity, and $\alpha = \omega_i S_{ij}\omega_j/\omega^2$ is the rate of

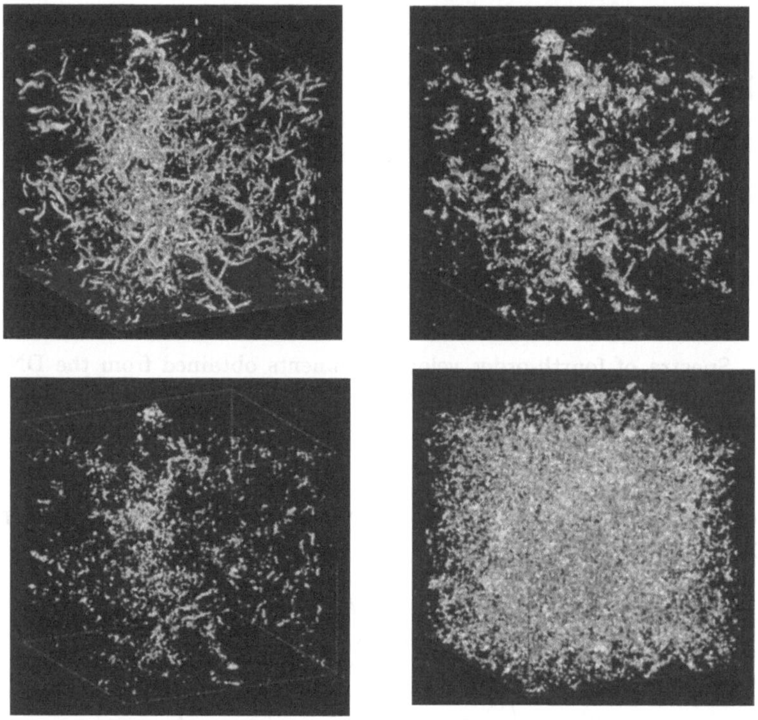

Figure 3. Iso-surface of the amplitude of vorticity ($\omega = \bar{\omega} + 3\sigma_\omega$) (top left), the total rate of strain ($s = \bar{s} + 3\sigma_s$) (top right), α ($\alpha = \bar{\alpha} + 3\sigma_\alpha$) (bottom left), and $\cos\theta_2 = 0.98$ (bottom right), where $\bar{a}$ and σ_a denotes the mean and the standard deviation of a.

enstrophy generation. Regions with large α are called *active* regions.[6] Let $\lambda_1 > \lambda_2 > \lambda_3$ be the eigenvalues of S_{ij} and $\mathbf{\Lambda}_i$ be the eigenvector of S_{ij} associated with λ_i. Then α can be expressed as $\alpha = \lambda_1 \cos^2\theta_1 + \lambda_2 \cos^2\theta_2 + \lambda_3 \cos^2\theta_3$, where θ_i is the angle between $\boldsymbol{\omega}$ and $\mathbf{\Lambda}_i$. Ashurst *et al.*[7] showed in the analysis of their DNS data that $\boldsymbol{\omega}$ is well aligned with $\mathbf{\Lambda}_2$ in high enstrophy regions.

Figure 3 shows the region of high ω, high $s = (S_{ij}S_{ij})^{1/2}$, high α, and the region where $\cos\theta_2 \approx 1$. Note that s is proportional to the local dissipation. A close inspection of the figures shows that (i) the location of high ω and that of high s are almost the same, whereas their shapes are different; in contrast to the worm-like shape of the high ω region, the shape of the high s region is sheet-like, (ii) the location of high α region is in the immediate neighbourhood of the high ω region and the shape of the high α region is say petal-like, and (iii) the regions with $\cos\theta_2 \approx 1$ do not always correspond to high enstrophy. Regarding (i) and (ii), similar observations were also reported in Ref. [1].

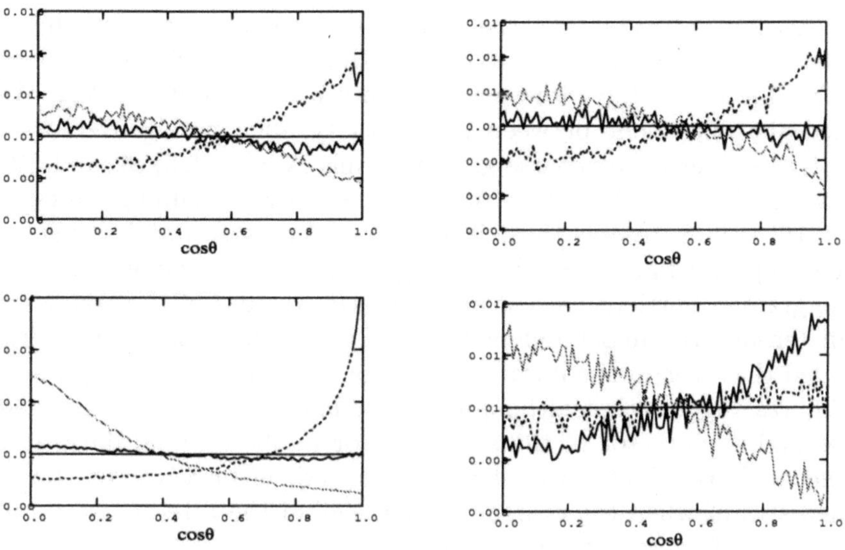

Figure 4. Pdf's of the cosine of the angle between $\boldsymbol{\omega}^{(M)}$ and Ith eigenvector of $S_{ij}^{(N)}$; $M = N = 8$ (top left), $M = N = 16$ (top right), $M = N = 0$ (bottom left), and $M = 0, N = 32$ (bottom right). Black solid lines are for $I = 1$, broken ones for $I = 2$, and gray for $I = 3$.

5. The alignment in coarse grained fields

In order to get some idea on the flow structure at the inertial subrange scales, we have studied the alignment between vorticity and eigenvectors of rate-of-strain tensor computed from several coarse-grained filterings of the same data-base. Replacing spatial derivatives by central differences with spacing $2N\Delta$, we define the central differences at coarse grained level 'N', i.e.,

$$\frac{\partial A(\mathbf{x})}{\partial x_1}^{(N)} \equiv \frac{A(x_1 + N\Delta, x_2, x_3) - A(x_1 - N\Delta, x_2, x_3)}{2(N\Delta)},$$

where $\Delta = 2\pi/512$ is the grid spacing of the DNS data. The vorticity and rate-of-strain tensors at coarse-grained level 'N' are defined by

$$\omega_i^{(N)} \equiv \epsilon_{ijk} \frac{\partial u_k}{\partial x_j}^{(N)}, \qquad S_{ij}^{(N)} \equiv \frac{1}{2}\left(\frac{\partial u_i}{\partial x_j}^{(N)} + \frac{\partial u_j}{\partial x_i}^{(N)}\right),$$

respectively, where ϵ_{ijk} is the alternating tensor and we have defined $\omega_i^{(0)} \equiv \omega_i$ and $S_{ij}^{(0)} \equiv S_{ij}$. Since x-derivative of $\sin kx$ is expressed as $Ck \cos kx$, where $C = (\sin k\delta)/(k\delta)$, in the central difference scheme with spacing δ, the scheme gives a good approximation when $k\delta < \epsilon\pi$ and $0 < \epsilon \ll 1$. This consideration suggests that the central difference scheme at the coarse-grained level 'N' resolves approximately the behavior of the x-derivative of $\exp(ikx)$ for $k < 128/N$, where we have put $\epsilon = 1/2$ and $\delta = N\Delta$. We have studied the alignment between the vorticity and the eigenvectors of the rate-of-strain tensor at coarse-grained levels $N = 0, 4, 8, 16$, and 32. Note that level '32' roughly resolve modes in energy containing ranges, while levels '16' and '8' resolve those in the inertial ranges ($8 < k < 18$).

The top two figures in Fig. 4 suggest that the alignment between the coarse-grained vorticity vector and the eigenvectors of the coarse-grained strain tensor are almost independent of the scale of the coarse graining, provided that it is in the inertial subrange. On the other hand, the bottom two figures suggest that although the vorticity vector is well aligned with the second eigenvector of the strain tensor in the original data, it is well aligned with the first, instead of the second, eigenvector of the coarse-grained strain, when the coarse-grained level of the rate-of-strain tensor is at the scale of energy containing eddies.

This work was supported by "Research for the Future" Program of the Japan Society for the Promotion of Science under the project JSPS-RETF97P01101. The computation was carried out by the VPP500/42 system at the computer center of Nagoya University.

References

1. Jiménez, J., Wray, A.A., Saffman, P.G. and Rogallo, R.S. (1993) The structure of intense vorticity in isotropic turbulence. *J. Fluid Mech* **255**, pp. 65–90.
2. She, Zhen-Su, Jackson, E. and Orszag. S.A. (1991) Structure and dynamics of homogeneous turbulence: models and simulation. *Proc. R. Soc. Lond. A* **434**, pp. 101–124.
3. Kaneda, Y., Gotoh, K. and Ishihara, T. (1998) Taylor expansions and Padé approximations of Lagrangian and Eulerian two-time velocity correlations in turbulence. *J. Phys. Soc. Jpn.* **67** pp. 1075–1078.
4. Kaneda, Y., Ishihara, T. and Gotoh, K. (1999) Taylor expansions in powers of time of Lagrangian and Eulerian two-point two-time velocity correlations in turbulence. *Phys. Fluids* **11**, pp. 2154–2166.
5. Chen, H., Herring, J.R., Kerr, R.M. and Kraichnan, R.H. (1989) Non-Gaussian statistics in isotropic turbulence. *Phys. Fluids A* **1**, pp. 1844–1854.
6. Tsinober, A. (1998) Is concentrated vorticity that important? *Eur. J. Mech. B, Fluids* **17** pp. 421–449.
7. Ashurst, W.T., Kerstein, A.R., Kerr, R.M. and Gibson, C.H. (1987) Alignment of vorticity and scalar gradient with strain rate in simulated Navier-Stokes turbulence. *Phys. Fluids* **30**, pp. 2343-2353.

ON UNIVERSALITY OF STATISTICS OF PRESSURE FIELD IN HOMOGENEOUS TURBULENCE

T. GOTOH

Department of Systems Engineering,
Nagoya Institute of Technology
Showa-ku, Nagoya 466, Japan

AND

K. NAGAYA

NEC Corporation, 7-1, Shiba 5-chome,
Minato-Ku, Tokyo 108-8001, Japan

1. Introduction

Since the source term of the Poisson equation for pressure is quadratic in the velocity gradient and the PDFs for the vorticity and velocity gradients are themselves different from Gaussian, it is quite natural to expect markedly non-Gaussian statistics for the pressure field. Moreover the pressure and its gradients are given by the volume integral of the source term weighted by the Poisson kernel, so that non-local effects in physical space as well as local ones affect the statistics. These facts suggest that small scales of pressure field are highly intermittent and may be influenced by large scales.

We consider how the small scales of the pressure field are scaled in length and Reynolds number and to what extent they are affected by large scales. Also relation bewteen pressure statistics and coherent structure in turbulence is explored. Our approach here is to examine the changes of the pressure and pressure gradient statistics to the Reynolds nmbers by DNS with different forcings and to see to what extent they are universal.

T. Kambe et al. (eds.), IUTAM Symposium on Geometry and Statistics of Turbulence, 139–144.
© 2001 *Kluwer Academic Publishers.*

2. Numerical computation

An incompressible fluid is assumed to obey the Navier Stokes and the continuity equations:

$$\frac{\partial \boldsymbol{u}}{\partial t} + \boldsymbol{u} \cdot \nabla \boldsymbol{u} = -\nabla p + \nu \nabla^2 \boldsymbol{u} + \boldsymbol{f}_u, \tag{1}$$

$$\nabla \cdot \boldsymbol{u} = 0, \tag{2}$$

where $\boldsymbol{u}$ is the velocity vector, p is the pressure, ν the kinematic viscosity and the density ρ is assumed to be unity without loss of generality. Two kinds of forcings are considered. One is the random force $\boldsymbol{f}$ which is additive, solenoidal, statistically homogenous and isotropic. And it is assumed to be white in time and Gaussian with zero mean and the second order moment given by

$$\langle f_i(\boldsymbol{x}, t) f_j(\boldsymbol{x}', s) \rangle = K_{ij}(\boldsymbol{x} - \boldsymbol{x}') \delta(t - s), \tag{3}$$

$$K_0 = K_{ii}(0) = \left\langle \boldsymbol{f}^2(\boldsymbol{x}, t) \right\rangle = \int_0^\infty E_F(k) dk, \tag{4}$$

$$E_F(k) = \begin{cases} c_f > 0 & \text{for low band} \\ 0 & \text{otherwise} \end{cases} \tag{5}$$

where $\langle \cdot \rangle$ denotes the ensemble average [2]. Another is an excitation of the Fourier modes by the instability coefficient applied at lower wavenumber band, that is, in the wavevector representation, to replace $\boldsymbol{f}_u(\boldsymbol{k}, t)$ in (1) by $c(k, t)\boldsymbol{u}(\boldsymbol{k})$ with

$$c(k, t) = \begin{cases} c(t) > 0 & \text{for low band} \\ 0 & \text{otherwise} \end{cases} \tag{6}$$

where $c(t)$ is determined to keep the $K_{max}\eta$ constant greater than unity during the computation [1, 3]. Yeung and Pope, and Vedula and Yeung used the additive forcing which obeys the Orstein-Uhlenbeck process [4, 5].

Runs with additive forcing are further separated into two groups. One is a group of data (group A) computed from a series of DNSs in which turbulence field at higher Reynolds number is succesively generated from that of lower Reynolds number. Another is a group of data (group B) which are generated through long time evolution from independent initial velocity fields. When the averaging time is sufficiently long, the statistics of two groups of data should be the same, but for the first group, practically, there remains some statistical correlation between large scales of two turbulent flow fields of adjacent Reynolds numbers. In the second group, on the other hand, there is no correlation, so that the large scale flow structures at different Reynolds numbers are different and independent.

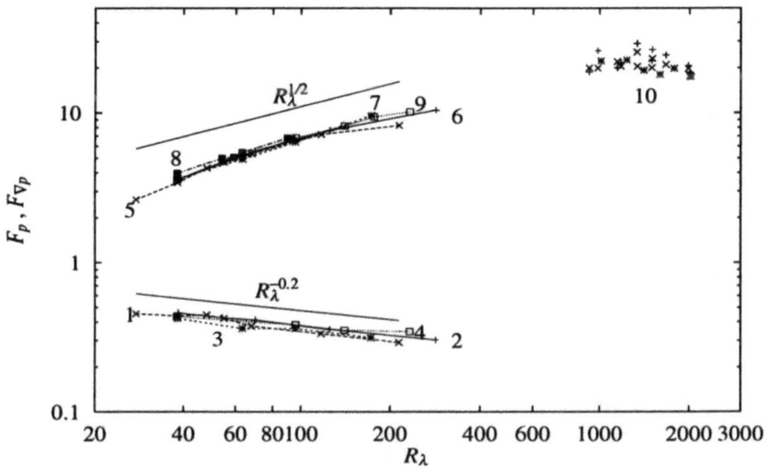

Figure 1. Comparison of DNS variances $F_p, F_{\nabla p}$ by various forcings. 1: F_p (group A), 2: F_p (group B), 3: F_p (Gotoh & Rogallo 1998), 4: F_p (Vedula & Yeung 1999), 5: $F_{\nabla p}$ (group A), 6: $F_{\nabla p}$ (group B), 7: $F_{\nabla p}$ (Gotoh & Rogallo 1998), 8; $F_{\nabla p}$ (Yeung & Pope 1989). 9; $F_{\nabla p}$ (Vedula & Yeung 1999). 10; $F_{\nabla p}$ (Voth *et al.* 1998). Batchelor's values (estimated using Gaussian approximation for the fourth order moment of the velocity gradients.) are $F_p^G = 0.15$ and $F_{\nabla p}^G = 3.9$, repsectively.

3. Normalized variances of pressure and pressure gradient

The normalized variance of the pressure F_p and the one for the pressure gradient $F_{\nabla p}$ are defined as

$$F_p = \frac{\langle p^2 \rangle}{(\langle \boldsymbol{u}^2 \rangle /2)^2}, \qquad F_{\nabla p} = \frac{\langle (\nabla p)^2 \rangle}{\bar{\epsilon}^{3/2} \nu^{-1/2}}, \tag{7}$$

where $\bar{\epsilon}$ is the average rate of energy dissipation. In the classical Kolmogorov scaling, both F_p and $F_{\nabla p}$ are independent of $\mathcal{R}_\lambda$. Figure 1 shows the variation of F_p and $F_{\nabla p}$ against $\mathcal{R}_\lambda$. F_p decreases slowly with $\mathcal{R}_\lambda$, approximately as ${\mathcal{R}_\lambda}^{-0.2}$, and $F_{\nabla p}$ increases roughly as ${\mathcal{R}_\lambda}^{1/2}$ for low to moderate $\mathcal{R}_\lambda$, but the rate of increase for $F_{\nabla p}$ becomes smaller at high end of $\mathcal{R}_\lambda$, irrespective of the type of forcing. This is consistent with recent DNS studies [1, 5, 7]. On the other hand, for $\mathcal{R}_\lambda$ between 1000 and 2000, the experimental data of $F_{\nabla p}$ are seen to be nearly independent of $\mathcal{R}_\lambda$ [6]. DNS values seem to approach the experimental values when the Reynolds number becomes large. It is not known whether or not F_p becomes independent of $\mathcal{R}_\lambda$ for large $\mathcal{R}_\lambda$.

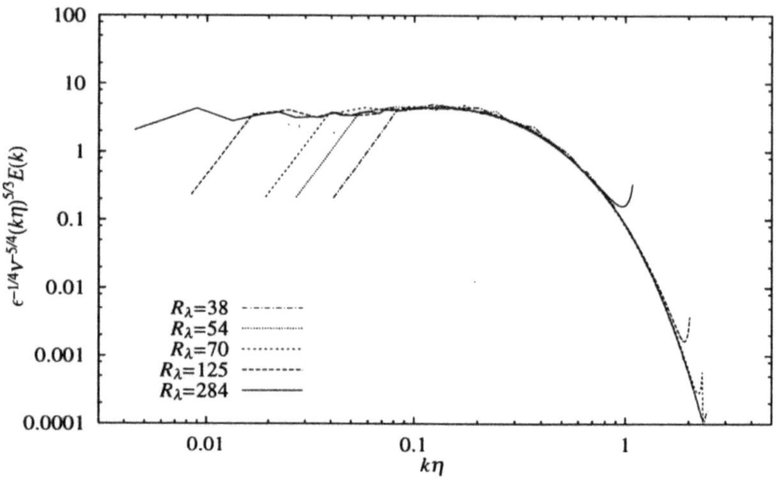

Figure 2. Kolmogorov Scaling of the Energy spectra. Data are group B.

4. Scaling of pressure spectrum

In the K41 the pressure spectrum, defined by $\langle p^2 \rangle = \int P(k)dk$, is scaled as $P(k) = \bar{\epsilon}^{3/4}\nu^{7/4}f_p(k\eta)$ and of the form $P(k) = C_p\bar{\epsilon}^{4/3}k^{-7/3}$ in the inertial range, where $f_p(x)$ is a nondimensional function and C_p is a constant of order one. However, it has been growing to observe the facts that the K41 scaling of the pressure spectrum $P(k)$ is not as good as the case of the energy spectrum even for the wavenumber range $k\eta = 0.1 \sim 1.0$ [5, 1, 7]. In fact, as seen from figures 2 and 3, collapse of energy spectrum by the K41 scaling from the data of group B is very satisfactory, while, but not good for the K41 scaling for the pressure spectra.

Here we propose a new scaling of the pressure spectrum as

$$P(k) = \bar{\epsilon}^{3/4}\nu^{7/4}F_{\nabla p}\phi(k\eta), \tag{8}$$

where η is the Kolmogorov length and ϕ is a nondimensional function. Figure 4 shows the pressure spectra by two kinds of forcings at various Reynolds numbers by using the scaling (8) and compensation $(k\eta)^{5/3}$. Collapse of the curves is satisfactory, supporting the scaling (8). Another point is that the curves extend toward lower wavenumbers horizontally as the Reynolds number increases, implying that $P(k)$ is proportional to $k^{-5/3}$, unlike the Kolmogorov scaling $k^{-7/3}$. The $k^{-5/3}$ scaling of the pressure spectrum has also been found in other recent DNS with different forcing [5, 7]. Physical explanation for this scaling has not been known [1]. These observations suggest that the pressure spectrum in the inertial to higher wavenumber range is universally scaled when $F_{\nabla p}$ is included for the range

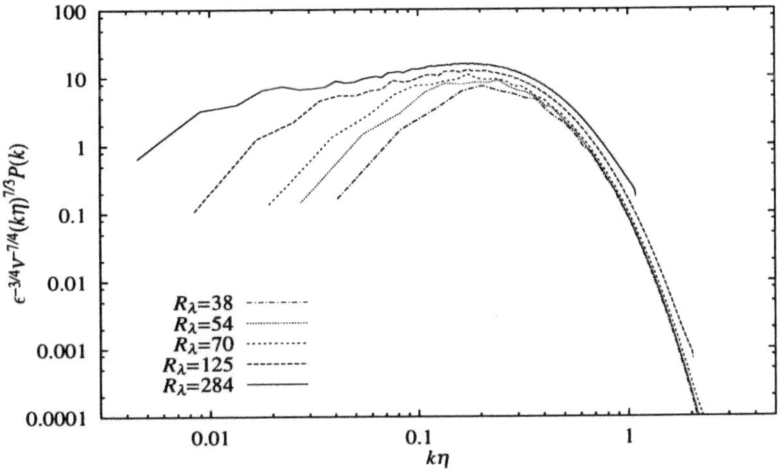

Figure 3. Kolmogorov Scaling of the pressure spectra (group B) with compensation by $(k\eta)^{7/3}$.

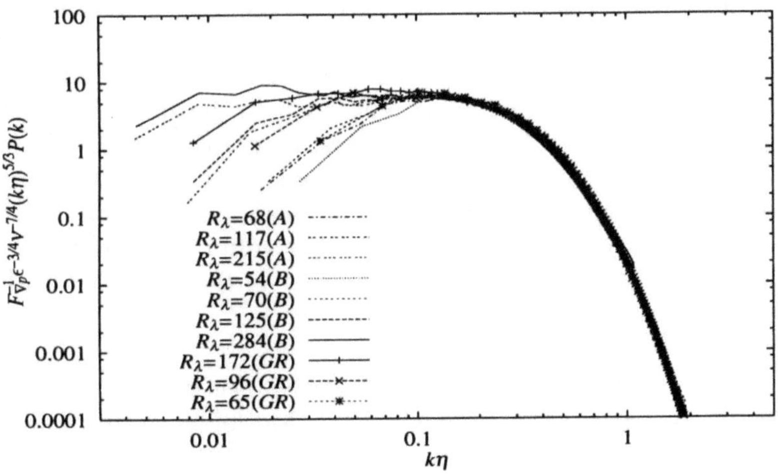

Figure 4. Scaling of the pressure spectra by using $F_{\nabla p}$ and compensation by $(k\eta)^{5/3}$. The data are from group A, B and Gotoh & Rogallo (1999).

of Reynolds numbers studied here. However, since the experimental data suggests that $F_{\nabla p}$ tends to be independent of $\mathcal{R}_\lambda$, which implies that the scaling of pressure spectrum by (8) becomes independent of $\mathcal{R}_\lambda$ so that universal scaling of the pressure spectrum is achieved at large Reynolds numbers. Further studies at large Reynolds numbers are necessary.

The above discussion means that the Reynolds number dependency of

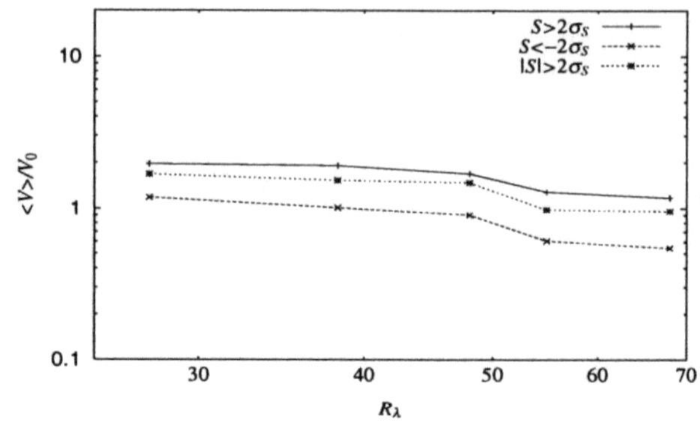

Figure 5. Average volume of the domains of the source term S for the Poisson equation for the pressure normalized by $V_0 = L\lambda\eta$ at $S_c = 2\sigma_s$. Data are group A.

$F_{\nabla p}$ is a key ingredient to the scaling of the pressure statistics, and it requires physical explanation. A physical argument using a model field for the source term of the Poisson equation for the pressure has recently been presented by Gotoh and Rogallo [1]. The model is given by a random distribution of dipoles consisting of a pair of positive and negative source terms. It is found that the Reynolds number dependence of $F_{\nabla p}$ is explained by this model as that of the effective volume over which the space integral for the pressure gradient is done. A direct measurement of the effective volume shown in figure 5 indeed shows that the average of the effective volume is of the order of $L\lambda\eta$, where L is the integral scale. In the limit of large Reynolds numbers, we expect that the effective volume would be of the order of $AL\eta^2$, where A is a constant of the order of ten, and this leads to independence of $F_{\nabla p}$ on $\mathcal{R}_\lambda$ [2].

The authors thank Mr. Fukayama for his assistance for numerical computation. This work was supported by Grant-in-Aid for Scientific Research (C) 09640260 by The Ministry of Education, Science, Sports and Culture of Japan, and by RIKEN.

References

1. T. Gotoh and R. S. Rogallo, 1999 J. Fluid Mech. **396**, 257.
2. T. Gotoh and K. Nagaya, submitted to Phys. Fluids.
3. T. Gotoh, R. S. Rogallo, J. R. Herring and R. H. Kraichnan, 1993 Phys. Fluids A **5**, 2846.
4. Yeung, P. K. & Pope, S. B. 1989 J. Fluid Mech. **207**, 531.
5. Vedula, P. and Yeung, P. K. 1999 Phys. Fluids **11**, 1208.
6. Voth, G. A., Satyanarayan K. and Bodenschatz, E. 1998 Phys. Fluids **10**, 2268.
7. Cao, N., Chen, S. and Doolen, G. D. 1999 Phys. Fluids **11**, 2235.

A STATISTICAL LAW OF VELOCITY CIRCULATIONS IN FULLY DEVELOPED TURBULENCE

K. YOSHIDA
Department of Physics, University of Tokyo
Hongo, Bunkyo-ku, Tokyo 113-0033, Japan †

AND

N. HATAKEYAMA
Institute of Fluid Science, Tohoku University,
Katahira, Aoba-ku, Sendai 980-8577, Japan

1. Introduction

Numerical simulations of the Navier-Stokes equations allow us to investigate the three-dimensional field of the fluid motion at $R_\lambda \leq 200$, where R_λ is the Reynolds number based on the Taylor microscale λ. It is observed in many DNSs that the region of the high amplitude vorticity has filament-like structures and the details of the geometry of the vortex filaments has been investigated (e.g., [3, 2]). The average radius of the cross section $\langle R \rangle$ is observed to be $3 \sim 6\eta$, where η is the Kolmogorov dissipation scale.

A question arises that how the fine vortex structures are reflected in the statistical properties of turbulence. The statistics of longitudinal velocity increments is usually discussed in the theory of isotropic turbulence since Kolmogorov [4]. However, the relation between the statistics and the geometric picture of vortex filaments would become clearer when we discuss the statistics of the velocity circulation $\Gamma(A)$, which is defined as

$$\Gamma(A) = \int_{\partial A} \mathbf{u} \cdot \mathrm{d}\mathbf{l} = \int_A \omega \cdot \mathrm{d}\mathbf{S},$$

where A is a plane region with a boundary ∂A in the whole domain D, $\mathrm{d}\mathbf{l}$ and $\mathrm{d}\mathbf{S}$ are the line and the surface elements respectively and $\omega = \nabla \times \mathbf{u}$ the vorticity. The velocity circulation is given by the vorticity over the region

†Current address: Department of Computational Science and Engineering, Nagoya University, Nagoya 464-8603, Japan

T. Kambe et al. (eds.), IUTAM Symposium on Geometry and Statistics of Turbulence, 145–150.
© 2001 *Kluwer Academic Publishers.*

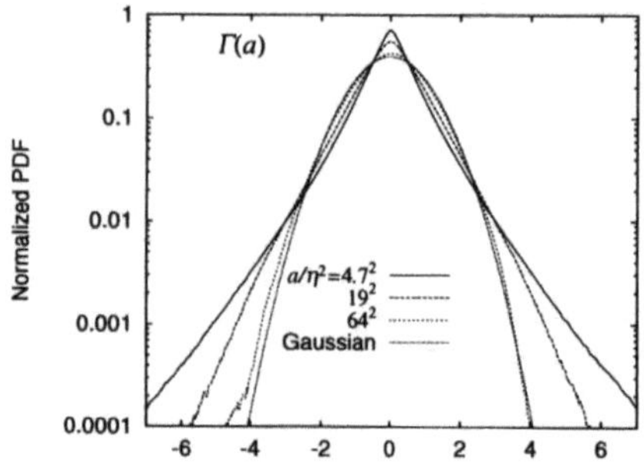

Figure 1. Normalized PDFs of the circulation $\Gamma(a)$

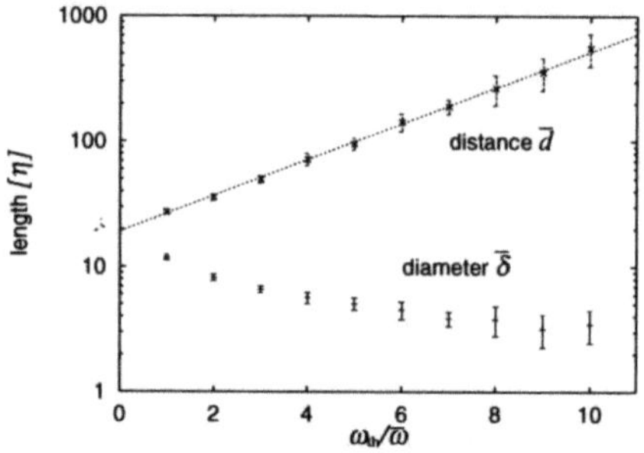

Figure 2. Typical scales of the connected components of the strong vorticity region, the average distance $\bar{d}$ and the average diameter $\bar{\delta}$. The line is the fitted function of the form $\exp(a + b(\omega_{\text{th}}/\bar{\omega}))$ to the average distance $\bar{d}$.

A. It takes a large magnitude when an intense vorticity structure intersects the region A. Cao *et al.* [1] studied the statistical properties of velocity circulation using data from DNS. They found that the PDF of velocity circulation depended on the area a of the region A and the form of PDF varied as a function of a and has significant exponential tails when $\sqrt{a}$ is in the inertial range.

In this context, we study the relation between the statistical properties of the circulation and the intense vorticity structure in turbulence using the data from DNS.

2. Statistics of $\Gamma(a)$

The DNS of the force-free Navier-Stokes equations was performed with the resolution of 256^3 in three dimensional torus by Yamamoto *et al.* [6, 5]. The sample space is generated by calculating the velocity circulations for the snap-shot velocity field at the maximum energy dissipation rate along all possible square contours on the grid for each fixed area a. The Taylor microscale Reynolds number at the time is $R_\lambda = 89.5$. η is 0.424 [grid] and Taylor microscale is $\lambda/\eta = 18.6$. The inertial range is located in $19 \leq r/\eta \leq 40$.

Figure 1 shows the PDFs of the velocity circulations $\Gamma(a)$ normalized with respect to the variance for various areas. The PDF has a substantial stretched tail compared to that of Gaussian when the area a is $O(\lambda^2)$ $(a/\eta^2 = 19^2)$ or $O(\eta^2)$ $(a/\eta^2 = 4.7^2)$. The PDF approximates Gaussian when the $\sqrt{a}$ is as large as the integral scale L_p $(a/\eta^2 = 64^2)$. The deviation from Gaussian is estimated by computing flatness $F_4 = \langle(\Gamma(a))^4\rangle/\langle(\Gamma(a))^2\rangle^2$. It is found that the flatness of the PDF of circulation is substantially larger than the Gaussian value 3 or that of the longitudinal velocity increments $\delta u_l(\sqrt{a})$ and closer to that of the transversal velocity increments $\delta u_t(\sqrt{a})$, yet still larger. This suggests that the statistics of the circulation is of different type from that of the velocity increments. The results are consistent with those of the DNS analysis of Cao *et al.* [1].

3. Decomposition of the vorticity field

Let us fix a plane $S \subset D$, where D is the whole cubic domain, and consider the scalar field $\omega_{\rm n}(\mathbf{x})(\mathbf{x} \in S)$ of the vorticity component normal to the plane S. We fix a positive threshold $\omega_{\rm th}$ and decompose the scalar field $\omega_{\rm n}$ into two components: the strong component $\omega_>$ and the weak component $\omega_<$. Their definitions are as follows:

$$\omega_>(\mathbf{x}) = \begin{cases} \omega_{\rm n}(\mathbf{x}) & (|\omega_{\rm n}(\mathbf{x})| \geq \omega_{\rm th}) \\ 0 & (|\omega_{\rm n}(\mathbf{x})| < \omega_{\rm th}) \end{cases}, \qquad \omega_<(\mathbf{x}) = \begin{cases} 0 & (|\omega_{\rm n}(\mathbf{x})| \geq \omega_{\rm th}) \\ \omega_{\rm n}(\mathbf{x}) & (|\omega_{\rm n}(\mathbf{x})| < \omega_{\rm th}). \end{cases}$$

The motivation of the decomposition is to investigate the contributions of the strong component often concentrated in small regions and the weak component to the statistics of the velocity circulation separately. The strong vorticity region $\{\mathbf{x} \in S : |\omega_{\rm n}(\mathbf{x})| \geq \omega_{\rm th}\}$ occupies rather a small area fraction.

The geometry of the connected components of the strong vorticity region $\{\mathbf{x} \in S : |\omega_{\rm n}(\mathbf{x})| \geq \omega_{\rm th}\}$ is studied. Six arbitrary planes which are perpendicular to the axes of the cube are chosen and the number of the connected components is counted. The average distance $\bar{d}$ and the average diameter $\bar{\delta}$, which are defined from the average area and the number density of the connected components respectively is shown in Figure 2 for various thresholds $\omega_{\rm th}$. It is found that $\bar{d}$ increases exponentially as the threshold $\omega_{\rm th}$ increases and that the average diameter $\bar{\delta} = 7 \sim 9\eta$ for $\omega_{\rm th} = 2 \sim 3\bar{\omega}$ where $\bar{\omega}$ is the r.m.s. of the normal components of the vorticity field $\omega_{\rm n}$. The average diameter $\bar{d}$ is consistent with the average diameter of the vortex filaments which are known from other DNS analyses. We expect that the strong component $\omega_>(\mathbf{x})$ corresponds to the intersections with vortex filaments for the appropriate threshold ($\omega_{\rm th} = 2 \sim 3\bar{\omega}$).

4. Statistics of $\Gamma_>(a)$ and $\Gamma_<(a)$

The circulations $\Gamma_>(A)$ and $\Gamma_<(A)$ associated with the strong vorticity $\omega_>$ and the weak vorticity $\omega_<$ respectively are defined by

$$\Gamma_>(A) = \int_A \omega_>(\mathbf{x})\mathrm{d}\mathbf{x}, \qquad \Gamma_<(A) = \int_A \omega_<(\mathbf{x})\mathrm{d}\mathbf{x},$$

where $A \subset S$. It is obvious that $\Gamma(A) = \Gamma_>(A) + \Gamma_<(A)$.

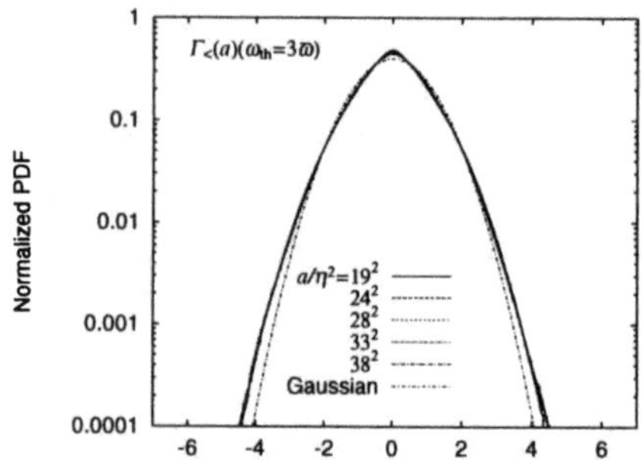

Figure 3. PDFs of the circulation $\Gamma_<(a)$ associated with the weak vorticity with $\omega_{\text{th}} = 3\bar{\omega}$

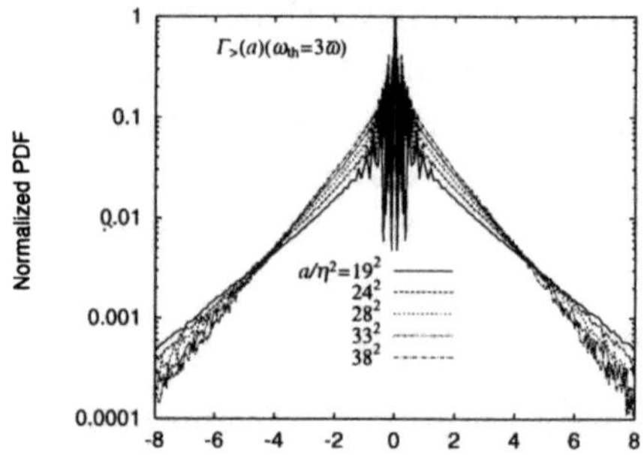

Figure 4. PDFs of the circulation $\Gamma_>(a)$ associated with the strong vorticity with $\omega_{\text{th}} = 3\bar{\omega}$

It is found that the normalized PDF of $\Gamma_<(a)$ is less dependent on the area a (Figure 3), that is, the statistics of $\Gamma_<(a)$ is approximately self-similar. The self-similarity of $\Gamma_<(a)$ may be related to the Kolmogorov's picture. This suggests that the fine vortex structure is responsible for the non self-similarity, namely intermittency, at least at this range of Reynolds number R_λ.

Indeed, the PDF of $\Gamma_>(a)$ depends strongly to the area a and it has stretched tail compared to Gaussian. The normalized PDFs are shown for $\omega_{\text{th}} = 3\bar{\omega}$ in Figure 4. The oscillating behavior of the PDF at the small amplitude of $|\Gamma_>(a)|$ is due to the truncation of the vorticity field and the finite grid size in the following reason. Let Δ be the grid size, then the minimum positive value of $|\Gamma_>(a)|$ is $\Delta^2\omega_{\text{th}}$. The existence of this finite minimum causes jumps in PDF at $\Gamma_> = n\Delta^2\omega_{\text{th}}$ for integers n of small absolute value.

The behaviors of the PDFs of $\Gamma_>(a)$ and $\Gamma_<(a)$ are qualitatively same as long as the average size of the connected components of $\omega_>$ corresponds to that of the intersections of the filaments ($\omega_{\text{th}} = 2 \sim 3\bar{\omega}$).

5. A model of the statistics of vortex filaments

The authors considered a simplified model of the statistics of velocity circulation caused by a random distribution of vortex filaments in turbulence in [7]. Here we neglect the weaker back ground vorticity field.

Let N vortex filaments in which the vorticity is concentrated is randomly distributed in the domain D. We choose a plane S in the domain D. It is assumed that each vortices intersects with S only once and that areas of intersections are small and approximated by points. Let $Y_j \in S(j = 1,\ldots,N)$ and $Z_j \in \mathbf{R}(j = 1,\ldots,N)$ be the intersection point with the plane S and the circulation of the j-th vortex respectively. By assuming the randomness and mutual independence of Y_j and Z_j, it is shown that the circulation in the model denoted by $\Gamma_{\rm m}(a)$ obeys a compound Poisson distribution $P(\theta a, \sigma)$ whose characteristic function is given by

$$\varphi_{\Gamma_{\rm m}(a)}(s) = \exp\left(\theta a \int_{-\infty}^{\infty} (e^{{\rm i}s\gamma} - 1)\sigma(\gamma){\rm d}\gamma\right), \tag{1}$$

where $\theta = N/|S|$ is the number density of the intersections and $\sigma(\gamma)$ is the common probability distribution of Z_j.

The tail of a compound Poisson distribution in general decays with the order $\mathsf{P}(|\Gamma_{\rm m}(a)| \geq x) \sim \exp(-x \log x/c)$ where c depends on σ and it decays slower than that of Gaussian distributions. Normalized PDF of $\Gamma_{\rm m}(a)$ depends on the area a and approximates Gaussian for large a. We assume the probability distribution σ of the circulation is symmetric. Then it is shown that the second moment $\langle(\Gamma_m(a))^2\rangle$ is proportional to the area.

It is easily seen from the equation (1) that the characteristic function of the circulation $\Gamma_{\rm m}(a)$ satisfies

$$(\varphi_{\Gamma_{\rm m}(a)}(s))^{\frac{C}{a}} = \varphi_{\Gamma_{\rm m}(C)}(s) \tag{2}$$

where C is an arbitrary positive constant, thus l.h.s. is independent of the area a. We call the PDF corresponds to the l.h.s. (i.e. the Fourier transform of the l.h.s. multiplied by a normalizing factor) the convoluted PDF. The convoluted PDF of $\Gamma_{\rm m}(a)$ is independent of the area a. The property is due to the fact that $\Gamma_{\rm m}(a)$ is an additive process in area a.

6. Comparison between $\Gamma_>(a)$ and $\Gamma_{\rm m}(a)$

Here we compare the statistics of the circulation associated with the strong vorticity $\Gamma_>(a)$ obtained from the data of DNS and the statistics of the circulation $\Gamma_{\rm m}(a)$ of the model in §5.

The scaling exponent of the second moment of $\Gamma_{\rm m}(a)$ is 1. The local exponent of the second moment of $\Gamma_>(a)$ asymptotically approaches to 1 when the threshold $\omega_{\rm th}$ becomes large or the area a becomes large.

The convoluted PDF of the circulation $\Gamma_{\rm m}(a)$ in the model is independent of the area a and is identical. The convoluted PDFs of the $\Gamma_>(a)$ are computed from the data of DNS for various a and are shown in Figure 5. They tend to converge for the larger areas. From the investigation of the second moments and the convoluted PDFs, it appears that the model $\Gamma_{\rm m}(a)$ represents an asymptotic behavior of $\Gamma_>(a)$ for the large area.

The discrepancy found in the smaller scale may be due to the finite size of the strong components. Since the Reynolds number achieved by the DNS is limited, the inertial range scale is not sufficiently separated from the size

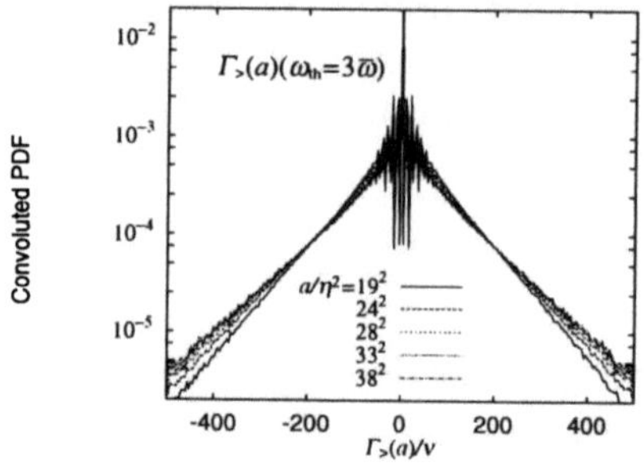

Figure 5. Convoluted PDFs of the circulation $\Gamma_>(a)$ ($\omega_{\rm th} = 3\bar{\omega}$) with the parameter $C = 19^2\eta^2$.

of the strong components and the model may be a over-simplification for the small scale statistics.

In the model, the probability distribution of the circulation of the filaments $\sigma(\gamma)$ is a free parameter and is not determined a priori. However, it is found that the PDF of the circulation $\Gamma_>(a)$ associated with strong component for the area larger than λ^2 is reproduced well by setting σ as a two-sided exponential distribution $\sigma(\gamma) = (1/2G)\exp(-|\gamma|/G)$, where G is the average circulation of the filaments. The distribution σ seems to be related to the dynamics and the stability of the filaments. The detailed analysis of the dynamics is beyond the scope of this paper. We just point out that the probability distribution of the circulation σ of filaments plays an important role in the statistics of velocity circulation $\Gamma(a)$ at the range of Reynolds number of the DNS ($R_\lambda \sim 90$).

References

1. Cao, N., S. Chen, and K. Sreenivasan: 1996, 'Properties of velocity circulation in three-dimensional turbulence'. *Phys. Rev. Lett.* **76**, 616–619.
2. Jiménez, J. and A. A. Wray: 1998, 'On the Characteristics of Vortex Filaments in Isotropic Turbulence'. *J. Fluid Mech.* **373**, 255–285.
3. Jiménez, J., A. A. Wray, P. G. Saffman, and R. S. Rogallo: 1993, 'The Structure of Intense Vorticity in Isotropic Turbulence'. *J. Fluid Mech.* **255**, 65–90.
4. Kolmogorov, A. N.: 1941, 'The local structure of turbulence in incompressible viscous fluid for very large Reynolds numbers'. *Dokl. Akad. Nauk SSSR* **30**, 301–305. (reprinted in *Proc. R. Soc. Lond.* A 434:9-13).
5. Takahashi, N.: 1997, 'Statistical laws of turbulence and simulation'. Master's thesis, Dept. Phys. Univ. Tokyo.
6. Yamamoto, K.: 1995, 'Parallel Computational Fluid Dynamics: New Algorithms and Applications'. In: N. Satofuka, J. Periaux, and A. Ecer (eds.): *Proc. Parallel CFD '94 Conf.*
7. Yoshida, K. and N. Hatakeyama: 1999, 'Statistical Laws of Velocity Circulation in Homogenous Turbulence'. *J. Phys. Soc. Jpn.* (submitted).

D

Passive Scalar Advections

LAGRANGIAN METHOD FOR MULTIPLE CORRELATIONS IN PASSIVE SCALAR ADVECTION

U. FRISCH
C.N.R.S., Observatoire de la Côte d'Azur,
B.P. 4229, F-06304 Nice Cedex 4, France

A. MAZZINO
INFM–Dipartimento di Fisica, Università di Genova,
I–16146 Genova, Italy

AND

A. NOULLEZ, M. VERGASSOLA
C.N.R.S., Observatoire de la Côte d'Azur,
B.P. 4229, F-06304 Nice Cedex 4, France

Reprinted by permission from *Phys. Fluids* **11**, 2178-2186 (1999).

Abstract. A Lagrangian method is introduced for calculating simultaneous n-point correlations of a passive scalar advected by a random velocity field, with random forcing and finite molecular diffusivity κ. The method, which is here presented in detail, is particularly well suited for studying the $\kappa \to 0$ limit when the velocity field is not smooth. Efficient Monte-Carlo simulations based on this method are applied to the Kraichnan model of passive scalar and lead to accurate determinations of the anomalous intermittency corrections in the fourth-order structure function as a function of the scaling exponent ξ of the velocity field in two and three dimensions. Anomalous corrections are found to vanish in the limits $\xi \to 0$ and $\xi \to 2$, as predicted by perturbation theory.

1. Introduction

Robert Kraichnan's model of passive scalar advection by a white-in-time velocity field has been particularly fertile ground for theoreticians trying to develop a theory of intermittency [1, 2, 3] (see also Refs. [4, 5]). Al-

T. Kambe et al. (eds.), IUTAM Symposium on Geometry and Statistics of Turbulence, 153–173.
© 2001 *Kluwer Academic Publishers.*

though the model leads to closed equations for multiple-point moments, only second-order moments can be obtained in closed analytic form [6]. Theoretical predictions differed as to the behavior of higher-order quantities, regarding in particular the survival or the vanishing of intermittency corrections (anomalies) in certain limits. Obtaining reliable numerical results was thus an important challenge. Until recently numerical simulations have been based on the direct integration of the passive scalar partial differential equation and have been limited to two dimensions [5, 7, 8]. Such calculations are delicate; to wit, the difficulty of observing for the second-order structure function the known high-Péclet number asymptotic scaling [6]. Also, the numerical scheme used in Refs. [5, 7] involves a slightly anisotropic velocity field which is not expected to give exactly the right scaling laws for the passive scalar [9].

Lagrangian methods for tackling the Kraichnan model and which require only the integration of ordinary differential equations were recently proposed independently by Frisch, Mazzino and Vergassola [10] and by Gat, Procaccia and Zeitak [11]. Our goal here is to give a detailed presentation of the Lagrangian method and to present new results.

In Section 2 we give the theoretical background of the Lagrangian method for general random velocity fields which need not be white-in-time. In Section 3 we investigate the limit of vanishing molecular diffusivity which depends crucially on how nearby Lagrangian trajectories separate. We then turn to the Kraichnan model which is given a Lagrangian formulation (Section 4). Then, we show how it can be solved numerically by a Monte-Carlo method (Section 5) and present results in both two and three space dimensions (Section 6). We make concluding remarks in Section 7.

2. The Lagrangian method

The (Eulerian) dynamics of a passive scalar field $\theta(\boldsymbol{r},t)$ advected by a velocity field $\boldsymbol{v}(\boldsymbol{r},t)$ is described by the following partial differential equation (written in space dimension d):

$$\partial_t\theta(\boldsymbol{r},t) + \boldsymbol{v}(\boldsymbol{r},t)\cdot\nabla\,\theta(\boldsymbol{r},t) = \kappa\nabla^2\theta(\boldsymbol{r},t) + f(\boldsymbol{r},t), \qquad (1)$$

where $f(\boldsymbol{r},t)$ is an external source (forcing) of scalar and κ is the molecular diffusivity. In all that follows we shall assume that $\theta(\boldsymbol{r},t) = 0$ at some distant time in the past $t = -T$ (eventually, we shall let $T \to \infty$).

In this section the advecting velocity $\boldsymbol{v}$ and the forcing f can be either deterministic or random. In the latter case, no particular assumption is made regarding their statistical properties.

In order to illustrate the basic idea of the Lagrangian strategy, let us first set $\kappa = 0$. We may then integrate (1) along its characteristics, the

Lagrangian trajectories of tracer particles, to obtain

$$\theta(\boldsymbol{r},t) = \int_{-T}^{t} f(\boldsymbol{a}(s;\boldsymbol{r},t),s)\,ds, \tag{2}$$

where $\boldsymbol{a}(s;\boldsymbol{r},t)$ is the position at time $s \le t$ of the fluid particle which will be at position $\boldsymbol{r}$ at time t. (Two-time Lagrangian positions of this type were also used in Kraichnan's Lagrangian History Direct Interaction theory [12].) This Lagrangian position, which will henceforth be denoted just $\boldsymbol{a}(s)$, satisfies the ordinary differential equation

$$\frac{d\boldsymbol{a}(s)}{ds} = \boldsymbol{v}\left(\boldsymbol{a}(s),s\right), \qquad \boldsymbol{a}(t) = \boldsymbol{r}. \tag{3}$$

For $\kappa > 0$ we shall now give a stochastic generalization of this Lagrangian representation. Roughly, θ will be the average of a random field ϕ which satisfies an advection–forcing equation with no diffusion term, in which the advecting velocity is the sum of $\boldsymbol{v}$ and a suitable white-noise velocity generating Brownian diffusion.

To be more specific we need to introduce some notation. We shall use a set of n d-dimensional time-dependent random vectors

$$\dot{\boldsymbol{w}}_i(s) = \{\dot{w}_{i,\alpha};\ i = 1,\ldots,n;\ \alpha = 1,\ldots,d\}, \tag{4}$$

which are Gaussian, identically distributed, independent of each other and independent of both $\boldsymbol{v}$ and f. The time dependence is assumed to be white noise:

$$\langle \dot{w}_{i,\alpha}(s)\dot{w}_{j,\beta}(s')\rangle_w = \delta_{ij}\delta_{\alpha\beta}\delta(s-s'). \tag{5}$$

The notation $\langle\cdot\rangle_w$ stands for "average over the $\dot{\boldsymbol{w}}_i$'s for a fixed realization of $\boldsymbol{v}$ and f". Similarly, $\langle\cdot\rangle_{vf}$ stands for "average over $\boldsymbol{v}$ and f for a fixed realization of the $\dot{\boldsymbol{w}}_i$'s". Unconditional averages are denoted just $\langle\cdot\rangle$. Clearly,

$$\langle\cdot\rangle = \langle\langle\cdot\rangle_w\rangle_{vf} = \langle\langle\cdot\rangle_{vf}\rangle_w\,. \tag{6}$$

The white noise, which is a random distribution, is here denoted by $\dot{w}(s)$ since it is the time derivative of the Brownian motion (or Wiener–Lévy) process. In the numerical implementation we shall work with increments of $w(s)$.

We can now state the main result which is at the basis of our Lagrangian method.

Let $\phi_i(\boldsymbol{r},t)$ ($i = 1,\ldots,n$) be the solutions of the following advection–forcing equations:

$$\begin{aligned} &\partial_t\phi_i(\boldsymbol{r},t) + \left(\boldsymbol{v}(\boldsymbol{r},t) + \sqrt{2\kappa}\,\dot{\boldsymbol{w}}_i(t)\right)\cdot\nabla\phi_i(\boldsymbol{r},t) = f(\boldsymbol{r},t).\\ &\phi_i(\boldsymbol{r},-T) = 0, \qquad i = 1,2,\ldots,n. \end{aligned} \tag{7}$$

For any $\boldsymbol{r}_1, \boldsymbol{r}_2, \boldsymbol{r}_3, \ldots,$ *we have*

$$\theta(\boldsymbol{r}_1) = \langle \phi(\boldsymbol{r}_1) \rangle_w , \quad \forall i \tag{8}$$

$$\theta(\boldsymbol{r}_1)\theta(\boldsymbol{r}_2) = \langle \phi_i(\boldsymbol{r}_1)\phi_j(\boldsymbol{r}_2) \rangle_w , \quad \forall i \neq j, \tag{9}$$

$$\theta(\boldsymbol{r}_1)\theta(\boldsymbol{r}_2)\theta(\boldsymbol{r}_3) = \langle \phi_i(\boldsymbol{r}_1)\phi_j(\boldsymbol{r}_2)\phi_k(\boldsymbol{r}_3) \rangle_w , \quad \forall i, j, k, \quad i \neq j \neq k \neq i, \tag{10}$$

$$\ldots\ldots\ldots\ldots\ldots\ldots\ldots\ldots\ldots\ldots\ldots,$$

where all the fields θ *and* ϕ *are evaluated at the same time* t.

The proof of (8) is obtained by taking the mean value of (7) (only over w) and noting that

$$\left\langle \sqrt{2\kappa}\dot{\boldsymbol{w}}_i \cdot \nabla \phi_i \right\rangle_w = -\kappa \nabla^2 \langle \phi_i \rangle_w . \tag{11}$$

(This is a standard result for linear stochastic equations having white-noise coefficients; it is derived using Gaussian integration by parts (see Ref. [13], Chap. 4). For similar derivations see Refs. [6, 14, 15].)

To prove (9), we first derive from (7), assuming $i \neq j$,

$$\begin{aligned} &\partial_t \left(\phi_i(\boldsymbol{r}_1)\phi_j(\boldsymbol{r}_2) \right) + [\boldsymbol{v}(\boldsymbol{r}_1) \cdot \nabla_1 + \boldsymbol{v}(\boldsymbol{r}_2) \cdot \nabla_2 \\ &+ \sqrt{2\kappa}(\dot{\boldsymbol{w}}_i \cdot \nabla_1 + \dot{\boldsymbol{w}}_j \cdot \nabla_2)]\phi_i(\boldsymbol{r}_1)\phi_j(\boldsymbol{r}_2) = \\ &f(\boldsymbol{r}_1)\phi_j(\boldsymbol{r}_2) + f(\boldsymbol{r}_2)\phi_j(\boldsymbol{r}_1), \end{aligned} \tag{12}$$

where ∇_1 and ∇_2 stand for $\nabla_{\boldsymbol{r}_1}$ and $\nabla_{\boldsymbol{r}_2}$. We then average (12) (over w) and use a relation similar to (11)

$$\begin{aligned} &\left\langle \sqrt{2\kappa} \left(\dot{\boldsymbol{w}}_i \cdot \nabla_1 + \dot{\boldsymbol{w}}_2 \cdot \nabla_2 \right) \phi_i(\boldsymbol{r}_1)\phi_j(\boldsymbol{r}_2) \right\rangle_w \\ &= -\kappa \left(\nabla_1^2 + \nabla_2^2 \right) \langle \phi_i(\boldsymbol{r}_1)\phi_j(\boldsymbol{r}_2) \rangle_w . \end{aligned} \tag{13}$$

(Notice that cross terms involving $\nabla_1 \cdot \nabla_2$ disappear because of the independence of $\dot{\boldsymbol{w}}_i$ and $\dot{\boldsymbol{w}}_j$.) The higher order equations in (10) and following are proved similarly.

Because of the absence of a diffusion operator in (7), its solution has an obvious Lagrangian representation

$$\phi_i(\boldsymbol{r}, t) = \int_{-T}^{t} f(\boldsymbol{a}_i(s), s)\, ds, \tag{14}$$

where

$$\frac{d\boldsymbol{a}_i(s)}{ds} = \boldsymbol{v}\left(\boldsymbol{a}_i(s), s\right) + \sqrt{2\kappa}\dot{\boldsymbol{w}}_i(s), \tag{15}$$

$$\boldsymbol{a}_i(t) = \boldsymbol{r}. \tag{16}$$

So far we have not used the random or deterministic character of $\boldsymbol{v}$ and f. In the random case, taking the average of (8)–(10) over $\boldsymbol{v}$ and f, we obtain

$$\begin{aligned}
&\langle\theta(\boldsymbol{r}_1)\rangle = \langle\phi(\boldsymbol{r}_1)\rangle\,, \quad \forall i && (17)\\
&\langle\theta(\boldsymbol{r}_1)\theta(\boldsymbol{r}_2)\rangle = \langle\phi_i(\boldsymbol{r}_1)\phi_j(\boldsymbol{r}_2)\rangle\,, \quad \forall i \neq j && (18)\\
&\langle\theta(\boldsymbol{r}_1)\theta(\boldsymbol{r}_2)\theta(\boldsymbol{r}_3)\rangle = \langle\phi_i(\boldsymbol{r}_1)\phi_j(\boldsymbol{r}_2)\phi_k(\boldsymbol{r}_3)\rangle\,, &&\\
&\forall i,j,k, \quad i \neq j \neq k \neq i, && (19)\\
&\cdots\cdots\cdots\cdots\cdots\cdots\cdots\cdots\cdots\cdots\cdots\cdots\cdots\cdots\cdots &&
\end{aligned}$$

Eqs. (14), (15), (16), together with (17), (18) and higher orders, constitute our Lagrangian representation for multiple-point moments of the passive scalar.

Let us stress that, for moments beyond the first order, it is essential to use more than one white noise. Indeed, we could have made use of just (8) and written

$$\theta(\boldsymbol{r}_1)\theta(\boldsymbol{r}_2) = \langle\phi_i(\boldsymbol{r}_1)\rangle_w\langle\phi_j(\boldsymbol{r}_2)\rangle_w, \quad \forall i, j. \tag{20}$$

(Including the case $i = j$.) It is however not possible to $(\boldsymbol{v}, f)$-average (20) because it involves a *product* of w-averages rather than just one average as in (9).

In the more restricted context of the Kraichnan model, a functional integral representation of nth order moments involving, as here, n white noises has already been given in Refs. [16, 17].

3. The limit of vanishing molecular diffusivity

Interesting pathologies occur when we bring two or more Eulerian space arguments of the moments to coincidence and *simultaneously* let $\kappa \to 0$.

This is already seen on the second-order moment $\langle\theta^2(\boldsymbol{r}_1)\rangle$. From (14) and (18), we have

$$\langle\theta^2(\boldsymbol{r}_1)\rangle = \int_{-T}^{t}\int_{-T}^{t}\langle f(\boldsymbol{a}_i(s), s) f(\boldsymbol{a}_j(s'), s')\rangle\, ds ds', \tag{21}$$

for any $i \neq j$. The differential equations for $\boldsymbol{a}_i(s)$ and $\boldsymbol{a}_j(s)$ involve different white noises. If, in the limit $\kappa \to 0$, we simply ignore the $\sqrt{2\kappa}\dot{\boldsymbol{w}}_i$ terms in (15), we find that all the $\boldsymbol{a}_i(s)$'s satisfy the same equation (3) and the same boundary condition $\boldsymbol{a}_i(t) = \boldsymbol{r}_1$. It is then tempting to conclude that all the $\boldsymbol{a}_i(s)$'s are identical, namely are just the Lagrangian trajectories $\boldsymbol{a}(s)$ of the unperturbed $\boldsymbol{v}$-flow. As a consequence, we can then rewrite (21) as

$$\langle\theta^2(\boldsymbol{r}_1)\rangle \quad = \quad \int_{-T}^{t}\int_{-T}^{t}\langle f(\boldsymbol{a}(s), s) f(\boldsymbol{a}(s'), s')\rangle\, ds ds',$$

$$= \left\langle \left(\int_{-T}^{t} f(\boldsymbol{a}(s), s)\, ds \right)^2 \right\rangle . \tag{22}$$

For a large class of random forcings f of zero mean the r.h.s. of (22) will diverge $\propto T$ as $T \to \infty$. This is for example the case when f is homogeneous, stationary and short- (or delta-) correlated in time as in the Kraichnan model. The reason of this divergence is that, although the forcing has zero mean, its integral (along the Lagrangian trajectory) over times long compared to the correlation time behaves like Brownian motion (in the T-variable) and, thus, has a variance $\propto T$. From this naïve procedure we would thus conclude that, when $T = \infty$, the scalar variance becomes infinite as $\kappa \to 0$.

This conclusion is actually correct when the $\boldsymbol{v}$-flow is smooth (differentiable in the space variable): this is the so-called Batchelor limit which has been frequently investigated [6, 16, 18, 19, 20, 21]. The conclusion however becomes incorrect when the $\boldsymbol{v}$-flow is only Hölder continuous, i.e. its spatial increments over a small distance ℓ vary as a fractional power of ℓ (e.g. $\ell^{1/3}$ in Kolmogorov 1941 turbulence). As pointed out in Ref. [16] (see also Ref. [4]), when $\boldsymbol{v}$ is not smooth the solution to (3) lacks uniqueness, so that two Lagrangian particles which end up at the same point $\boldsymbol{r}_1$ at time t may have different past histories. This is exemplified with the one-dimensional model

$$\frac{dx}{ds} = -x^{1/3}, \quad -T \le s \le t, \quad x(t) = \epsilon \ge 0, \tag{23}$$

where x is a deputy for $\boldsymbol{a}_i - \boldsymbol{a}_j$, the separation between two nearby Lagrangian particles. For $\epsilon > 0$, the solution of (23) is

$$x(s) = \left[\epsilon^{2/3} + \frac{2}{3}(t-s) \right]^{3/2} \tag{24}$$

If we now set $\epsilon = 0$ in (24) or consider times s such that $|t-s| \gg \epsilon^{2/3}$, we obtain

$$x(s) = \left[\frac{2}{3}(t-s) \right]^{3/2} \tag{25}$$

This is indeed *a* solution of (23) with $\epsilon = 0$, but there is another trivial one, namely $x(s) = 0$. Related to this non-uniqueness is the fact that, when ϵ is small the solution given by (24) becomes independent of ϵ, namely is given by the non-trivial solution (25) for $\epsilon = 0$.

Whenever the flow $\boldsymbol{v}$ is just Hölder continuous, the separation of nearby Lagrangian particles proceeds in a similar way, becoming rapidly independent of the initial separation. Such a law of separation is much more explosive than would have been obtained for a smooth flow with sensitive dependence of the Lagrangian trajectories. In the latter case we would have $x(s) = \epsilon\, e^{\lambda(t-s)}$ $(\lambda > 0)$, which grows exponentially with $t-s$ but still tends to zero with ϵ.

We propose to call this explosive growth a Richardson walk, after Lewis Fry Richardson who was the first to experimentally observe this rapid separation in turbulent flow and who was also much interested in the role of non-differentiability in turbulence [22]. It is this explosive separation which prevents the divergence of $\langle\theta^2(\boldsymbol{r}_1)\rangle$ when $\kappa \to 0$ (and more generally of moments with several coinciding points). Indeed, as the time s moves back from $s = t$, even an infinitesimal amount of molecular diffusion will slightly separate, say, by an amount ϵ, the Lagrangian particles $\boldsymbol{a}_i(s)$ and $\boldsymbol{a}_j(s)$ which coincide at $s = t$. Then, the Richardson walk will quickly bring the separations to values independent of ϵ and, thus of κ. Hence, the double integral in (21), which involves points $\boldsymbol{a}_1(s)$ and $\boldsymbol{a}_2(s)$ with uncorrelated forces when $|s-s'|$ is sufficiently large, may converge for $T \to \infty$. (For it to actually converge more specific assumptions must be made about the space and time correlations of $\boldsymbol{v}$ and f, which are satisfied, e.g., in the Kraichnan model.)

An alternative to introducing a small diffusivity is to work at $\kappa = 0$, with "point splitting". For this one replaces $\langle\theta^2(\boldsymbol{r}_1)\rangle$ by $\langle\theta(\boldsymbol{r}_1)\theta(\boldsymbol{r}'_1)\rangle$, where $\boldsymbol{r}_1$ and $\boldsymbol{r}'_1$ are separated by a distance ϵ. Eventually, $\epsilon \to 0$. In practical numerical implementations we found that point splitting works well for second-order moments but is far less efficient than using a small diffusivity for higher-order moments.

4. The Kraichnan model

The Kraichnan model [6, 1] is an instance of the passive scalar equation (1) in which the velocity and the forcing are Gaussian white noises in their time dependence. This ensures that the solution is a Markov process in the time variable and that closed moment equations, sometimes called "Hopf equations", can be written for single-time multiple-space moments such as $\langle\theta(\boldsymbol{r}_1,t)\ldots\theta(\boldsymbol{r}_n,t)\rangle$. The equation for second-order moments was published for the first time by Kraichnan [6] and, for higher-order moments, by Shraiman and Siggia [20]. Note that Hopf's work [23] dealt with the characteristic functional of random flow; it had no white-noise process and no closed equations, making the use of his name not so appropriate in the context of white-noise linear stochastic equations. The fact that closed moment equations exist for such problems has been known for a long time (see, e.g., Ref. [14] and references therein). We shall not need the moment equations and shall not write them here (for an elementary derivation, see Ref. [15]).

The precise formulation of the Kraichnan model as used here is the following. The velocity field $\boldsymbol{v} = \{v_\alpha, \alpha = 1, \ldots, d\}$ appearing in (1) is incompressible, isotropic, Gaussian, white-noise in time; it has homogeneous increments with power-law spatial correlations and a scaling exponent ξ in the range $0 < \xi < 2$:

$$\langle [v_\alpha(\boldsymbol{r},t) - v_\alpha(\boldsymbol{0},0)][v_\beta(\boldsymbol{r},t) - v_\beta(\boldsymbol{0},0)]\rangle = 2\delta(t) D_{\alpha\beta}(\boldsymbol{r}), \tag{26}$$

where,

$$D_{\alpha\beta}(\boldsymbol{r}) = r^\xi \left[(\xi + d - 1)\,\delta_{\alpha\beta} - \xi \frac{r_\alpha r_\beta}{r^2} \right]. \tag{27}$$

Note that since no infrared cutoff is assumed on the velocity its integral scale is infinite; this is not a problem since only velocity increments matter for the dynamics of the passive scalar. Note also that when a white-in-time velocity is used in (15), the well known Ito–Stratonovich ambiguity could appear [24]. This ambiguity is however absent in our particular case, on account of incompressibility.

The random forcing f is independent of $\boldsymbol{v}$, of zero mean, isotropic, Gaussian, white-noise in time and homogeneous. Its covariance is given by:

$$\langle f(\boldsymbol{r},t)\, f(\boldsymbol{0},0)\rangle = F\,(r/L)\;\delta(t), \tag{28}$$

with $F(0) > 0$ and $F\,(r/L)$ decreasing rapidly for $r \gg L$, where L is the (forcing) integral scale.

In principle, to be a correlation, the function $F\,(r/L)$ should be of positive type, i.e. have a non-negative d-dimensional Fourier transform. In our numerical work we find it convenient to work with the step function $\Theta_L(r)$ which is equal to unity for $0 \le r \le L$ and to zero otherwise. Hence, the injection rate of passive scalar variance is $\varepsilon = F(0)/2 = 1/2$. The fact that the function Θ is not of positive type is no problem. Indeed, let its Fourier transform be written $E(k) = E_1(k) - E_2(k)$, where $E(k) = E_1(k)$ whenever $E(k) \geq 0$. Using the step function amounts to replacing in (1) the real forcing f by the complex forcing $f_1 + i f_2$ where f_1 and f_2 are independent Gaussian random functions, white-noise in time and chosen such that their energy spectra (in the space variable) are respectively $E_1(k)$ and $E_2(k)$. Since the passive scalar equation is linear, the solution may itself be written as $\theta_1 + i\theta_2$ where θ_1 and θ_2 are the (independent) solutions of the passive scalar equations with respective forcing terms f_1 and f_2. Using the universality with respect to the functional form of the forcing [25], it is then easily shown that the scaling laws for the passive scalar structure functions are the same as for real forcing.

We shall be interested in the passive scalar structure functions of even order $2n$ (odd order ones vanish by symmetry), defined as

$$S_{2n}(r;L) \equiv \left\langle (\theta(\boldsymbol{r}) - \theta(\boldsymbol{0}))^{2n} \right\rangle . \tag{29}$$

From Ref. [6] (see also Ref. [15]) we know that, for $L \gg r \gg \eta \sim \kappa^{1/\xi}$, the second-order structure function is given by

$$S_2(r;L) = C_2\, \varepsilon\, r^{\zeta_2}, \quad \zeta_2 = 2 - \xi, \tag{30}$$

where C_2 is a dimensionless numerical constant. If there were no anomalies, we would have, for $n > 1$,

$$S_{2n}(r;L) = C_{2n}\, \varepsilon^n\, r^{n\zeta_2}. \tag{31}$$

Note that (30) and (31) do not involve the integral scale L. Actually, for $n > 1$, we have *anomalous* scaling with $S_{2n}(r;L) \propto r^{\zeta_{2n}}$ and $\zeta_{2n} < n\zeta_2$. More precisely, we have

$$S_{2n}(r;L) = C_{2n}\, \varepsilon^n\, r^{n\zeta_2} \left(\frac{L}{r}\right)^{\zeta_{2n}^{\text{anom}}}, \tag{32}$$

$$\zeta_{2n}^{\text{anom}} \equiv n\zeta_2 - \zeta_{2n}, \tag{33}$$

where the structure function now displays a dependence on the integral scale L. Our strategy will be to measure the dependence of $S_{2n}(r;L)$ on L while the separation r and the injection rate ε are kept fixed and, thereby, to have a direct measurement of the anomaly ζ_{2n}^{anom}.

Let us show that, in principle, this can be done by the Lagrangian method of Section 2 using $2n$ tracer (Lagrangian) particles whose trajectories satisfy (15). The structure function of order $2n$ can be written as a linear combination of moments of the form $\langle \theta(\boldsymbol{r}_1)\theta(\boldsymbol{r}_2)\ldots\theta(\boldsymbol{r}_{2n})\rangle$, where p of the points are at location $\boldsymbol{r}$ and $2n-p$ at location $\boldsymbol{0}$ ($p = 0,\ldots,2n$). Because of the symmetries of the problem, p and $2n-p$ give the same contribution. Thus, we need to work only with the $n+1$ *configurations* corresponding to $p = 0,\ldots,n$. For example, we have

$$S_4(r;L) = 2\left\langle \theta^4(\boldsymbol{r})\right\rangle - 8\left\langle \theta^3(\boldsymbol{r})\theta(\boldsymbol{0})\right\rangle + 6\left\langle \theta^2(\boldsymbol{r})\theta^2(\boldsymbol{0})\right\rangle . \tag{34}$$

Let us first consider the case of the two-point function (second-order moment). Using (14), (18) and the independence of $\boldsymbol{v}$ and f, we have

$$\langle \theta(\boldsymbol{r}_1)\theta(\boldsymbol{r}_2)\rangle = \left\langle \int_{-T}^{t}\int_{-T}^{t} \langle f(\boldsymbol{a}_1(s_1),s_1) f(\boldsymbol{a}_2(s_2),s_2)\rangle_f \, ds_1 ds_2 \right\rangle_{vw} . \tag{35}$$

Here, $\langle\cdot\rangle_f$ is an average over the forcing and $\langle\cdot\rangle_{vw}$ denotes averaging over the velocity and the $\dot{\boldsymbol{w}}$'s, and $\boldsymbol{a}_1(s_1)$ and $\boldsymbol{a}_2(s_2)$ satisfy (15) with the "final" conditions $\boldsymbol{a}_1(t) = \boldsymbol{r}_1$and $\boldsymbol{a}_2(t) = \boldsymbol{r}_2$, respectively.

In (35) the averaging over f can be carried out explicitly using (28). With our step-function choice for F, we obtain

$$\langle\theta(\boldsymbol{r}_1)\,\theta(\boldsymbol{r}_2)\rangle = \langle T_{12}(L)\rangle_{vw}, \tag{36}$$

where

$$T_{12}(L) = \int_{-T}^{t} \Theta_L(|\boldsymbol{a}_1(s) - \boldsymbol{a}_2(s)|)\, ds \tag{37}$$

is the amount of time that two tracer particles arriving at $\boldsymbol{r}_1$ and $\boldsymbol{r}_2$ and moving backwards in time spent with their mutual distance $|\boldsymbol{a}_1(s) - \boldsymbol{a}_2(s)| < L$. Whether the particles move backwards or forward in time is actually irrelevant for the Kraichnan model since the velocity, being Gaussian, is invariant under reversal.

For the four-point function, we proceed similarly and use the Wick rules to write fourth-order moments of f in terms of sums of products of second-order moments, obtaining

$$\begin{aligned}\langle\theta(\boldsymbol{r}_1)\theta(\boldsymbol{r}_2)\theta(\boldsymbol{r}_3)\theta(\boldsymbol{r}_4)\rangle &= \langle T_{12}(L)T_{34}(L)\rangle_{vw} \\ &+ \langle T_{13}(L)T_{24}(L)\rangle_{vw} \\ &+ \langle T_{14}(L)T_{23}(L)\rangle_{vw}.\end{aligned} \tag{38}$$

Expressions similar to Eqs. (36) and (38) are easily derived for higher order correlations.

We see that the evaluation of structure functions and moments has been reduced to studying certain statistical properties of the random time that pairs of particles spend with their mutual distances less than the integral scale L. Generally, the distance between pairs of particles tends to increase with the time elapsed but, occasionally, particles may come very close and stay so; this will be the source of the anomalies in the scaling.

5. Numerical implementation of the Lagrangian method

In Section 4 we have shown that structure functions of the passive scalar are expressible in terms of $\boldsymbol{v}$-averages of products of factors $T_{ij}(L)$. For the structure function of order $2n$, these products involve configurations of $2n$ particles, p of which end at locations $\boldsymbol{r}$ at time $s = t$ and the remaining $2n - p$ at location $\boldsymbol{0}$. In the Kraichnan model $\boldsymbol{v}$ and f are stationary, so that after relaxation of transients, θ also becomes stationary. We may thus calculate our structure functions at $t = 0$. Time-reversal invariance of the $\boldsymbol{v}$ field and of the Lagrangian equations allows us to run the s-time forward rather than backward. Also, $T_{ij}(L)$ is sensitive only to differences in particle separations, whose evolution depends only on the difference of the velocities at $\boldsymbol{a}_i$ and $\boldsymbol{a}_j$. Furthermore, the $\boldsymbol{v}$ field has homogeneous increments. All this allows us to work with $2n-1$ particle separations, namely

$$\tilde{\boldsymbol{a}}_i(s) \equiv \boldsymbol{a}_i(s) - \boldsymbol{a}_{2n}(s), \quad i = 1, \ldots, 2n-1, \tag{39}$$

$$\tilde{\boldsymbol{a}}_i(t) = \tilde{\boldsymbol{r}}_i \equiv \boldsymbol{r}_i - \boldsymbol{r}_{2n}. \tag{40}$$

Using (15) we find that the quantities $\tilde{\boldsymbol{a}}_i(s)$ satisfy $2n-1$ (vector-valued) differential equations which involve the differences of velocities $\boldsymbol{v}(\boldsymbol{a}_i(s), s) - \boldsymbol{v}(\boldsymbol{a}_{2n}(s), s)$. The statistical properties of the solutions remain unchanged if we subtract $\boldsymbol{a}_{2n}(s)$ from all the space arguments. We thus obtain the following Lagrangian equations of motion for the $2n-1$ particle separations:

$$\frac{d\tilde{\boldsymbol{a}}_i(s)}{ds} = \tilde{\boldsymbol{v}}_i(s) + \sqrt{2\kappa}\tilde{\dot{\boldsymbol{w}}}_i(s) \tag{41}$$

$$\tilde{\boldsymbol{v}}_i(s) \equiv \boldsymbol{v}(\tilde{\boldsymbol{a}}_i(s), s) - \boldsymbol{v}(\boldsymbol{0}, s), \tag{42}$$

$$\tilde{\dot{\boldsymbol{w}}}_i(s) \equiv \dot{\boldsymbol{w}}_i - \dot{\boldsymbol{w}}_{2n}. \tag{43}$$

For numerical purposes (41) is discretized in time using the standard Euler–Ito scheme of order one half [24]

$$\tilde{\boldsymbol{a}}_i(s+\Delta s) - \tilde{\boldsymbol{a}}_i(s) = \sqrt{\Delta s}\ (\mathcal{V}_i + \sqrt{2\kappa}\ \mathcal{W}_i), \tag{44}$$

where Δs is the time step and the $\mathcal{V}_i$'s and the $\mathcal{W}_i$'s ($i = 1, \ldots, 2n-1$) are d-dimensional Gaussian random vectors chosen independently of each other and independently at each time step and having the appropriate correlations, which are calculated from (5) and (26), namely

$$\langle \mathcal{V}_{i,\alpha} \mathcal{V}_{j,\beta} \rangle = D_{\alpha\beta}(\tilde{\boldsymbol{a}}_i) + D_{\alpha\beta}(\tilde{\boldsymbol{a}}_j) - D_{\alpha\beta}(\tilde{\boldsymbol{a}}_i - \tilde{\boldsymbol{a}}_j)\ , \tag{45}$$

$$\langle \mathcal{W}_{i,\alpha} \mathcal{W}_{j,\beta} \rangle = (1+\delta_{ij})\delta_{\alpha\beta}, \tag{46}$$

where $D_{\alpha\beta}(\boldsymbol{r})$ is defined in (27).

To actually generate these Gaussian random variables, we use the symmetry and positive definite character of covariance matrices like (45) and (46). Indeed, any such matrix can be factorized as a product (taking $\mathcal{V}$ as an example) [26]

$$\langle \mathcal{V} \otimes \mathcal{V} \rangle = \mathcal{L}\mathcal{L}^T\ , \tag{47}$$

where $\mathcal{L}$ is a nonsingular lower triangular matrix and $\mathcal{L}^T$ is its transpose. $\mathcal{L}$ can be computed explicitly using the Cholesky decomposition method [26], an efficient algorithm to compute the triangular factors of positive definite matrices. It nevertheless takes $O([(2n-1)d]^3/3)$ flops to get $\mathcal{L}$ from $\langle \mathcal{V} \otimes \mathcal{V} \rangle$ and this is the most time-consuming operation at each time step. Once $\mathcal{L}$ is obtained, a suitable set of variables $\mathcal{V}$ can be obtained by applying the linear transformation

$$\mathcal{V} = \mathcal{L}\,\mathcal{N} \tag{48}$$

to a set of *independent* unit-variance Gaussian random variables $\mathcal{N}_{i,\alpha}$ coming from a standard Gaussian random number generator, that is with $\langle \mathcal{N}_{i,\alpha} \mathcal{N}_{j,\beta} \rangle = \delta_{ij}\delta_{\alpha\beta}$. The resulting variables $(\mathcal{V}_i)_\alpha$ then have the required covariances.

From the $\tilde{\boldsymbol{a}}_i(s)$'s we obtain the quantities $T_{ij}(L)$ for all the desired values of the integral scale L, typically, a geometric progression up to the maximum value $L_{\max}$. The easiest way to evaluate $S_{2n}(r; L)$ is to evolve simultaneously the $n+1$ configurations corresponding to $p = 0, \ldots, n$, stopping the current realization when *all* inter-particle distances in *all* configurations are larger than some appropriate large-scale threshold L_{th}, to which we shall come back. The various moments appearing in the structure functions are then calculated using expression such as (38) in which the (v, w)-averaging is done by the Monte-Carlo method, that is over a suitably large number of realizations. We note that the expressions for the structure functions of order higher than two involve heavy cancellations between the terms corresponding to different configurations of particles. For instance, the three terms appearing in the expression (34) for the fourth-order function, all have dominant contributions scaling as $L^{2(2-\xi)}$ for large L and a first subdominant correction scaling as $L^{2-\xi}$. The true non-trivial scaling $\propto L^{\zeta_4^{\text{anom}}}$ emerges only after cancellation of the dominant and first subdominant contributions. For small ξ the dominant contributions are particularly large. In the presence of such cancellations, Monte-Carlo averaging is rather difficult since the relative errors on individual terms decrease only as the inverse square root of the number of realizations. In practice, the number of realizations is increased until clean non-spurious scaling emerges. In three dimensions, for $S_4(r; L)$, between one and several millions realizations (depending on the value of ξ) are required. In two dimensions even more realizations are needed. For example, to achieve comparable quality of scaling for $\xi = 0.75$, in three dimensions 4×10^6 realizations are needed but 14×10^6 are needed in two dimensions.

Now, some comments on the choice of parameters.

The threshold L_{th} must be taken sufficiently large compared to largest integral scale of interest $L_{\max}$ to ensure that the probability of returning within $L_{\max}$ from L_{th} is negligible. But choosing an excessively large L_{th} is too demanding in computer resources. In practice, the choice of L_{th} depends both on the space dimension and on how far one is from the limit $\xi = 0$. In three dimensions, it is enough to take $L_{\text{th}} = 10\,L_{\max}$. In two dimensions there is a new difficulty when ξ is small. At $\xi = 0$ the motion of Lagrangian particles and also of separations of pairs of particles is exactly two-dimensional Brownian motion. As it is well-known, in two dimensions, Brownian motion is recurrent (see, e.g., Ref. [27]). Hence, with probability one a pair of particles will eventually achieve arbitrarily small separations. As a consequence, at $\xi = 0$ in two dimensions, the mean square value of θ is infinite. For very small positive ξ this mean square saturates but most of the contribution comes from scales much larger than the integral scale. This forces to choose extremely large values of $L_{\rm th}$ when ξ is small. In practice, for $0.6 \le \xi \le 0.9$ we take $L_{\rm th} = 4 \times 10^3\, L_{\rm max}$ and beyond $\xi = 0.9$ we take $L_{\rm th} = 10\, L_{\rm max}$. (The range $\xi < 0.6$ has not yet been explored.) In view of the accuracy of our results, we have verified that the use of larger values for $L_{\rm th}$ does not affect in any significant way the values of the scaling exponents.

The molecular diffusivity κ is chosen in such a way that r is truly in the inertial range, namely, we demand (i) that the dissipation scale $\eta \sim \kappa^{1/\xi}$ should be much smaller than the separation r and (ii) that the time a pair of particle spends with a separation comparable to η, which is $\sim \eta^2/\kappa \sim \eta^{2-\xi}$ should be much smaller than the time needed for this separation to grow from r to L, which is $\sim L^{2-\xi} - r^{2-\xi}$. The latter condition becomes very stringent when ξ is close to 2.

Finally, the time step Δs is chosen small compared to the diffusion time η^2/κ at scale η .

6. Results

We now present results for structure functions up to fourth order. The three-dimensional results have already been published in Ref. [10]. The two-dimensional results are new. Some results for structure functions of order six have been published in Ref. [28] and shall not be repeated here (more advanced simulations are in progress).

A severe test for the Lagrangian method is provided by the second-order structure function $S_2(r; L)$, whose expression is known analytically [1]. Its behavior being non-anomalous, a flat scaling in L should be observed. The L-dependence of S_2, measured by the Lagrangian method, is shown in Fig. 1 for $\xi = 0.6$ and $d = 3$ (all structure functions are plotted in log–log coordinates). The measured slope is 10^{-3} and the error on the constant is 3%. (These figures are typical also for other values of ξ studied.) We observe that, for separation r much larger than the integral scale L, correlations between $\theta(\boldsymbol{r})$ and $\theta(\boldsymbol{0})$ are very small; hence, the scaling for the second-order structure function is essentially given by the L-dependence of $\langle\theta^2\rangle$, namely $L^{2-\xi}$; the transition to the constant-in-L behavior around $r = L$ is

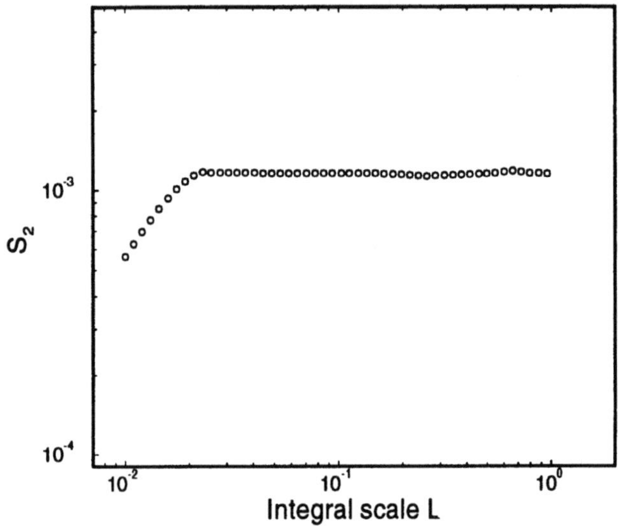

Figure 1. 3-D second-order structure function S_2 *vs* L for $\xi = 0.6$. Separation $r = 2.7 \times 10^{-2}$, diffusivity $\kappa = 1.115 \times 10^{-2}$, number of realizations 4.5×10^6.

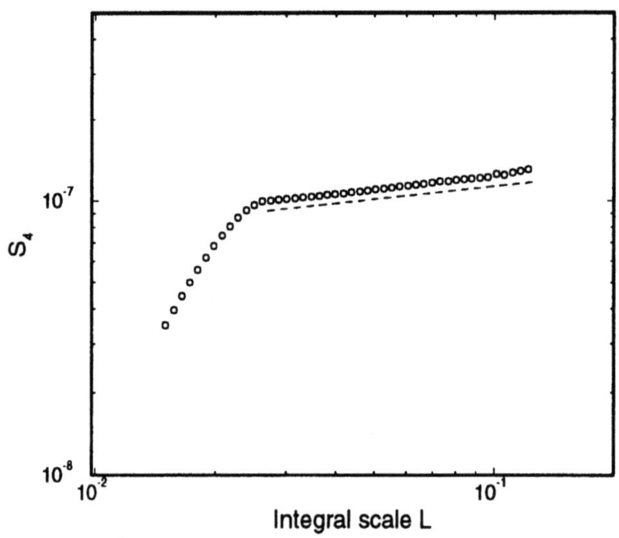

Figure 2. 3-D fourth-order structure function S_4 *vs* L for $\xi = 0.2$. Separation $r = 2.7 \times 10^{-2}$, diffusivity $\kappa = 0.247$, number of realizations 15×10^6.

very sharp, on account of the step function chosen for F.

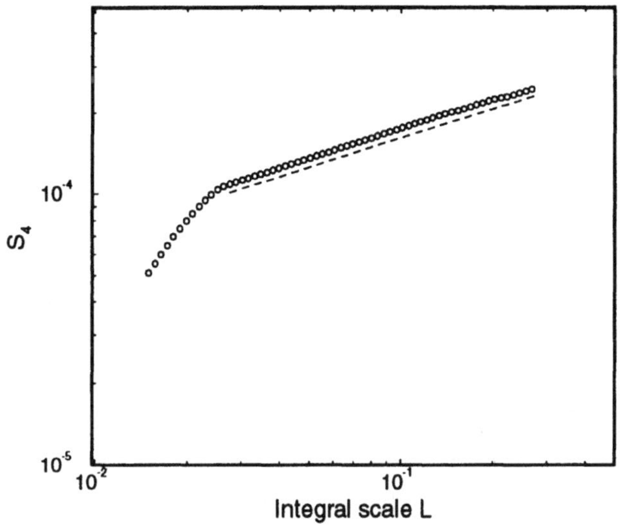

Figure 3. Same as in Fig. 2 for $\xi = 0.9$. Parameters: $r = 2.7 \times 10^{-2}$, $\kappa = 4.4 \times 10^{-4}$, number of realizations 8×10^6.

Figs. 2, 3 and 4 show the L-dependence of the fourth-order structure function in three dimensions for $\xi = 0.2$, 0.9 and 1.75, respectively. In each case the scaling region (which is basically $L > r$) is indicated by a dashed straight line whose slope is the anomaly. Note that, to obtain a similar high-quality scaling as shown on these figures, a much larger number of realizations is needed for small ξ; this is required to permit cancellation of leading contributions to (34), as explained in Section 5.

The two-dimensional case, which is numerically more difficult for reasons explained near the end of Section 5, is shown in Figs. 5, 6 and 7 for $\xi = 0.6$, 0.9 and 1.75, respectively. Fig. 6 also shows the data obtained for various values of the number of realizations. Note that if only 150×10^3 realizations are used, the anomaly (that is the slope obtained, e.g., by a least square fit) is grossly overestimated.

Fig. 8 shows a plot of the anomaly $\zeta_4^{\rm anom}$ *vs* ξ in both two and three dimensions. The error bars (shown in 2-D only for $\xi \leq 1.1$ to avoid crowding) are obtained by analyzing the fluctuations of local scaling exponents over octave ratios of values for L, a method which tends to overestimate errors.

Let us now comment on the results. In three dimensions, the error bars near $\xi = 2$ are exceedingly small and the data have a good fit (shown as dashed line) of the form $\zeta_4^{\rm anom} = a\gamma + b\gamma^{3/2}$ with $\gamma = 2 - \xi$ (the parameters are $a = 0.06$ and $b = 1.13$). This is compatible with an expansion in powers

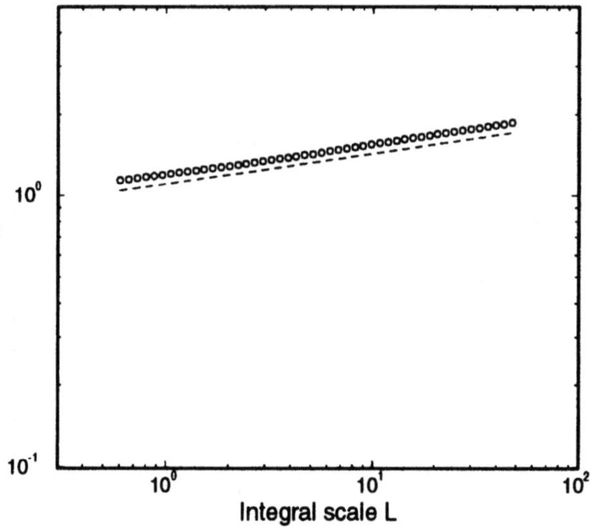

Figure 4. Same as in Fig. 2 for $\xi = 1.75$. Parameters: $r = 2.7 \times 10^{-2}$, $\kappa = 10^{-9}$, number of realizations 1.5×10^6.

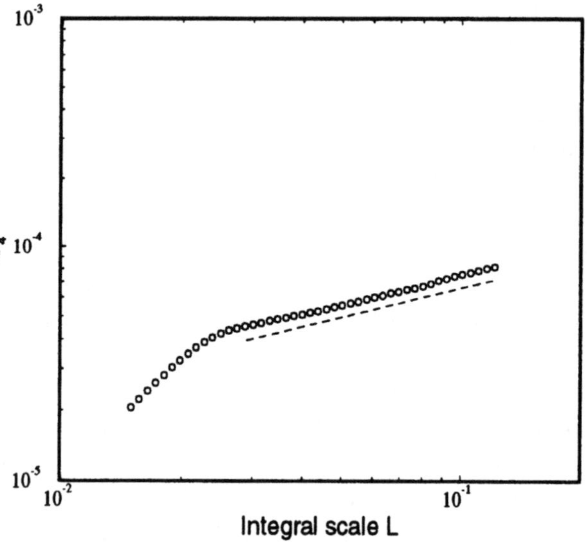

Figure 5. 2-D fourth-order structure function S_4 *vs* L for $\xi = 0.6$. Parameters: $r = 2.7 \times 10^{-2}$, $\kappa = 1.1 \times 10^{-2}$, number of realizations 5×10^6.

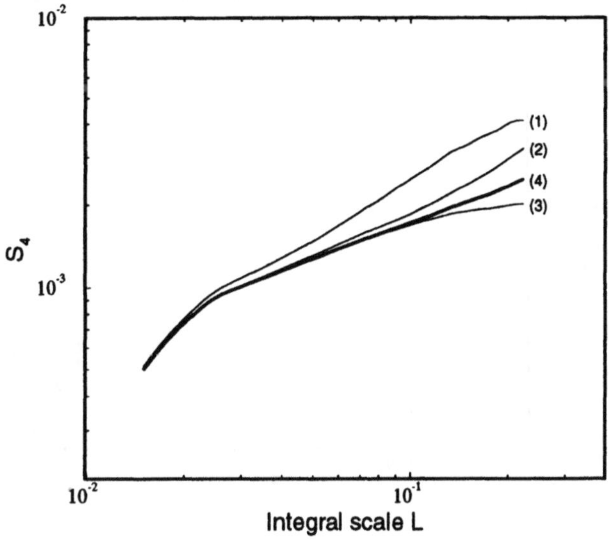

Figure 6. Same as in Fig. 5 for $\xi = 0.9$. Parameters are $r = 2.7 \times 10^{-2}$, $\kappa = 4.4 \times 10^{-4}$. To illustrate convergence, various numbers of realizations are shown: (1) 150×10^3, (2) 1.5×10^6, (3) 3.4×10^6, (4) from 4.8×10^6 to 7×10^6 (several curves superposed).

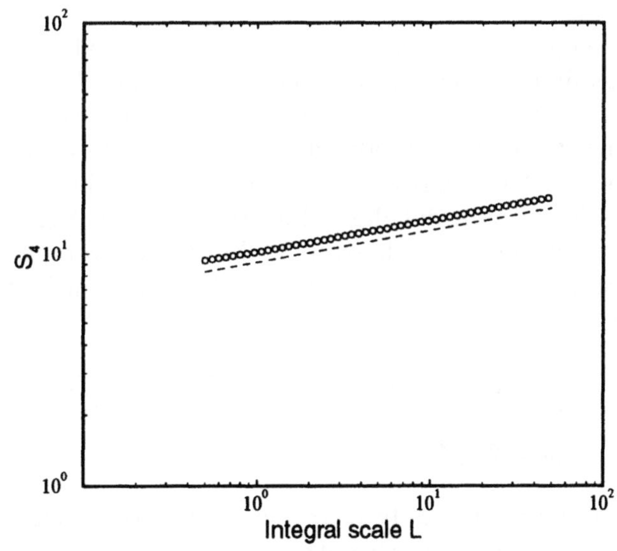

Figure 7. Same as in Fig. 5 for $\xi = 1.75$. Parameters: $r = 2.7 \times 10^{-2}$, $\kappa = 10^{-9}$, number of realizations 2.4×10^6.

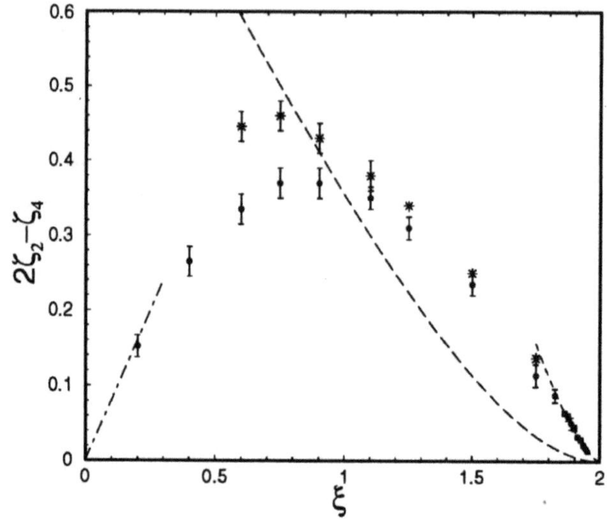

Figure 8. The anomaly $2\zeta_2 - \zeta_4$ for the fourth-order structure function in two dimensions (stars, upper graph) and three dimensions (circles, lower graph). Error bars in 2-D shown only for $\xi \leq 1.1$. The dashed line is the three-dimensional linear ansatz prediction (49).

of $\sqrt{\gamma}$ [29] in which a term $\propto \sqrt{\gamma}$ is ruled out by the Hölder inequality $\zeta_4 \leq 2\zeta_2 = 2\gamma$.

Of particular significance is that, when ξ is decreased from 2 to 0, the anomaly grows at first, achieves a maximum and finally decreases. In three dimensions, where small-ξ simulations are easier than in two dimensions, we have good evidence that the anomaly vanishes for $\xi \to 0$ as predicted by the perturbation theory of Gawędzki and Kupiainen [2], whose leading order $\zeta_4^{\text{anom}} = 4\xi/5$ is shown as a dot-dashed straight line on Fig. 8. Note that the next-order correction $\propto \xi^2$ is known [30] but the convergence properties of the ξ-series are not clear. The "linear ansatz" prediction for the anomaly given in Refs. [1, 7], that is

$$\zeta_4^{\text{anom}} = \frac{3\zeta_2 + d}{2} - \frac{1}{2}\sqrt{8d\zeta_2 + (d - \zeta_2)^2}, \tag{49}$$

$$\zeta_2 = 2 - \xi, \tag{50}$$

is consistent with our results only near $\xi = 1$, the point farthest from the two limits $\xi = 0$ and $\xi = 2$, which both have strongly nonlocal dynamics. This suggests a possible relation between deviations from the linear ansatz and locality of the interactions [31]. Whether this consistency for $\xi = 1$ persists for moments of order higher than four is an open problem.

The fact that the anomalies are stronger in two than in three dimensions is consistent with their vanishing as $d \to \infty$ [3]. The fact that the maximum anomaly occurs for a value of ξ smaller in two than in three dimensions can be tentatively interpreted as follows. Near $\xi = 0$ the dynamics is dominated by the nearly ultraviolet-divergent eddy diffusion, whereas near $\xi = 2$ it is dominated by the nearly infrared-divergent stretching. The former increases with d, but not the latter. The maximum is achieved when these two effects balance.

7. Concluding remarks

Compared to Eulerian simulations of the partial differential equation (1) our Lagrangian method has various advantages. When calculating moments of order $2n$ we do not require the complete velocity field at each time step but only $2n - 1$ random vectors, basically, the set of velocity differences between the locations of Lagrangian tracer particles which are advected by the flow and subject to independent Brownian diffusion. As a consequence the complexity of the computation (measured in number of floating point operations) grows polynomially rather than exponentially with the dimension d. Furthermore, working with tracers naturally allows to measure the scaling of the structure functions $S_{2n}(r; L)$ *vs* the integral scale L of the forcing. Physically, this means that the injection rate of passive scalar variance (which equals its dissipation rate) and the separation r are kept fixed while the integral scale L is varied. Anomalies, that is discrepancies from the scaling exponents which would be predicted by naïve dimensional analysis, are measured here directly through the scaling dependence on L of the structure functions.

Actually, direct Eulerian simulations and the Lagrangian method are complementary. Very high-resolution simulations of the sort found in Ref. [5] are really not practical in more than two dimensions but do give access to the entire Eulerian passive scalar field. Hence, they can and have been used to address questions about the geometry of the scalar field and about probability distributions.

Although we have presented here the numerical implementation of the Lagrangian method only for the Kraichnan model, it is clear that the general strategy presented in Sections 2 and 3 is applicable to a wide class of random flows, for example, with a finite correlation time.

An important aspect of the results obtained for the Kraichnan model is that they agree with perturbation theory [2] for $\xi \to 0$. As is now clear from theory and simulations, anomalies in the Kraichnan model and other passive scalar problems arise from zero modes in the operators governing the Eulerian dynamics of n-point correlation functions [2, 3, 32]. It is likely that some form of zero modes is also responsible for anomalous scaling in nonlinear turbulence problems. An instance are the "fluxless solutions" to Markovian closures based on the Navier–Stokes equations in dimension d close to two, which have a power-law spectrum with an exponent depending continuously on d [33]. Attempts to capture intermittency effects in three-dimensional turbulence are being made along similar lines (see, e.g., Ref. [34]). The Kraichnan model for passive scalar intermittency puts us on a trail to understanding anomalous scaling in turbulence.

ACKNOWLEDGMENTS

We are most grateful to Robert Kraichnan for innumerable interactions on the subject of passive scalar intermittency over many years. We acknowledge useful discussions with M. Chertkov, G. Falkovich, O. Gat, K. Gawędzki, F. Massaioli, S.A. Orszag, I. Procaccia, A. Wirth, V. Yakhot and R. Zeitak. Simulations were performed in the framework of the SIVAM project of the Observatoire de la Côte d'Azur. Part of them were performed using the computing facilities of CASPUR at Rome University, which is gratefully acknowledged. Partial support from the Centre National de la Recherche Scientifique through a "Henri Poincaré" fellowship (AM) and from the Groupe de Recherche "Mécanique des Fluides Géophysiques et Astrophysiques" (MV) is also acknowledged.

References

1. R.H. Kraichnan, "Anomalous scaling of a randomly advected passive scalar," Phys. Rev. Lett. **72**, 1016 (1994).
2. K. Gawędzki and A. Kupiainen, "Anomalous scaling of the passive scalar," Phys. Rev. Lett. **75**, 3834 (1995).
3. M. Chertkov, G. Falkovich, I. Kolokolov, and V. Lebedev, "Normal and anomalous scaling of the fourth-order correlation function of a randomly advected passive scalar," Phys. Rev. E **2**, 4924 (1995).
4. K. Gawędzki, "Intermittency of passive advection," in Advances in Turbulence VII, edited by U. Frisch (Kluwer Academic Publishers, London, 1998).
5. S. Chen and R.H. Kraichnan, "Simulations of a Randomly Advected Passive Scalar Field," Phys. Fluids **10**, 2867 (1998).
6. R.H. Kraichnan, "Small-scale structure of a scalar field convected by turbulence," Phys. Fluids **11**, 945 (1968).
7. R.H. Kraichnan, V. Yakhot, and S. Chen, "Scaling relations for a randomly advected passive scalar," Phys. Rev. Lett. **75**, 240 (1995).
8. A.L. Fairhall, B. Galanti, V.S. L'vov, and I. Procaccia, "Direct numerical simulations of the Kraichnan model: scaling exponents and fusion rules," Phys. Rev. Lett. **79**, 4166 (1997).
9. U. Frisch and A. Wirth, "Inertial-diffusive range for a passive scalar advected by a white-in-time velocity field," Europhys. Lett. **35**, 683 (1996).
10. U. Frisch, A. Mazzino, and M. Vergassola, "Intermittency in passive scalar advection," Phys. Rev. Lett. **80**, 5532 (1998).
11. O. Gat, I. Procaccia, and R. Zeitak, "Anomalous scaling in passive scalar advection: Monte Carlo Lagrangian trajectories," Phys. Rev. Lett. **80**, 5536 (1998).

12. R.H. Kraichnan, "Lagrangian-history closure approximation for turbulence," Phys. Fluids **8**, 575 (1965).
13. U. Frisch, *Turbulence. The Legacy of A. N. Kolmogorov* (Cambridge Univ. Press, Cambridge, 1995).
14. A. Brissaud and U. Frisch, "Solving linear stochastic differential equations," J. Math. Phys. **15**, 524 (1974).
15. U. Frisch and A. Wirth, "Intermittency of passive scalars in delta-correlated flow: introduction to recent work," in Proceedings of *Turbulence Modeling and Vortex Dynamics*, (Istanbul, Turkey, 2–6 September, 1996). *Springer Lect. Notes Phys.* **491**, 53, edited by O. Boratav, A. Eden and A. Erzan (Springer, Berlin, 1997).
16. D. Bernard, K. Gawędzki, and A. Kupiainen, "Slow modes in passive advection," J. Stat. Phys. **90**, 519 (1998).
17. M. Chertkov, "Instanton for random advection," Phys. Rev. E **55**, 2722 (1997).
18. G.K. Batchelor, "Small scale variation of convected quantities like temperature in turbulent fluid," J. Fluid Mech. **5**, 113 (1959).
19. R.H. Kraichnan, "Convection of a passive scalar by a quasi-uniform random straining field," J. Fluid Mech. **64**, 737 (1974).
20. B.I. Shraiman and E.D. Siggia, "Lagrangian path integrals and fluctuations in random flow," Phys. Rev. E **49**, 2912 (1994).
21. M. Chertkov, G. Falkovich, I. Kolokolov, and V. Lebedev, "Statistics of a passive scalar advected by a large-scale two-dimensional velocity field: analytic solution," Phys. Rev. E **51**, 5609 (1995).
22. L.F. Richardson, *Collected Papers,* vol. 1, edited by P.G. Drazin, (Cambridge University Press, 1993).
23. E. Hopf, "Statistical hydrodynamics and functional calculus," J. Ratl. Mech. Anal. **1**, 87 (1952).
24. P.E. Kloeden and E. Platen, *Numerical Solution of Stochastic Differential Equations* (Springer, Berlin, 1992).
25. K. Gawędzki and A. Kupiainen, "Universality in Turbulence: an Exactly Soluble Model," in *Lecture Notes in Physics* **469**, 71, edited by H. Grosse and L. Pittner (Springer, Berlin, 1996).
26. A. Ralston and P. Rabinowitz, *A First Course in Numerical Analysis* (McGraw–Hill, New York, 1978).
27. W. Feller, *An Introduction to Probability Theory and its Applications*, Vol. 1, (J. Wiley & Sons, New York, 1950).
28. U. Frisch, A. Mazzino, and M. Vergassola, "Lagrangian dynamics and high-order moments intermittency in passive scalar advection," Phys. Chem. Earth (in press).
29. A. Pumir, B.I. Shraiman and E.D. Siggia, "Perturbation theory for the δ-correlated model of passive scalar advection near the Batchelor limit," Phys. Rev. E **55**, R1263, (1997).
30. L.T. Adzhemyan, N.V. Antonov, and A.N. Vasil'ev, "Renormalization group, operator product expansion, and anomalous scaling in a model of advected passive scalar," Phys. Rev. E **58**, 1823 (1998).
31. R.H. Kraichnan, private communication (1998).
32. B.I. Shraiman and E.D. Siggia, "Anomalous scaling of a passive scalar in turbulent flow," C. R. Acad. Sci. Paris, série II **321**, 279 (1995).
33. U. Frisch, M. Lesieur, and P.-L. Sulem, "On crossover dimensions for fully developed turbulence," Phys. Rev. Lett. **37**, 895 (1976); U. Frisch and J.-D. Fournier, "d-dimensional turbulence," Phys. Rev. A **17**, 747 (1978).
34. V.I. Belinicher, V.S. L'vov, A. Pomyalov, and I. Procaccia, "Computing the scaling exponents in fluid turbulence from first principles: demonstration of multiscaling," J. Stat. Phys. **93**, 797 (1998). (see also chao-dyn/9708004.)

ANOMALOUS SCALING IN PASSIVE SCALAR ADVECTION AND LAGRANGIAN SHAPE DYNAMICS

ITAI ARAD AND ITAMAR PROCACCIA
Department of Chemical Physics, The Weizmann Institute of Science, Rehovot 76100, Israel

Abstract. The problem of anomalous scaling in passive scalar advection, especially with δ-correlated velocity field (the Kraichnan model) has attracted a lot of interest since the exponents can be computed analytically in certain limiting cases. In this paper we focus, rather than on the evaluation of the exponents, on elucidating the *physical mechanism* responsible for the anomaly. We show that the anomalous exponents ζ_n stem from the Lagrangian dynamics of shapes which characterize configurations of n points in space. Using the shape-to-shape transition probability, we define an operator whose eigenvalues determine the anomalous exponents for all n, in all the sectors of the SO(3) symmetry group.

1. Introduction

In the lecture at the IUTAM symposium the work of our group on the consequences of anisotropy on the universal statistics of turbulence has been reviewed. This material is available in print, and the interested reader can find it in [1, 2, 3, 4, 5, 6]. A short review is available in the proceedings of "Dynamics Days Asia" [7]. In this paper we review some recent work aimed at understanding the *physical mechanism* responsible for the anomalous exponents that characterize the statistics of passive scalars advected by turbulent velocity fields. We will consider isotropic advecting velocity fields, but will allow anisotropy in the forcing of the passive scalar. In such case the statistical objects like structure functions and correlation functions are not isotropic. Instead, they are composed of an isotropic and non-isotropic parts. We overcome this complication by characterizing these functions in terms of the SO(3) irreducible representations. Any such function can be

T. Kambe et al. (eds.), IUTAM Symposium on Geometry and Statistics of Turbulence, 175–184.
© 2001 *Kluwer Academic Publishers.*

written as a linear combination of parts which belong to a given irreducible representation of SO(3). We will show that each part is characterized by a set of *universal* scaling exponents. The weight of each part however will turn out to be non-universal, set by the boundary conditions.

The SO(3) classification will appear to be natural once we focus on the physics of Lagrangian trajectories in the flow. We will see that one can offer a satisfactory understanding of the physics of anomalous scaling by connecting the the statistics of the passive scalar to the Lagrangian trajectories. This connection provides a very clear understanding of the physical origin of the anomalous exponents, relating them to the dynamics in the space of *shapes* of groups of Lagrangian particles.

2. The Kraichnan Model of Passive Scalar Advection

The model of passive scalar advection with rapidly decorrelating velocity field was introduced by R.H. Kraichnan [8] already in 1968. In recent years [9, 10, 11, 12, 13, 14] it was shown to be a fruitful case model for understanding multiscaling in the statistical description of turbulent fields. The basic dynamical equation in this model is for a scalar field $T(\boldsymbol{r},t)$ advected by a random velocity field $\boldsymbol{u}(\boldsymbol{r},t)$:

$$[\partial_t - \kappa_0 \nabla^2 + \boldsymbol{u}(\boldsymbol{r},t)\cdot\nabla]T(\boldsymbol{r},t) = f(\boldsymbol{r},t)\ . \tag{1}$$

In this equation $f(\boldsymbol{r},t)$ is the forcing and κ_0 is the molecular diffusivity. In Kraichnan's model the advecting field $\boldsymbol{u}(\boldsymbol{r},t)$ as well as the forcing field $f(\boldsymbol{r},t)$ are taken to be Gaussian, time and space homogeneous, and delta-correlated in time:

$$\Big\langle (u^\alpha(\boldsymbol{r},t) - u^\alpha(\boldsymbol{r}',t))(u^\beta(\boldsymbol{r},t') - u^\beta(\boldsymbol{r}',t'))\Big\rangle_{\boldsymbol{u}} = h^{\alpha\beta}(\boldsymbol{r}-\boldsymbol{r}')\delta(t-t')\ , \tag{2}$$

where the "eddy-diffusivity" tensor $h^{\alpha\beta}(\boldsymbol{r})$ is defined by

$$h^{\alpha\beta}(\boldsymbol{r}) = \left(\frac{r}{\Lambda}\right)^{\xi}\left(\delta^{\alpha\beta} - \frac{\xi}{d-1+\xi}\frac{r^\alpha r^\beta}{r^2}\right)\ , \quad \eta \ll r \ll \Lambda\ . \tag{3}$$

Here η and Λ are the inner and outer scale for the velocity fields, and the coefficients are chosen such that $\partial_\alpha h^{\alpha\beta} = 0$. The averaging $\langle \ldots \rangle_{\boldsymbol{u}}$ is done with respect to the realizations of the velocity field.

The forcing f is also taken white in time and Gaussian:

$$\langle f(\boldsymbol{r},t) f(\boldsymbol{r}',t')\rangle_f = \Xi(\boldsymbol{r}-\boldsymbol{r}')\delta(t-t')\ . \tag{4}$$

Here the average is done with respect to realizations of the forcing. The forcing is taken to act only on the large scales, of the order of L (with a compact support in Fourier space). This means that the function $\Xi(r)$ is nearly constant for $r \ll L$ but is decaying rapidly for $r > L$.

From the point of view of the statistical theory one is interested mostly in the scaling exponents characterizing the structure functions

$$S_{2n}(\boldsymbol{r}_1, \boldsymbol{r}_2) = \left\langle (T(\boldsymbol{r}_1, t) - T(\boldsymbol{r}_2, t))^{2n} \right\rangle_{\boldsymbol{u},f} . \tag{5}$$

For isotropic forcing one expects S_{2n} to depend only on the distance $R \equiv |\boldsymbol{r}_1 - \boldsymbol{r}_2|$ such that in the scaling regime

$$S_{2n}(R) \propto R^{\zeta_{2n}} \propto [S_2(R)]^n \left(\frac{L}{R}\right)^{\delta_{2n}} . \tag{6}$$

In this equation we introduced the "normal" $(n\zeta_2)$ and the anomalous (δ_{2n}) parts of the scaling exponents $\zeta_{2n} = n\zeta_2 - \delta_{2n}$. The first part can be obtained from dimensional considerations, but the anomalous part cannot be guessed from simple arguments.

When the forcing is anisotropic, the structure functions depend on the vector distance $\boldsymbol{R} = \boldsymbol{r}_1 - \boldsymbol{r}_2$. In this case we can represent them in terms of spherical harmonics,

$$S_{2n}(\boldsymbol{R}) = \sum_{\ell,m} a_{\ell,m}(R) Y_{\ell,m}(\hat{\boldsymbol{R}}) , \tag{7}$$

where $\hat{\boldsymbol{R}} \equiv \boldsymbol{R}/R$. This is a case in which the statistical object is a scalar function of one vector, and the appropriate irreducible representation of the SO(3) symmetry group are obvious. We are going to explain in the next section that the coefficients $a_{\ell,m}(R)$ are expected to scale with a universal leading scaling exponent $\zeta_{2n}^{(\ell)}$. The exponent will turn out to be ℓ dependent but not m dependent.

Theoretically it is natural to consider correlation functions rather than structure functions. The $2n$-order correlation functions are defined as

$$F_{2n}(\boldsymbol{r}_1, \ldots, \boldsymbol{r}_{2n}) \equiv \langle T(\boldsymbol{r}_1) T(\boldsymbol{r}_2) \ldots T(\boldsymbol{r}_{2n}) \rangle_{\boldsymbol{u},f} . \tag{8}$$

For separations $r_{ij} \to 0$ the correlation functions converges to $\langle T^{2n} \rangle$, whereas for $r_{ij} \to L$ decorrelation leads to convergence to $\langle T \rangle_{\boldsymbol{u},f}^n$. For all $r_{ij} \approx O(r) \ll L$ one expects a behaviour according to

$$F_{2n}(\boldsymbol{r}_1, \ldots, \boldsymbol{r}_{2n}) = L^{n(2-\xi)}(c_0 + \cdots + c_k (r/L)^{\zeta_{2n}} \tilde{F}_{2n}(\tilde{\boldsymbol{r}}_1, \ldots, \tilde{\boldsymbol{r}}_{2n}) + \cdots) , \tag{9}$$

where $\tilde{F}_{2n}$ is a scaling function depending on $\tilde{\boldsymbol{r}}_i$ which denote a set of dimensionless coordinates describing the configuration of the $2n$ points.

The exponents and scaling functions are expected to be universal, but not the c coefficients, which depend on the details of forcing.

It has been shown [11] that the anomalous exponents ζ_{2n} can be obtained by solving for the zero modes of the exact differential equations which are satisfied by F_{2n}. The equations for the zero modes read

$$[-\kappa\sum_i \nabla_i^2 + \hat{\mathcal{B}}_{2n}]\mathcal{F}_{2n}(\mathbf{r}_1, \mathbf{r}_2, ..., \mathbf{r}_{2n}) = 0 \ . \tag{10}$$

The operator $\hat{\mathcal{B}}_{2n} \equiv \sum_{i>j}^{2n} \hat{\mathcal{B}}_{ij}$, and $\hat{\mathcal{B}}_{ij}$ are defined by

$$\hat{\mathcal{B}}_{ij} \equiv \hat{\mathcal{B}}(\boldsymbol{r}_i, \boldsymbol{r}_j) = h^{\alpha\beta}(\boldsymbol{r}_i - \boldsymbol{r}_j)\partial^2/\partial r_i^\alpha \partial r_j^\beta \ . \tag{11}$$

3. Lagrangian Trajectories, Correlation Functions and Shape Dynamics

An elegant approach to the correlation functions is furnished by Lagrangian dynamics [7, 15, 16, 17, 18]. In this formalism one recognizes that the actual value of the scalar at position $\boldsymbol{r}$ at time t is determined by the action of the forcing along the Lagrangian trajectory from $t = -\infty$ to t:

$$T(\boldsymbol{r}_0, t_0) = \int_{-\infty}^{t_0} dt\, \langle f(\boldsymbol{r}(t), t)\rangle_{\boldsymbol{\eta}} \ , \tag{12}$$

with the trajectory $\boldsymbol{r}(t)$ obeying

$$\begin{aligned} \boldsymbol{r}(t_0) &= \boldsymbol{r}_0 \ , \\ \partial_t \boldsymbol{r}(t) &= \boldsymbol{u}(\boldsymbol{r}(t), t) + \sqrt{2\kappa}\boldsymbol{\eta}(t) \ , \end{aligned} \tag{13}$$

and $\boldsymbol{\eta}$ is a vector of zero-mean independent Gaussian white random variables, $\left\langle \eta^\alpha(t)\eta^\beta(t')\right\rangle = \delta^{\alpha\beta}\delta(t-t')$. With this in mind, we can rewrite F_{2n} by substituting each factor of $T(\boldsymbol{r}_i)$ by its representation (12). Performing the averages over the random forces, we end up with

$$F_{2n}(\boldsymbol{r}_1, \ldots, \boldsymbol{r}_{2n}, t_0) = \left\langle \int_{-\infty}^{t_0} dt_1 \cdots dt_n \Big[\Xi(\boldsymbol{r}_1(t_1) - \boldsymbol{r}_2(t_1)) \cdots \right. \tag{14}$$

$$\left. \times \Xi(\boldsymbol{r}_{2n-1}(t_n) - \boldsymbol{r}_{2n}(t_n)) + \text{permutations}\Big] \right\rangle_{\boldsymbol{u}, \{\boldsymbol{\eta}_i\}} , \tag{15}$$

To understand the averaging procedure recall that each of the trajectories $\boldsymbol{r}_i$ obeys an equation of the form (13), where $\boldsymbol{u}$ as well as $\{\boldsymbol{\eta}_i\}_{i=1}^{2n}$ are independent stochastic variables whose correlations are given above. Alternatively, we refer the reader to section II of [18], where the above analysis is carried out in detail.

In considering Lagrangian trajectories of *groups* of particles, we should note that every initial configuration is characterized by a *center of mass*, say $\boldsymbol{R}$, a *scale* s (say the radius of gyration of the cluster of particles) and a *shape* $\boldsymbol{Z}$. In "shape" we mean here all the degrees of freedom other than the scale and $\boldsymbol{R}$: as many angles as are needed to fully determine a shape, in addition to the Euler angles that fix the shape orientation with respect to a chosen frame of coordinates. Thus a group of $2n$ positions $\{\boldsymbol{r}_i\}$ will be sometime denoted below as $\{\boldsymbol{R}, s, \boldsymbol{Z}\}$.

One component in the evolution of an initial configuration is a rescaling of all the distances which increase on the average like t^{1/ζ_2}; this rescaling is analogous to Richardson diffusion. The exponent ζ_2 which determines the scale increase is also the characteristic exponent of the second order structure function [8]. This has been related to the exponent ξ of (3) according to $\zeta_2 = 2 - \xi$. After factoring out this overall expansion we are left with a normalized 'shape'. It is the evolution of this shape that determines the anomalous exponents.

Consider a final shape $\boldsymbol{Z}_0$ with an overall scale s_0 which is realized at $t = 0$. This shape has evolved during negative times. We fix a scale $s > s_0$ and examine the shape when the configuration reaches the scale s for the last time before reaching the scale s_0. Since the trajectories are random the shape $\boldsymbol{Z}$ which is realized at this time is taken from a distribution $\gamma(\boldsymbol{Z}; \boldsymbol{Z}_0, s \to s_0)$. As long as the advecting velocity field is scale invariant, this distribution can depend only on the ratio s/s_0.
Next, we use the shape-to-shape transition probability to define an operator $\hat{\gamma}(s/s_0)$ on the space of functions $\Psi(\boldsymbol{Z})$ according to

$$[\hat{\gamma}(s/s_0)\Psi](Z_0) = \int d\boldsymbol{Z}\gamma(\boldsymbol{Z}; \boldsymbol{Z}_0, s \to s_0)\Psi(\boldsymbol{Z}) \qquad (16)$$

We will be interested in the eigenfunction and eigenvalues of this operator. This operator has two important properties. First, for an isotropic statistics of the velocity field the operator is isotropic. This means that this operator commutes with all rotation operators on the space of functions $\Psi(\boldsymbol{Z})$. In other words, if $\mathcal{O}_\Lambda$ is the rotation operator that takes the function $\Psi(\boldsymbol{Z})$ to the new function $\Psi(\Lambda^{-1}\boldsymbol{Z})$, then

$$\mathcal{O}_\Lambda\hat{\gamma} = \hat{\gamma}\mathcal{O}_\Lambda \ . \qquad (17)$$

This property follows from the obvious symmetry of the Kernel $\gamma(\boldsymbol{Z}; \boldsymbol{Z}_0, s \to s_0)$ to rotating $\boldsymbol{Z}$ and $\boldsymbol{Z}_0$ simultaneously. Accordingly the eigenfunctions of $\hat{\gamma}$ can be classified according to the irreducible representations of SO(3) symmetry group. We will denote these eigenfunctions as $B_{q\ell m}(\boldsymbol{Z})$. Here $\ell = 0, 1, 2, \ldots$, $m = -\ell, -\ell + 1, \ldots \ell$ and q stands for a running index if there is more than one representation with the same ℓ, m. The fact that the $B_{q\ell m}(\boldsymbol{Z})$ are classified according to the irreducible representations of SO(3) in manifested in the action of the rotation operators upon them:

$$\mathcal{O}_\Lambda B_{q\ell m} = \sum_{m'} D^{(\ell)}_{m'm}(\Lambda) B_{q\ell m'} \tag{18}$$

where $D^{(\ell)}_{m'm}(\Lambda)$ is the SO(3) $\ell \times \ell$ irreducible matrix representation.

The second important property of $\hat{\gamma}$ follows from the δ-correlation in time of the velocity field. Physically this means that the future trajectories of n particles are statistically independent of their trajectories in the past. Mathematically, it implies for the kernel that

$$\gamma(\boldsymbol{Z};\boldsymbol{Z}_0, s \to s_0) = \int d\boldsymbol{Z}_1 \gamma(\boldsymbol{Z};\boldsymbol{Z}_1, s \to s_1)\gamma(\boldsymbol{Z}_1;\boldsymbol{Z}_0, s_1 \to s_0)\,, \quad s > s_1 > s_0 \tag{19}$$

and in turn, for the operator, that

$$\hat{\gamma}(s/s_0) = \hat{\gamma}(s/s_1)\hat{\gamma}(s_1/s_0)\,. \tag{20}$$

Accordingly, by a successive application of $\hat{\gamma}(s/s_0)$ to an arbitrary eigenfunction, we get that the eigenvalues of $\hat{\gamma}$ have to be of the form $\alpha_{q,\ell} = (s/s_0)^{\zeta^{(q,\ell)}_{2n}}$:

$$\left(\frac{s}{s_0}\right)^{\zeta^{(q,\ell)}_{2n}} B_{q\ell m}(\boldsymbol{Z}_0) = \int d\boldsymbol{Z} \gamma(\boldsymbol{Z};\boldsymbol{Z}_0, s \to s_0) B_{q\ell m}(\boldsymbol{Z}) \tag{21}$$

Notice that the eigenvalues are not a function of m. This follows from Schur's lemmas [19], but can be also explained from the fact that the rotation operator mixes the different m's (18): Take an eigenfunction $B_{q\ell m}(\boldsymbol{Z})$, and act on it once with the operator $\mathcal{O}_\Lambda\hat{\gamma}(s/s_0)$ and once with the operator $\hat{\gamma}(s/s_0)\mathcal{O}_\Lambda$. By virtue of (17) we should get that same result, but this is only possible if all the eigenfunctions with the same ℓ *and* the same q share the same eigenvalue.

To proceed we want to introduce into the averaging process in (15) by averaging over Lagrangian trajectories of the $2n$ particles. This will allow us to connect the shape dynamics to the statistical objects. To this aim consider any set of Lagrangian trajectories that started at $t = -\infty$ and end up at time $t = 0$ in a configuration characterized by a scale s_0 and center of mass $\boldsymbol{R}_0 = 0$. A full measure of these have evolved through the scale L or larger. Accordingly they must have passed, during their evolution from time $t = -\infty$ through a configuration of scale $s > s_0$ at least once. Denote now

$$\mu_{2n}(t, R, \boldsymbol{Z}; s \to s_0, \boldsymbol{Z}_0)\, dt d\boldsymbol{R}\, d\boldsymbol{Z} \tag{22}$$

as the probability that this set of $2n$ trajectories crossed the scale s for the last time before reaching $s_0, \boldsymbol{Z}_0$, between t and $t+dt$, with a center of mass between $\boldsymbol{R}$ and $\boldsymbol{R}+d\boldsymbol{R}$ and with a shape between $\boldsymbol{Z}$ and $\boldsymbol{Z}+d\boldsymbol{Z}$.

In terms of this probability we can rewrite Eq.(15) (displaying, for clarity, $\boldsymbol{R}_0 = 0$ and $t=0$) as

$$
\begin{aligned}
F_{2n}(\boldsymbol{R}_0 = 0, s_0, \boldsymbol{Z}_0, t=0) &= \int d\boldsymbol{Z} \int_{-\infty}^{0} dt \int d\boldsymbol{R}\, \mu_{2n}(t, R, \boldsymbol{Z}; s \to s_0, \boldsymbol{Z}_0) \\
&\times < \int_{-\infty}^{0} dt_1 \cdots dt_n \Big[\Xi(\boldsymbol{r}_1(t_1) - \boldsymbol{r}_2(t_1)) \cdots \Xi(\boldsymbol{r}_{2n-1}(t_n) - \boldsymbol{r}_{2n}(t_n)) \\
&+ \text{ perms } \Big] \Big|(s; \boldsymbol{R}, \boldsymbol{Z}, t) >_{\boldsymbol{u}, \boldsymbol{\eta}_i} . \qquad (23)
\end{aligned}
$$

The meaning of the conditional averaging is an averaging over all the realizations of the velocity field and the random $\boldsymbol{\eta}_i$ for which Lagrangian trajectories that ended up at time $t=0$ in $\boldsymbol{R} = 0, s_0, \boldsymbol{Z}_0$ passed through $\boldsymbol{R}, s, \boldsymbol{Z}$ at time t.

Next, the time integrations in Eq.(23) are split to the interval $[-\infty, t]$ and $[t, 0]$ giving rise to 2^n different contributions:

$$
\int_{-\infty}^{t} dt_1 \cdots \int_{-\infty}^{t} dt_n + \int_{t}^{0} dt_1 \int_{-\infty}^{t} dt_2 \cdots \int_{-\infty}^{t} dt_n + \ldots \qquad (24)
$$

Consider first the contribution with n integrals in the domain $[-\infty, t]$. It follows from the delta-correlation in time of the velocity field, that we can write

$$
\begin{aligned}
&< \int_{-\infty}^{t} dt_1 \cdots dt_n \Big[\Xi(\boldsymbol{r}_1(t_1) - \dot{\boldsymbol{r}}_2(t_1)) \cdots \Xi(\boldsymbol{r}_{2n-1}(t_n) - \boldsymbol{r}_{2n}(t_n)) \\
&\qquad + \text{ perms } \Big] \Big|(s; \boldsymbol{R}, \boldsymbol{Z}, t) >_{\boldsymbol{u}, \boldsymbol{\eta}_i} \\
&= < \int_{-\infty}^{t} dt_1 \cdots dt_n \Big[\Xi(\boldsymbol{r}_1(t_1) - \boldsymbol{r}_2(t_1)) \cdots \Xi(\boldsymbol{r}_{2n-1}(t_n) - \boldsymbol{r}_{2n}(t_n)) \\
&\qquad + \text{ perms } \Big] >_{\boldsymbol{u}, \boldsymbol{\eta}_i} \\
&= F_{2n}(\boldsymbol{R}, s, \boldsymbol{Z}, t) = F_{2n}(s, \boldsymbol{Z}) . \qquad (25)
\end{aligned}
$$

The last equality follows from translational invariance in space-time. Accordingly the contribution with n integrals in the domain $[-\infty, t]$ can be written as

$$
\int d\boldsymbol{Z} F_{2n}(s, \boldsymbol{Z}) \int_{-\infty}^{0} dt \int d\boldsymbol{R} \;\; \mu_{2n}(t, R, \boldsymbol{Z}; s \to s_0, \boldsymbol{Z}_0) . \qquad (26)
$$

We identify the shape-to-shape transition probability:

$$
\gamma(\boldsymbol{Z}; \boldsymbol{Z}_0, s \to s_0) = \int_{-\infty}^{0} dt \int d\boldsymbol{R} \;\; \mu_{2n}(t, R, \boldsymbol{Z}; s \to s_0, \boldsymbol{Z}_0) . \qquad (27)
$$

Finally, putting all this added wisdom back in Eq.(23) we end up with

$$F_{2n}(s_0, \boldsymbol{Z}_0) = I + \int d\boldsymbol{Z}\gamma(\boldsymbol{Z}; \boldsymbol{Z}_0, s \to s_0) F_{2n}(s, \boldsymbol{Z}) \ . \tag{28}$$

Here I represents all the contributions with one or more time integrals in the domain $[t, 0]$. The key point now is that only the term with n integrals in the domain $[-\infty, t]$ contains information about the evolution of $2n$ Lagrangian trajectories that probed the forcing scale L. Accordingly, the term denoted by I cannot contain information about the leading anomalous scaling exponent belonging to F_{2n}, but only of lower order exponents. The anomalous scaling dependence of the LHS of Eq.(28) has to cancel against the integral containing F_{2n} without the intervention of I.

Representing now

$$\begin{aligned} F_{2n}(s_0, \boldsymbol{Z}_0) &= \sum_{q\ell m} a_{q,\ell m}(s_0) B_{q\ell m}(\boldsymbol{Z}_0) \ , \\ F_{2n}(s, \boldsymbol{Z}) &= \sum_{q\ell m} a_{q,\ell m}(s) B_{q\ell m}(\boldsymbol{Z}) \ , \\ I &= \sum_{q\ell m} I_{q\ell m} B_{q\ell m}(\boldsymbol{Z}_0) \end{aligned} \tag{29}$$

and substituting on both sides of Eq.(28) and using Eq.(21) we find, due to the linear independence of the eigenfunctions $B_{q\ell m}$

$$a_{q,\ell m}(s_0) = I_{q\ell m} + \left(\frac{s}{s_0}\right)^{\zeta_{2n}^{(q,\ell)}} a_{q,\ell m}(s) \tag{30}$$

To leading order the contribution of $I_{q\ell m}$ is neglected, leading to the conclusion that *the spectrum of anomalous exponents of the correlation functions is determined by the eigenvalues of the shape-to-shape transition probability operator*. Calculations show that the leading exponent in the isotropic sector is always smaller than the leading exponents in all other sectors. This gap between the leading exponent in the isotropic sector to the rest of the exponents determines the rate of decay of anisotropy upon decreasing the scale of observation.

4. Concluding remarks

The derivation presented above has used explicitly the properties of the advecting field, in particular the δ-correlation in time. Accordingly it cannot be immediately generalized to more generic situations in which there exist time correlations. Nevertheless we find it pleasing that at least in the present case we can trace the physical origin of the exponents anomaly, and connect it to the underlying dynamics. In more generic cases the mechanisms may be more complicated, but one should still keep the lesson in mind - higher order correlation functions depend on many coordinates, and these define a configuration in space. The scaling properties of such functions may very well depend on how such configurations are reached by the dynamics. Focusing on static objects like structure functions of one variable may be insufficient for the understanding of the physics of anomalous scaling.

Acknowledgments

This work has been supported in part by the German-Israeli Foundation, The European Commission under the TMR program and the Naftali and Anna Backenroth-Bronicki Fund for Research in Chaos and Complexity.

References

1. I. Arad, V.S. L'vov and I. Procaccia, "Correlation Functions in Isotropic and Anisotropic Turbulence: the Role of the Symmetry Group", Phys. Rev. E, **59**, 6753 (1999).
2. I. Arad, B. Dhruva, S. Kurien, V.S. L'vov, I. Procaccia and K.R. Sreenivasan, "The extraction of anisotropic contributions in turbulent flows", Phys. Rev. Lett., **81**, 5330 (1998).
3. I. Arad, L. Biferale, I. Mazzitelli and I. Procaccia, "Disentangling Scaling Properties in Anisotropic and Inhomogeneous Turbulence", Phys. Rev. Lett., **82**, 5040 (1999).
4. S. Kurien, V. S. L'vov, I. Procaccia and K.R. Sreenivasan. "The Scaling Structure of the Velocity Statistics in Atmospheric Boundary Layer", Phys. Rev. E, in press.
5. I. Arad, L. Biferale and I. Procaccia, "Nonperturbative Spectrum of Anomalous Scaling Exponents in the Anisotropic Sectors of Passively Advected Magnetic Fields", Phys. Rev. E, in press.
6. I. Arad, V.S. L'vov, E. Podivilov and I. Procaccia , "Anomalous Scaling in the Anisotropic Sectors of the Kraichnan Model of Passive Scalar Advection", Phys Rev. E, submitted.
7. I. Arad, V.S. L'vov and I. Procaccia, "Anomalous Scaling in Anisotropic Turbulence", Proceedings of "Dynamics Days Asia, 1999" B. Hu, Editor.
8. R. H. Kraichnan. "Small-scale structure of passive scalar advected by turbulence." Phys. Fluids, **11**, 945, (1968).
9. R. H. Kraichnan, Phys. Rev. Lett. **72** 1016 (1994).
10. V.S. L'vov, I. Procaccia and A.L. Fairhall, Phys. Rev E **50**, 4684 (1994).
11. K. Gawedzki and A. Kupiainen, "Anomalous scaling of the passive scalar", Phys. Rev. Lett., **75**, 3834 (1995).
12. M. Chertkov, G. Falkovich, I. Kolokolov and V. Lebedeev, "Normal and anomalous scaling of the fourth-order correlation function of a randomly advected passive scalar", Phys. Rev E **52** 4924 (1995)
13. A. Fairhall, O. Gat, V.S. L'vov and I. Procaccia, "Anomalous scaling in a model of passive scalar advection: Exact results",
14. D. Bernard, K. Gawedcki and A. Kupiainen. Phys. Rev. E **54**, 2624 (1996).
15. B.I. Shraiman and E.D. Siggia, "Lagrangian path integrals and fluctuations of random flows", Phys. Rev. E **49**, 2912 (1974).
16. O. Gat and R. Zeitak, Phys. Rev. E **57**,5331 (1998).

17. O. Gat, I. Procaccia and R. Zeitak, "Anomalous scaling in passive scalar advection: Monte-Carlo Lagrangian trajectories", Phys. Rev. Lett., 80, 5536 (1998)
18. U. Frisch, A. Mazzino, A. Noullez and M. Vergassola, "Lagrangian method for multiple correlations in passive scalar advection", Phys. Fluids, **11**, 2178, (1999).
19. M. Hamermesh, *Group Theory and its Applications to Physical Problems* (Addison-Wesley, Reading, Ma. 1962).

CORRELATION BETWEEN ENERGY AND TEMPERATURE DISSIPATION RATES IN TURBULENT FLOWS

R. A. ANTONIA, T. ZHOU AND G. XU
Department of Mechanical Engineering,
University of Newcastle, N.S.W., 2308, Australia

Abstract.

The correlation coefficient ρ between the locally averaged energy and temperature dissipation rates, ϵ and ϵ_θ respectively, is studied using measurements in a plane wake, decaying grid turbulence and a round jet. In the first two flows, ϵ and ϵ_θ are approximated by quantities, denoted by ϵ_{ap} and $\epsilon_{\theta ap}$ respectively, derived from measurements using a probe comprising a combination of 4 hot wires and 2 cold wires. In contrast to the wake, the magnitude of ρ in grid turbulence calculated using ϵ_{ap} and $\epsilon_{\theta ap}$ are comparable to those obtained using ϵ_{iso} and $\epsilon_{\theta iso}$, the isotropic counterparts of ϵ and ϵ_θ. In a particular flow, ρ decreases both with increasing wire separation and as the Taylor microscale Reynolds number R_λ increases. The major effect on ρ appears to be associated with R_λ although a possible dependence of ρ on the flow type cannot be discounted.

1. Introduction

By analogy to the refined similarity hypothesis (RSH) proposed by Kolmogorov (1962) and Obukhov (1962) for the turbulent velocity field, the RSH for a passive scalar (RSHP) states that the nth-order moment of the temperature increment $\delta\theta \equiv \theta(x+r) - \theta(x)$ between two points separated by a distance r may be expressed as (e.g. Korchaskin, 1970; Van Atta, 1971; Antonia and Van Atta, 1975; Stolovitzky et al., 1995; Zhu et al., 1995)

$$\langle(\delta\theta)^n\rangle \sim \langle\epsilon_{\theta_r}^{n/2}\epsilon_r^{-n/6}\rangle r^{n/3} \sim r^{\zeta_\theta(n)} \quad , \qquad (1)$$

when the separation r is in the inertial range (IR), i.e. $\eta \ll r \ll L$, where $\eta \equiv (\nu^3/\langle\epsilon\rangle)^{1/4}$ is the Kolmogorov length scale, L is the integral length

T. Kambe et al. (eds.), IUTAM Symposium on Geometry and Statistics of Turbulence, 185–190.
© 2001 *Kluwer Academic Publishers.*

scale, ν is the kinematic viscosity of the fluid, $\langle\epsilon\rangle$ is the mean energy dissipation rate, ϵ_r and ϵ_{θ_r} represent the locally averaged dissipation rates of turbulent energy and temperature variance, $\zeta_\theta(n)$ is the IR scaling exponent of $\langle(\delta\theta)^n\rangle$. Clearly, the correlation between $\epsilon_{\theta_r}^{n/2}$ and $\epsilon_r^{-n/6}$ is important to the scaling of $\langle(\delta\theta)^n\rangle$ in RSHP.

Van Atta (1971) assumed that the probability density functions (pdfs) of ϵ_r and ϵ_{θ_r} for IR scales are lognormally distributed and their joint pdf is bivariate lognormal with a correlation coefficient ρ defined as

$$\rho = \frac{\langle(\alpha - \langle\alpha\rangle)(\beta - \langle\beta\rangle)\rangle}{\sigma_\alpha \sigma_\beta} \ , \tag{2}$$

where α and β represent $\ln \epsilon_r$ and $\ln \epsilon_{\theta_r}$ respectively, σ_α and σ_β are the standard deviations of α and β. With this assumption, the scaling exponent $\zeta_\theta(n)$ is related directly to ρ (e.g. Antonia and Van Atta, 1975; Meneveau et al., 1990). The correlation between ϵ_r and ϵ_{θ_r} is therefore important in terms of quantifying the scaling of $\langle(\delta\theta)^n\rangle$, and perhaps in a more physical sense, ascertaining the interdependence between the two dissipative fields through their relation to the vortical structure of the flow. Van Atta (1971) argued that ρ should vary only weakly with r across the IR. Previous experimental results obtained by simply approximating ϵ and ϵ_θ with $\epsilon_{iso} \equiv 15\nu(\partial u/\partial x)^2$ and $\epsilon_{\theta iso} \equiv 3\kappa(\partial\theta/\partial x)^2$, where κ is the thermal diffusivity of the fluid, have yielded apparently disparate values for ρ in different flows (e.g. Antonia and Van Atta, 1975; Antonia and Chambers, 1980; Meneveau et al., 1990). However, the validity of ϵ_{iso} and $\epsilon_{\theta iso}$ as surrogates for ϵ and ϵ_θ has been questioned. The magnitude of ρ may depend on a number of factors, e.g. the Reynolds number, the separation between the hot and cold wires as well as the type of approximation used for ϵ and ϵ_θ. Meneveau et al. (1990) noted that the apparent discrepancy of ρ may be due to the influence of the hot wire on the cold wire signal; this effect may be significant in a turbulent jet because of the large turbulent intensity of this flow. However, this explanation may not be adequate, especially when the separation between the hot and cold wires is sufficient to avoid any interference from the hot wire signal. The main aim of this paper is to quantify the dependence between ϵ_r and ϵ_{θ_r}, using measurements in a plane wake, decaying grid turbulence and a round jet. The dependence of ρ on different approximations to ϵ and ϵ_θ is studied in the first two flows. In the round jet, parallel single hot and cold wires provide "isotropic" estimates for ϵ and ϵ_θ; the wires are mounted on separated traversing mechanisms. The separation between the wires is varied in order to assess its influence on ρ. The dependence of ρ on the Taylor microscale Reynolds number R_λ ($\equiv \langle u^2\rangle^{1/2}\lambda/\nu$, where $\lambda \equiv \langle u^2\rangle^{1/2}/\langle u_{1,1}^2\rangle^{1/2}$ is the Taylor microscale) is also examined in the jet ($R_\lambda = 386 - 550$).

2. Experimental Details

Three types of flows: grid turbulence, a cylinder wake and a circular jet are investigated. Grid turbulence is generated by a square mesh (M = 24.76 mm) grid, located at the entrance of the test section. A screen, located at a distance $x_\theta = 1.5M$ downstream of the grid, introduces heat into the flow. The measurements are made at $x/M = 40$. The wake is generated by an aluminium circular cylinder with a diameter $d = 6.35$ mm. The measurement location is $x/d = 240$, i.e. in the self-preserving region. Details of the probe and its performance check can be found in Zhou and Antonia (1999).

The jet which issues from an axisymmetric nozzle of diameter $d = 55$ mm, is supplied by a variable centrifugal blower. The air is heated by an electrical fan heater located at the blower entrance. The measurement location is at $x/d = 40$, where the flow is approximately self-preserving. Details of the jet facility can be found in Xu (1999).

3. Experimental Results and Discussions

3.1. DEPENDENCE OF ρ ON THE APPROXIMATIONS TO ϵ AND ϵ_θ

In the present wake and grid experiments, the true values of ϵ and ϵ_θ are not measured. The six-wire probe allows three of the nine velocity derivatives which feature in ϵ and two of the three temperature derivatives which appear in ϵ_θ to be measured. Approximations to ϵ and ϵ_θ can then be obtained using isotropy and continuity (Zhou and Antonia, 1999)

$$\epsilon_{ap} \simeq \nu \left[6 \left(\frac{\partial u}{\partial x}\right)^2 + 3 \left(\frac{\partial u}{\partial y}\right)^2 + 2 \left(\frac{\partial v}{\partial x}\right)^2 + 2 \left(\frac{\partial u}{\partial y}\right) \left(\frac{\partial v}{\partial x}\right) \right] \tag{3}$$

and

$$\epsilon_{\theta ap} \simeq \kappa \left[\left(\frac{\partial \theta}{\partial x}\right)^2 + 2 \left(\frac{\partial \theta}{\partial y}\right)^2 \right] . \tag{4}$$

The validity of Eqs. (3) and (4) has been checked extensively in grid turbulence (Zhou and Antonia, 1999). The values of ρ can then be quantified using different approximations for ϵ and ϵ_θ, for example (ϵ_{ap}, $\epsilon_{\theta ap}$) or (ϵ_{iso}, $\epsilon_{\theta iso}$). It has been shown (Zhou and Antonia, 1999) that, for grid turbulence, important differences exist between the statistics of either ϵ_{ap} and ϵ_{iso} or $\epsilon_{\theta ap}$ and $\epsilon_{\theta iso}$. Since ϵ_{ap} and $\epsilon_{\theta ap}$ represent ϵ and ϵ_θ more reliably than ϵ_{iso} and $\epsilon_{\theta iso}$, the correlation coefficient ρ based on (ϵ_{ap}, $\epsilon_{\theta ap}$) may be different from that based on (ϵ_{iso}, $\epsilon_{\theta iso}$). The values of ρ calculated using either (ϵ_{ap}, $\epsilon_{\theta ap}$) or (ϵ_{iso}, $\epsilon_{\theta iso}$) in grid and wake turbulence are shown in Figure 1 (hereafter, the asterisk represents normalization by the Kolmogorov length scale η). For grid turbulence ($R_\lambda = 52$), the values of ρ, obtained either with (ϵ_{ap}, $\epsilon_{\theta ap}$) or (ϵ_{iso}, $\epsilon_{\theta iso}$), agree quite well in the scaling range. They increase steadily and reach a maximum at the centre of this range with a magnitude of about 0.1, indicating that (ϵ_{iso}, $\epsilon_{\theta iso}$) seems to represent (ϵ, ϵ_θ) well enough for the purpose of evaluating ρ. This value is in close agreement with that reported by Mydlarski and Warhaft (1999) at $R_\lambda = 582$ for turbulence generated by an active grid; apparently suggesting that ρ is approximately independent of R_λ, at least in decreasing shearless turbulence. For forced box turbulence, the DNS data of Wang et al. (1999) indicated that ρ increases slightly with R_λ. For the plane wake ($R_\lambda = 40$), however, the magnitude of ρ, obtained using (ϵ_{ap}, $\epsilon_{\theta ap}$), is about 20% larger in the scaling range than when ($\epsilon_{iso}, \epsilon_{\theta iso}$) is used. This suggests that, for this flow, the isotropic surrogates ($\epsilon_{iso}, \epsilon_{\theta iso}$) cannot represent ($\epsilon, \epsilon_\theta$) as reliably as in the grid flow. The magnitude of ρ in the wake is about twice as large as that in the grid. The different behaviours of ($\epsilon_{ap}, \epsilon_{\theta ap}$) and ($\epsilon_{iso}, \epsilon_{\theta iso}$) in the wake and grid flows may reflect the influence of the mean shear and of the structural organization in the wake. They may also indicate that local isotropy is better satisfied in grid turbulence than in the turbulent wake.

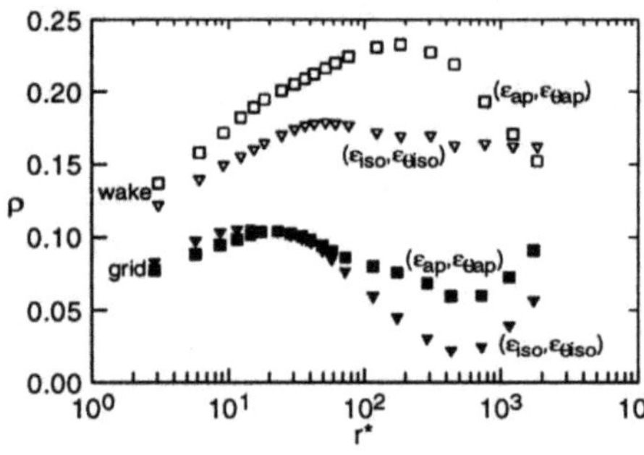

Figure 1. Comparison of estimates of ρ obtained using different approximations to ϵ and ϵ_θ. Wake: □, ($\epsilon_{ap}, \epsilon_{\theta ap}$); ∇, ($\epsilon_{iso}, \epsilon_{\theta iso}$). Grid: ■, ($\epsilon_{ap}, \epsilon_{\theta ap}$); ▼, ($\epsilon_{iso}, \epsilon_{\theta iso}$).

3.2. DEPENDENCE OF ρ ON WIRE SEPARATION S^*

Intuitively, one may expect ρ to increase as the separation between the hot and cold wires decreases. Too small a separation may result in the hot wire contaminating the cold wire signal. Values of ρ for different separations ($s^* = 2 - 11$) are presented in Figure 2a for the jet ($R_\lambda = 550$). The dependence of ρ on s^* is evident even though it is not as significant as one might have expected. For example, when s^* decreases from 11 to 2, ρ increases by only 10%. Using different combinations of single hot and cold wires, ρ was evaluated for two different separations in the two flows (Figure

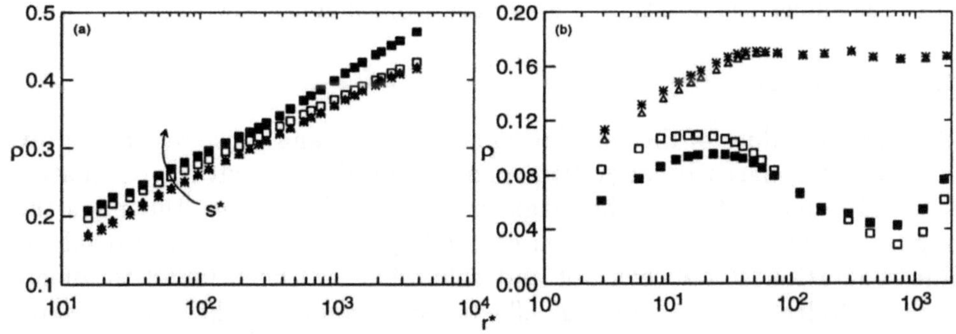

Figure 2. Dependence of ρ on the wire separation. (a) jet: ■, $s^* = 2$; □, 5.2; △, 8.6; $*$, 11. (b) grid: □, $s^* = 1$; ■, 7. wake: $*$, $s^* = 0.4$; △, 3.

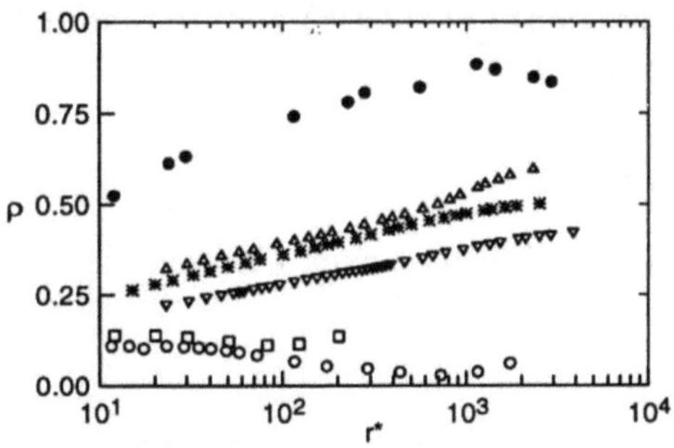

Figure 3. Dependence of ρ on R_λ in the jet and comparison with other data. Present jet: △, $R_\lambda = 386$; $*$, 410; ▽, 550. Present grid: ○, 52. Meneveau et al.: □, 160. Antonia and Van Atta: ●. 240.

2b). For grid turbulence, when s^* increases from 1 to 7, ρ decreases by about 20% across the scaling range; for the wake, ρ decreases by about 6% when s^* increases from 0.4 to 3, in qualitative agreement with the results of Figure 2a.

3.3. DEPENDENCE OF ρ ON R_λ

To gain some insight into the R_λ dependence of ρ, measurements are made on the jet axis for three values of R_λ over the range $386 \leq R_\lambda \leq 550$. The results are shown in Figure 3 together with the values from Antonia and Van Atta (1975) for $R_\lambda = 240$ in a co-flowing jet. The magnitude of ρ increases approximately linearly with $\ln r^*$ in the scaling range, but decreases significantly with the increase of R_λ. The influence of R_λ on ρ is much larger than that of s^* (Figure 2). The smaller values of ρ in the grid flow ($R_\lambda = 52$) than in the wake ($R_\lambda = 160$) of Meneveau et al. (1990) and in the jet flows suggest that ρ may vary from flow to flow.

4. Conclusions

The present values of the correlation coefficient ρ and those reported previously show apparently large discrepancies. While different approximations to ϵ and ϵ_θ may influence the magnitude of ρ in shear flows, the magnitude of ρ is only reduced slightly when ϵ_{iso} and $\epsilon_{\theta iso}$ are used in grid turbulence. The separation s^* between hot and cold wires has been found to affect ρ slightly. Neither of the previous effects can account for the above discrepancies. However, the evidence of Figure 3 indicates that the effect of R_λ cannot be ignored. Further, the possibility that ρ can depend on the nature of the flow or initial conditions cannot be dismissed. The dependencies of ρ on R_λ and flow type indicate that, with the assumption made by Van Atta (1971), the scaling exponents of $\langle(\delta\theta)^n\rangle$ should also reflect a dependence on both R_λ and the nature of the flow.

Acknowledgements

The support of the Australian Research Council is gratefully acknowledged.

References

Antonia, R. A. and Chambers, A. J. *Boundary-layer Meteorol.*, **18**, 399 (1980).
Antonia, R. A. and Van Atta, C. W. *J. Fluid Mech.*, **67**, 273 (1975).
Kolmogorov, A. N. *J. Fluid Mech.*, **13**, 82 (1962).
Korchaskin, N. N. *Izv. Atmos. Ocean. Phys.*, **6**, 947 (1970).
Meneveau, C, Sreenivasan, K. R., Kailasnath, P. and Fan, M. *Phys. Rev. A*, **41**, 894 (1990).
Mydlarski, L. and Warhaft, Z. *J. Fluid Mech.*, **358**, 135 (1999).
Obukhov, A. M. *J. Fluid Mech.*, **13**, 77 (1962).
Stolovitzky, G., Kailasnath, P. and Sreenivasan, K. R. *J. Fluid Mech.*, **297**, 275 (1995).
Van Atta, C. W. *Phys. Fluids*, **14**, 1803 (1971).
Wang, L-P., Chen, S. and Brasseur, J. G. *J. Fluid Mech.*, **400**, 163 (1999).
Xu, G. Ph.D. thesis, The University of Newcastle (1999).
Zhou, T. and Antonia, R. A. *Phys. Fluids*, **12**, 335 (2000).
Zhu, Y., Antonia, R. A. and Hosokawa, I. *Phys. Fluids*, **7**, 1637 (1995).

SIMULATION OF SCALAR FLUCTUATION FIELD BY A NON-BUOYANT PLUME IN GRID-GENERATED TURBULENCE BY RANDOM FOURIER MODES METHOD

Y.SAKAI
Department of Mechano-Informatics and Systems, Nagoya University, Furocho, Chikusaku, Nagoya 464-8603, Japan

T.SUZUKI
Mechanical Engineering Laboratory, Agency of Industrial Science and Technology, MITI,
Namiki 1-2, Tsukuba 305-8564, Japan

AND

I.NAKAMURA
Department of Mechano-Informatics and Systems, Nagoya University, Furocho, Chikusaku, Nagoya 464-8603, Japan

1. Introduction

It is well known that the characteristics of the diffusion field of a scalar quantity like the concentration of matter and temperature in turbulent flow are remarkably different from ones in laminar flow. This fact is very important and interesting not only in fluid mechanics as a typical nature of turbulence, but also in engineering (e.g., mixing problems with chemical reactions). A lot of theoretical and experimental studies have been done in the past, and in recent years, numerical studies have been extensively made with the development of computer systems.

In this study, a new Lagrangian numerical scheme to calculate scalar fluctuation fields by a non-buoyant plume in grid-generated turbulence has been developed. The Lagrangian velocities of fluid particles are simulated by a summation of unsteady random Fourier modes[1, 2]. The previous random Fourier modes methods developed by Fung et al.[2] is only useful to simulate the non-decaying homogeneous isotropic turbulence. So, firstly we improved their random Fourier modes method to simulate the downstream decaying turbulence. Next, the two-particle backward diffusion technique[4] was applied to calculate the scalar statistics up to the second order mo-

T. Kambe et al. (eds.), IUTAM Symposium on Geometry and Statistics of Turbulence, 191–196.
© 2001 *Kluwer Academic Publishers.*

ments of scalar probability density function. The simulations of two types of scalar fluctuation fields (a point source plume of dye solution and a line source thermal plume) have been performed, and the simulation results are compared with the experimental results.

2. The outline of the model

2.1. RANDOM FOURIER MODES METHOD

The details of the random Fourier modes method are shown in references[1, 2]. Here the outline of the method and only the improved points are shown. The velocity vector $\boldsymbol{u}$ is represented as a Fourier series in the limits N,

$$\boldsymbol{u}(\boldsymbol{x},t) = \sum_{n=1}^{N}[(\boldsymbol{a}_n \times \hat{\boldsymbol{k}}_n)\cos(\boldsymbol{k}_n\cdot\boldsymbol{x} - \omega_n t) + (\boldsymbol{b}_n \times \hat{\boldsymbol{k}}_n)\sin(\boldsymbol{k}_n\cdot\boldsymbol{x} - \omega_n t)], \quad (1)$$

where $\boldsymbol{k}$ is a wavenumber vector, ω is an angular frequency. Subscript n shows a value concerned with the nth Fourier mode, $\hat{\boldsymbol{k}}$ is a unit vector which has the same direction as $\boldsymbol{k}$, $\hat{\boldsymbol{k}}_n = \boldsymbol{k}_n/|\boldsymbol{k}_n|$. $\boldsymbol{a}_n$, $\boldsymbol{b}_n$, $\boldsymbol{k}_n$ are chosen independently with each other, and these vectors are also independent upon the position $\boldsymbol{x}$, then $\boldsymbol{u}$ is satisfied with the continuity condition automatically. $\boldsymbol{k}_n$ are decided so that their directions are distributed isotropically. $\boldsymbol{a}_n$ and $\boldsymbol{b}_n$ are three-dimensional random vectors which have zero average and have the distribution of the mean square value expressed as

$$\overline{|\boldsymbol{a}_n|^2} = \overline{|\boldsymbol{b}_n|^2} = 3\int_{k_n}^{k_{n+1}} E(k)dk, \quad (2)$$

where $k = |\boldsymbol{k}|$. The overlines show the ensemble average. $E(k)$ is an arbitrarily prescribed three-dimensional energy spectrum. The angular frequencies ω_n in Eq.(1) determine the unsteadiness associated with the nth Fourier mode, which is modeled here by $\omega_n = k_n\cdot\nu_n$, where

$$\nu_n^2 = \frac{2}{3}\int_{k_n}^{k_{n+1}} E(k)dk. \quad (3)$$

The present distribution of wave numbers k_n is on algebraic progression (Vassilicos et al.[3])

$$k_n = k_1 + \frac{n(n-1)}{2}\cdot\frac{2\,(k_N - k_1)}{N\,(N-1)}. \quad (4)$$

With regard to the energy spectrum, we chose one for the low Reynolds number turbulence proposed by Kraichnan [1]

$$E(k) = 16(2/\pi)^{\frac{1}{2}}u'^2k^4k_0^{-5}\exp(-2k^2/k_0^2), \quad (5)$$

where u' is a r.m.s. value of one component of velocity fluctuation $\boldsymbol{u}$, k_0 is a wave number that the spectrum shows a peak, which is given by $k_0 = \sqrt{2\pi}/L_E$, and L_E is the integral length scale of the Eulerian velocity field.

The grid-generated turbulence is not homogeneous, and the intensity of the turbulence decreases and the length scale increase in the downstream direction from the grid. So, in order to simulate the downstream decaying turbulence like the grid-generated turbulence, we need incorporate the spatial variation of the turbulence intensity and length scale into the model. Fortunately, the Kraichnan's energy spectrum(Eq.(5)) contains these parameters in the simple form. By using this energy spectrum, the downstream decaying turbulence can be easily modeled by adjusting only two parameters , i.e. u' and L_E.

The positions of particles $\boldsymbol{x}(t)$ are calculated by the following equations,

$$\boldsymbol{x}(t+\Delta t) = \boldsymbol{x}(t) + \boldsymbol{v}(\boldsymbol{x}(t), t)\Delta t + \sqrt{2\kappa\Delta t}\chi, \tag{6}$$

where κ is a moleculer diffusion coefficient, χ is a normalized white noise vector. We should note that the influence of molecule diffusion is considered in the last term of Eq.(6).

2.2. TWO-PARTICLE MODEL

The statistics of the scalar field are calculated by tracking the fluid particles that transport the scalar quantity. The mean scalar value Γ at the position $\boldsymbol{x}$ and time t is calculated by the following equation,

$$\Gamma(\boldsymbol{x}, t) = \int P_1(\boldsymbol{x}', 0; \boldsymbol{x}, t) s(\boldsymbol{x}') d\boldsymbol{x}', \tag{7}$$

where P_1 is the transition probability density function[4], $s(\boldsymbol{x})$ is a function of the scalar source distribution. $P_1(\boldsymbol{x}', t'; \boldsymbol{x}, t)$ means the probability density that the particle reaches $\boldsymbol{x}'$ at time t' on the condition that it existed in the position $\boldsymbol{x}$ at time t. We also consider another transition probability density function $P_2(\boldsymbol{x}'^{(1)}, \boldsymbol{x}'^{(2)}, t'; \boldsymbol{x}^{(1)}, \boldsymbol{x}^{(2)}, t)$ for the two particles, where the superscripts (1) and (2) show the quantity concerned with particles 1 and 2. P_2 is the probability density that the particles arrive at the position $\boldsymbol{x}^{(1)'}$ and $\boldsymbol{x}^{(2)'}$ at time t' on the condition that these existed at the position $\boldsymbol{x}^{(1)}$ and $\boldsymbol{x}^{(2)}$ at time t. The mean square value of the scalar fluctuation $\langle\gamma^2\rangle$ is similarly given by the following equation with P_2,

$$\langle\gamma^2(\boldsymbol{x}, t)\rangle = \int P_2(\boldsymbol{x}'^{(1)}, \boldsymbol{x}'^{(2)}, 0; \boldsymbol{x}, \boldsymbol{x}, t) s(\boldsymbol{x}'^{(1)}) s(\boldsymbol{x}'^{(2)}) d\boldsymbol{x}'^{(1)} d\boldsymbol{x}'^{(2)}. \tag{8}$$

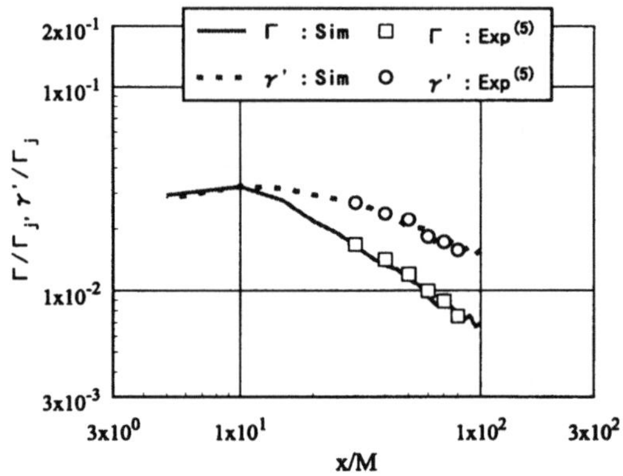

Figure 1. Downstream variation of Γ and γ' on plume centerline for the continuous point source.

3. Simulation results

As the targets of the simulation, two kinds of scalar diffusion fields are chosen, i.e. one is the concentration field of dye solution from the continuous point source in grid-generated turbulence by Nakamura et al.[5], and another is the thermal diffusion field from the heated wire in grid-generated turbulence by Stapountzis et al.[6].

In both case, the number of simulation trials is 10000. The turbulent velocity fluctuation, integral length scale and time scale which are necessary to make the simulation are determined on the basis of the experimental data in references[5, 6]

Figure 1 presents the downstream variations of the mean concentration Γ and r.m.s. value γ' of the concentration fluctuation on the plume centerline in case of the continuous point source. Γ and γ' are normalized by the injection concentration Γ_j. The abscissa shows the distance x from the source, and it is normalized by the grid size M. In the figure, "Sim" shows the simulation results by the model suggested in this study, and "Exp" shows the experimental results by Nakamura et al.[5]. The simulation results of the both Γ and γ' give good agreements with experimental data.

Figure 2 and Figure 3 show the radial distributions of the mean concentration Γ and the r.m.s. value of the concentration fluctuation γ' in case of the continuous point source, respectively. In these figures, the abscissa represents the distance from the centerline z, and it is normalized by the half width b_Γ of the mean concentration distribution. The ordinate is normalized by the value on the plume centerline $\Gamma(0)$ or $\gamma'(0)$. For the mean concentration, the simulation results obey well the Gaussian curve $\Gamma(z)/\Gamma(0) = \exp(-(\ln 2)\xi^2)$, $\xi = z/b_\Gamma$ which is shown as a "Gaussian" in

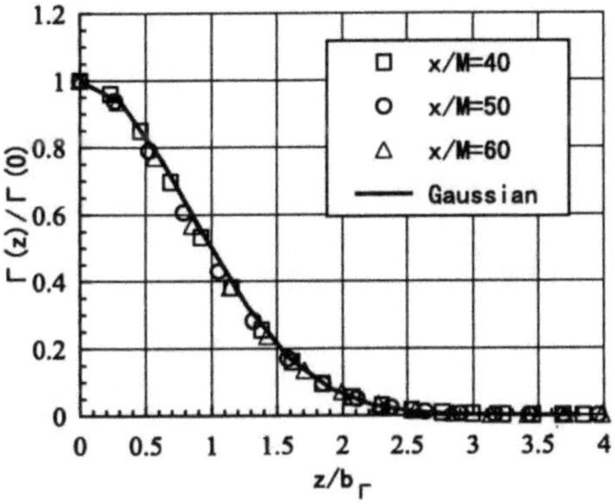

Figure 2. Normalized mean concentration against cross-stream distance for the continuous point source.

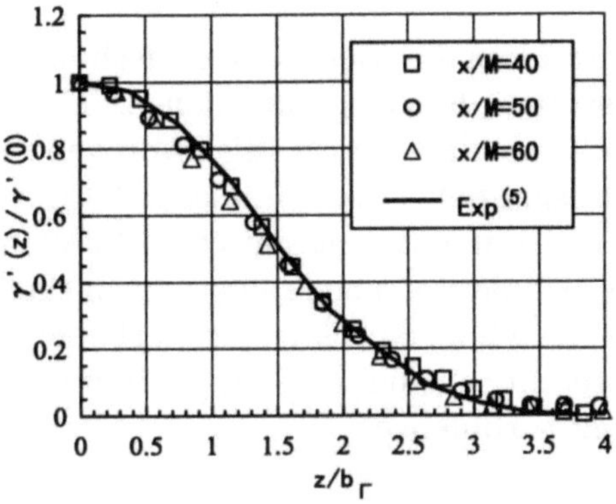

Figure 3. Normalized r.m.s. value of the concentration fluctuation against cross-stream distance for the continuous point source.

the figure. This Gaussian curve is known to be a good approximation of the distribution of experimental data. With regard to the r.m.s. value of concentration fluctuation, the simulated distribution agrees well with the experimental results shown as "Exp".

With regard to the line source thermal plume, it was found that the model predictions of Γ and γ' are consistent with the experimental results. However, in the near region of the source, we observed some deviation between the simulation and experimental results of γ' (Figure 4). This deviation may be caused by the "meandering effect" but the modeling of this meandering effect is a difficult task at present, so we remains this work in the future.

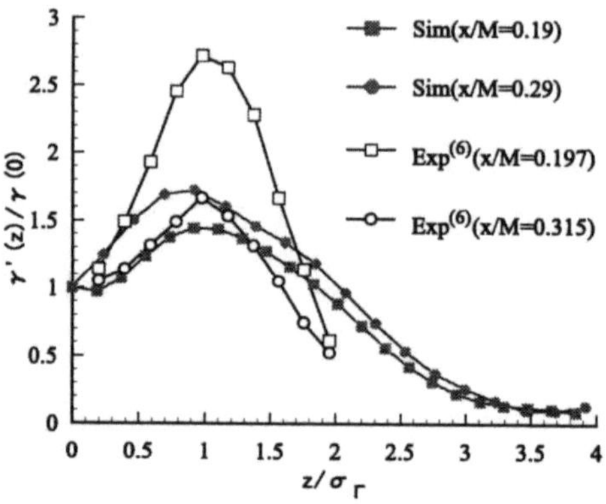

Figure 4. Radial profile of the r.m.s. value of the temperature fluctuation for the line source thermal plume.

4. Concluding remarks

In this research, a random Fourier modes method has been improved to simulate the downstream decaying turbulence. The Lagrangian velocities of fluid particles ware calculated with this improved random Fourier modes method, and the diffusion fields were simulated by the two-particle diffusion model in cases of a continuous point source and a continuous line source in the grid-gernerated turbulence. The predictions of the mean and fluctuation r.m.s. values are on the whole consistent with the experimental data. So we conclude that the present scheme is useful to simulate the scalar fluctuation fields in the decaying turbulence.

References

1. Kraichnan, R.H. (1970) Diffusion by a random velocity field, *Phys. Fluids* **13**, 22–31.
2. Fung, J.C.H., Hunt, J.C.R., Malik, N.A. and Perkins, R.J. (1992) Kinematic simulation of homogeneous turbulence by unsteady random fourier modes *J. Fluid Mech.* **236**, 281-318.
3. Vassilicos, J.C. and Fung, J.C.H. (1995) The self-similar topology of passive interfaces advected by two-dimensional turbulent-like flows *Phys. Fluids* **7**, 1970-1998.
4. Durbin, P.A. (1980) A stochastic model of two-particle dispersion and concentration fluctuations in homogeneous turbulence, *J. Fluid Mech.* **100**, 279–302.
5. Nakamura, I., Sakai, Y. and Miyata, M. (1987) Diffusion of matter by a non-buoyant plume in grid-generated turbulence, *J. Fluid Mech.* **178**, 379–403.
6. Stapountzis, H., Sawford, B.L., Hunt, J.C.R. and Britter, R.E. (1991) Structure of the temperature field downwind of a line source in grid turbulence, *J. Fluid Mech.* **165**, 401-424.

E

Vortices, Vorticity, and Strain Dynamics

SPATIOTEMPORAL STRUCTURE OF TUBULAR VORTICES IN TURBULENCE

S. KIDA, H. MIURA
Theory and Computer Simulation Research Center,
National Institute for Fusion Science,
Oroshi-cho 322-6, Gifu 509-5292, Japan

AND

T. ADACHI
Faculty of Science and Engineering,
Saga University, 1 Honjo-machi, Saga 840-8502, Japan

Abstract.
Vortical structure and its dynamics in turbulence are investigated by the use of the low-pressure vortex. The double spiral structure of cross-axial vorticity is formed around a tubular vortex on which the axial vorticity is dominant. Vortex axes are being re-linked by reconnections either of tip-type or T-type. An automatic tracking scheme of vortex axes is successfully applied to a turbulent field.

1. Introduction

It is widely recognized that long-lived tubular vortices commonly exist in many kinds of turbulence. They are characterized not only by strong axial vorticity along their axes but also by double spiral layers of cross-axial vorticity wrapping around them. These spiral layers may play significant roles in turbulence dynamics such as enhancement of mixing, diffusion and drag, activation of turbulent motion itself. There are two dynamical mechanisms of generation of spiral layers. One is the Kelvin-Helmholtz instability through which a vortex layer rolls up into numbers of thin tubular vortices. The original vortex layer forms into double spiral structures around tubular vortices. Vorticity in the layers is parallel to the tube. A strong

T. Kambe et al. (eds.), IUTAM Symposium on Geometry and Statistics of Turbulence, 199–204.
© 2001 *Kluwer Academic Publishers.*

tubular vortex thus generated tends to wrap and stretch the surrounding background vorticity into a double spiral [2]. This is the second mechanism of spiral formation, in which vorticity in spiral layers is perpendicular to the tube.

It is not easy to characterize the time-dependent three-dimensional vortical structures. Actually there have been proposed so far a lot of representation methods by emphasizing different aspects of vortical structures. The low-pressure vortex we have recently introduced picks up tubular swirling vortices irrespective of their intensity and define their central axes and core regions [3-5]. The axes of low-pressure vortices can be used for a concise representation of the flow structures.

In this paper we investigate numerical turbulence by the use of the low-pressure vortex. The spiral structure around a low-pressure vortex is analyzed in §2. A couple of typical reconnection mechanisms are taken up in §3. A new automatic tracking scheme of vortex axes is described in §4. Section 5 is devoted to concluding remarks.

2. Spiral Structure around a Vortex Axis

The numerical turbulence we employ in this section has evolved from a random velocity field of relatively large scales with isotropic energy spectrum $E(k) \propto (k/k_0)^4 \exp[-2(k/k_0)^2]$, where $k_0 = 4$. The flow is periodic in the three orthogonal directions with period 2π, and the simulation is performed by the Fourier spectral method with 256^3 modes. The temporal evolution of enstropy is shown in figure 1. It takes a single peak at $t \approx 195$. The micro-scale Reynolds number at $t = 200$ is 106.

Initially, vorticity-concentrated blobs are scattered due to statistical fluctuation of vorticity, but the intensity is relatively low. As time progresses, these blobs are distorted and elongated more and more into layers with intensified vorticity, some of which subsequently roll up into many

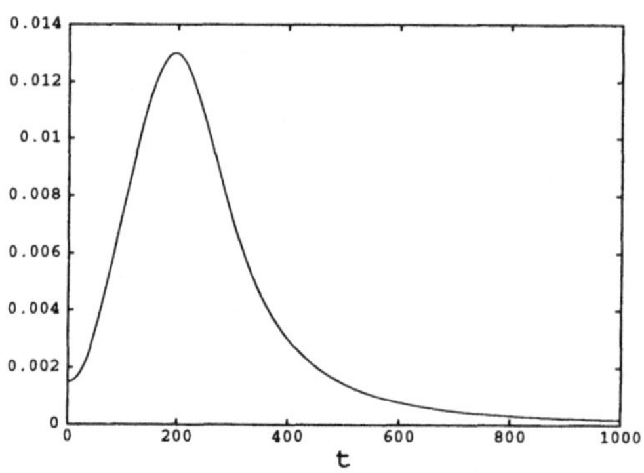

Fig.1 Temporal evolution of enstrophy.

tubular vortices [4]. In general they are accompanied with spiral vortical structure. In figure 2, we draw the vorticity distribution on a cross-section of a low-pressure vortex chosen arbitrarily at $t = 200$ (see figure (d)). Figures (a), (b), and (c) show contours of the magnitude, the axial component, and

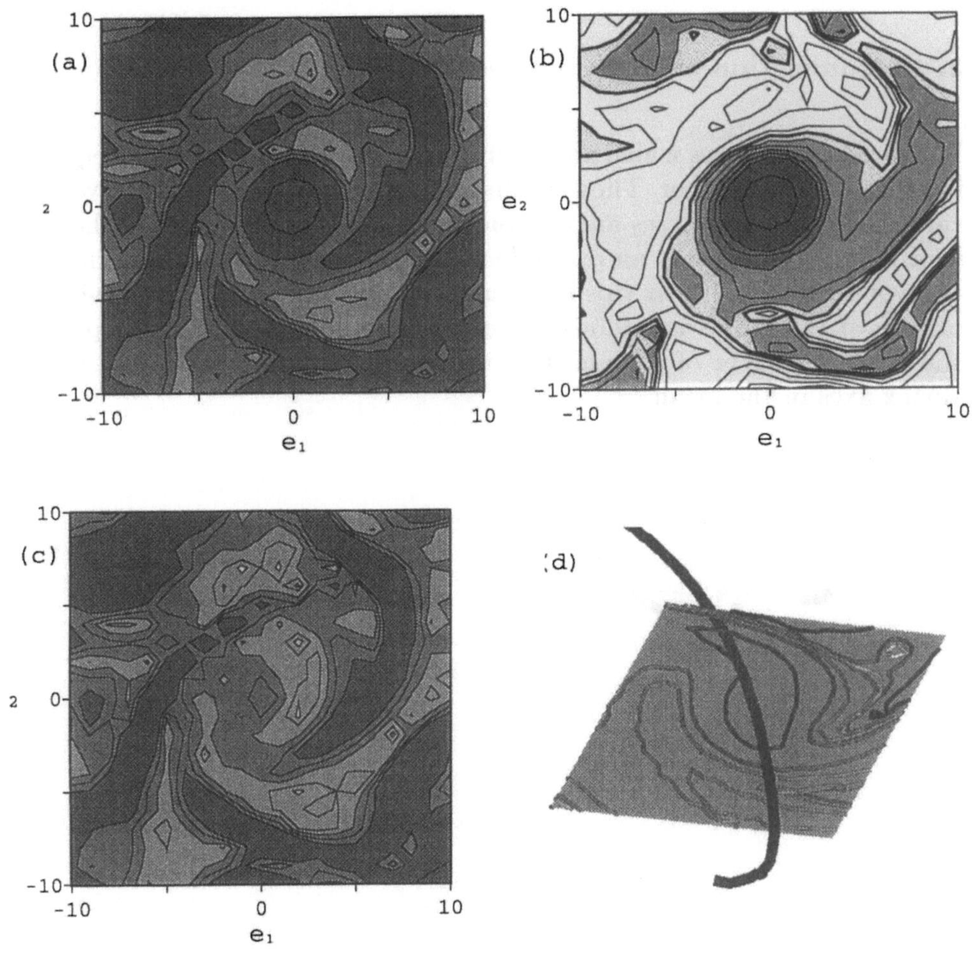

Fig.2 Spiral vortical structure around a low-pressure vortex. Contours of (a) magnitude, (b) axial component, and (c) cross-axial component of vorticity in a cross-section shown in (d). Darker shade implies larger values in (a) and (c). Vorticity is pointed into (or out of) the paper in gray (or white) area in (b). The coordinates of the cross-section, e_1 and e_2, are measured in the unit of the grid width taken in the numerical simulation.

the cross-axial component of vorticity, respectively. The vortex axis is located at the center. The axial component of vorticity is concentrated circularly on the axis, while the cross-axial component forms a double spiral wrapping around it. A similar double spiral is observed in a simple sheared turbulence as well [2].

3. Reconnection of Vortex Axes

Low-pressure vortices which are moving about in a turbulent flow frequently interact with each other. They are disconnected and reconnected. Two typical reconnection processes are conspicuous. One is of tip-type and the other is of T-type. Figure 3 is an example of the tip-type reconnection, in which two vortices link at their tips in a numerical turbulence obtained with resolution of 128^3. This is reminiscent of the relinking of vortices by pressure waves [6]. This type of reconnection contributes to increase of the length of vortex axes in the mean.

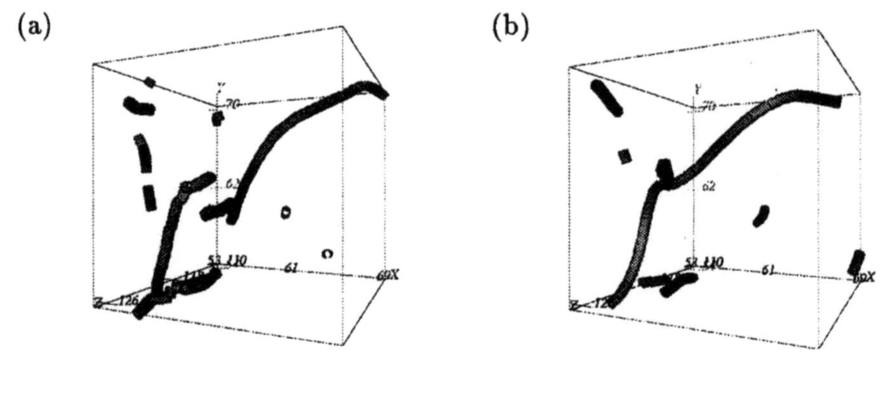

Fig.3 Tip-type reconnection of low-pressure vortices. Two vortex axes are being linked at their tips. (a) $t = 265$, (b) 290.

Another type of reconnection, T-type, is shown in figure 4. There are two long vortex axes. The one which is running from the top of the (X, Y)-plane to the left in figure (a) has reconnected at the middle with another one which is going to the right in figure (b). This T-type reconnection is observed quite a few over the whole flow field (see figure 5).

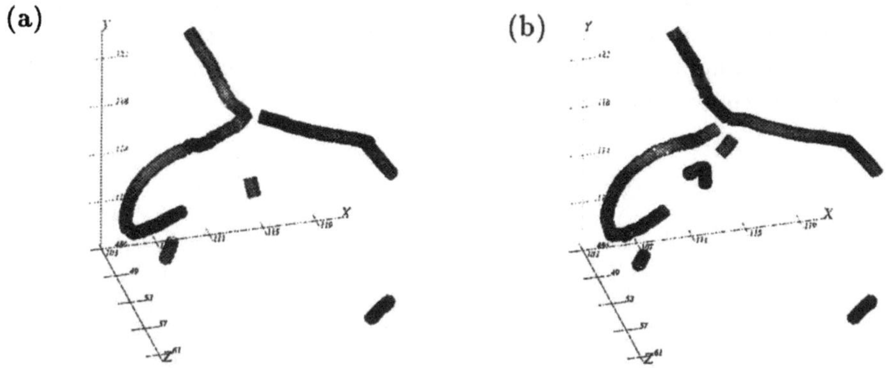

Fig.4 T-type reconnection of low-pressure vortices.
(a) $t = 602$, (b) 611.

4. Automatic Vortex Tracking

Since there are so many vortices in a turbulent flow, it is not easy to grasp their movement as a whole. Therefore it may be useful if a particular vortex can be followed automatically. Very recently we have invented an automatic tracking scheme of vortex axes and identified successfully the motion of vortex axes in a collision of two orthogonal pairs of vortex tubes [1].

Here, we apply it to a decaying isotropic turbulence of the initial micro-scale Reynolds number 34. Enstrophy decays monotonically and becomes half around $t = 2$. In figure 5, we show the temporal evolution of a particular vortex axis chosen arbitrarily at $t = 10$ (figure (a)) which is followed by the present tracking scheme. The vortex axis in question is highlighted by black thick lines. It reconnects with other vortex axes in course of time. Those portions of axes linked by reconnection are depicted by gray thick lines. Many other vortex axes are drawn with thin lines. Vortex axes are almost straight at most parts and kinked strongly at several reconnection points where other axes are attached to make Y-branches.

5. Concluding Remarks

We have shown that the axis of the low-pressure vortex is convenient to represent the vortical structure of complex flow fields. The spiral structure of cross-axial vorticity around a tubular vortex has been discovered with reference to it. The topology of vortex axes changes through the tip-type and the T-type reconnections. Their dynamical role in various turbulent transport processes as well as their own physical characteristics such as the Reynolds number dependence of shape and size of the core are interesting future subjects. The automatic tracking scheme of vortex axes is expected to be useful for these studies.

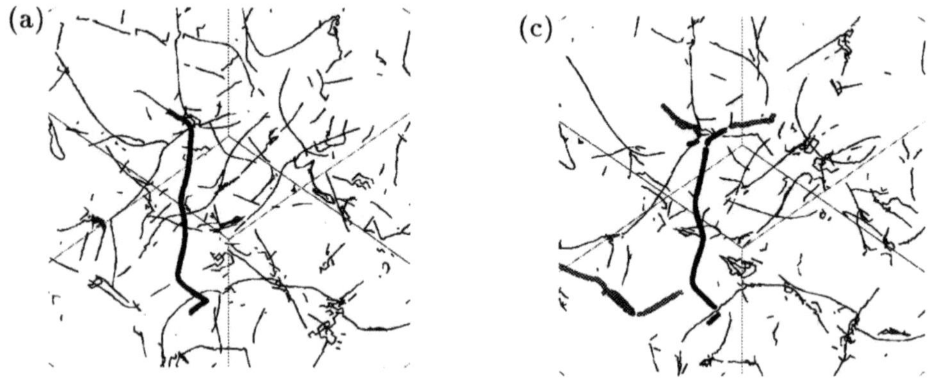

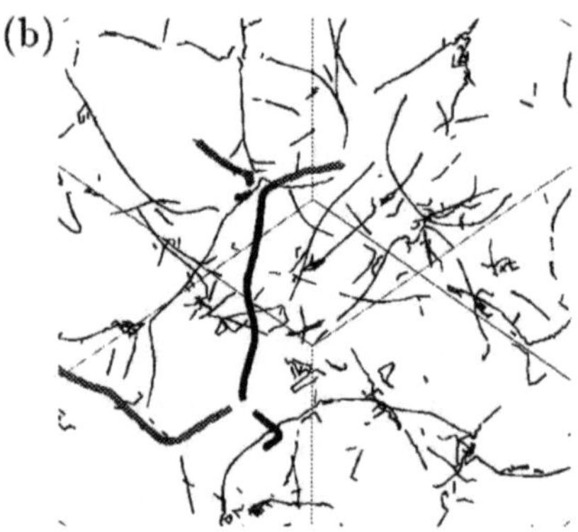

Fig.5 Vortex tracking. A vortex axis chosen arbitrarily in figure (a) is followed by the automatic tracking scheme. Thick lines denote the union of the chosen vortex axis (shown by black) and those linked with it (shown by gray). There are many other vortex axes drawn with thin lines. (a) $t = 10$, (b) 12, (c) 14.

References

1. T. Adachi, H. Miura and S. Kida: Automatic chase of low-pressure vortex axes. *J. Plasma Fusion Res. SERIES* (1998) (in press).
2. G. Kawahara, S. Kida, M. Tanaka and S. Yanase: J. Fluid Mech. **353** (1997) 115.
3. S. Kida and H. Miura: Identification and analysis of vortical structures. *Euro. J. Mech. B/Fluids*, **17** (1998) 471-488.
4. S. Kida and H. Miura: Swirl condition in low-pressure vortices. *J. Phys. Soc. Jpn.*, **67** (1998) 2166-2169.
5. H. Miura and S. Kida: Identification of tubular vortices in turbulence. *J. Phys. Soc. Jpn.*, **66** (1997) 1331–1334.
6. R. Verzicco, J. Jiménez and P. Orlandi: On steady columnar vortices under local compression. *J. Fluid Mech.*, **299** (1995) 367-388.

GEOMETRIC ASPECTS OF THE INTERACTION OF VORTICITY AND STRAIN IN HOMOGENEOUS SHEARED TURBULENCE

K.K. NOMURA
Department of Mechanical and Aerospace Engineering
University of California, San Diego
La Jolla, CA 92093, U.S.A.

1. Introduction

The interaction of the vorticity vector ω and the rate-of-strain tensor S through vortex stretching is considered a primary mechanism in the generation and associated dynamics of the small scales in three-dimensional incompressible turbulence. As described by the evolution equation for ω,

$$\frac{D\omega_i}{Dt} = S_{ik}\omega_k + \nu\frac{\partial^2\omega_i}{\partial x_k \partial x_k}, \quad (1)$$

$S_{ik}\omega_k$ represents the response of ω to S. Implicit is the coupled, nonlocal nature of the interaction which is indicated by the equation for S,

$$\frac{DS_{ij}}{Dt} = -S_{ik}S_{kj} - \frac{1}{4}(\omega_i\omega_j - \delta_{ij}\omega_k\omega_k) - \frac{1}{\rho}\Pi_{ij} + \nu\frac{\partial^2 S_{ij}}{\partial x_k \partial x_k}. \quad (2)$$

The quantity, $\Pi_{ij} = \partial^2 p/\partial x_i \partial x_j$, is the pressure Hessian and represents local and nonlocal action through the pressure field.

Experiments and simulations have revealed the presence of distinct small-scale structures and various geometric properties of ω and S. Results from direct numerical simulations (DNS) of homogeneous turbulence show that intense ω tends to concentrate into tube-like or filamentary structures (She et al., 1991; Jimenez et al., 1993; Tanaka and Kida, 1993; Kida and Tanaka, 1994). As observed by Kida and Tanaka (1994) (hereinafter referred to as KT94), in homogeneous shear flow, these structures tend to exhibit distinct spatial orientation. Preferential alignment of ω with the intermediate principal strain eigenvector, $\mathbf{e}_\beta$, is also observed (Ashurst et al., 1987; Nomura and Elghobashi, 1992). This can be explained in terms

T. Kambe et al. (eds.), IUTAM Symposium on Geometry and Statistics of Turbulence, 205–210.
© 2001 *Kluwer Academic Publishers.*

of the coupled dynamics of ω and S in the principal strain basis (Nomura and Post, 1998). In this reference frame, vortex stretching and rotation of the principal S axes, $\dot{\Omega}$, are effectively distinguished. The latter mechanism is associated with misaligned ω with respect to the S axes and includes both local and nonlocal effects. By applying the assumptions of the restricted Euler problem (denoted here as RE) which considers isotropic Π (Vieillefosse, 1984), the strain basis equations show that local (misaligned) ω rotates the S axes such that ω orients *away from* $\mathbf{e}_\alpha$ and *towards* $\mathbf{e}_\beta$ or $\mathbf{e}_\gamma$. Results from DNS of isotropic turbulence (Nomura and Post, 1998) show the significance of this mechanism in the dynamics of the flow. Nonlocally-induced $\dot{\Omega}$ through Π is found to be influenced by the proximate spatial structure. In this paper, we consider the interaction of ω and S in homogeneous sheared turbulence. DNS provide data for the analysis. Results further demonstrate the strain basis dynamics and illustrate various geometric aspects of the interaction. Comparison with linear simulations establish nonlinear aspects of the structure and dynamics.

2. Results and Discussion

Results from two DNS of homogeneous sheared turbulence are presented. These are parameterized by the initial values of the turbulent Reynolds number, $Re_\lambda = v\lambda/\nu$, and Shear number, $Sh = Sv^2/\varepsilon$, where S is the mean shear. Note, $r_\omega = Re_\lambda/Sh = \langle\omega'\rangle/S$, gives the ratio of fluctuating vorticity to mean vorticity ($\overline{\omega} = S$). In RUN1, $Re_{\lambda_o} = 24$, $Sh_\circ = 1.5$; thus $r_{\omega_o} = 16$. In RUN2, $Re_{\lambda_o} = 29$, $Sh_\circ = 16.0$; thus $r_{\omega_o} = 1.8$.

Due to the imposed mean shear, ω and the principal axes of S tend to exhibit preferential spatial orientation. Here, the S axes are denoted by $\mathbf{e}_\alpha$, $\mathbf{e}_\beta$, and $\mathbf{e}_\gamma$, corresponding to the principal eigenvalues, $\alpha \geq \beta \geq \gamma$, respectively. In general, the orientation of a vector may be described by the angle pair, $(\theta_{pitch}, \theta_{yaw})$, where θ_{pitch} is the angle of the vector from its projection in the $x-y$ (horizontal) plane and θ_{yaw} is the angle of the projection from the positive $y-$ (spanwise) axis (Fig. 1a). In the early development of the flow, as observed by KT94, amplification of ω' in the direction of the extensional strain of the mean shear, $\overline{\mathbf{e}}_\alpha$ $(\pm 45°, \pm 90°)$, prevails. The presence of $\overline{\omega}$ establishes a *predominant misalignment* of ω with respect to the S axes (Fig. 1b). This results in a locally-induced rotation of the S axes as described by the first term on the right-hand side of,

$$\dot{\Omega}_{\mathbf{e}_\alpha - \mathbf{e}_\beta} \equiv \frac{D\mathbf{e}_\alpha}{Dt} \cdot \mathbf{e}_\beta = -\frac{1}{\alpha - \beta}\Big(\underbrace{\tfrac{1}{4}\omega_\alpha\omega_\beta}_{local} + \underbrace{\widetilde{\Pi}_{\alpha,\beta}}_{nonlocal} \Big). \qquad (3)$$

If the eigenvectors of S initially coincide with those of the mean shear, i.e., $\overline{\mathbf{e}}_\alpha$, $\overline{\mathbf{e}}_\beta$, and $\overline{\mathbf{e}}_\gamma$, the induced rotation, $D\mathbf{e}_\alpha/Dt \cdot \mathbf{e}_\beta$, will reorient $\mathbf{e}_\alpha$ and $\mathbf{e}_\beta$

in the plane comprised of $\overline{\mathbf{e}}_\alpha$ and $\overline{\mathbf{e}}_\beta$. The sense of rotation is such that $\mathbf{e}_\beta$ is directed towards ω (RE dynamics). Further amplification of ω' will occur along the reoriented S axes. If these local effects are significant, ω, $\mathbf{e}_\alpha$, and $\mathbf{e}_\beta$ will occur primarily in the $\overline{\mathbf{e}}_\alpha - \overline{\mathbf{e}}_\beta$ plane. Note, this plane corresponds to a reverse "S"-shaped curve on a $(\sin(\theta_{pitch}), \theta_{yaw})$ plot (Fig. 1c). In general, the presence of ω' in other directions (isotropic initial conditions) will cause $\dot{\Omega}$ to vary and the simultaneous reorientation of all three axes may result in rather complex orientations in S and ω.

Joint probability distributions (jpd) of $(\sin\theta_{pitch}, \theta_{yaw})$ for ω and ω' (not shown) indicate spatial orientations which are similar to those observed by KT94 and consistent with the described dynamics. Figure 2 shows probable orientations for $\mathbf{e}_\alpha$. Here, we consider the total flow and three conditional samples based on the second invariant of the velocity gradient tensor, $II = (\omega^2/2 - S^2)/2$, which indicates the relative significance of rotation versus strain. In each case, the shape of the distributions generally follows the $\overline{\mathbf{e}}_\alpha - \overline{\mathbf{e}}_\beta$ curve (Fig. 1c). Since the rate of locally-induced rotation is proportional to the vorticity components (3), these effects may be particularly significant in high-amplitude, rotation-dominated $(II > 0)$ regions (Fig. 2b). However, in regions characterized by significant strain (Figs. 2c,d), strain generation acts to promote positive β which then tends to approach α. This will effectively enhance the induced rotation through the prefactor in (3) thereby retaining its significance in these regions. In regions where both ω and S are significant $(II \sim 0$; Fig. 2c), reorientation of the S axes will contribute to amplification of spanwise ω which, in turn, will affect the relative orientation of ω. The S axes tend to equilibrate in these regions as a result of this feedback (Fig. 2c). In strain-dominated $(II < 0)$ regions (Fig. 2d), the orientation of the S axes is particularly sensitive to the flow parameter r_ω. Results for flows with differing r_ω indicate that the extent of rotation of $\mathbf{e}_\alpha$ and $\mathbf{e}_\beta$ in these regions depends on the local geometry which is associated with the relative significance of $\langle\omega'\rangle$ to $\overline{\omega}$ (Fig. 1b).

In regards to the term associated with $\widetilde{\Pi}$ in (3), we consider this as a nonlocally-induced rotation of the S axes. Recall in RE, Π is assumed isotropic. In this case, the strain basis equations are autonomous and effectively describe the local dynamics of a fluid particle. The anisotropic component of Π, which contributes to $\dot{\Omega}$, then represents deviation from RE and, in this sense, is regarded as nonlocal. As shown in Fig. 3a, the nonlocally-induced rotation of the S axes through $\widetilde{\Pi}_{\alpha,\beta}$ tends to counteract locally-induced rotation, $\frac{1}{4}\omega_\alpha\omega_\beta$, in high-amplitude rotation-dominated regions $(II > 0)$. As discussed in (Nomura and Post, 1998), these regions are characterized by tube-like spatial structure as indicated by the strong alignment of ω with the eigenvector of Π, $\mathbf{f}_3$, corresponding to its smallest (nearly zero) eigenvalue. Similar structures and properties are observed in the shear flow. Fig. 3c shows the relative orientation of ω with the eigenvectors of S and Π. As indicated, there is a strong tendency for ω to orient in the direction of $\mathbf{f_3}$ for high positive II. Note, this tendency is much stronger than that associated with the alignment of ω with $\mathbf{e}_\beta$ in these regions. As in isotropic turbulence, the corresponding spatial structure of these regions is tube-like (Fig. 4a).

Simulation results of the linearized Navier-Stokes equations exhibit similar local characteristics at early time. However, they lack the distinct nonlocal features present in the nonlinear flow. In particular, characteristics of Π differ significantly. The strain basis components exhibit nearly symmetric distributions about zero. The anisotropic component $\widetilde{\Pi}_{\alpha,\beta}$ is no longer significant in rotation-dominated regions (Fig. 3b). This suggests the absence of nonlocal effects as represented by the RE problem in which isotropic Π acts to preserve the volume of a fluid element with no directional distinction. In high amplitude rotation-dominated regions in fully nonlinear flow, Π exhibits distinct directionality with respect to ω due to the presence of spatial structure. In the corresponding regions of the linear flow, such directionality does not exist (Fig. 3d). This indicates that the (nonlinear) influence of the surrounding flow is absent, i.e., there is no spatial *coherence*. Visualization of high amplitude rotation-dominated regions in the linear flow (Fig. 4b) show elongated structures which do exhibit inclination from the streamwise (x_1) direction as in the nonlinear flow (Fig. 4a). However, based on the characteristics of Π, these structures are not consistent with vortex *tubes*. It is likely that these isosurfaces represent regions of concentrated ω^2 reorganized by mean advection rather than what are considered to be *coherent* structures.

References

Ashurst, W. T., Kerstein, A. R., Kerr, R. M., and Gibson, C. H. (1987). Alignment of vorticity and scalar gradient with strain rate in simulated Navier-Stokes turbulence. *Phys. Fluids*, 30:2343–2353.

Jimenez, J., Wray, A. A., Saffman, P. G., and Rogallo, R. S. (1993). The structure of intense vorticity in isotropic turbulence. *J. Fluid Mech.*, 255:65–90.

Kida, S. and Tanaka, M. (1994). Dynamics of vortical structures in a homogeneous shear flow. *J. Fluid Mech.*, 274:43–68.

Nomura, K. K. and Elghobashi, S. E. (1992). Mixing characteristics of an inhomogeneous scalar in isotropic and homogeneous sheared turbulence. *Phys. Fluids*, 4:606–625.

Nomura, K. K. and Post, G. K. (1998). The structure and dynamics of vorticity and rate of strain in incompressible homogeneous turbulence. *J. Fluid Mech.*, 377:65–97.

She, Z.-S., Jackson, E., and Orszag, S. A. (1991). Structure and dynamics of homogeneous turbulence: models and simulations. *Proc. R. Soc. Lond. A*, 434:101–124.

Tanaka, M. and Kida, S. (1993). Characterization of vortex tubes and sheets. *Phys. Fluids*, A5:2079–2082.

Vieillefosse, P. (1984). Internal motion of a small element of fluid in an inviscid flow. *Physica*, 125A:150–162.

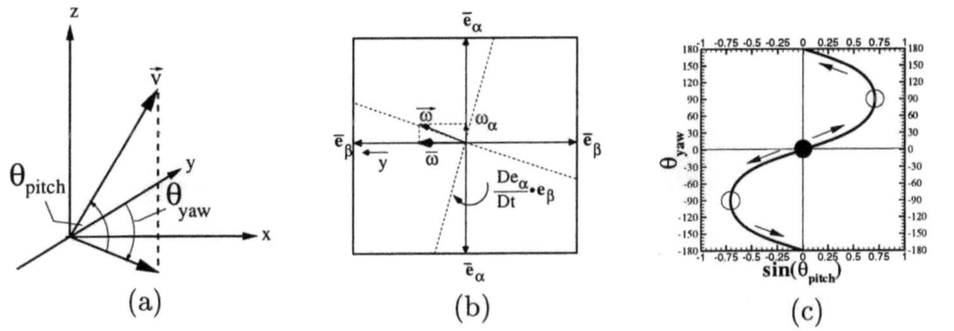

Figure 1. (a) Definition of the angles of orientation $(\theta_{pitch}, \theta_{yaw})$ of a vector in 3-D cartesian coordinates. (b) Diagram of $\overline{\mathbf{e}}_\alpha - \overline{\mathbf{e}}_\beta$ plane indicating locally-induced rotation of the principal strain axes, (c) Curve representing $\overline{\mathbf{e}}_\alpha - \overline{\mathbf{e}}_\beta$ plane on $(\sin(\theta_{pitch}), \theta_{yaw})$ plot. Note: '∘' indicates orientation of $\overline{\mathbf{e}}_\alpha$, '•' indicates orientation of $\overline{\mathbf{e}}_\beta$, and arrows indicate trajectories of principal axes.

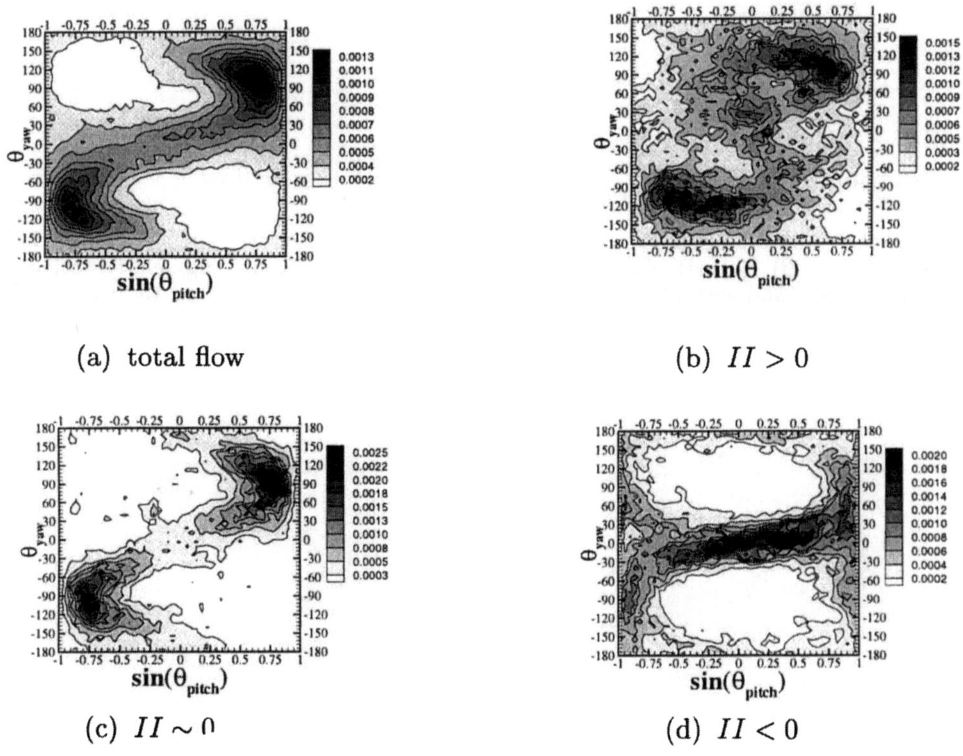

Figure 2. Joint probability distributions of $(\sin(\theta_{pitch}), \theta_{yaw})$ indicating orientation of eigenvector $\mathbf{e}_\alpha$ for (a) total flow and high-amplitude (b) rotation-dominated $(II > 0)$, (c) comparable rotation-strain $(II \sim 0, \omega^2 \geq 2\langle\omega^2\rangle)$, (d) strain-dominated $(II < 0)$ conditional samples (RUN1, $St = 6$).

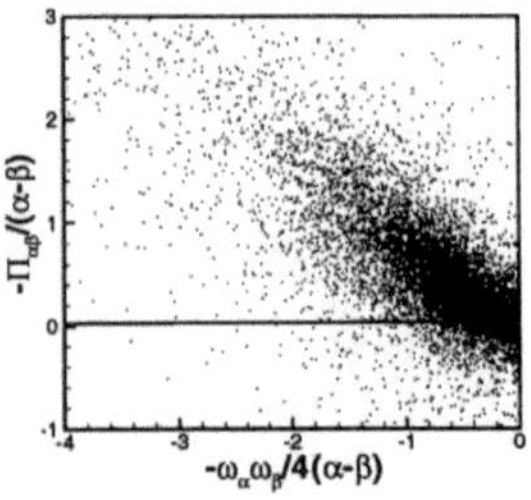

(a) nonlinear flow

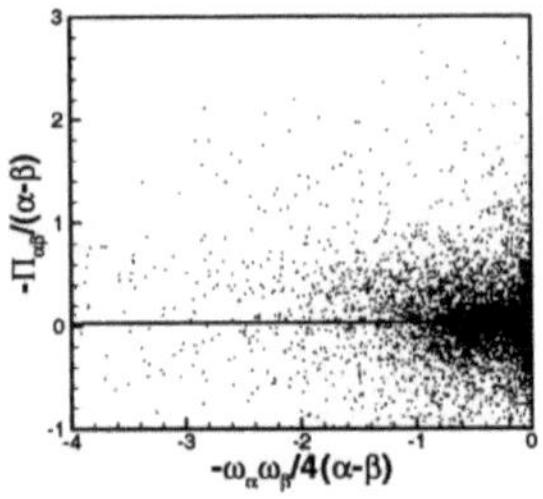

(b) linear flow

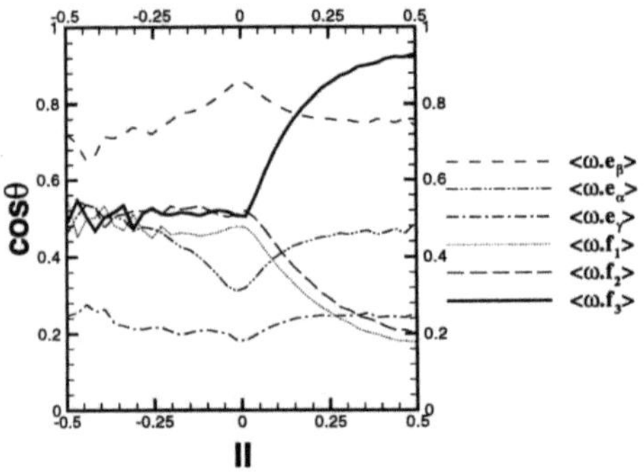

(c) nonlinear flow

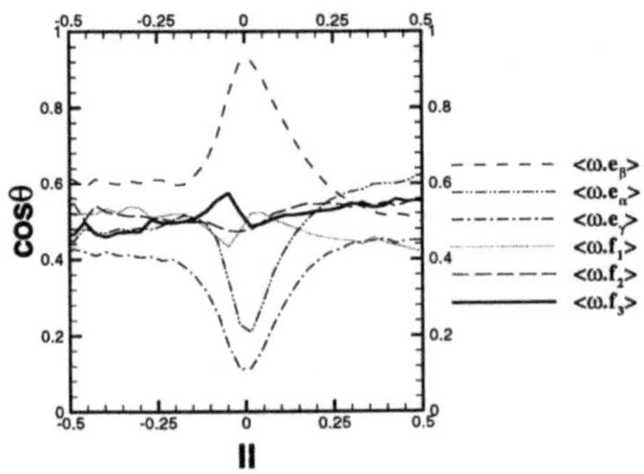

(d) linear flow

Figure 3. (a),(b) Scatter plots showing correlation between $-\widetilde{\Pi}_{\alpha\beta}/(\alpha-\beta)$ and $-\omega_\alpha\omega_\beta/4(\alpha-\beta)$, (c)-(d) Conditional expectation of $\cos\theta$ indicating relative orientation between ω and each of the eigenvectors of S and Π (conditioned on the value of II with $III < 0$) (a),(c) nonlinear and (b),(d) linearized shear flow (RUN2 - $St = 6.0$).

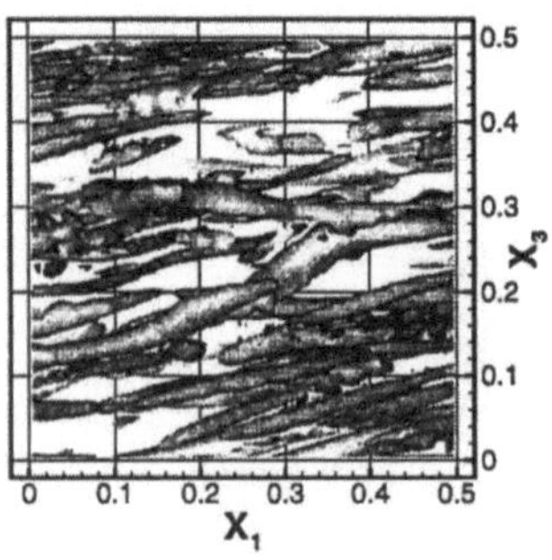

(a) nonlinear flow

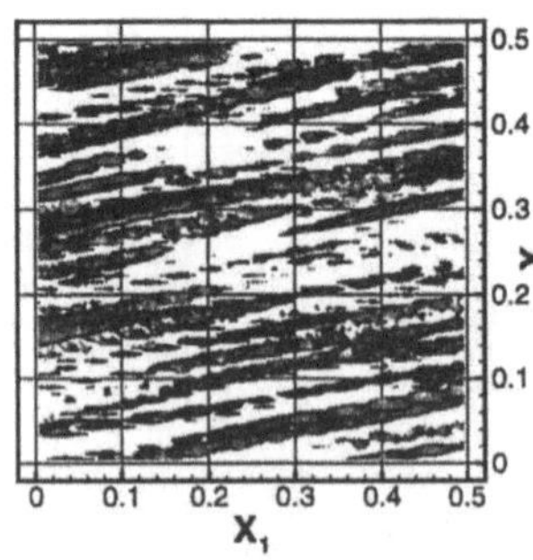

(b) linear flow

Figure 4. Three-dimensional visualizations (64^3 subdomain of computational grid) of high-amplitude rotation-dominated ($II > 0$) regions in (a) nonlinear and (b) linearized shear flow (RUN2 - $St = 6.0$) (streamwise (x_1) and vertical (x_3) directions shown).

MOTION OF A CURVED VORTEX FILAMENT: HIGHER-ORDER ASYMPTOTICS

YASUHIDE FUKUMOTO
Graduate School of Mathematics,
Kyushu University 33, Fukuoka 812–8581, Japan

Abstract. Three-dimensional motion of a slender vortex tube, embedded in an inviscid incompressible fluid, is investigated based on the Euler equations. Using the method of matched asymptotic expansions in a small parameter ϵ, the ratio of core radius to curvature radius, the velocity of a vortex filament is derived to $O(\epsilon^3)$, whereby the influence of elliptical deformation of the core due to the self-induced strain is taken into account. In the *localized induction approximation*, this is reducible to a completely integrable evolution equation among the *localized induction hierarchy.*

1. Introduction

The traveling speed U for a thin axisymmetric vortex ring in an inviscid incompressible fluid of infinite extent is known, to third (virtually fourth) order in a small parameter $\epsilon = \sigma/R_0$, the ratio of core radius σ to the ring radius R_0, as

$$U = \frac{\Gamma}{4\pi R_0}\left\{\log\left(\frac{8}{\epsilon}\right) - \frac{1}{4} - \frac{3\epsilon^2}{8}\left[\log\left(\frac{8}{\epsilon}\right) - \frac{5}{4}\right] + O(\epsilon^4 \log\epsilon)\right\}. \qquad (1)$$

Here Γ is the circulation carried by the ring and a particular distribution of vorticity is assumed (Dyson 1893). The first two terms are Kelvin's formula which are considered as the first order. Dyson's correction takes account of an *elliptical deformation* of cross-section of the core caused by the self-induced straining field.

In practical flows, viscous diffusion of vorticity swells the core with time. Extending the Saffman's first-order formula (Saffman 1970), Fukumoto & Moffatt (1999) succeeded in constructing a formula for the third-order correction to the traveling speed of a viscous vortex ring.

T. Kambe et al. (eds.), IUTAM Symposium on Geometry and Statistics of Turbulence, 211–216.
© 2001 *Kluwer Academic Publishers.*

In contrast, it is not easy to render the motion of a curved vortex filament amenable to a systematic analysis. The simplest asymptotic theory is the so called '*localized induction approximation (LIA)*' put forward by Da Rios (1906). By limiting the domain of Biot-Savart *line*-integral to the neighboring segment of length $2L$ and, in addition, introducing a short cut-off σ to avoid the logarithmic infinity, we are led to the following evolution equation for the filament curve $\boldsymbol{X} = \boldsymbol{X}(s,t)$, expressed as functions of the arclength s and the time t:

$$\boldsymbol{X}_t = C\kappa \boldsymbol{b}\,; \qquad C = \frac{\Gamma}{4\pi}\log\left(\frac{2L}{\sigma}\right), \tag{2}$$

where κ is the curvature, $\boldsymbol{b}$ is the binormal vector, and a subscript denotes a differentiation with respect to the indicated variable.

A distinguishing feature is that (2) is a completely integrable evolution equation equivalent to a cubic nonlinear Schrödinger equation (Hasimoto 1972). Langer & Perline (1991) unveiled the bi-Hamiltonian structure of (2) behind this integrability, and manipulated a recursion operator to generate successively an infinite sequence of commuting vector fields, called the '*localized induction hierarchy (LIH)*'. The first three of them are

$$\boldsymbol{V}^{(1)} = \kappa \boldsymbol{b}\,, \qquad \boldsymbol{V}^{(2)} = \frac{1}{2}\kappa^2 \boldsymbol{t} + \kappa_s \boldsymbol{n} + \kappa\tau \boldsymbol{b}\,, \tag{3}$$

$$\boldsymbol{V}^{(3)} = \kappa^2\tau \boldsymbol{t} + (2\kappa_s\tau + \kappa\tau_s)\boldsymbol{n} + (\kappa\tau^2 - \kappa_{ss} - \frac{1}{2}\kappa^3)\boldsymbol{b}\,, \tag{4}$$

where τ is the torsion and $(\boldsymbol{t}, \boldsymbol{n}, \boldsymbol{b})$ are the Frenet-Serret vectors. Observe that, when specialized to a circle with constant curvature $\kappa \neq 0$ and $\tau = 0$, a superposition of $\boldsymbol{V}^{(1)}$ and $\boldsymbol{V}^{(3)}$ is no other than (1).

This unexpected coincidence inspires us to pursue the higher-order velocity of a vortex filament in three dimensions. Fukumoto & Miyazaki (1991) showed that a vortex filament with axial velocity in the core is governed by an evolution equation comprising a summation of (3) (cf. Moore & Saffman 1972). In the present investigation, we rule out axial flow at leading order, and make an attempt at an extension of matched asymptotic expansions to $O(\epsilon^3)$.

2. Asymptotic development of the Biot-Savart law

The velocity $\boldsymbol{v}(\boldsymbol{x})$ of the fluid at a position $\boldsymbol{x}$ is uniquely determined from the vorticity $\boldsymbol{\omega}(\boldsymbol{x})$ by the Biot-Savart law. In order to evaluate this volume integral at points near the core, it is expedient to introduce local coordinates $(\tilde{x}, \tilde{y}, \xi)$ moving with the filament. Here ξ parameterizes the central curve of the filament. Given a point $\boldsymbol{x}$ sufficiently close to the core, there corresponds uniquely the nearest point $\boldsymbol{X}(\xi,t)$ on the centerline of filament. Then $\boldsymbol{x}$ is expressed as $\boldsymbol{x} = \boldsymbol{X}(\xi,t) + \tilde{x}\boldsymbol{n}(\xi,t) + \tilde{y}\boldsymbol{b}(\xi,t)$. We conveniently use the cylindrical coordinates (r,ϕ,ξ) such that $\tilde{x} = r\cos\phi$ and $\tilde{y} = r\sin\phi$. They are rendered orthogonal by replacing ϕ with θ:

$$\theta(s,t) = \phi - \int_{s_0}^{s} \tau(s',t)\,ds'\,. \tag{5}$$

Though incomplete, we begin with the following ansatz for vorticity:

$$\boldsymbol{\omega} = \zeta(\tilde{x},\tilde{y},t)\boldsymbol{t}(\xi,t)\,. \tag{6}$$

Using the shift-operator technique, the vector potential $\boldsymbol{A}$ for velocity, with the vorticity being substituted from (6), is rewritten as

$$\begin{aligned}\boldsymbol{A}(\boldsymbol{x}) &= \frac{1}{4\pi}\iiint \zeta(\tilde{x},\tilde{y})\frac{\boldsymbol{t}(s)}{|\boldsymbol{x}-\boldsymbol{X}-\tilde{x}\boldsymbol{n}-\tilde{y}\boldsymbol{b}|}(1-\kappa\tilde{x})\,ds d\tilde{x} d\tilde{y}\\ &= \frac{1}{4\pi}\int ds\left\{\iint d\tilde{x}d\tilde{y}\zeta(\tilde{x},\tilde{y})(1-\kappa\tilde{x})\,e^{-\tilde{x}(\boldsymbol{n}\cdot\nabla)-\tilde{y}(\boldsymbol{b}\cdot\nabla)}\right\}\frac{\boldsymbol{t}(s)}{R}\,. \end{aligned}\tag{7}$$

where

$$R = |\boldsymbol{x} - \boldsymbol{X}(s)|\,. \tag{8}$$

The exponential function is formally expanded in powers of $\tilde{x}$ and $\tilde{y}$ as

$$\begin{aligned}\boldsymbol{A} &= \frac{1}{4\pi}\int ds\bigg\{\iint d\tilde{x}d\tilde{y}\,\zeta\Big(1-\kappa\tilde{x}-\tilde{x}(\boldsymbol{n}\cdot\nabla)-\tilde{y}(\boldsymbol{b}\cdot\nabla)+\frac{1}{2}[\tilde{x}^2(\boldsymbol{n}\cdot\nabla)^2\\ &+2\tilde{x}\tilde{y}(\boldsymbol{n}\cdot\nabla)(\boldsymbol{b}\cdot\nabla)+\tilde{y}^2(\boldsymbol{b}\cdot\nabla)^2]+\kappa\tilde{x}^2(\boldsymbol{n}\cdot\nabla)+\kappa\tilde{x}\tilde{y}(\boldsymbol{b}\cdot\nabla)+\cdots\Big)\bigg\}\frac{\boldsymbol{t}}{R}\,.\end{aligned}\tag{9}$$

We shall know from the inner expansion that the axial component ζ of vorticity has the following dependence on ϕ:

$$\zeta(\tilde{x},\tilde{y}) = \zeta_0(r) + \zeta_{11}(r,\xi,t)\cos\phi + \zeta_{12}(r,\xi,t)\sin\phi + \zeta_{21}(r,\xi,t)\cos 2\phi + \cdots\,, \tag{10}$$

where

$$\zeta_0 = \zeta^{(0)}(r) + \kappa^2\hat{\zeta}_0^{(2)}(r) + \cdots\,,\quad \zeta_{11} = \kappa\hat{\zeta}_{11}^{(1)}(r) + \cdots\,,\quad \zeta_{21} = \kappa^2\hat{\zeta}_{21}^{(2)}(r) + \cdots\,. \tag{11}$$

The transversal components emerge at $O(\epsilon^2)$ in the form:

$$\omega_r = \hat{\omega}_r(\kappa\tau\cos\phi + \kappa_s\cos\phi)\,,\quad \omega_\theta = \hat{\omega}_\theta(\kappa_s\cos\phi - \kappa\tau\cos\phi)\,. \tag{12}$$

Substituting (10) into (9), we get

$$\boldsymbol{A}(\boldsymbol{x}) = \boldsymbol{A}_m(\boldsymbol{x}) + \boldsymbol{A}_d(\boldsymbol{x}) + \cdots\,, \tag{13}$$

where

$$\boldsymbol{A}_m(\boldsymbol{x}) = \frac{\Gamma}{4\pi}\int \frac{\boldsymbol{t}(s)}{R}ds\,, \tag{14}$$

$$\boldsymbol{A}_d(\boldsymbol{x}) = -\frac{1}{8\pi}[\pi\int_0^\infty r^3\zeta^{(0)}dr]\int\frac{\kappa_s\boldsymbol{n}+\kappa\tau\boldsymbol{b}}{R}ds - \frac{d^{(1)}}{2}\int ds[\kappa(\boldsymbol{n}\cdot\nabla)+\kappa^2]\frac{\boldsymbol{t}}{R}, \tag{15}$$

with

$$d^{(1)} = \frac{1}{2\pi}\left\{[\pi\int_0^\infty r^2\hat{\zeta}_{11}^{(1)}dr] - \frac{1}{2}[\pi\int_0^\infty r^3\zeta^{(0)}dr]\right\}. \tag{16}$$

The second term $\boldsymbol{A}_d$ corresponds to the flow field induced by a line of 'dipoles', arranged on the centerline, with their axes in the binormal direction.

In the LIA, the dipole velocity field $\boldsymbol{v}_d = \nabla\times\boldsymbol{A}_d$ simplifies to

$$\begin{aligned}\boldsymbol{v}_d \;=\; & -\frac{d^{(1)}}{2}\left\{\frac{2\kappa}{r^2}[\sin\phi\,\boldsymbol{e}_r - \cos\phi\,\boldsymbol{e}_\theta] - \frac{\kappa^2}{r}\cos 2\phi\,\boldsymbol{e}_\theta\right.\\ & \left. +\log\left(\frac{2L}{r}\right)\left[(2\kappa_s\tau+\kappa\tau_s)\boldsymbol{n} + (\kappa\tau^2 - \kappa_{ss} - \frac{1}{4}\kappa^3)\right]\boldsymbol{b}\right\} + \cdots,\end{aligned} \tag{17}$$

where non-singular terms are omitted. The first terms stand for the dipole field. Intriguingly, the logarithmic terms are identical with the third vector field $\boldsymbol{V}^{(3)}$ of the LIH (4), except for the coefficient of κ^3. We are reminded of the fact that, for the speed of a vortex ring, the logarithmic terms at $O(\epsilon^3)$ arise also from the inner solution. This remarkable coincidence invites a further investigation of the inner solution to higher orders.

3. Inner solution

The inner solution is addressed by solving the Euler equations in the moving coordinates. We introduce the dimensionless variables; the radial distance r is normalized by σ, the core radius, time by R_0^2/Γ with R_0 being a measure of the curvature radius, relative velocity by Γ/σ and the centerline velocity by $\Gamma/R^{(0)}$. To eliminate pressure from the Euler equations, it is advantageous to handle the vorticity equation by taking the curl of the Euler equations. We introduce the relative velocity $\boldsymbol{v} = \dot{\boldsymbol{X}}(\xi,t) + u\boldsymbol{e}_r + v\boldsymbol{e}_\theta + w\boldsymbol{t}$, where a dot stands for a derivative in t with fixing ξ, and $\boldsymbol{e}_r$ and $\boldsymbol{e}_\theta$ are the unit vectors in the radial and azimuthal directions respectively.

Suppose that the leading-order flow consists only of the azimuthal component $v^{(0)}$ possessing both rotational and translational symmetry about the local central axis $\boldsymbol{t}$: $v^{(0)} = v^{(0)}(r)$. We can conveniently introduce the streamfunction ψ for the local flow in the plane transversal to the $\boldsymbol{t}$-direction.

The solution at $O(\epsilon)$, meeting the condition that the relative velocity is finite at $r = 0$, is written out as follows:

$$\psi^{(1)} = \kappa \left[\hat{\psi}_{11}^{(1)} - \frac{1}{\kappa} (\dot{\boldsymbol{X}}^{(0)} \cdot \boldsymbol{b}) \right] \cos\phi \,, \tag{18}$$

where

$$\hat{\psi}_{11}^{(1)} = v^{(0)} \left\{ \frac{r^2}{2} + \int_0^r \frac{dr'}{r'[v^{(0)}(r')]^2} \int_0^{r'} r'' \left[v^{(0)}(r'') \right]^2 dr'' \right\} + c_{11}^{(1)} v^{(0)} \,, \tag{19}$$

with $c_{11}^{(1)}$ being a constant. The matching condition then relates C in (2) to the flow structure in the core (Widnall, Bliss & Zalay 1971).

Fortunately $p^{(1)}$ is straightforwardly available:

$$p^{(1)} = \kappa \left\{ v^{(0)} \frac{\partial \hat{\psi}_{11}^{(1)}}{\partial r} - \zeta^{(0)} \hat{\psi}_{11}^{(1)} - r [v^{(0)}(r)]^2 \right\} \cos\phi \,. \tag{20}$$

The gradient of $p^{(1)}$, in turn, drives axial flow at $O(\epsilon^2)$:

$$w^{(2)} = \hat{w}_1^{(2)} (-\kappa\tau \cos\phi + \kappa_s \sin\phi) \,, \tag{21}$$

where

$$\hat{w}_1^{(2)} = Cr - r \frac{\partial \hat{\psi}_{11}^{(1)}}{\partial r} + \frac{r\zeta^{(0)}}{v^{(0)}} \hat{\psi}_{11}^{(1)} + r^2 v^{(0)} \tag{22}$$

Notice that torsion and arcwise variation of curvature is vital for the presence of pressure gradient and thus of axial velocity. The streamfunction $\psi^{(2)}$ at $O(\epsilon^2)$ for flow in the transversal plane is built in parallel with the case of a circular vortex ring.

We are now ready to deduce the *third-order* velocity. The vorticity equation in the axial direction has much in common with that for a circular vortex ring. The effect of τ and κ_s, which is missing in the latter case, makes its appearance only in the following few terms:

$$\dot{\zeta}^{(1)} + A \left(\tau^2 - \frac{\kappa_{ss}}{\kappa} \right) \frac{\partial \zeta^{(1)}}{\partial \theta} + \frac{\partial}{\partial t} \left[\int_0^{s(\xi,t)} \tau ds' \right] \frac{\partial \zeta^{(1)}}{\partial \theta} + \frac{v^{(0)}}{r} \frac{\partial \zeta^{(3)}}{\partial \theta} + u^{(3)} \frac{\partial \zeta^{(0)}}{\partial r}$$
$$+ \cdots = \frac{\zeta^{(0)}}{\eta} \frac{\partial w^{(2)}}{\partial \xi} + \cdots ; \qquad \eta = |\partial \boldsymbol{X} / \partial \xi| \,. \tag{23}$$

The third-order velocity $\dot{\boldsymbol{X}}^{(2)}$ under question is included in $\zeta^{(3)}$ and $u^{(3)}$. Relevant to the traveling speed is the terms proportional to $\cos\phi$ and $\sin\phi$. The relevant part of the matching condition is

$$\kappa\psi^{(3)} \sim \left(\frac{3}{32\pi} r^3 - \frac{d^{(1)}}{2} r \right) \log\left(\frac{2L}{\epsilon r} \right) [(\kappa_{ss} - \kappa\tau^2) \cos\phi$$
$$+ (2\kappa_s \tau + \kappa\tau_s) \sin\phi] + \cdots \quad \text{as } r \to \infty \,. \tag{24}$$

After some manipulation, we eventually arrive at

$$\frac{\partial \boldsymbol{X}}{\partial t}(s,t) = \frac{\Gamma}{4\pi}\log\left(\frac{2L}{\sigma}\right)\left\{\kappa\boldsymbol{b} + \left[\frac{\pi}{\Gamma}\int_0^\infty r^3\zeta^{(0)}dr\right]\left[(2\kappa_s\tau + \kappa\tau_s)\boldsymbol{n} + (\kappa\tau^2 - \kappa_{ss})\boldsymbol{b} + \kappa^2\tau\boldsymbol{t}\right] - \sigma^2 C_b\kappa^3\boldsymbol{b}\right\}. \quad (25)$$

Determination of the coefficients C_b requires a further task.

As an example, we consider a specific vorticity distribution at $O(\epsilon^0)$ of constant vorticity in the circular domain of unit radius surrounded by an irrotational flow, known as the Rankine vortex:

$$v^{(0)} = r/(2\pi) \text{ for } r \le 1, \quad v^{(0)} = 1/(2\pi r) \text{ for } r > 1. \quad (26)$$

With this choice, $\int_0^\infty r^3\zeta^{(0)}dr = 1/4$ and

$$C_b = \frac{3}{8} + \left(c_{11}^{(1)} - \frac{5}{8}\right). \quad (27)$$

The disposable parameter $c_{11}^{(1)}$ in (19) has bearing with the freedom of choosing the local origin $r = 0$, to $O(\epsilon)$, in the $\boldsymbol{n}$-direction (Fukumoto & Moffatt 1999). When $c_{11}^{(1)} = 5/8$, the origin $r = 0$ sits at the form center. It is remarkable that the choice of $c_{11}^{(1)} = 3/8$ reduces (25) to a summation of $\boldsymbol{V}^{(1)}$ and $\boldsymbol{V}^{(3)}$ of the LIH provided by (3) and (4). This implies that there is a line in the core obeying the integrable equation. Since the stagnation point is located at the distance $(5/8)\epsilon\sigma$, this 'integrable line' is located between the form center-line and the stagnation line.

References

Da Rios, L. S. 1906 On the motion of an unbounded fluid with a vortex filament of any shape. [in Italian] *Rend. Circ. Mat. Palermo* **22**, 117-135.

Dyson, F. W. 1893 The potential of an anchor ring. – part II. *Phil. Trans. R. Soc. Lond. A* **184**, 1041–1106.

Fukumoto, Y. & Moffatt, H. K. 1999 Motion and expansion of a viscous vortex ring. Part 1. A higher-order asymptotic formula for the velocity. submitted to *J. Fluid Mech.*

Fukumoto, Y. & Miyazaki, T. 1991 Three-dimensional distortions of a vortex filament with axial velocity. *J. Fluid Mech.* **222**, 369–416.

Hasimoto, H. 1972 A soliton on a vortex filament. *J. Fluid Mech.* **51**, 477–485.

Langer, J. & Perline, R. 1991 Poisson geometry of the filament equation. *J. Nonlinear Sci.* **1**, 71–93.

Moore, D. W. & Saffman, P. G. 1972 The motion of a vortex filament with axial flow. *Phil. Trans. R. Soc. Lond. A* **272**, 403–429.

Saffman, P. G. 1970 The velocity of viscous vortex rings. *Stud. Appl. Math.* **49**, 371–380.

Widnall, S. E., Bliss, D. B. & Zalay, A. 1971 Theoretical and experimental study of the stability of a vortex pair. In *Aircraft Wake Turbulence and its Detection* (eds. Olsen, Goldberg & Rogers), pp. 305–338, Plenum.

STATISTICS OF TURBULENCE AROUND A COLUMNAR VORTEX

TAKESHI MIYAZAKI,
HIROYUKI ISHII
AND JULIAN. C. R. HUNT*
Department of Mechanical and Control Engineering,
University of Electro-Communications, Chofu, Tokyo 182-8585, Japan
* *Department of Geological Sciences,*
University College London, Gower Street, LONDON, WC1E 6BT, UK.

1. Introduction

Interactions between intense elongated vortices and surrounding turbulent motions are often produced artificially in engineering and environmental flows, and they occur naturally in most homogeneous and sheared turbulent flows. Several experimental and computational studies have been undertaken on the instability of vortices interacting with background turbulence (Sreedhar & Ragab (1994) and Risso et al. (1995)). These interactions are thought to play an important role in the three- dimensionalization of the flow fields. It is, also, of practical interest to estimate the lifetime of trailing vortices under the influence of atmospheric turbulence.

2. Direct Numerical Simulation

Melander & Hussain (1993) conducted an idealized DNS at moderate Reynolds numbers, and showed the followings; (1) azimuthal alignment of small-scale vorticity, (2) merger and axisymmetrization of these vortex rings, (3) excitation of bending waves on the columnar vortex. In this paper, the statistics of initially isotropic homogeneous turbulence interacting with a Lamb-Oseen vortex (with circulation $\Gamma = 40r_0^2\omega_{rms} = 41.23$ and initial radius $r_0 = 0.5 = \frac{1}{3}L_{INT}$) are investigated numerically by extending the work by Melander & Hussain. Unlike their artificial initial conditions, the

T. Kambe et al. (eds.), IUTAM Symposium on Geometry and Statistics of Turbulence, 217–222.
© 2001 *Kluwer Academic Publishers.*

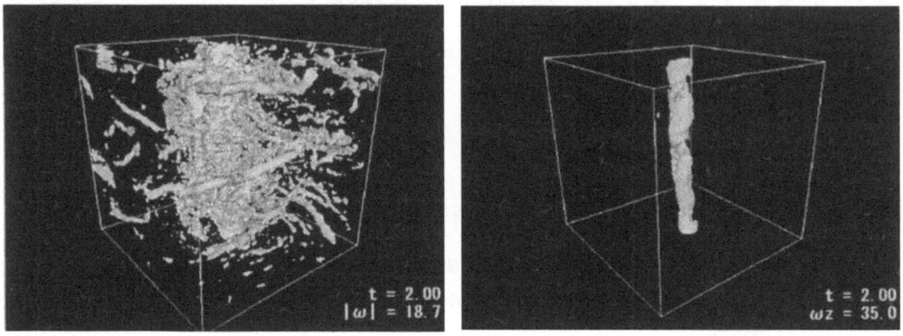

Figure 1. (a)Distorted turbulent eddies, (b)Distortion of the columnar vortex.

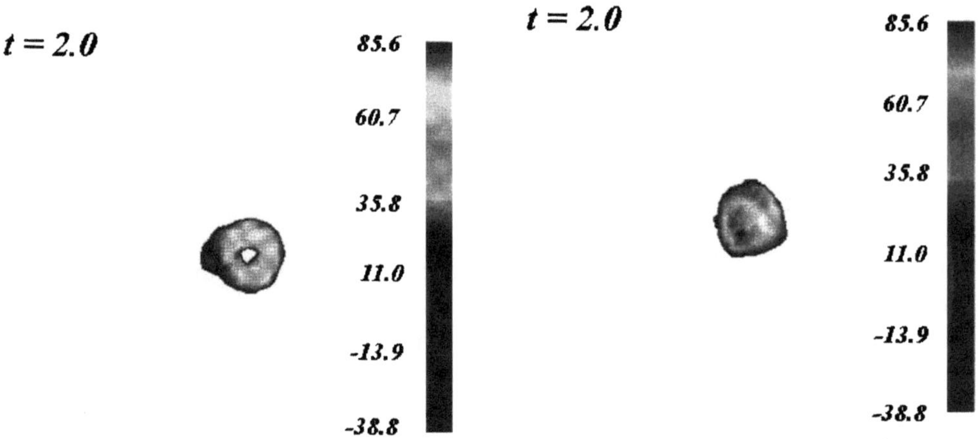

Figure 2. (a)Vorticity contours inside the vortex core at $z = \pi/2$. (b)at $z = 5\pi/6$.

initial homogeneous isotropic turbulence $Re_\lambda = 126$, itself, is produced by a direct numerical simulation of decaying turbulence. Statistical properties are computed by averaging 20 cases, and they are compared with the results of a linear RDT.

After the columnar vortex is introduced in the fully developed turbulence, the vorticity field is distorted. Intense vortices are wrapped around the vortex core (Fig.1a). This is because the differential rotation in the mean flow causes the azimuthal component of turbulent vorticity to grow algebraically, in proportion to time, while the other components do not grow.

We show, in Fig.1(b), the iso-surface of ω_z (axial vorticity). This corresponds to the surface of the columnar vortex. As time goes on, the surface is distorted and it becomes like a spiral, indicating that bending waves are excited on the vortex core.

Figures 2a,b show contours of ω_z at two horizontal cross-sections of the

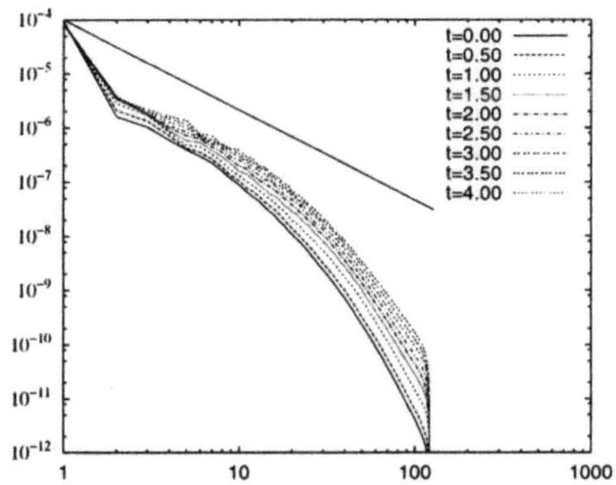

Figure 3. One-dimensional energy spectrum, $E(k_z)$.

columnar vortex at $t = 2.0$. We can see that axisymmetric waves (with "bubbles" and "hot spots" of vorticity) are excited on the vortex core.

3. Statistics of turbulence

Next, let us consider the statistical properties of the flow field. As a result of the distortion of the turbulence by the mean flow, it becomes inhomogeneous and anisotropic. The flow field, however, remains statistically homogeneous in the azimuthal and axial directions. The one-dimensional energy spectrum, where the horizontal axis is k_z, is shown in Fig.3. The energy of higher wavenumbers is piled up and an inverse energy cascade can be observed, although this tendency is weak.

The velocity correlation functions can be Fourier-decomposed in the azimuthal and axial directions, whereas the coefficients have spatial dependence on the radial coordinate, so that the two-point energy spectrum tensors $\Phi_{ij}(r_1, r_2, t; k, m)$, i.e.,

$$\langle u_i(r_1, \theta_1, z_1, t) u_j(r_2, \theta_2, z_2, t) \rangle =$$
$$\sum_{m=-\infty}^{\infty} \int_{-\infty}^{\infty} dk \Phi_{ij}(r_1, r_2, t; k, m) \exp[ik(z_1 - z_2) + im(\theta_1 - \theta_2)]$$

are computed by averaging "20 realizations"($Re_\lambda = 126$) produced numerically.

Figures 4a,b show the time-evolution of axisymmetric axial and radial components, respectively. We find two peaks in Fig.4a, i.e., at the center and at the core-surface of the Lamb-Oseen vortex. The central peak corresponds to the excitation of axisymmetric vortex waves, whereas the peak at the vortex surface indicates a blocking effect, i.e., the outer turbulent eddies

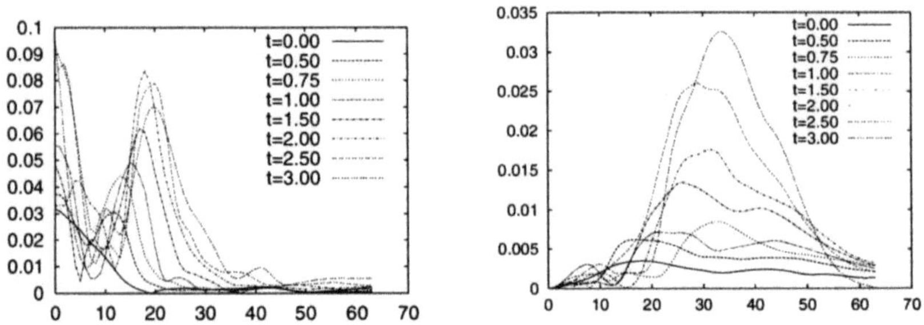

Figure 4. (a)Axisymmetric axial component, (b)Axisymmetric radial component.

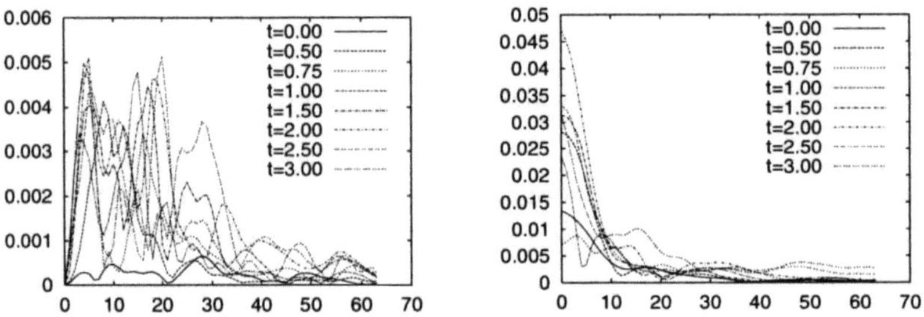

Figure 5. (a)Bending axial component, (b)Bending radial component.

are blocked by the columnar vortex and they cannot penetrate inside the core. In Fig.4b, the axisymmetric radial component grows with time, again. Comparing two figures, we find that the axial velocity dominates near the core surface ($r \approx 1$: 'blocking' effect) but the radial velocity dominates away from the vortex core ($r \approx 1.5 - 2.0$), corresponding to the formation of dipole vortex rings wrapped around the columnar vortex.

Figures 5a,b show the bending ($m = 1$) component. These are nonzero, only inside the vortex core, where the bending component oscillate with time, indicating the excitation of bending vortex waves. Higher ($m \geq 2$) components are presented only inside the core, too, and they are small compared with the axisymmetric and bending components. Thus only the axisymmetric component grows with time outside the vortex core and the velocity field of small scale eddies near the vortex becomes statistically axisymmetric, within a period of two or three revolutions of the columnar vortex.

Figure 6a shows the time-evolution of the maximum values of axisymmetric axial component (at the vortex surface) for several k_z, showing the strength of the blocking effect. They grow with time initially, but saturate after t= about 1.5. Similarly, the axisymmetric radial component grow with time, as we can see in Fig.6b. They are not saturated, although saturation may occur later. This is because the maximum values of the radial

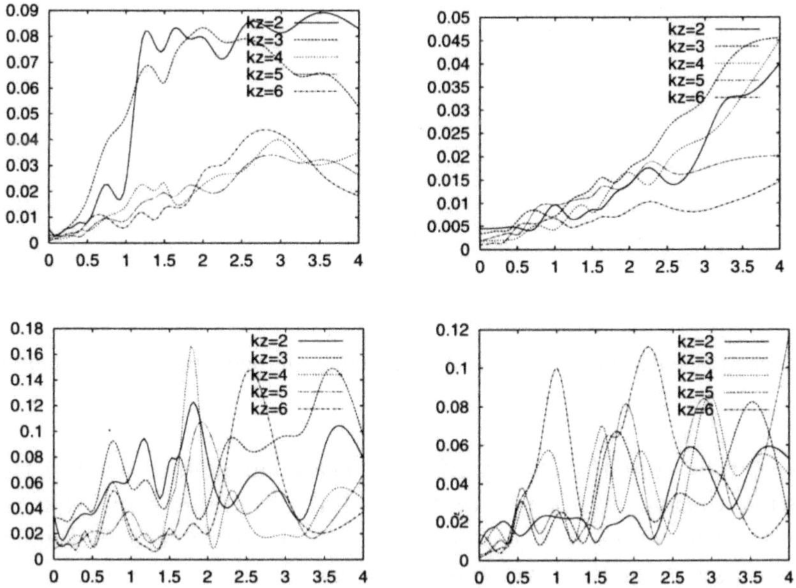

Figure 6. Peak values of (a)axisymmetric axial component, (b)axisymmetric radial component, (c)axisymmetric axial component at the center, (d)bending radial component.

component are attained away from the vortex surface, where the effect of differential rotation is weaker than at the core surface. Higher (larger k_z) components seem to be saturated faster than lower components.

In Figs.6c,d, the peak values at the vortex center of axisymmetric axial component and those of bending radial component are shown. They oscillate with time, representing the vortex axisymmetric and bending waves excited by the turbulence outside the vortex core, respectively. The amplitude of these oscillations seem to increase gradually, leading to vortex breakdown.

4. Comparison with RDT

These findings are compared with the theoretical results based on the linear rapid distortion theory (RDT: Miyazaki & Hunt (1999)). For simplicity, we consider a "solid cylinder vortex", i.e., a model vortex whose core is replaced by a solid cylinder rotating with a constant angular velocity. Then, quantitative comparisons with the results of DNS, inside the core and near the core surface, are difficult (Fig.7a shows the axial component near the core- surface). However, a fairly good agreement is found, in Fig.7b, for the axisymmetric radial component, which grows away from the vortex core surface.

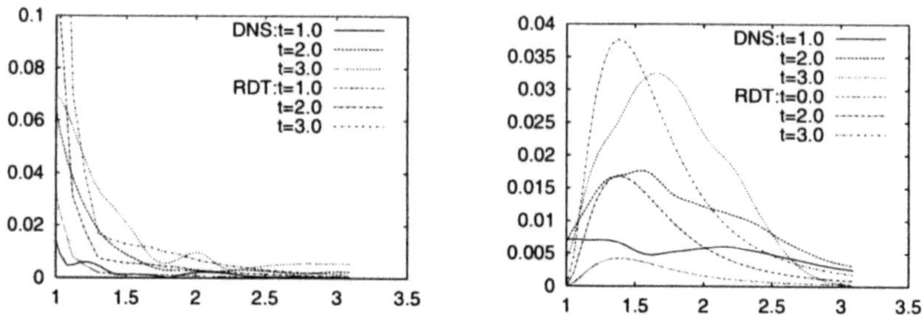

Figure 7. Comparison with RDT (a)axisymmetric axial component, (b)radial component.

RDT theory tells that the energy of small scale turbulence around the columnar vortex decays about r^{-5} in the radial direction, grows like t^2 in time, in the earlier stage of the simulation. As we have seen in Fig.6a, nonlinear effects, such as, the inverse energy cascade due to the merger of "vortex rings" and the breakdown of the columnar vortex caused by the amplification of vortex waves, are observed at later times.

5. Summary

In this paper, statistical properties of turbulence around a columnar vortex is investigated numerically. The followings are explored.

1.The intense vortices are wrapped around the columnar vortex core.

2.Bending and axisymmetric vortex waves are excited in the vortex core.

3.The turbulent eddies around the vortex core become statistically axisymmetric, within a period of two or three revolutions of the columnar vortex.

4.The outer turbulence is blocked by the columnar vortex, whereas vortex waves are excited inside the core.The axial velocity dominates near the core surface and radial velocity dominates away from the core surface.

5.The linear RDT can explain the behavior of the two-point energy spectrum tensors in the earlier stage of the simulation, but nonlinear effects, such as, the inverse energy cascade, are observed later.

References

1. Melander & Hussain (1993) *Phys. Rev. ,E48 (4), 2669.*
2. Miyazaki & Hunt (1999) *J. Fluid Mech. in press.*
3. Risso et al.(1995) *AIAA Paper 96-0820.*
4. Sreedhar & Ragab(1994) *Phys. Fluids 6,2501.*

STUDY OF THE GEOMETRY OF FLOW PATTERNS IN A COMPRESSIBLE WAKE

On the behavior of velocity gradient tensor invariants

H. MAEKAWA
Dept. of Mechanical Engineering and Intelligent Systems,
Univ. of Electro-Communications
Chofu-city Tokyo 182-8585 JAPAN

1. Introduction

Direct numerical simulations of turbulence provide better understanding of fine scale motions and indicate that the dissipation motion in incompressible turbulence has a universal character. The fine scale motion by the evolution equation of the velocity gradient tensor A_{ij} in incompressible turbulence has been investigated by Vielliefosse [1] and others. In compressible turbulence, there is another kind of dissipation motion, so called dilatation dissipation due to $div\mathrm{v} \neq 0$. Maekawa [2](1998) extended the incompressible evolution equations analyzed by Cantwell[3] (1992) for compressible flow and found that the velocity gradient tensor was governed by a linear second order system whose coefficients are expressed in terms of the invariant P and its derivatives and the invariant Q. This fact suggests that analytical feature of A_{ij} is characterized by the analytical features of the invariants P and Q. Maekawa and Matsuo [4](1998) applied the solutions of the evolution equations to explain the statistical robust tendency of homogenous isotropic compressible turbulence for initial high turbulent Mach numbers. In what follows, flow structures of inhomogeneous compressible flow such as a wake in the invariant space are investigated. The solutions of the evolution equations are applied to understanding the definite trends presented in the invariant space.

2. Numerical simulations

2.1. FLOW DESCRIPTION

The initial conditions for the time-developing wake simulations consists of linear eigenfunctions obtained from a linear stability analysis superimposed

T. Kambe et al. (eds.), IUTAM Symposium on Geometry and Statistics of Turbulence, 223–228.
© 2001 *Kluwer Academic Publishers.*

on the laminar mean flow. For a perfect gas with a Prandtl number of unity, the Crocco-Busemann relation is employed to relate the mean temperature to the mean velocity. The most unstable 2-D antisymmetric mode with a pair of equal and opposite oblique 3-D modes are superimposed on the mean velocity, temperature, and density profiles at the beginning of a simulation. The code was validated by comparison between the numerical result and the linear theory. Simulations addressed the range of wavenumber leading to secondary transition. Then, the spanwise wavenumber was chosen to correspond to pair of oblique waves inclined at $\theta = \pm 45^0$ relative to the streamwise direction, because the modal energies for the oblique modes grow well after saturation of the fundamental mode.

2.2. SIMULATION TECHNIQUE

The technique used in this study is direct numerical solution of the compressible Navier-Stokes equations without modeling assumptions. This is naturally very expensive, even for a very low Reynolds number wake, but this method is uniquely capable of providing the desired detailed look at the physics. Because fully resolved information is available for the fine-scale motions, it offers an ideal means of studying the flow dynamics. To compute the flow solution, the fully compressible equations are solved on Cartesian grid systems having $72 \times 150 \times 48$ points in the streamwise, transversal and spanwise directions respectively. The algorithm employed to solve the equations used sixth-order compact finite difference schemes. A fourth-order Runge-Kutta algorithm was employed to integrate the equations forward in time. For time-developing wake, periodic boundary conditions can be applied in the streamwise and spanwise directions. The major gradient direction is finite with NSCBC applied at $y = \pm 10$. The length of the computational domain in the streamwise and spanwise directions is determined from linear stability theory. In the direct numerical simulations of the temporally evolving wake, the governing equations were nondimensionalized by the characteristic physical scales such as the initial half width δ and free stream velocity U.

3. Numerical results

3.1. EVOLUTION OF MODAL ENERGY

To illustrate the effect of Mach number, comparisons were made for simulations performed at M=1 and M=2 for an initial velocity profile of Re=200. Simulations for Re=600 were performed to minimize the effects of viscosity on the flow field. The modal energy history of the fundamental and oblique waves is shown in Fig.1 for Re=400, and Re=600 wakes. The modal energy is defined as $E_k = \int \hat{u}_i(\alpha, y, \beta)\hat{u}_i^*(\alpha, y, \beta)dy$ where summation over i from 1 to 3 is imposed, $\hat{u}(\alpha, y, \beta)$ is the Fourier transform of u_i in the $x-z$ plane. Several simulation data at varying time shown by arrows in Fig.1 beyond the region of exponential growth are analyzed. When the oblique modes grow well, the streamwise fine structures are observed.In this paper, the results for M=1.2 and Re=600 are described.

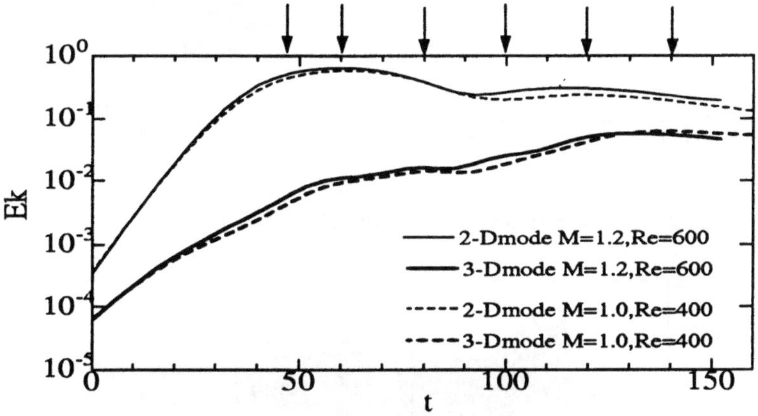

Fig.1 Evolution of modal energies of the fundamental and oblique modes.

3.2. FLOW STRUCTURE IN PHYSICAL/INVARIANT SPACE

The main points to note, observed in physical space, are that high values of the enstrophy and high rates of dissipation occur in relatively isolated regions which tend to be highly elongated in the streamwise/transversal direction. These observations are similar to those of incompressible mixing layers. These structures are composed by the flow patterns characterized by stable-focus-stretching and unstable-focus-compressing and also a few flow patterns due to compressibility such as stable-focus-compressing. In terms of the topology there has been a change from stable-focus-stretching to unstable-focus-compressing and vise versa in these structures. This fact is investigated by the solutions of the evolution equations for the invariants. Figure 2 is the very pronounced nearly straight ridgeline extending into the upper left quadrant. As will be examined, however, only a small fraction of the data collects along the ridgeline. A significant fraction of the data lies in an elliptical region about the origin in the Q-R space. In the P-R space, as shown in Fig.3, the predominance of points collects along a straight ridgeline of R=0. Around the region of modal energy saturation of the 2-D fundamental disturbance, a small fraction of the data shows the pronounced plot in the Q-R space.

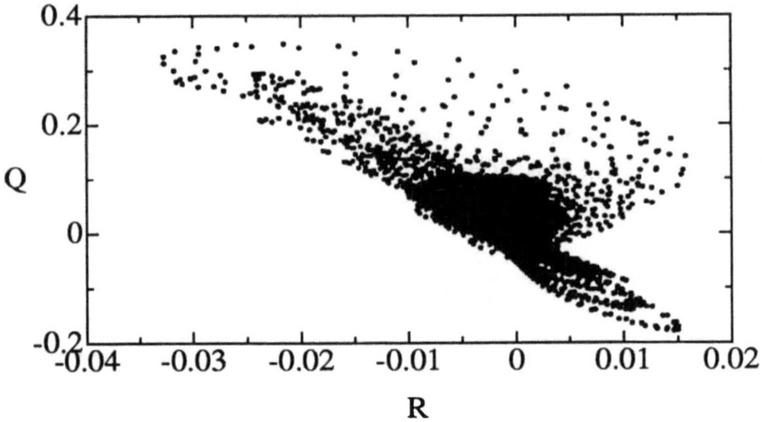

Fig.2 Scatter plots of Q versus R at $tU/\delta = 120$.

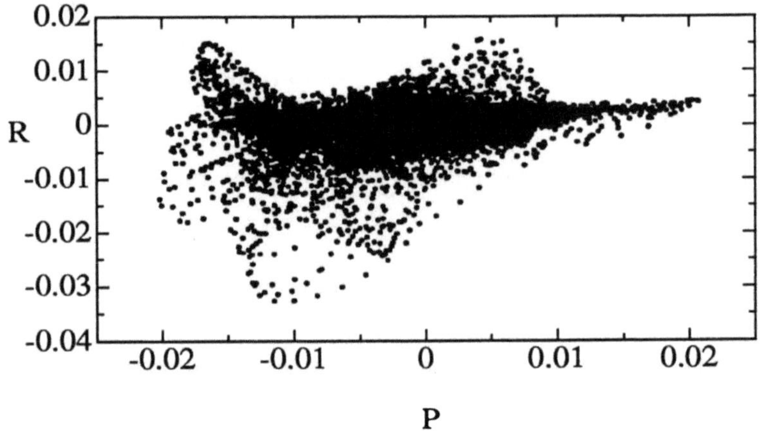

Fig.3 Scatter plots of P versus R at $tU/\delta = 120$.

4. Homogeneous model for compressible flow

Differentiating the Euler equations with respect to x_j leads to the transport equation for the velocity gradient tensor A_{ij}. The evolution equations for the invariants P,Q and R are given by

$$\frac{dQ}{dt} - \frac{4}{3}P\frac{dP}{dt} + \frac{2}{3}P^3 - \frac{7}{3}PQ + 3R = -A_{ik}H_{ki} \tag{1}$$

$$\frac{dR}{dt} - \frac{1}{3}(Q+P^2)\frac{dP}{dt} + \frac{1}{3}P^2Q - PR - \frac{2}{3}Q^2 = -A_{in}A_{nm}H_{mi} + A_{ik}H_{ki}P \tag{2}$$

$$\frac{dP}{dt} + P^2 - 2Q = -\frac{\partial}{\partial x_i}\left(\frac{1}{\rho}\frac{\partial p}{\partial x_i}\right) \tag{3}$$

where $H_{ij} = -(\frac{\partial}{\partial x_j}(\frac{1}{\rho}\frac{\partial p}{\partial x_i}) - \frac{\partial}{\partial x_k}(\frac{1}{\rho}\frac{\partial p}{\partial x_k})\frac{\delta_{ij}}{3})$. The evolution equations for imcompressible flo ·re obtained from the condition of the invariant $P = 0$ by incompressibility. A homogeneous model, imposing the right hand sides of the evolution equations (1) and (2) to be 0, is employed in this paper.

4.1. STATISTICS OF RIGHT HAND SIDES OF EVOLUTION EQUATIONS

Figure 4 shows the histogram of the right hand side of $A_{ik}H_{ki}$ of the evolution equation (1). Significant fraction of the plots is concentrated around the origin. A small fraction of the data in Fig.4 lies on the large magnitudes. This feature suggests that the homogenous model should not be applied for the fraction of data. The Laplacian of p/ρ is a local quantity which is directly related to the vorticity and shear. An examination of the histogram of $-(\frac{\partial}{\partial x_i}(\frac{1}{\rho}\frac{\partial p}{\partial x_i}))$ shows that the magnitudes are close to -2Q in the simulated results for Re=600. We can propose a homogeneous model for compressible flow that the right hand side of the equation (3) equals to -2Q even for the particle lying far from the origin.

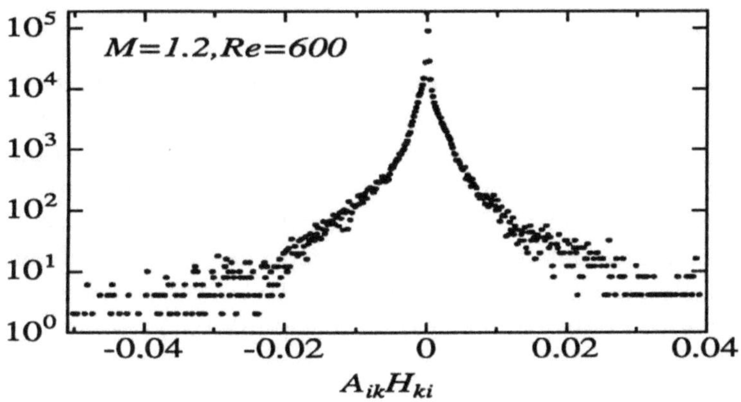

Fig.4 Histogram of $A_{ik}H_{ki}$ for the compressible wake at $tU/\delta = 120$.

4.2. FAMILIY OF SOLUTION

Figure 5 shows the solutions of the compressible evolution equations where DNS data at $tU/\delta = 120$ are employed as the initial conditions. Homogenous model can predict characteristics of evolving around the origin of the invariant space. Trajectories are indicative of a slowing down of the evolution of invariants for particles close to the origin and the regions where the curvatures of the solution trajectories have highly changed. When P is very small, the solution family projected in the Q-R space is very close to the incompressible case. In the P-R space, a plot remains for a long time around the P axis. However, further examinations of the right hand sides of the evolution equations for the points far from the origin suggest that a small correction will be required for the homogeneous model.

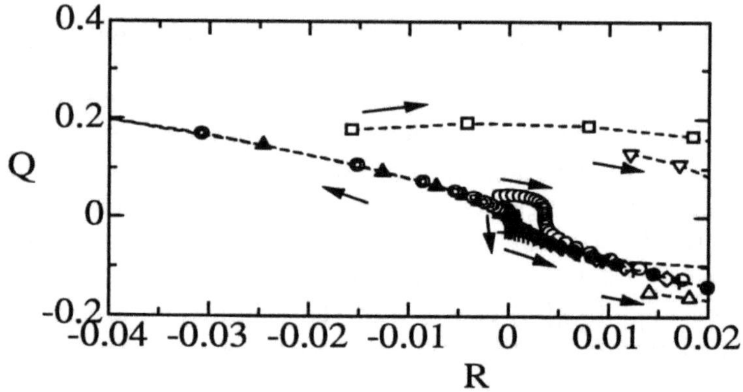

Fig.5 Trajectories of the invariants of the velocity gradient tensor.

5. Conclusions

When the streamwise fine-scale structures appear in a simulated compressible wake, corresponding to the saturation of oblique modes, the predominance of points collects in an elliptical region about the origin with its major axis aligned with upper left and lower right quadrants in the Q-R space. Significant number of plots is described by the evolution equations of the velocity gradient tensor with a homogeneous model of compressible turbulence. Points far from the origin tend to be associated with relatively large rates of the right hand sides of the evolution equations

Acknowledgements

This work was supported with a grant-in-aid (No. 10650163) from the Japanese Ministry of Education and Culture.

References

1. P.Vieillefosse, (1984) Internal motion of a small element of fluid in an inviscid flow, *Physica* **125** A pp.150-162.
2. H. Maekawa, (1998) Topology of fine scale motions in compressible turbulence, *NAGARE* **17** pp.411-416.
3. B. J. Cantwell, (1992) Exact solution of a restricted Euler equation for the velocity gradient tensor, *Phys. of Fluids* A **4** (4) pp.782-793
4. H. Maekawa and Y. Matsuo, (1998) Study of the geometry of flow patterns in compressible isotropic turbulence, *Proc. of 13th U.S. National Congress of Applied Mechanics*, FB1.

INTERACTION OF A VORTEX RING LINKED WITH A VORTEX FILAMENT

KATSUHIRO SUZUKI, MOTONORI ABE, MASAYUKI HIRANO
AND OSAMU SANO
Department of Applied Physics,
Tokyo University of Agriculture and Technology,
Saiwai-cho 3-5-8, Fuchu, Tokyo 183-0054, Japan

Abstract. In the laboratory, we produced a system of vortices with link helicity, *i.e.*, a vortex ring linked with a vortex filament. Observations based on flow visualization reveal several processes of interactions including the topological change by reconnection.

1. Introduction

Conservation of helicity in an ideal fluid leads to a concept of topological invariants of the field, such as the winding number of vortex lines [1]. The idea is important for understanding not only the dynamics of an ideal fluid but also that of nondissipative MHD [2, 3]. However, reconnection will cause the topological change of field lines in viscous fluid, which makes the dynamics much more complicated than that in ideal one. There are a lot of numerical while a few experimental works in which the processes of reconnection are reported [5, 6, 7]. Especially, to the author's knowledge, the dynamics of vortex tubes with link helicity has never been investigated experimentally. The aim of the present work is to realize linked vortices in the laboratory and study their dynamics.

2. Experimental Apparatus

Commonly, a vortex ring is produced by impulsive ejection of fluid from a nozzle (or an orifice), while a vortex filament by strong suction. We combine these methods to generate a system of vortices with link helicity.

T. Kambe et al. (eds.), IUTAM Symposium on Geometry and Statistics of Turbulence, 229–234.
© 2001 *Kluwer Academic Publishers.*

As a working medium, water in a rectangular tank of $50 \times 50 \times 70\ \ \mathrm{cm}^3$ is used. In order to maintain a thin vortex filament, a pump sucks the water through a pipe at the center of the upper surface of the tank (Figure 1). An acrylic cylinder with a diameter of 30 cm and a height of 60 cm is installed in the tank to keep the axial symmetry of the flow. The circulation Γ_{flmt} of the filament is controlled by the volume flow rate Q_f of the pump.

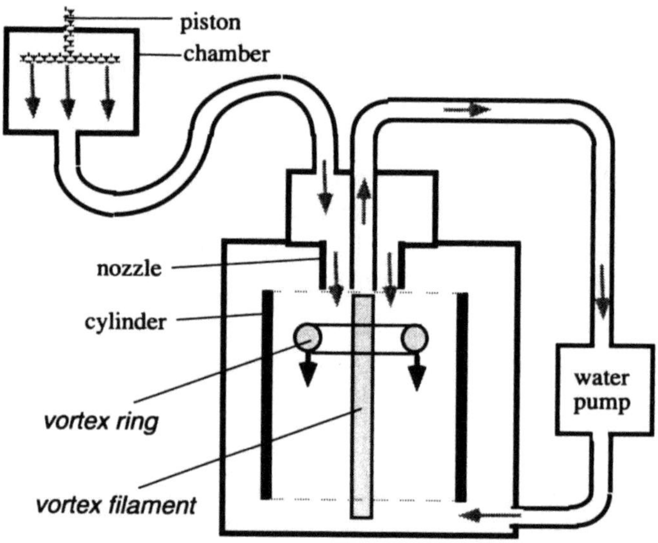

Figure 1. General schematic of the apparatus.

The vortex ring is produced by ejecting a given volume of the fluid from a nozzle (diameter D =10.4 cm) at the upper surface of the tank. Water impulsively expelled by a piston from a separate chamber is led to the nozzle through six hoses. This geometry enhances the reproducibility of the vortices because it prevents the harmful oscillations generated by the piston from affecting the test section. The strength Γ_{ring} of the ring is determined by the ejection time and the ejection volume, which are controlled by a cam and a spring driving the piston. Γ_{ring} takes discrete values, depending on the form of the cam.

The "linked" topology of the vortices is achieved by a configuration illustrated in Figure 2: the suction pipe for the maintenance of the steady filament passes through the center of the nozzle. The inner tube is installed in order that the fluid ejected from the nozzle may not influence the filament.

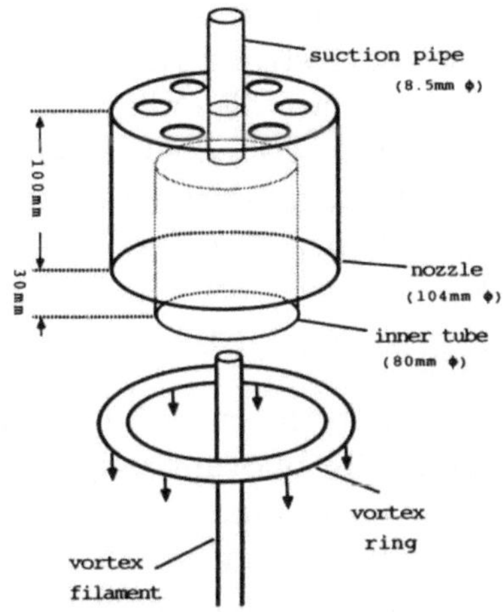

Figure 2. Geometry of the nozzle and the suction pipe.

3. Observations

3.1. MEASUREMENT OF THE CIRCULATION

A laser-Doppler velocimeter (LDV) system (DANTEC, FlowLite Optics/ Flow Velocity Analyzer/ FLOware system) is used to measure the velocity field induced by each vortex. When only the vortex filament exists, the velocity field is almost steady and axisymmetric for all the tested ranges of Q_f. According to the measurement by the LDV, the radial distribution of the azimuthal velocity V_θ (in a cylindrical coordinate system), has a peak typical of the flow induced by a straight vortex filament, from which both the core radius $\varepsilon_{\text{flmt}}$ and the strength Γ_{flmt} are evaluated. It is found that $\varepsilon_{\text{flmt}}$ ranges from 5.5 mm to 10 mm, and Γ_{flmt} from 1.1×10^{-3} m^2 s^{-1} to 1.9×10^{-2} m^2 s^{-1}.

Concerning the velocity field induced by the vortex ring, it is unsteady because the ring moves downwards from the nozzle to the bottom of the tank. Owing to this property, the core radius $\varepsilon_{\text{ring}}$ of the ring can be evaluated from LDV signals obtained at a single point, where the center of the vortex core passes. In fact, a $t - V_r$ curve measured at the point can be converted to $z - V_r$ one, by multiplying t by the translational velocity U of the ring. Then the distance between a positive maximal peak and a negative minimum one in the latter curve corresponds to $\varepsilon_{\text{ring}}$. Similarly, the circulation Γ_{ring} of the ring can be calculated by measuring V_r and V_z. Utilizing the radial distribution of V_r and the vertical one of V_z converted from $t - V_z$ curves, one can execute the line integral of the velocity field along a rectangular path around the core. Two kinds of vortex rings with $(\varepsilon_{\text{ring}}, \Gamma_{\text{ring}}) = (5.0\text{ mm}, 1.9 \times 10^{-3}\text{ m}^2\text{ s}^{-1})$ and $(\varepsilon_{\text{ring}}, \Gamma_{\text{ring}}) = (5.1$ mm,

2.4×10^{-2} m^2 s^{-1}) are employed in the present experiments. In the following, we simply call the former "strong ring", and the latter "weak ring". The major radii of both rings are 55 millimeters immediately after their formation. The measurement described so far indicates that the Reynolds number $Re = \Gamma_\alpha/\nu$ (α = flmt, ring) ranges from 1.1×10^3 to 2.4×10^4, where ν is the kinematic viscosity of water.

3.2. INTERACTION OF THE VORTICES

Interaction of the filament with the ring is examined by flow visualization using tracer particles (neutrally buoyant polystyrene beads) and dye. We utilized the particles for strong filaments ($\Gamma_{\text{flmt}} = O(10^{-2}$ m^2 s^{-1})) while dye for weak ones ($\Gamma_{\text{flmt}} = O(10^{-3}$ m^2 s^{-1})). The ring is visualized by the particles loosely sticked inside the nozzle with glue. It is found that the glue itself is efficient to make the ring visible, since it is well concentrated into the core.

Based on the visualization, the following processes are observed:

1. If the strength of one of the vortices is much stronger than that of the other, the structure with weaker circulation is broken down immediately after the outset of the interaction.
2. If only the weak ring exists, a wave with a wavenumber $n \approx 8$ begins to appear on it about 1.5 s after its formation (Figure 3). The wavenumber is consistent with Maxworthy's work [4] because the *ring* Reynolds number is about 5×10^3 in this case. Though the wave does not propagate along the ring, it slowly rotates around the core at a frequency of 0.5 Hz. After moving downwards for about 60 s, the ring is broken just before it arrives at the bottom plane. Provided that the filament with $\Gamma_{\text{flmt}} = 1.2 \times 10^{-3}$ m^2 s^{-1} ($\approx \Gamma_{\text{ring}}$) coexists with this ring, the swirling flow induced by the filament strongly distorts and amplifies the wave, which causes faster decay of the ring at $t \approx 12$ s. The filament is also broken in this case. Therefore, the vortex system consists of vortices with weak strength (of the same order) is unstable.
3. After the formation of the strong ring alone, it proceeds downwards to reach the bottom of the tank within 10 s, without showing instability. If the filament with the strength $\Gamma_{\text{flmt}} = 1.9 \times 10^{-2}$ m^2 s^{-1} ($\approx \Gamma_{\text{ring}}$) is simultaneously generated with this stable ring, two kinds of interactions are observed.

   (a) If the cylindrical symmetry of the system is well maintained, the part of the filament ahead of the ring becomes thick because the flow induced by the ring compresses it. (On the contrary, the part behind the ring will be stretched, though it is not well visualized in this work.) It is also observed that some tracer particles swirling around the filament are entrained to the ring, which may correspond to the transportation of the vorticity from the filament to the ring.

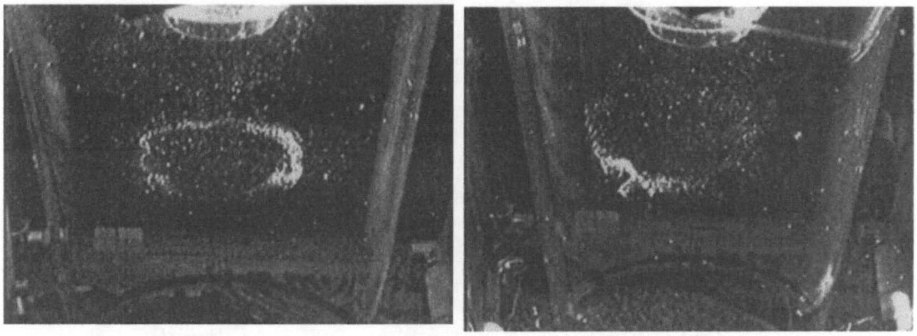

(a) (b)

Figure 3. The collapse of a vortex ring caused by a vortex filament. (a) When only the weak ring ($\Gamma_{\rm ring} = 1.1 \times 10^{-3}$ m^2 s^{-1}) exists, a wave with wavenumber $n \approx 8$ appears on it. ($t = 15.8$ s after the formation of the ring.) (b) When the ring and a filament exist ($\Gamma_{\rm flmt} = 1.2 \times 10^{-3}$ m^2 s^{-1}), the wave is amplified to break at $t = 11.8$ s. (The filament is not visualized.)

(b) We can intentionally break the cylindrical symmetry of the setup, because the position of the suction pipe can be slightly changed. It is also observed that the symmetry can be spontaneously broken during the process of interaction. In fact, if the filament is relatively thick ($\varepsilon_{\rm flmt} \approx 10$mm), the filament is not compressed but bended by the flow induced by the ring. In either cases of broken symmetry, visualization suggests the occurrence of reconnection of the vortices. In Figure 4, typical process is illustrated. After such a stage that the ring becomes wavy and the filament is bended, it seems that a new helical filament is generated through reconnection.

4. Summary

We produced a system of vortices with link helicity in the laboratory, and investigated the interactions of it by flow visualization using tracers and dye. When $\Gamma_{\rm flmt} \approx \Gamma_{\rm ring} = O(10^{-3}$ m^2 s$^{-1})$, both the vortices are unstable. When $\Gamma_{\rm flmt} \approx \Gamma_{\rm ring} = O(10^{-2}$ m^2 s$^{-1})$, the compression of the filament is observed in cylindrically symmetric case, and the topological change of vortices caused by reconnection is suggested if the symmetry is broken. For the sake of more quantitative and precise study of the processes observed in the present work, we are planning to introduce PIV and execute numerical simulations in the near future.

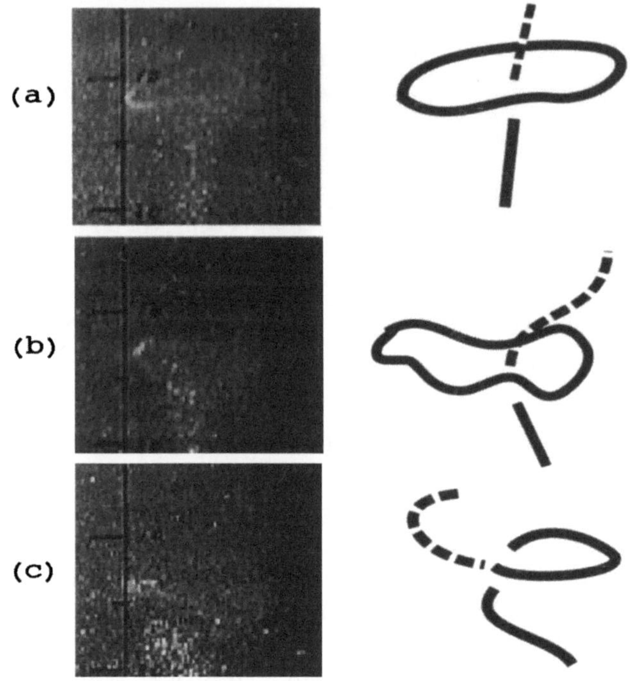

Figure 4. Topological change observed when the symmetry is broken ($\Gamma_{\rm flmt} \approx \Gamma_{\rm ring} = O(10^{-2}\ {\rm m}^2\ {\rm s}^{-1})$). (a) The vortex ring proceeds downwards, keeping the linked topology. (b) The filament approaches the ring, deforming its shape. (c) It seems that two vortices are reconnected to form a helical structure.

References

1. Moffatt, H. K. (1969) The Degree of Knottedness of Tangled Vortex Lines, *J. Fluid Mech.* **35**, 117–29.
2. Moffatt, H. K. (1990) The Energy Spectrum of Knots and Links, *Nature* **347**, 367–9.
3. Marsh, G. E. (1993) Helicity, Topology and Force-Free Fields, *Phys. Rev. E* **47**, 3607–11.
4. Maxworthy, T. (1977) Some Experimental Studies of Vortex Rings, *J. Fluid Mech.* **81**, 465–95.
5. Kida, S. and Takaoka, M. (1994) Vortex Reconnection, *Ann. Rev. Fluid Mech.* **26**, 169–89.
6. Oshima, Y. and Asaka, S. (1977) Interaction of Two Vortex Rings along Parallel Axis in Air, *J. Phys. Soc. Jpn.* **42**, 708–13.
7. Schatzle, P. R. (1987) An Experimental Study of Fusion of Vortex Rings, *Thesis, California Institute of Technology.*
8. Suzuki, K., Abe, Y., Hirano, M., and Sano, O. (1998) Experiments on Interaction of a Vortex Filament with a Vortex Ring, *Proc. Third Int. Conf. Fluid Mech. (Beijing)*, 376–79.

MECHANISM OF FLAME PROPAGATION ALONG A VORTEX TUBE

T. HASEGAWA, S. NISHIKI AND S. MICHIKAMI
Nagoya Institute of Technology
Gokiso-cho, Showa-ku, Nagoya 466-8555, Japan

1. Introduction

Vortex tubes are recognized as sinews of turbulence: The length of the vortex tube represents the integral scale of turbulence, the spacing of the nodes of the vortex tube represents the Taylor micro scale, and the diameter of the vortex tube represents the 10 times the Kolmogorov scale (Tanahashi and Miyauchi, 1999). Thus the interaction between a fine vortex tube and a flame seems to be an essential process in turbulent combustion. The aim of this work is to study the mechanism of flame propagation along a fine vortex tube of a premixed gas when the vortex tube interacts perpendicularly with the flame (Chomiak, 1976). It is well known that a premixed flame propagates rapidly with a velocity similar to the maximum circumferential velocity of the vortex core. It is also known that the premixed flame can propagate along the vortex tube when the maximum circumferential velocity is faster than the burning velocity and the core diameter is larger than the flame thickness (Hasegawa *et al.*, 1995). However, the mechanism of the flame propagation is still not clear though several models have been proposed. In this study, the flame propagation along a fine vortex tube is numerically simulated and the mechanism that provokes the flame to propagate along the vortex tube is discussed in terms of the vorticity transport equation.

2. Numerical Simulation

Three-dimensional conservation equations for compressible reacting flows are solved on the basis of the following assumptions: i) The chemical reaction is an one-step irreversible one with heat release, where the molecular

T. Kambe et al. (eds.), IUTAM Symposium on Geometry and Statistics of Turbulence, 235–240.
© 2001 *Kluwer Academic Publishers.*

weights of the reactants and products are the same. ii) The bulk viscosity, the Soret and the Dufour effects and the pressure gradient diffusion are neglected. iii) The specific heat at constant pressure and the specific heat ratio are constant. iv) The equation of the state of the burned and unburned gases is that of an ideal gas. The physical parameters are given as $Q = 2.446\times10^6$ J/kg, $C_p = 1.25\times10^3$J/kg·K, $\lambda = 7.4\times10^{-2}$ W/m·K, $\gamma = 1.4$, $Pr = 1.0$ and $Le = 1.0$. The initial pressure is 0.5 MPa and the initial temperature is 300 K, thus the adiabatic flame temperature is 2260 K and the density ratio of the flame is 7.53. The frequency factor of the one-step irreversible reaction of $B = 3.47\times10^6$ 1/s·K and the activation temperature of $\theta = 19600$ K give the laminar burning velocity of $u_L = 0.236$ m/s and the flame thickness of $\delta = 0.0734$ mm.

A straight vortex tube is placed at the center of a volume of 1 mm × 1 mm × 2.5 mm as shown in Figure 1. The side boundaries are periodic, but longitudinal boundaries are non-periodic and posed by non-reflecting boundary conditions (Poinsot and Lele, 1992, Hasegawa *et al.*, 1999). The initial vortex tube is assumed to have a Gaussian vorticity distribution and the circumferential velocity field is given by the following equation:

$$V(r) = \frac{\Gamma_0}{2\pi r}\left[1 - \exp\left(-\frac{r^2}{(\sigma_0/2)^2}\right)\right], \quad \Gamma_0 = \pi\sigma_0 V_{m0}$$

The initial parameters of the vortex tube are $\sigma = 0.156$ mm and $V_{m0} = 12.4$ m/s, and thus the initial Reynolds number is $Re = 152$. A laminar flame is placed at $x = 1$ mm to interact with the vortex tube.

Fourier spectral collocation method is used for y, z directions and 6th-order central finite difference method is used for x direction to treat non-periodic boundary conditions. The number of grid points is 128 in y, z directions, while it is 640 in x direction to keep the accuracy similar to the spectral method. 3rd-order Runge-Kutta method is used in time integration and the time step is selected to satisfy the CFL condition. A vector-parallel computer (Fujitsu VPP 500) with 16 PEs is used for the numerical simulation. The total memory used is 10.5 GBytes.

3. Results and Discussions

Contour lines of temperature, total vorticity and azimuthal component of vorticity in the center cross section at $t = 31$ μs are illustrated in Figure 2. It is found that the flame propagates rapidly along the vortex tube, and that the azimuthal component of vorticity appears near the flame front. This azimuthal vorticity can make the flame to propagate along the vortex tube (Hasegawa and Nishikado, 1996).

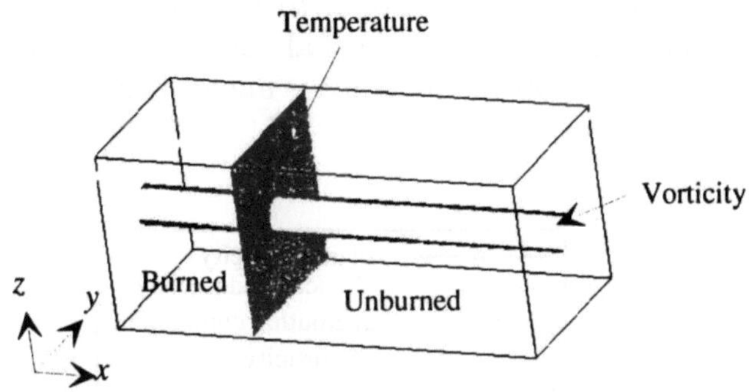

Figure 1. Initial conditions.

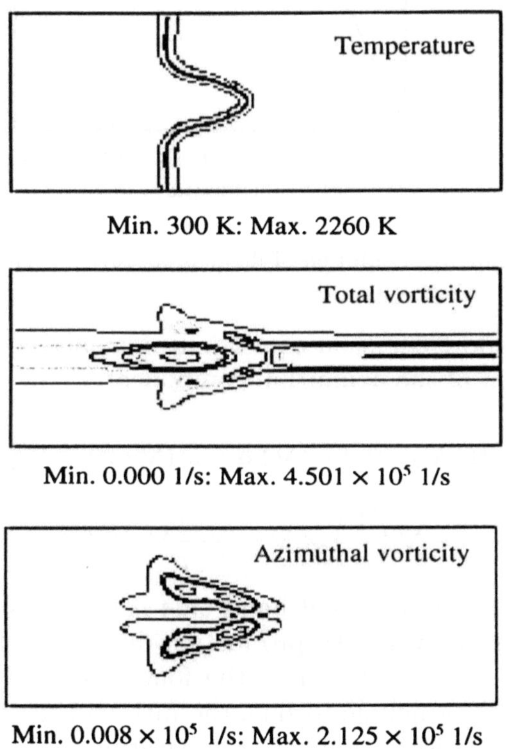

Figure 2. Production of azimuthal vorticity.

As shown in Figure 3, The proportional factor of the propagation velocity to the maximum circumferential velocity of the vortex tube increases at the initial stage, but it becomes a constant value of 1.2. This means that the propagation velocity is proportional to the maximum circumferential velocity with a factor of 1.2. The azimuthal component of vorticity induces a similar velocity at the top of the convex flame, though the factor has a higher value.

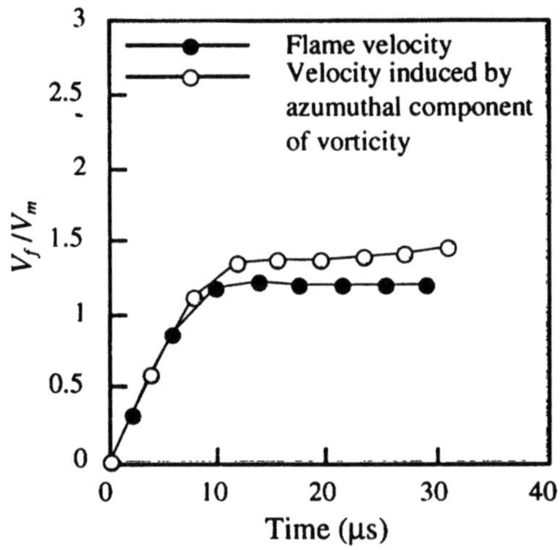

Figure 3. Evolution of flame propagation velocity.

Next the transport equation of vorticity is analyzed to clarify the production mechanism of the azimuthal component.

$$\frac{\partial \omega}{\partial t} = -(\mathbf{u} \cdot \nabla)\omega + (\omega \cdot \nabla)\mathbf{u} - \omega(\nabla \cdot \mathbf{u}) - \nabla\left(\frac{1}{\rho}\right) \times \nabla p$$
$$+\nabla\left(\frac{1}{\rho}\right) \times \nabla \cdot \tau + \frac{1}{\rho}\nabla \times \nabla \cdot \tau$$

where the first term of the right-hand side is the convection of vorticity, the second term is the vorticity production due to stretch, the third term is vorticity decay due to dilatation, the fourth term is the production of vorticity due to the baroclinic torque, the fifth and the sixth terms are the viscous diffusion and dissipation. After integrating the transport equation in time, the azimuthal component of vorticity produced by each term is shown in Figure 4. It is obvious that the convection term has a main effect on the appearance of azimuthal component of vorticity in front of the convex flame. This means that the initial vortex tube is expanded by combustion and as a result it is twisted due to the conservation of the angular momentum. The twisted vortex in front of the convex flame induces the velocity that pulls the flame into the unburned vortex tube. The stretch of the vortex also appears simultaneously during the expansion and enhances the vorticity that pulls the flame. Besides the baroclinic effect produces the vorticity behind the convex flame that pushes the flame into the unburned vortex tube. However, the production of negative vorticity behind the convex flame due to the convection and stretch terms compensates the baroclinic effect. The dilatation effect and the viscous dissipation reduce the vorticity behind the flame but they play a small part. Therefore the main mechanism of the production of azimuthal component of vorticity is attributed to the convection and the stretch in front of the flame.

4. Conclusions

1) Azimuthal component of vorticity appears in front of the flame. 2) Flame propagation velocity is proportional to the maximum circumferential velocity and is similar to the velocity induced by the azimuthal component of vorticity. 3) Convection and stretch of axial vorticity produces positive azimuthal component in front of the flame and produces negative azimuthal vorticity behind the flame. This negative vorticity cancels the vorticity production due to the baroclinic effect. As a result, the azimuthal vorticity in front of the flame, which is produced by convection and stretch, provokes the flame propagation.

5. Acknowledgments

This work is partially supported by the Grant-in-Aid for Scientific Research (40164818).

References

Tanahashi, M., Iwase, S., Uddin, Md.A. and Miyauchi, T. (1999) Three-dimensional features of coherent fine scale eddies in turbulence, *Proceedings of First International Symposium on Turbulence and Shear Flow Phenomena.*

Chomiak, J. (1976) Dissipation fluctuations and the structure and propagation of turbulent flames in premixed gases at high Reynolds number, *Sixteenth Symposium (International) on Combustion*, The Combustion Institute, Pittsburgh, pp. 1665-1673.

Hasegawa, T., Nishikado, K. and Chomiak, J. (1995) Flame propagation along a fine vortex tube, *Combustion Science and Technology*,**108**, 67-80.

Poinsot, T. and Lele, S.K. (1992) Boundary conditions for direct simulations of compressible viscous flows, *Journal of Computational Physics***101**, 104-129.

Hasegawa, T., Morifuji, T. and Borghi, R. (1999) Direct numerical simulation of a turbulent premixed flame, *Proceedings of the 5th ASME/JSME Joint Thermal Engineering Conference*, ASME, AJTE99-6316.

Hasegawa, T. and Nishikado, K. (1996) Effect of density ratio on flame propagation along a vortex tube, *Twenty-Sixth Symposium (International) on Combustion*, The Combustion Institute, Pittsburgh, pp. 291-297.

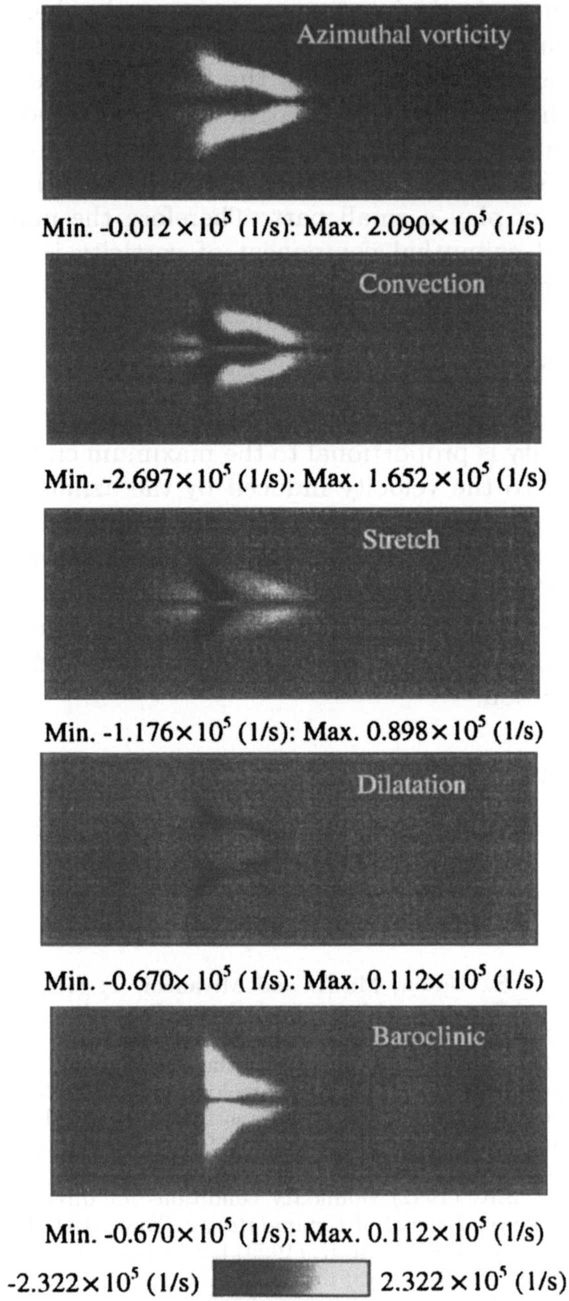

Figure 4. Production of azimuthal vorticity due to convection, stretch, dilatation and baroclinic terms.

F

Large Scale Motions, LES, and Closure

MAPPING CLOSURE FOR NON-GAUSSIAN VELOCITY FIELDS

M. TAKAOKA
Division of Physics and Astronomy, Graduate School of Science, Kyoto University, Sakyo-ku, Kyoto 606-8502, Japan

1. Introduction

In our recent paper [1] (henceforth referred to as T99), an attempt was made to assess a mapping closure for turbulent velocity fields with non-Gaussian probability distribution functions (PDF's). The *a priori* assumption of the Gaussian velocity statistics is *not* made in our mapping closure, which generalizes the results for the Burgers turbulence by Kraichnan [2] and Gotoh and Kraichnan [3]. Our results show that further generalization is required to reproduce the sub-Gaussian PDF (its tails fall below the Gaussian distribution) observed in the direct numerical simulation (DNS) of the Burgers equation. It is indispensable to investigate the properties related to the sub-Gaussian PDF in order to get a clue of such generalization.

The Burgers equation for the velocity u is written as

$$\frac{Du}{Dt} = \nu \frac{\partial^2 u}{\partial x^2}, \tag{1}$$

and the governing equation for the velocity gradient $\xi = \partial_x u$ is

$$\frac{D\xi}{Dt} = -\xi^2 + \nu \frac{\partial^2 \xi}{\partial x^2}, \tag{2}$$

where ν is the kinematic viscosity and $D/Dt = (\partial/\partial t) + u(\partial/\partial x)$ is the substantial derivative. The first and the second terms in the r.h.s. of (2) represent stretching and dissipation respectively.

We also apply the mapping closure to a generalized equation of the Burgers equation, which is replaced the r.h.s. of equation (1) by $\sum_j \nu_j \frac{\partial^j u}{\partial x^j}$. The generalized equation includes the Korteweg-de Vries equation, the Kuramoto-Sivashinsky equation and a hyper viscous equation. But here we restrict ourselves to the results for the sub-Gaussian velocity case of the Burgers equation for lack of space.

T. Kambe et al. (eds.), IUTAM Symposium on Geometry and Statistics of Turbulence, 243–248.
© 2001 *Kluwer Academic Publishers.*

2. Formulation of Mapping Closure

In T99, it is shown that the velocity and its gradient PDF's relax to a unique distribution in the Burgers turbulence even if it start from non-Gaussian velocity distribution. In the limiting distribution, the velocity PDF is sub-Gaussian and the velocity gradient PDF has a Λ-shape. The present mapping closure has succeeded to predict that the PDF's started from non-Gaussian distribution relax to a unique PDF. However, it fails to reproduce the sub-Gaussian structure in velocity PDF and the far-negative tail of velocity gradient PDF. The former discrepancy is also noticed in [3].

The linear-dependence assumption $u(x,t) = r(t)u_0(z)$ made in [2, 3] is equivalent to assume the Gaussianity of velocity PDF. We generalized the map for u in T99 to investigate non-Gaussian velocity PDF,

$$u(x,t) = X(u_0(z),t), \tag{3}$$

$$\frac{\mathrm{d}z}{\mathrm{d}x} = J(\xi_0(z),t), \tag{4}$$

$$\xi(x,t) = \frac{\partial X(u_0,t)}{\partial u_0}\xi_0 J(\xi_0,t) \equiv Y(u_0,\xi_0,t), \tag{5}$$

where u_0 and ξ_0 are the multivariate Gaussian reference fields. The functional form of the coordinate map J is the same as that in [2, 3].

Once the functional form of maps are determined, the governing equations of the maps are readily obtained as

$$\frac{\partial}{\partial t}X(u_0,t) = \nu\left\langle \frac{\partial^2 u}{\partial x^2}\middle| u_0\right\rangle, \tag{6}$$

$$\frac{\partial}{\partial t}Y(u_0,\xi_0,t) = -Y^2(u_0,\xi_0,t) + \nu\left\langle \frac{\partial^2 \xi}{\partial x^2}\middle| u_0,\xi_0\right\rangle + \frac{1}{Q(u_0,\xi_0,t)}\int_{-\infty}^{\xi_0} K(u_0,\xi_0',t)\mathrm{d}\xi_0', \tag{7}$$

where

$$K(u_0,\xi_0',t) = \frac{\partial Y(u_0,\xi_0',t)}{\partial \xi_0'}Q(u_0,\xi_0',t)\left(\frac{\partial}{\partial t}\ln\frac{N(t)}{J(\xi_0,t)} - Y(u_0,\xi_0',t)\right) - \left(\frac{\partial X}{\partial u_0}\right)^{-1}\frac{\partial(Z(u_0,\xi_0',t),Y(u_0,\xi_0',t))}{\partial(\quad u_0 \quad , \quad \xi_0' \quad)},$$

$$Z(u_0,\xi_0,t) = \nu Q(u_0,\xi_0,t)\left(\left\langle \frac{\partial^2 u}{\partial x^2}\middle| u_0\right\rangle - \left\langle \frac{\partial^2 u}{\partial x^2}\middle| u_0,\xi_0\right\rangle\right)$$

Here $\langle\cdot|\, u_0\rangle$ and $\langle\cdot|\, u_0,\xi_0\rangle$ denote respectively the ensemble mean conditional on given values of u_0 and (u_0,ξ_0), and $\frac{\partial(f\ ,g\ )}{\partial(u_0,\xi_0)} = \frac{\partial f}{\partial u_0}\frac{\partial g}{\partial \xi_0} - \frac{\partial f}{\partial \xi_0}\frac{\partial g}{\partial u_0}$ represents the Jacobian of f and g.

3. PDF Tail and sub-Gaussian Velocity Model

Although the discrepancies are found only in the tail part and the PDF's take very small values there, the quality of closure *assumption* does appear in this region. It is important to clarify the reasons for this discrepancy in order to make an assessment of the mapping closure.

3.1. SUB-GAUSSIAN VELOCITY MODEL

We found by the DNS in T99 that if the initial velocity has appropriate sub-Gaussian distribution then the normalized velocity PDF almost remains in the initial sub-Gaussian. Our interest here is to examine the improvement of the discrepancy in the far-negative tail of velocity gradient PDF by changing the velocity PDF to be sub-Gaussian in the mapping closure. Following the method for the evolution of Gaussian velocity PDF's in [2, 3], the shape of the normalized velocity PDF, now the appropriate sub-Gaussian, is fixed. And its evolution is calculated by using the integral quantities; energy and energy dissipation, both of which are determined by the mapping closure. The velocity gradient PDF, on the other hand, evolves under the same equation in T99.

In the pioneering work by Kraichnan [2], who is the creator of mapping closure and has the best knowledge of it, he obtained the evolution equation of $r(t)$, the map for u, by the energy balance equation:

$$\frac{\mathrm{d}\langle u^2\rangle}{\mathrm{d}t} = -2\nu\langle\xi^2\rangle, \tag{8}$$

$$\Longrightarrow \quad \frac{\mathrm{d}r(t)}{\mathrm{d}t} = -\nu\frac{\langle\xi^2\rangle}{\langle u^2\rangle}r(t), \tag{9}$$

$$\Longrightarrow \quad \frac{\partial}{\partial t}X(u_0,t) = -\nu\frac{\langle\xi^2\rangle}{\langle u^2\rangle}X(u_0,t), \tag{10}$$

where use has been made of the relation $X(u_0,t) = r(t)u_0(z)$.

In our sub-Gaussian velocity model, we use (10) as the evolution equation for $X(u_0,t)$ instead of (6). The evolution of $Y(u_0,\xi_0,t)$, or $J(\xi_0,t)$, is determined by the same equation as (7). The proportionality of $\frac{\partial}{\partial t}X(u_0,t)$ to $X(u_0,t)$ guarantees the invariance of the dependence of $X(u_0,t)$ on u_0. That is, the normalized velocity PDF is unchanged in the time evolution. If we set the initial velocity PDF as the appropriate sub-Gaussian, it remains the sub-Gaussian as observed in DNS.

Figures 1(a) and (b) show the velocity PDF and velocity gradient PDF, respectively. The normalized velocity PDF in (a) remains almost in the

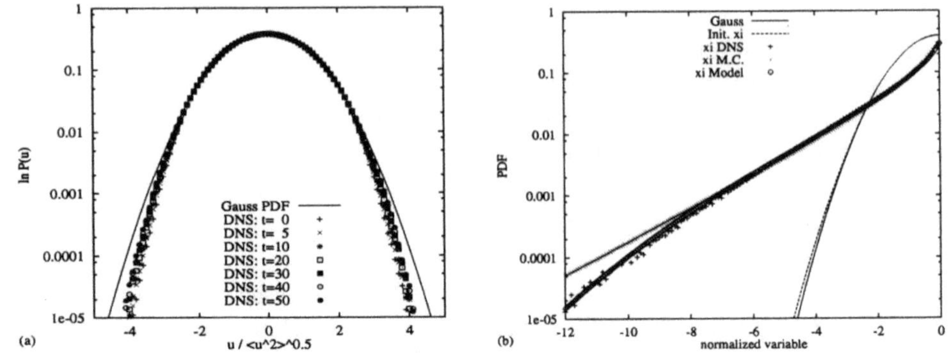

Figure 1. Time evolution of velocity PDF (a) and PDF tail of the velocity gradient in far-negative region at $t = 20$ (b). Solid and dashed lines respectively represent Gaussian PDF and the initial PDF of velocity gradient.

initial sub-Gaussian distribution, which is made by a simple map:

$$u = X(u_0, 0) = -\frac{a}{c^2} + u_0 + \frac{a + su_0}{u_0^2 + c^2} \tag{11}$$

with parameter assignments $(c, a, s) = (3, 0, 4)$. We show only the tails of the velocity gradient PDF in far-negative region, which correspond to Fig. 4(b) in T99. $+$ symbols in (b) represent the tail obtained from DNS of the freely decaying Burgers equation. $\times$ and $\circ$ respectively show the results by the mapping closure in T99 and the model assuming the sub-Gaussian velocity PDF as explained above.

This result confirms the conjecture by Avellaneda *et al.* [4] that the discrepancy in the PDF tail for velocity gradient is due to the *a priori* assumption of Gaussian velocity statistics in [2, 3]. Although most works on the Burgers turbulence are focused on the PDF of gradient or of higher-order derivative fields, this example shows the importance of developing the analysis for the sub-Gaussian structure in the velocity field.

4. Asymptotic Tails of Velocity PDF

Several authors have characterized the *asymptotic tails* ($|u| \gg 1$) of velocity PDF's [4, 5, 6, 7, 8]. The PDF tails of the velocity depend on the PDF tails of external forcing or on the order of differentiation of initial Gaussian fields. In both cases of white-in-time external forcing and of white-in-space initial field, they predict the cubic tail: $P(u) \sim \exp(-Cu^3)$.

Avellaneda *et al.* [4] characterize the tails for the cases of Gaussian initial data with self-similar spectrum and with "ultraviolet-regularized" spectrum. The latter corresponds to the DNS's in [1, 2, 3], in which case

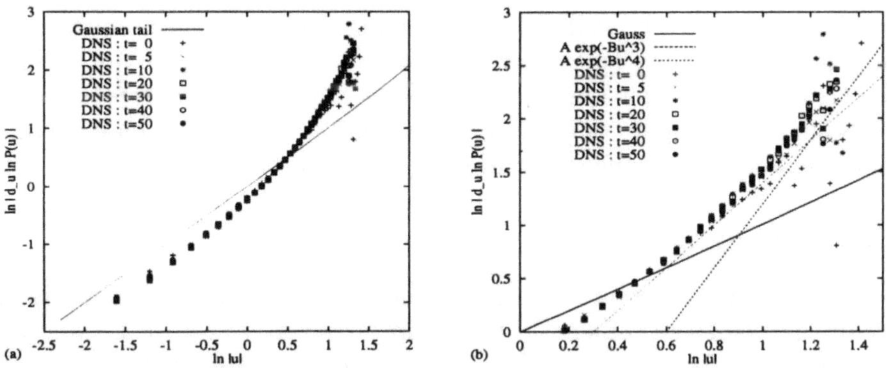

Figure 2. Time evolution of asymptotic tails of velocity PDF. The abscissa is $\ln\left|\frac{u}{u_{\mathrm{rms}}}\right|$, the ordinate is $\ln\left|\frac{\partial \ln P(u)}{\partial u}\right|$. (b) is the close-up of the tail part in (a) with lines showing Gaussian tail (solid line) and cubic tail (dashed line).

they predict $P(u) \sim \exp(-Ct^2u^4)$. Falkovich and Lebedev [6] showed that a Gaussian random force having correlation scale and correlation time produces velocity PDF tails: $P(u) \sim \exp(-Cu^4)$, and Gaussian velocity would, in particular, correspond to the force PDF decaying exponentially.

Our DNS of the Burgers turbulence shows there exists the cubic tail before the asymptotic tail corresponding to the "ultraviolet-regularized" spectrum. Fig. 2 shows the asymptotic tails of velocity PDF observed in the DNS starting from sub-Gaussian distribution shown in Fig. 1 (a). This cubic tail is also observed in the DNS starting from Gaussian distribution. We have checked that the cubic tail becomes longer for the higher Reynolds number. This might be understood as follows. Since the "ultraviolet-regularized" part in the spectrum has only small energy in later time, the body of the velocity PDF are determined by the "unregularized" part in the spectrum.

5. Conditional means

We lastly examine the detailed comparison of conditional means. The mapping closure maps a statistically known reference field to a surrogate field which satisfies the same evolution equation as the true field [9]. The governing equations of the maps are closed by letting the conditional means of the surrogate field be equal to that of the true field. Although no adjustable parameters appear in the mapping closure, there is a vast room in choosing the functional form of the map and the reference field. The quality of the closure is determined by the adopted maps. Comparing the conditional means, one might find the clues for choosing the appropriate functional form of the map.

It is shown that the present map can reproduce the structure of the conditional mean fairly well for all the time. The graphs are omitted here. The small discrepancy emerges in the large deviation regions of u and ξ, when the energy dissipation rate takes peak value. Both the stretching and the dissipation compete actively at that time. The present map seems to underestimate the enhancement of dissipation due to stretching.

6. Concluding Remarks

By using the sub-Gaussian velocity model, it is clarified that the discrepancy in the velocity-gradient PDF tail in far-negative region is caused by the discrepancy in velocity PDF tail. It is important to develop a mapping closure which is able to reproduce sub-Gaussian velocity PDF.

The statistics of second- and higher-order derivative terms are also investigated, but not shown here. The generalized mapping closure in T99 successfully predicts the transition from Gaussian distribution to non-Gaussian one in velocity PDF, though it is hard to obtain long time solutions. These results show the potential ability of the mapping closure.

To correctly capture the sub-Gaussian property of velocity PDF and the far-negative tail structure of velocity gradient PDF, we should use the map which can express fusion in folded regions due to the convective term. Difficulty in obtaining a long time solution for higher-order derivatives, pointed out in T99, might be related to the lack of reproducibility of dispersion effect. It is an important future work to characterize the mechanism of enhancement of dissipation and dispersion by stretching.

References

1. Takaoka, M. (1999) Application of Mapping closure to non-Gaussian velocity fields, *Phys. Fluids*, **11**, pp. 2205–2214.
2. Kraichnan, R.H. (1990) Models of intermittency in hydrodynamic turbulence, *Phys. Rev. Lett.*, **65**, pp. 575–578.
3. Gotoh, T. and Kraichnan, R.H. (1993) Statistics of decaying Burgers turbulence, *Phys. Fluids*, **A5**, pp. 445–457.
4. Avellaneda, M., Ryan, R. and E, W. (1995) PDFs for velocity and velocity gradients in Burgers' turbulence, *Phys. Fluids*, **7**, pp. 3067-3071.
5. Balkovsky, E., Falkovich, G., Kolokolov, I. and Lebedev, V. (1997) Intermittency of Burgers' turbulence, *Phys. Rev. Lett.*, **78**, pp. 1452–1455.
6. Falkovich, G. and Lebedev, V. (1997) Single-point velocity distribution in turbulence, *Phys. Rev. Lett.*, **79**, pp. 4159–4161.
7. Molchan, G.M. (1997) Burgers Equation with self-Similar Gaussian Initial Data: Tail Probabilities, *J. Stat. Phys.* **88**, pp. 1139-1150.
8. Ryan, R. (1998) Large-Deviation Analysis of Burgers Turbulence with White-Noise Initial Data, *Comm. Pure Appl. Math.*, **L1**, pp. 47-75
9. Chen, H., Chen, S. and Kraichnan, R.H. (1989) Probability distribution of a statistically advected scalar field, *Phys. Rev. Lett.*, **63**, pp. 2657-2660.

Electronic mail: takaoka@kyoryu.scphys.kyoto-u.ac.jp

COHERENT STRUCTURE AND SUBGRID-SCALE ENERGY TRANSFER IN TURBULENCE

KIYOSI HORIUTI
Department of Mechano-Aerospace Engineering,
Tokyo Institute of Technology,
2-12-1 O-okayama, Meguro-ku, Tokyo 152-8552, Japan

Abstract. Correlation of the transfer and dissipation of turbulent energy with the coherent structures is investigated. A classification method is developed to distinguish the tube and sheet structures. Two different kinds are considered for sheet structures. The grid-scale and subgrid-scale (SGS) energy exchange in large-eddy simulation (LES) is utilized to estimate the energy transfer. It is found that the cylindrical sheet region around the tube is organized, shielding the rather incoherent core region of tube. The energy transfer and dissipation were shown to take place primarily in the region similar to the Burgers' vortex layer. The backward scatter primarily occurred along the compressed vortex tube. The role reversal of the vorticities along eigenvalues for the strain rate tensor is discussed.

1. Proposal of classification method and its assessment

Coherent structures are roughly divided into the two groups: the vortex tube and sheet, but these structures are not uniquely defined, e.g., for the vortex sheet structures, there are (at least) two kinds. One is that similar to the sheet generated in the background flow with the plane strain (Burgers' layer), and the other one is the cylindrical sheet surrounding the vortex tube. In order to distinguish these sheet and tube structures, a method in which these structures are accurately classified is necessary. Furthermore, this method should be capable of an estimation for the flatness and curvature of the object, enabling it to distinguish the two kinds of sheet structures mentioned above. We considered the eigenvalue solutions for the strain rate tensors, S_{ij}, and the λ_2 method [1], because it was felt that the rigorous

T. Kambe et al. (eds.), IUTAM Symposium on Geometry and Statistics of Turbulence, 249–254.
© 2001 *Kluwer Academic Publishers.*

use of the off-diagonal elements for the products of the velocity gradient tensors is indispensable to estimate the flatness of the objects. Although it may be inappropriate to rely heavily on the Burgers' vortex models, we can derive some insight for the classification by using the eigenvalue solutions of the Burgers' tube and layer models. The three eigenvalues for the λ_2 method can be given for the Burgers' vortex tube as

$$\lambda_{\pm} = \frac{\alpha^2}{4}\left[\left(1 \pm Re_{\Gamma}[\frac{4\nu}{\alpha r^2}\{1-\exp(-\frac{\alpha r^2}{4\nu})\} - \exp(-\frac{\alpha r^2}{4\nu})]\right)^2 - Re_{\Gamma}^2 \exp(-\frac{\alpha r^2}{2\nu})\right], \lambda_z = \alpha^2, \quad (1)$$

where α denotes the stretching parameter, and $Re_{\Gamma} = \Gamma/(4\pi\nu)$ is the Reynolds number based on the circulation around the tube, Γ. It can be seen that the eigenvalues scale with α^2 for low Reynolds numbers and with α^2 and Re_{Γ}^2 for high Reynolds numbers [2]. Because of the crossover of the eigenvalues, an alignment of the vorticity vector with the eigenvector for the intermediate eigenvalue often occurs [3, 4]. We reordered the eigenvalues so that the eigenvalue, the eigenvector of which is most aligned with the vorticity vector, is chosen as λ_z, and the largest eigenvalue among the remaining two eigenvalues as λ_+, and the smallest one as λ_-. Similar reordering was used to define the eigenvalues for S_{ij} as $(\sigma_z, \sigma_+, \sigma_-)$ [5] and the corresponding eigenvectors are denoted as $\mathbf{e}_z, \mathbf{e}_+, \mathbf{e}_-$, respectively. The λ_2 method was used because it was found that eigenvalue crossover less frequently occurred in the λ_2 method, and the strain eigenvalue method yielded an overestimation of the cylindrical sheet region and an underestimation of the tube core region.

It was shown that the eigenvalues $(\lambda_z, \lambda_+, \lambda_-)$ and $(\sigma_z, \sigma_+, \sigma_-)$ can be equated as $\lambda_{\pm} = \sigma_{\mp}^2$ in the cylindrical vortex sheet region around the

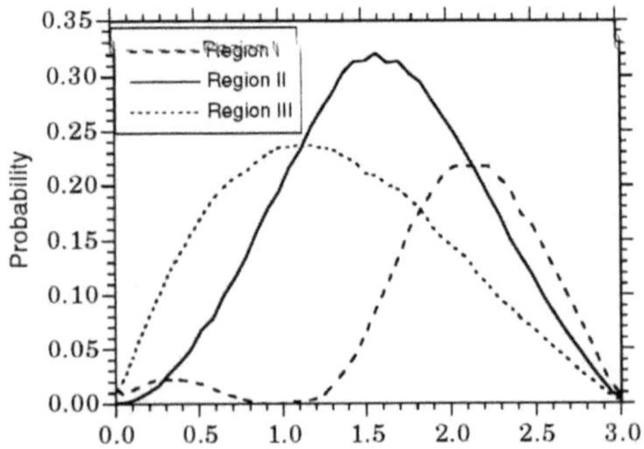

Fig.1 Pdf profiles for strain parameter

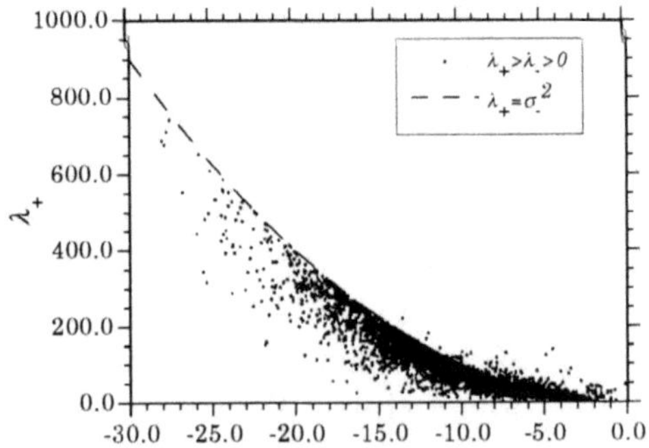

Fig.2 Scatter plot between σ_- and λ_+

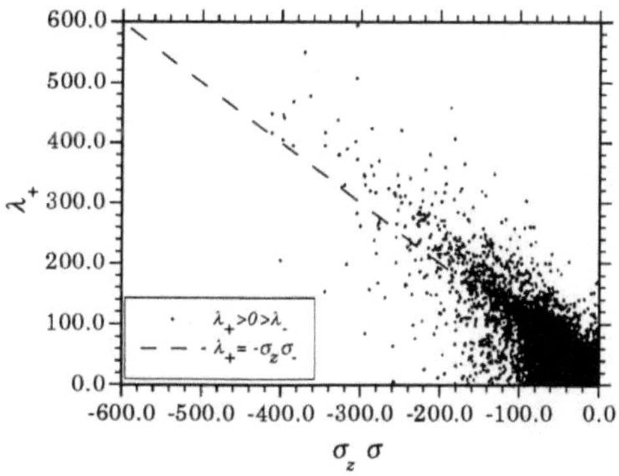

Fig.3 Scatter plot between σ_- and λ_+

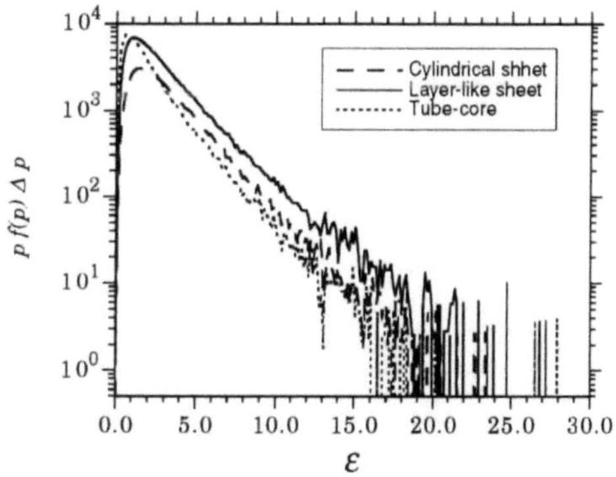

Fig.4 Pdfs for the dissipation rate ε

tube, which leads to the inequality, $\lambda_+ \geq \lambda_- > 0$. This scaling of $\lambda_\pm$ and $\sigma_\mp$ is a consequence of a curvature in the cylindrical sheet. They can be equated as $\lambda_\pm = \alpha\,\sigma_\mp$ in the Burgers' vortex layer, leading to the inequality, $\lambda_+ \geq 0 \geq \lambda_-$. This scaling of $\lambda_\pm$ and $\sigma_\mp$ was obtained since the layer-like sheet is flat. We note that $\lambda_z = \sigma_z^2$ for both tube and layer. Using these results, we classify the cylindrical sheet and the Burgers' layer like sheet by imposing the constraints as ($\lambda_+ \geq \lambda_- > 0$) (Region I) and ($\lambda_+ \geq 0 \geq \lambda_-$) (Region II), respectively. The core region of the vortex tube is classified as $\lambda_+ < 0$ (Region III).

We have utilized the incompressible decaying homogeneous isotropic turbulence Direct numerical simulation (DNS) data which were generated with 128, 128 and 128 grid points, at the instant when the Taylor microscale Reynolds number ≈ 30. The profiles for the probability density functions (pdf) for the strain parameter [6], $s_p = ((\sigma_- - \sigma_+)/(\sigma_- + \sigma_+))$, for the Regions I, II and III are shown in Fig. 1. The pdfs for Regions I, II and III are mostly concentrated in characteristically different three regions, respectively, in the ranges $1.5 < s_p < 3$ (biaxial strain), $1 < s_p < 2$ (biaxial strain close to plane strain), and $0 < s_p < 1$ (axial strain), approximately educing the desired objects. It should be noted that the present classification criterion is merely necessary for the equations, $\lambda_\pm = \sigma_\mp^2$, $\lambda_\pm = \alpha\,\sigma_\mp$, to hold. To examine if it is also sufficient to hold these relations, we show the scatter plot between σ_- and λ_+ in Figs. 2 and 3. Plot between σ_- and λ_+ obtained by imposing the constraint $\lambda_+ \geq \lambda_- > 0$ is shown in Fig. 2. The plotted points are clustered along the parabolic profile, $\lambda_+ = \sigma_-^2$. The deviation from this parabola indicates some difference from the pure Burgers' vortex tube model [6]. The plot between $\alpha\sigma_-$ and λ_+ obtained by imposing the constraint $\lambda_+ \geq 0 \geq \lambda_-$ is shown in Fig. 3, in which α is estimated by using σ_z. Overall, the plotted points are aligned with the linear

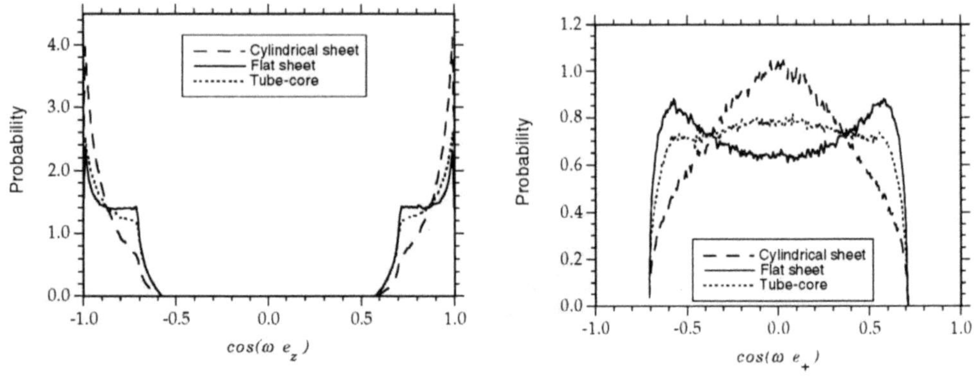

Fig.5 Pdfs for the cosine of the angle between the vorticity vector and the eigenvectors (a) ω and σ_z, (b) ω and σ_+.

profile, $\lambda_+ = \alpha\ \sigma_-$. Significant deviation, however, is discernible, because the Burgers' vortex layer is susceptible to small disturbances.

2. Classification of statistics in the DNS data

Figure 4 shows the (weighted-)pdfs for the dissipation rate of the turbulent energy, ε. To show the actual strength of the terms, we multiplied pdf, $f(p)$, by its bin value, p, and the bin interval, Δp. Intense dissipation primarily occurs in the sheet region, which is close to the Burgers' vortex layer. The dissipation generated in the tube region is rather weak, and the dissipation generated in the cylindrical sheet region is intermediate between them. Figure 5 shows the pdf distributions of the cosine of the angle between the vorticity vector, ω, and the eigenvectors, $\mathbf{e}_z$ and $\mathbf{e}_+$. Strong tendency for alignment between ω and $\mathbf{e}_z$ can be seen in the cylindrical sheet region, and $\mathbf{e}_+$ is placed almost perpendicular to ω in this region. These tendencies are weak in the Burgers' layer-like region and the tube-core region. The flat distributions of cosine values in the tube-core and layer-like regions indicate that the velocity fields in these regions are close to a random Gaussian field [5]. These results implies that the cylindrical sheet region is coherent, and shielding a rather incoherent core region of the tube. We consider that the vortex tube persists for a long time pf period [6] due to this shielding effect.

3. Analysis of the grid-scale and SGS energy transfer

We examine the contribution of the structures for the energy cascade by estimating the grid scale - SGS energy transfer in LES, because in this method, we can estimate energy cascade from the given scale down to the

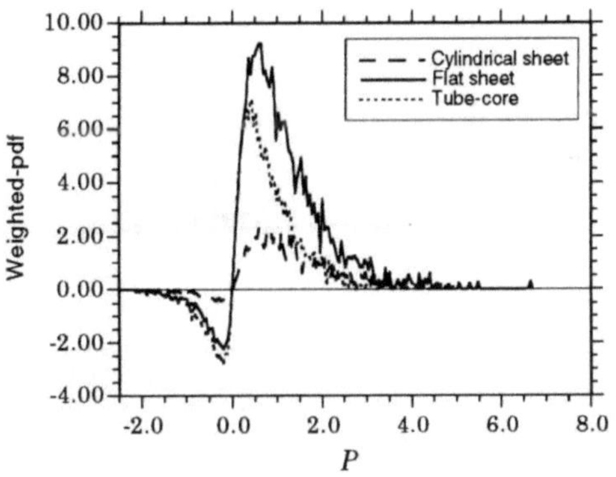

Fig.6 Pdf profiles for the P term

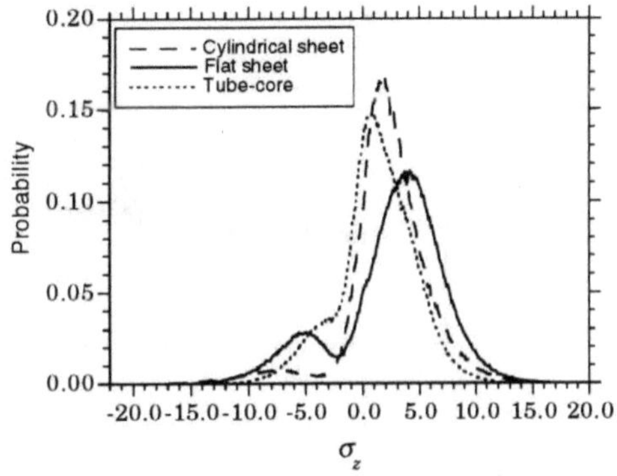

Fig.7 Pdf profiles for σ_z

smaller scales. The energy transfer occurs through the SGS production term, $P = -\tau_{ij}\partial\overline{u}_i/\partial x_j$ (τ_{ij} is the SGS stress tensor). Positive P implies the forward scatter of the grid-scale energy into the SGS, while negative P implies the backward scatter. We obtained an estimate of the P term by applying the Gaussian filter to the homogeneous isotropic DNS data, reducing 128 grid points to 64 grid points in each direction. Similar results were obtained when 128 grid points were reduced to 32 points.

Figure 6 shows the pdfs ($pf(p)\Delta p$) of the P term. In good correspondence with the decomposition of the dissipation rate (Fig. 2), the energy cascade (forward scatter) primarily arises in the layer-like region. These results are consistent with the previous result that the sheet structures are mostly responsible for the cascade of the turbulent energy [7]. It should be noted that the backward scatter primarily occurs in the tube-core region. Considerable occurrence of negative σ_z was noticed in the tube-core

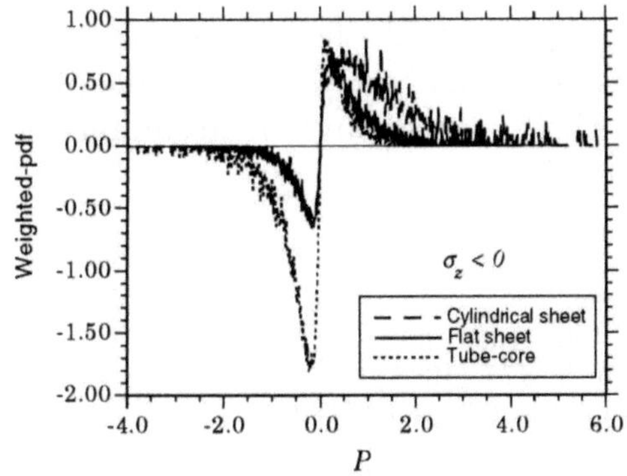

Fig.8 Pdfs for the P term with $\sigma_z < 0$

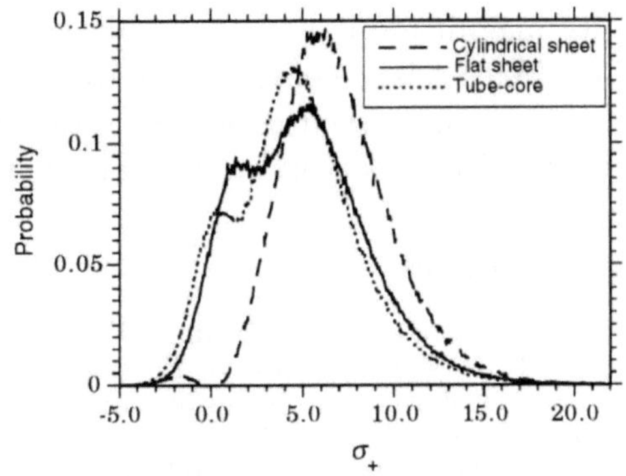

Fig.9 Pdf profiles for σ_+

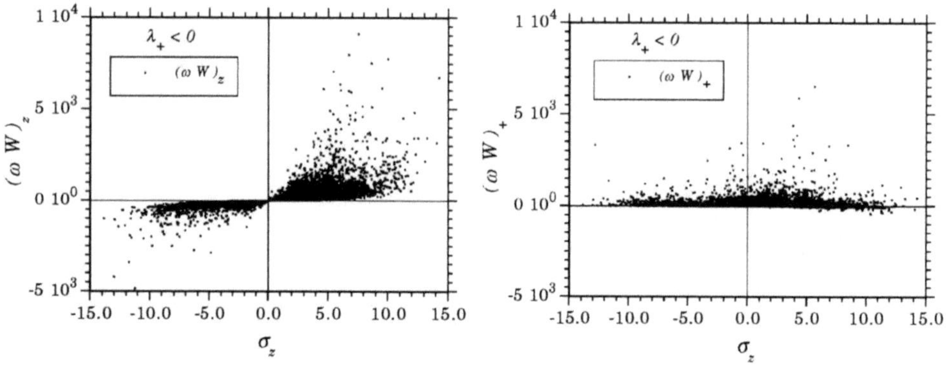

Fig.10 Scatter plot between σ_z and the vortex-stretching terms. (a) $\sigma_z(\mathbf{e}_z \cdot \omega)^2$, (b) $\sigma_+(\mathbf{e}_+ \cdot \omega)^2$.

and cylindrical sheet regions, as is shown in Fig. 7 for the pdfs of σ_z. It was found that the backward scatter events were highly correlated with the events with negative σ_z, indicating that the backward scatter predominantly occurs along the compressed vortex tube. In fact, Fig. 8 for the pdfs $(pf(p)\Delta p)$ of the SGS production term, which was conditionally sampled with $\sigma_z < 0$, shows that the P term is mostly negative in the tube-core region when $\sigma_z < 0$. Figure 9 shows the pdf for σ_+, and σ_+ is mostly positive, implying that, when $\sigma_z < 0$, the amplitude of ω_z is reduced, while the amplitude of ω_+ is increased due to the backward scatter.

This can be seen in Fig. 10, in which the scatter plot between σ_z and the vortex-stretching term, $\omega \cdot \mathbf{W}$, obtained in the tube-core region of the (unfiltered) DNS data is shown. $\mathbf{W}$ denotes the vortex-stretching vector,

$$\mathbf{W} = \sigma_z(\mathbf{e}_z \cdot \omega)\mathbf{e}_z + \sigma_+(\mathbf{e}_+ \cdot \omega)\mathbf{e}_+ + \sigma_-(\mathbf{e}_- \cdot \omega)\mathbf{e}_-. \qquad (2)$$

It can be seen that the component along $\mathbf{e}_+$, is predominantly positive, and its magnitude is not negligibly small compared with the component along $\mathbf{e}_z$. With a lapse of time, the role reversal of ω_z and ω_+ [2, 6] will take place in the compressed tube. We note that this reversal can't be accomplished in the Burgers' vortex tube and layer models.

References

1. J. Jeong and F. Hussain: J. Fluid Mech. **285** (1995) 69.
2. K.K. Nomura and G.K. Post: J. Fluid Mech. **377** (1998) 65.
3. S. Kida: Lecture Notes in Numerical Applied Analysis **12** (1993) 137.
4. J. Jiménez: Phys. Fluids A **4** (1992) 652.
5. B. Andreotti: Phys. Fluids **9** (1997) 735.
6. H.K. Moffatt, S. Kida and K. Ohkitani: J. Fluid Mech. **259** (1994) 241.
7. W.T. Ashurst, A.R. Kerstein, R.M. Kerr and C.H. Gibson: Phys. Fluids **30**, 2343 (1987).

A DYNAMICAL MODEL FOR TURBULENCE

J-P. LAVAL
DAPNIA/SAp, CE Saclay, 91191 Gif sur Yvette Cedex, France
B. DUBRULLE
NCAR, P.O. Box 3000, Boulder CO 80307-3000, USA
CNRS, URA 285, Observatoire Midi-Pyrénées, 14 avenue Belin, F-31400 Toulouse, France
AND
S. NAZARENKO
Mathematics Institute University of Warwick COVENTRY CV4 7AL, UK
University of Arizona, Department of Mathematics, Tucson AZ 85721, USA

In classical turbulence theory, the closure problem arises when considering the statistical or filter average of the forced Navier-Stokes equations:

$$\partial_t \bar{u}_i + \partial_j \overline{u_i u_j} = -\partial_i \bar{p} + \nu \Delta \bar{u}_i + \bar{f}_i. \tag{1}$$

Here, $\overline{u_i u_j}$ are the Reynolds stresses, which need to be prescribed in order to close the system. Most turbulence models try to close the system by prescribing directly the shape of the Reynolds stresses using, e.g. dimensional arguments and arbitrary parameters which need to be calibrated. A famous example is $\overline{u_i u_j} = \nu^{(t)}_{ijkl} \partial_k \bar{u}_l$, where $\nu^{(t)}_{ijkl}$ is the eddy-viscosity tensor. We have used a novel approach, in which the Reynolds stresses are computed by an estimate of the r.m.s. or sub-grid velocities $u'_i = u_i - \bar{u}_i$. The estimate is done via a dynamical equation linking the average or resolved velocity and the r.m.s. or sub-grid velocities. This equation is not postulated, but derived from the constitutive equations using a simplifying assumption. In this model, only the non-local interactions between the small random scales and the mean large scale flow are exactly taken into account [5]. It is thus referred to as a "non-local model" of turbulence. It is in fact a generalization of the Rapid Distortion Theory (RDT) [3]. This new model allows both analytical predictions and numerical implementa-

T. Kambe et al. (eds.), IUTAM Symposium on Geometry and Statistics of Turbulence, 255–260.
© 2001 *Kluwer Academic Publishers.*

tion [6]. We present a numerical application for 2D decaying turbulence and theoretical results about a special 3D case as plane flows.

1. Two-dimensional turbulence

We consider a two-dimensional incompressible inviscid fluid obeying the Euler equations. The large $(\mathbf{U}, \Omega)$ and the small-scale part $(\mathbf{u}', \omega')$ of the velocity and the vorticity fields are defined via a filtering procedure.

$$\mathbf{U} = \overline{\mathbf{u}} = \int G(\mathbf{x} - \mathbf{x}')\mathbf{u}(\mathbf{x}')d\mathbf{x}' \tag{2}$$

$$\Omega = \overline{\omega} = \int G(\mathbf{x} - \mathbf{x}')\omega(\mathbf{x}')d\mathbf{x}' \tag{3}$$

$$\mathbf{u}' = \mathbf{u} - \overline{\mathbf{u}} \tag{4}$$

$$\omega' = \omega - \overline{\omega}$$

where G is the filtering function which is defined by the numerical method used to solve the equations. Inserting this decomposition in the Euler equation and separating the large and the small scale part, we get a set of coupled equations. The base of our model is to neglect the terms corresponding to local interactions at small scales (terms involving $\mathbf{u}'\omega'$). The validity of such hypothesis is detailed in [5]. In this paper the authors compared the weight of each terms of the equations. It is shown that even additional terms can be neglected at the first order. After such approximations, the equations are reduced to:

$$\partial_t \Omega + \operatorname{div} \overline{\mathbf{U}\Omega} + \operatorname{div} \overline{\mathbf{U}\omega'} = 0, \tag{5}$$

$$\partial_t \omega' + \operatorname{div}(\mathbf{U}\omega') = -(\operatorname{div}(\mathbf{U}\Omega) - \operatorname{div} \overline{\mathbf{U}\Omega}) + \operatorname{div} \overline{\mathbf{U}\omega'} \equiv F \tag{6}$$

where F correspond to a forcing term from large scales on the smallest one. We implemented numerically this model for 2D periodical flows using a pseudo-spectral code for the large scales. Because of our modelisation, the small scale equation is linear, thereby providing room for computational time reduction via semi–Lagrangian methods, using a time step related to the large scales. We choose to use a Garbor decomposition, which provides a localized description while allowing theoretical manipulations similar to that obtained with Fourier modes. The equation of motion of the small-scale wave-packets is obtained by applying a Gabor Transform (GT) to the linearized equation of motion of small scales in the physical space. This equation is the following :

$$D_t \, GT(\omega'(\mathbf{x}, t)) = GT(F(\mathbf{x}, t)) \tag{7}$$

where D_t is a total derivative defined as :

$$
\begin{aligned}
D_t &= \partial_t + \dot{\mathbf{x}} \cdot \nabla + \dot{\mathbf{k}} \partial_k, && (8)\\
\dot{\mathbf{x}} &= \partial_k \Phi, && (9)\\
\dot{\mathbf{k}} &= -\nabla \Phi, && (10)\\
\Phi &= \mathbf{U} \cdot \mathbf{k}, && (11)
\end{aligned}
$$

and $F(\mathbf{x}, t)$ is the forcing of small scales correspondly to the creation of small scales from large-scales. A Particles In Cell (PIC) method is used to advect the small-scale wave-packets.

We performed several simulations with our model to compare with DNS. As it is not possible to do DNS with an infinite Reynolds number we performed simulations with a Reynolds number as high as possible. The comparison was done for two typical situations. We present only the comparison for decaying turbulence. The energy of the initial field is concentrated at very large scales. Small scales are created under the effect of enstrophy cascade. The DNS was performed on a grid 1024^2 with a viscosity $\nu = 1.8\ 10^{-4}$ The corresponding Reynolds number is approximately 10^4. The separation scale used in our model was $k_c = 21$. The large scales evolve on a 64^2 grid and the Reynolds Stress is computed on a grid twice as big as the large-scale one in order to compute the forcing term $F(\mathbf{x}, t)$ involved in the small-scale equation (7). The results are compared after approximately 50 turnover times The large scales vorticity fields are represented fig. 1. The large scales field are obtained by a truncature at $k < 21$ of the total vorticity field. These scales correspond to the smallest scales directly resolved in our modelisation. The total vorticity field of our modelisation is obtained by the addition of the resolved vorticity field to the small-scales field rebuilt from wave-packets. In this case we used between 2×10^4 and 6×10^4 wave-packets for the small scales discretisation. There is a good agreement between the result of our model and the DNS one. The largest structures are very similar and they are well localized even after 50 turnover times. We also compared our model to two other models. The first one is a hyperviscous direct simulation on a 64^2 grid. The results are quite different from the large DNS. The vortices are much less localized than for our model. The second LES model of comparison is the APVM model developed by Sadourny and Basdevant [4]. The result of this model is better than the HDNS 64^2 but not as accurate as our model in this particular case. Using the wave-packets, on can rebuild the small-scale fields. The good discretisation of small scales is obvious in the comparison of the small scale vorticity field between our simulation and the DNS 1024^2 (fig. 2). The corresponding energy spectra of the 4 simulations fig. 1 are represented on the same graph fig. 3. This comparison confirms the good agreement between DNS and our modelisation. The simulation with our model develop a k^{-3} energy spectrum very close to the DNS when the equivalent spectrum of the HDNS 64^2 and the APVM 64^2 seems to be steeper.

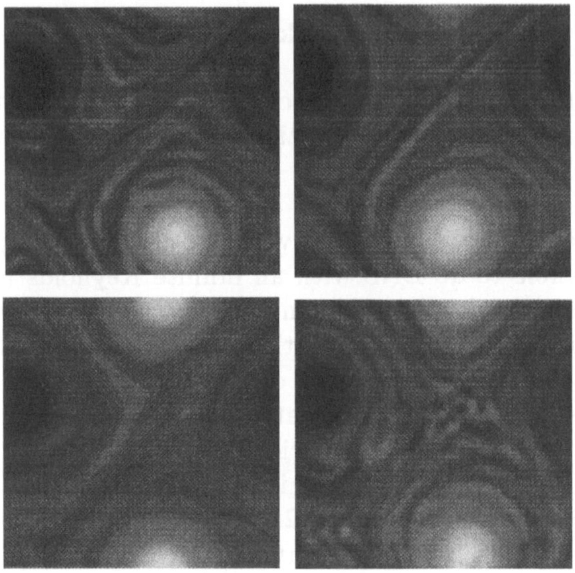

Figure 1. Large scales vorticity field ($k \leq 21$) after 50 turnover times for a decaying turbulence. **Left**: DNS on a 1024^2 grid and a viscosity $\nu = 1.810^{-4}$, **middle**: our model with a separation scale at $k = 21$, **right**: simulation with hyper-viscosity on a 64^2 grid.

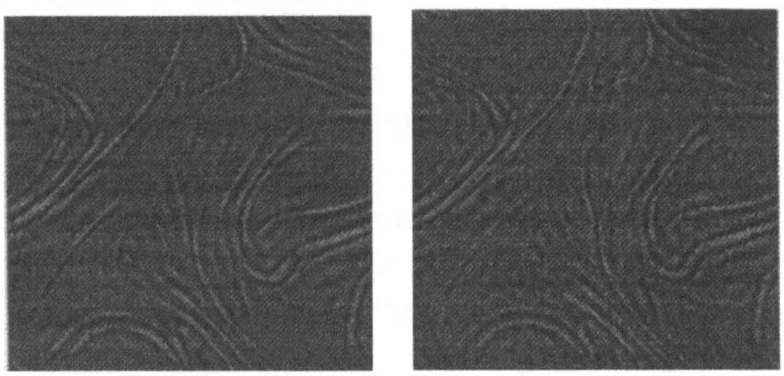

Figure 2. Small Scales vorticity field after 15 turnover time for a decaying turbulence. Result from a DNS on a 1024^2 grid and a viscosity $\nu = 1.810^{-4}$ (left) and from our model with a separation scale at $k = 21$.

2. Equilibrium profiles of channel flows

In plane parallel geometry, our model can be used to derive analytically the equilibrium profiles both in the near-wall and core regions [1]. Using an

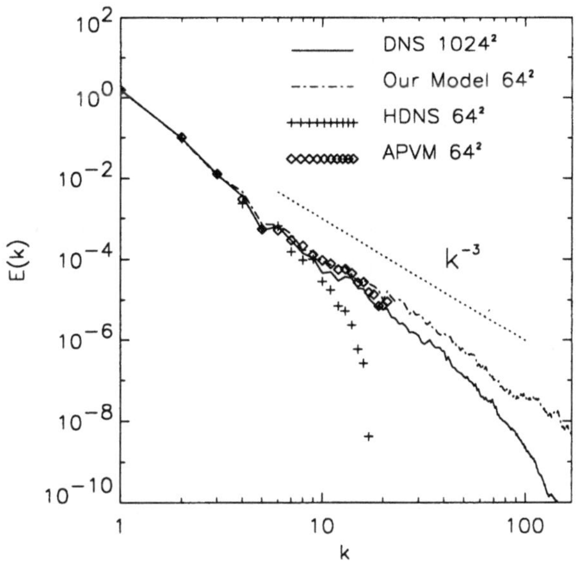

Figure 3. Comparison of the energy spectra obtain from the total vorticity field for decaying turbulence.

ideal filter to separate large and small scales, we can introduce the same decomposition into large and small scale equations. As in 2D, the small scale equation involves a forcing term which is responsible for the energy exchange between the two scales. Because of the linearity of the small scale equation, one can solve it with respect to the large scale field and the forcing term. Thanks to a reasonable hypothesis on the statistics of the forcing, we deduce the expression of the Reynolds sub-grid stress involved in the large scale equation. In the near-wall region we derive a general expression for the velocity profile which is linear in the viscous layer and logarithmic outside. This expression involves two physical parameters: the von Karman constant and the size of the viscous layer (which can be computed via a numerical implementation of our model). Fits of experimental profiles using our general formula provides reasonable values of these parameters ($\kappa = 0.4$ to $\kappa = 0.45$, size of the viscous layer about 15 wall units). In the core region, we find that the shape of the profile depends on the geometry of the flow: it ranges from algebraic in channel flow, to exponential in the bulk of boundary layers, or linear in plane Couette flow. This classification is consistent with Oberlack's system [2], which is based on symmetry arguments. Fits of boundary layer flows or channel flows at different Reynolds number over the whole flow region are performed using our results, and are found to be in very good agreement with available data (see fig. 4).

3. Conclusion

We presented two applications of our new model which is based on a simple hypothesis about the small scale evolution. This approach is different from traditional turbulent models since our model provides an expression of the turbulent Reynolds sub-grid stress via estimate of the sub-grid velocities rather than velocities correlations. This model leads to a reduction of simulation time with respect to DNS as the small scales are computed with a linearized equation using a semi-Lagrangian method.

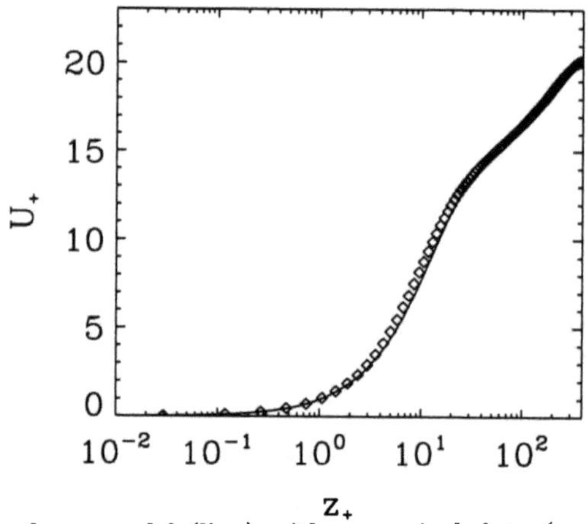

Figure 4. Comparison of our model (line) with numerical data (symbols) of a channel flow at $Re = 587.19$.

References

1. Dubrulle, B. and Laval, J.-P. and Nazarenko, S. V. and Kevlahan, N. K.-R. (1999) Derivation of equilibrium profiles in plane parallel flows using a dynamic subgrid-scale model, *Submitted to Phys. Fluids*
2. Oberlack, M., Unified theory for symmetries in plane parallel turbulent shear flows Center for Turbulence Research, Stanford University/NASA Ames *manuscript no. 163, under review in J. Fluid Mech.*
3. Townsend, A. A., The structure of turbulent shear flow (second edition), Cambridge university press, Cambridge
4. Sadourny, R. and Basdevant, C., Parameterization of subgrid scale barotropic eddies in quasi-geotrophic models: Anticipated Potential vorticity method, *J. Atm. Sci.*,**Vol. no. 42, pp. 1353-1363**
5. Laval, J.-P. and Dubrulle, B. and Nazarenko, S., Nonlocality of interaction of scales in the dynamics of 2D incompressible fluids",*Phys. Rev. Lett.*,**Vol. no. 83, pp. 4061-4064**
6. Laval, J.-P., Développement d'un nouveau modèle dynamique pour la turbulence: application aux écoulement bidimensionnels et aux écoulements plans, *PhD thesis, Université Paris VI*

TWO- AND THREE- DIMENSIONAL BEHAVIOR OF THE LARGE SCALES IN ROTATING TURBULENCE

ROBERT RUBINSTEIN
Institute for Computer Applications in Science & Engineering
NASA Langley Research Center
Hampton, Virginia
USA

Abstract. Weak turbulence theory is treated as an approximate solution of the DIA equations for rotating turbulence. The consistency of weak turbulence theory with numerical simulations of rotating turbulence in small aspect-ratio domains is evaluated. Finally, the possibility of four-wave interactions in this problem and in the problem of compressible rotating turbulence is discussed.

1. DIA and weak turbulence theory for rotating turbulence

Following Leslie (1973), an approximate solution of the DIA closure equations for rotating turbulence can be constructed by introducing the Markovian *ansatz*

$$G_{ij}(\mathbf{k},\tau) = e^{-\mu_k\tau}\{\cos(\omega_k\tau)P_{ij}(\mathbf{k}) + \sin(\omega_k\tau)\xi_{ij}(\mathbf{k})\}H(\tau)$$
$$Q_{ij}(\mathbf{k},\tau) = e^{-\mu_k\tau}\{\cos(\omega_k\tau)P_{ij}(\mathbf{k}) + \sin(\omega_k\tau)\xi_{ij}(\mathbf{k})\}Q(\mathbf{k}) \qquad (1)$$

for the DIA response and correlation tensors; compare also Carnevale and Martin (1982). In Eq. (1), τ denotes time separation, $P_{im} = \delta_{im} - k^{-2}k_ik_m$ and ξ_{im} is the (unique) antisymmetric tensor formed from the vectors of the local Craya-Herring or toroidal-poloidal basis (Cambon and Jacquin, 1989). Integrating the DIA response equation over time separation, and separating the coefficients of P and ξ, there results

$$1 - 2\frac{\Omega_k\omega_k}{\mu_k^2+\omega_k^2} = \{P_{imn}(\mathbf{k})P_{mrs}(\mathbf{p})P_{ns}(\mathbf{q})I_1 + P_{imn}(\mathbf{k})\xi_{mrs}(\mathbf{p})\xi_{ns}(\mathbf{q})I_2$$
$$+P_{imn}(\mathbf{k})\xi_{mrs}(\mathbf{p})P_{ns}(\mathbf{q})I_3 + P_{imn}(\mathbf{k})P_{mrs}(\mathbf{p})\xi_{ns}(\mathbf{q})I_4\}Q(\mathbf{q}) \times$$

T. Kambe et al. (eds.), IUTAM Symposium on Geometry and Statistics of Turbulence, 261–266.
© 2001 *Kluwer Academic Publishers.*

$$[P_{ir}(\mathbf{k})\frac{\mu_k}{\mu_k^2+\omega_k^2}+\xi_{ir}(\mathbf{k})\frac{\omega_k}{\mu_k^2+\omega_k^2}] \tag{2}$$

and

$$2\frac{\Omega_k\mu_k}{\mu_k^2+\omega_k^2}=\{P_{imn}(\mathbf{k})P_{mrs}(\mathbf{p})P_{ns}(\mathbf{q})I_1+P_{imn}(\mathbf{k})\xi_{mrs}(\mathbf{p})\xi_{ns}(\mathbf{q})I_2$$
$$+P_{imn}(\mathbf{k})\xi_{mrs}(\mathbf{p})P_{ns}(\mathbf{q})I_3+P_{imn}(\mathbf{k})P_{mrs}(\mathbf{p})\xi_{ns}(\mathbf{q})I_4\}Q(\mathbf{q})\times$$
$$[\xi_{ir}(\mathbf{k})\frac{\mu_k}{\mu_k^2+\omega_k^2}-P_{ir}(\mathbf{k})\frac{\omega_k}{\mu_k^2+\omega_k^2}] \tag{3}$$

where

$$\begin{aligned}
I_1 &= \frac{\mu_p+\mu_q}{(\mu_p^2+\mu_q^2)+(\omega_p^2+\omega_q^2)}+\frac{\mu_p+\mu_q}{(\mu_p^2+\mu_q^2)+(\omega_p^2-\omega_q^2)}\\
I_2 &= -\frac{\mu_p+\mu_q}{(\mu_p^2+\mu_q^2)+(\omega_p^2+\omega_q^2)}+\frac{\mu_p+\mu_q}{(\mu_p^2+\mu_q^2)+(\omega_p^2-\omega_q^2)}\\
I_3 &= \frac{\omega_p+\omega_q}{(\mu_p^2+\mu_q^2)+(\omega_p^2+\omega_q^2)}-\frac{\omega_p-\omega_q}{(\mu_p^2+\mu_q^2)+(\omega_p^2-\omega_q^2)}\\
I_4 &= \frac{\omega_p+\omega_q}{(\mu_p^2+\mu_q^2)+(\omega_p^2+\omega_q^2)}+\frac{\omega_p-\omega_q}{(\mu_p^2+\mu_q^2)+(\omega_p^2-\omega_q^2)}
\end{aligned} \tag{4}$$

In Eqs. (2)-(3), integration over wavevector triads $\mathbf{k}=\mathbf{p}+\mathbf{q}$ is understood, the inertial wave frequency is

$$\Omega_k=\Omega(\mathbf{k})=\mathbf{k}\cdot\boldsymbol{\Omega}/k \tag{5}$$

where $\boldsymbol{\Omega}$ is the angular velocity of the flow domain, $\mu_k=\mu(\mathbf{k})$ is the eddy damping, and $\omega_k=\omega(\mathbf{k})$ is the effective rotation rate. The closure is completed by an equation for $Q(\mathbf{k})$ which is the EDQNM form of the DIA energy balance equation. Note that the effective rotation rate ω_k and the eddy damping μ_k are to be determined simultaneously.

Basic limits of these equations are defined by the relative strengths of rotation and nonlinearity. In the absence of rotation, the equations reduce to Leslie's (1973) Markovianized DIA, and when rotation is nonzero but small, $\Omega_k << \eta_k$ where η_k is the Kolmogorov frequency proportional to $\varepsilon^{1/3}k^{2/3}$, rotation is a small perturbation of nonlinearity. A theory of this limit applicable to mean flow quantities has been developed by Shimomura and Yoshizawa (1986).

The opposite case, in which $\Omega_k >> \eta_k$ is analyzed instead by ignoring both eddy damping and renormalization of the rotation rate. This is the limit of *weak turbulence* (Zakharov *et al*, 1992) in which rotating turbulence is the turbulence of weakly interacting *inertial waves*

$$\mathbf{u}_i(\mathbf{x},t)=\mathbf{U}_i\exp(i\mathbf{k}\cdot\mathbf{x}+i\Omega_k t) \tag{6}$$

This instance of weak turbulence theory is somewhat special because of the existence of non-oscillating two-dimensional modes for which $\Omega_k = 0$. The dynamics of these modes can only be derived correctly on the basis of a theory which includes eddy damping, like Eqs. (2)-(3). The characteristic time-scale of this theory is the relaxation time of resonant three-wave interactions

$$\Theta(\mathbf{k}, \mathbf{p}, \mathbf{q}) = \int_0^\infty d\tau \exp i\tau\{\Omega_k \pm \Omega_p \pm \Omega_q\} = \frac{1}{\Omega}\delta(\Omega_k \pm \Omega_p \pm \Omega_q) \tag{7}$$

where it is found that the principal value terms in the integral do not contribute at this level of approximation (J. Scott, private communication). Double resonances, singlarities of the surfaces $\Omega_k \pm \Omega_p \pm \Omega_q = 0$, which would cause corrections to Eq. (7) (Newell and Aucoin, 1971), can be shown to be irrelevant (Rubinstein, 1999).

From the viewpoint of Eqs. (2)-(4), weak turbulence theory is the lowest order solution of DIA in the limit of very strong rotation. It is therefore natural to attempt to solve DIA perturbatively about this limit (Carnevale and Martin, 1982). At the next order, damping corrections of order

$$\mu \sim k\sqrt{\varepsilon/\Omega} \tag{8}$$

are obtained; the resonance conditions allow this correction even for two-dimensional modes. Thus, although these modes cannot exchange energy with inertial waves through three-wave processes, energy transfer through higher order interactions is possible. The damping correction of Eq. (8) may be related to the Newell quartet mechanism (Newell, 1969) discussed in the context of rotating turbulence by Smith and Waleffe (1999).

The main result of this theory of three-wave interactions in rotating turbulence is that energy is transferred toward the plane perpendicular to the rotation axis (Cambon and Jacquin, 1989; Waleffe, 1993). A typical consequence is the growth of the integral scale parallel to the rotation axis during decay.

2. Rotating turbulence in small aspect ratio domains

Simulations of turbulence in small aspect ratio domains have been reported by L. Smith and collaborators. By removing scales of motion in one direction, the possibility of a continuous transition from three- to two-dimensional behavior is raised (Smith *et al*, 1996). Suppose that the energy flux depends continuously on aspect ratio. Then, as in the d-dimensional turbulence of Fournier and Frisch (1978), at a critical aspect ratio, the forward transfer of energy ceases; at smaller aspect ratios, the energy transfer is toward large scales as in fully two-dimensional turbulence. This reversed cascade should be possible even though the turbulence remains three-dimensional, and enstrophy is not conserved.

Numerical simulations (Smith *et al*, 1996) were only partially consistent with this picture. In sufficiently small aspect ratio, an inverse cascade could be generated by two-dimensional forcing. This establishes a weaker result, namely the stability of small aspect ratio two-dimensional turbulence to three-dimensional perturbations.

More recent simulations (Smith and Waleffe, 1999) have treated rotating turbulence in small aspect ratio domains. In this case, it is found that a two-dimensional inverse cascade can be sustained at a fixed aspect ratio if the rotation rate is sufficiently large.

A qualitative explanation can be based on three-wave interactions. In the simulations, the forcing is isotropic and of a scale smaller than the length of the domain along the rotation axis. Resonant triads (for example, equilateral triangles) exist at these scales and transfer energy toward the perpendicular plane. In the absence of three dimensional scales, energy in this plane is naturally transferred towards large scales. The development of two-dimensionality in the large scales is brought about by the triads at scales comparable to the forcing scale.

It is stressed by Smith and Waleffe (1999) that since resonant interaction between two inertial waves and a planar Fourier mode transfers no energy to the planar mode, energy cannot be transferred to the perpendicular plane by resonant triad interactions. However, since the set of modes dominated by nonlinearity always includes a region surrounding the perpendicular plane, this explanation only requires transfer *toward* this plane.

Smith and Waleffe also report the development of a two-dimensional k^{-3} spectrum in some cases. Since this spectrum does not correspond to the flux of a conserved quantity, it must be unsteady.

3. Four wave interactions in rotating turbulence

An important consequence of small aspect ratio was noted by V. Yakhot, who argued that if $k_z >> k_x, k_y$ with $k_z \approx L_z^{-1}$ then the three-wave dispersion relation cannot be satisfied. In fact then,

$$\Omega_k = \Omega \frac{k_z}{k} \approx \Omega\{1 - \frac{1}{2}(\frac{k_\perp}{k_z})^2 \pm \cdots\} \tag{9}$$

which can be compared to the dispersion relation of Langmuir turbulence (Zakharov *et al*, 1992)

$$\Omega(\mathbf{k}) = \Omega_P \sqrt{1 + 3(kr_D)^2} \approx \Omega_P\{1 + \frac{3}{2}(r_D k)^2 + \cdots\} \tag{10}$$

where Ω_P is the plasma frequency and r_D is the Debye length.

To leading order, the three-wave resonance condition cannot be satisfied and the theory leads to four-wave resonances. This has an important implication for the large scales: such resonances conserve *wave action* as well as energy; therefore, a spectrum carrying constant flux of wave action to large scales exists of the form

$$E = C\varepsilon_a^{1/3}\Omega^{4/3}L_z^{2/3}k_\perp^{-5/3}f(k_z) \tag{11}$$

This spectrum is essentially three-dimensional. Thus, four-wave interactions prevent the two-dimensionalization of the velocity field. It is an open question whether four-wave interactions, like three-wave interactions, transfer energy toward the plane $\Omega_k = 0$.

However, this spectrum has not been observed in simulations. The inverse cascade is always observed to be two-dimensional. A possible explanation is that, as noted above, small scale triads can exist at scales larger than the forcing scale. Such triads should also be eliminated in order to achieve a dynamics dominated by four-wave interactions. Possible ways to eliminate these triads include

1. small aspect-ratio viscosity $\nu_x = \nu_y >> \nu_z$
2. small aspect-ratio lattice $\Delta k_x = \Delta k_y << \Delta k_z$
3. small aspect-ratio force $\langle f(\mathbf{x}+\mathbf{r})f(\mathbf{x})\rangle \sim 1 - a(x^2+y^2) - bz^2 + \cdots$ with $b >> a$.

Simulations using various combinations of small aspect ratio forcing in small aspect ratio domains are in progress to test this possibility (L. Smith, private communication).

4. Four-wave interactions in rotating compressible turbulence

The linear part of the motion

$$\begin{aligned} \dot{\rho} + \rho_0 \nabla\cdot\mathbf{u} &= 0 \\ \dot{\mathbf{u}} + \boldsymbol{\Omega}\times\mathbf{u} &= -c^2\nabla\rho \end{aligned} \tag{12}$$

leads to the dispersion relation for coupled inertial and sound waves. For large k

$$\begin{aligned} \omega &= \pm c_o k\{1 + \frac{1}{2}(\frac{\Omega}{c_0 k})^2\cos^2\theta + \cdots\} \\ \omega &= \pm\Omega\sin\theta\{1 + \frac{1}{2}(\frac{\Omega}{c_0 k})^1(1+\sin^2\theta) + \cdots\} \end{aligned} \tag{13}$$

where $\cos\theta = k_\perp/k, \sin\theta = k_z/k$ so that the wave systems decouple to leading order. Rotating compressible turbulence in this limit is sound in a background of inertial waves. But for small k,

$$\begin{aligned} \omega &= \Omega\{1+\frac{1}{2}(\frac{c_0 k_\perp}{\Omega})^2+\cdots\} \\ \omega &= \pm\sin\theta c_0 k\{1-\frac{1}{2}(\frac{c_0 k}{\Omega})^2\cos^2\theta \end{aligned} \tag{14}$$

so that long inertial waves have the dispersion relation of Langmuir waves with $r_D = c_0/\Omega$. Interaction among these waves does not admit three-wave resonances.

References

Cambon, C. and Jacquin, L., 1989. "Spectral approach to non-isotropic turbulence subjected to rotation," *J. Fluid Mech.* **202**, p. 295.

Carnevale, G. F. and Martin, P. C. 1982. "Field theoretical techniques in statistical fluid dynamics: with application to nonlinear wave dynamics," *Geophys. Astrophys. Fluid Dyn.* **20**, p. 131.

Fournier, J.-D. and Frisch, U. 1978. "*d*-dimensional turbulence," *Phys. Rev. A* **17**, p. 747.

Leslie, D. C. 1973. *Developments in the theory of turbulence*, Oxford, Clarendon Press.

Newell, A. C. 1969. "Rossby wave packet interactions," *J. Fluid Mech.* **35**, p. 255.

Newell, A. C. and Aucoin, P. J. 1971. "Semidispersive wave systems," *J. Fluid Mech.* **49**, p. 593.

Rubinstein, R. 1999. "Double resonances and spectral scaling in the weak turbulence theory of rotating and stratified turbulence," ICASE Report no. 99-7.

Shimomura, Y. and Yoshizawa, A., 1986. "Statistical analysis of anisotropic turbulent viscosity in a rotating system," *J. Phys. Soc. Japan* **55**, p. 1904.

Smith, L. M., Chasnov, J. and Waleffe, F. 1996. "Crossover from two- to three-dimensional turbulence," *Phys. Rev. Lett.* **77**, p. 2467.

Smith, L. M. and Waleffe, F. 1999. "Transfer of energy to two-dimensional large scales in forced, rotating three-dimensional turbulence," *Phys. Fluids* **11**, p. 1608.

Waleffe, F., 1993. "Inertial transfers in the helical decomposition," *Phys. Fluids* **5**, p. 677.

Zakharov, V.E., L'vov, V.S., and Falkovich, G., 1992. Kolmogorov Spectra of Turbulence I, Springer.

G

Thermal Turbulence, Stratified and Rotating Turbulence

Comments on high Rayleigh number convection

J.J. Niemela, L. Skrbek, K.R. Sreenivasan[†], and R.J. Donnelly
University of Oregon, Eugene, OR 97403
[†] *Yale University, New Haven, CT 06520*

Abstract.
We have recently conducted a series of experiments on turbulent convection in the range of Rayleigh numbers between 10^6 and 10^{17} (Niemela *et al.* 1999). The working fluid is cryogenic helium gas. The eleven decades of dynamic range enable us to make a few conclusive observations. Among them, the following aspects are noteworthy.

1. Scaling of the heat transport

The Nusselt number, Nu, scales with Ra according to $Nu = 0.124 Ra^{0.309 \pm 0.0043}$, essentially over all eleven decades of Ra (Fig. 1)[1]. The data can also be fitted equally well by a $3/10^{\text{ths}}$ power of Ra with logarithmic corrections, $Nu \sim (Ra^{3/2} \ln Ra^{3/2})^{1/5}$. The form of this latter expression is derived from a weakly nonlinear theory for conditions just past the onset of convection. This theory, due in various stages to the efforts of Howard, Roberts, Stewartson, and Herring (see also Toomre *et al.* 1977) consists of calculating the steepest variation of the Nu by a single- mode solution of weakly nonlinear convection. The functional form fits the data very well, but the coefficient in front is measured to be smaller by a factor of about 4. Regardless of its perceived applicability, we note that the number of adjustable parameters in this expression is only one (namely the prefactor); even the simple power-law fit has two unknown coefficients. The Rayleigh number span of the data is large enough to rule out, for the present convection cell, the classical $1/3^{\text{rd}}$ power (Malkus 1954, Priestly 1959) and the more recent $2/7^{\text{th}}$ power (Castaing *et al.* 1989, Shraiman and Siggia 1990). In particular, we do not observe a transition to Kraichnan's (1962) asymptotic regime (see also Howard 1963, Doering and Constantin 1996). This observation is consistent with the recent finding of Glazier et al. 1999. Over the Rayleigh number range covered in the experiment, it is possible to discern the differences between the formula culled from the weakly nonlinear theory and the upperbound result of Constantin

[1] The precise value of the exponent depends on how well one knows the physical properties of helium gas. We have used the most recently available standard data. See Niemela et al. (2000) for some details.

T. Kambe et al. (eds.), IUTAM Symposium on Geometry and Statistics of Turbulence, 269–277.
© 2001 *Kluwer Academic Publishers.*

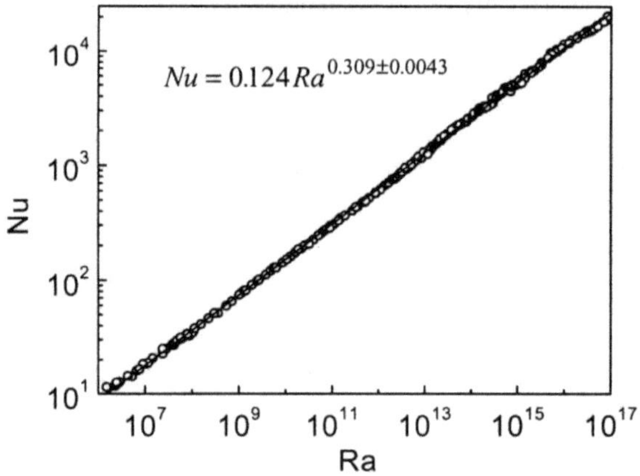

Figure 1. Log-log plot of the Nusselt number versus Rayleigh number. The line through the data is the least squares fit over the entire range of Ra.

and Doering (1999) for the case of very large Prandtl numbers, namely $Nu \leq Ra^{1/3}(1 + \log Ra)^{2/3}$

2. The mean wind

The so-called mean wind is the strong recirculating motion in the convection cell. At low Rayleigh numbers it does seem to exist in the form of unidirectional circulation, but the situation is complex at high Ra. If averaged over a suitable intervals of time, a semblance of the mean wind can be observed in this latter regime as well, but it is small compared to the free-fall velocity by a factor of 10 to 30 (depending on details of how the mean wind is estimated). In particular, this mean wind seems to alternate its direction quite frequently (Fig. 2a). This conclusion is not based on a direct measurement of velocity, but on correlating signals from two neighboring temperature probes. By necessity, the measurement technique does not discern velocity fluctuations whose time scales are smaller than the averaging time scale. If the averaging time becomes smaller, the distinctly bimodal nature of the distribution disappears (see Figs. 2b and 2c). Thus, a realistic picture may be one of weak large-scale circulation, upon which strong small-scale velocity fluctuations are superimposed. This does not necessarily mean that the theories invoking mean wind are incorrect: for their purposes, the

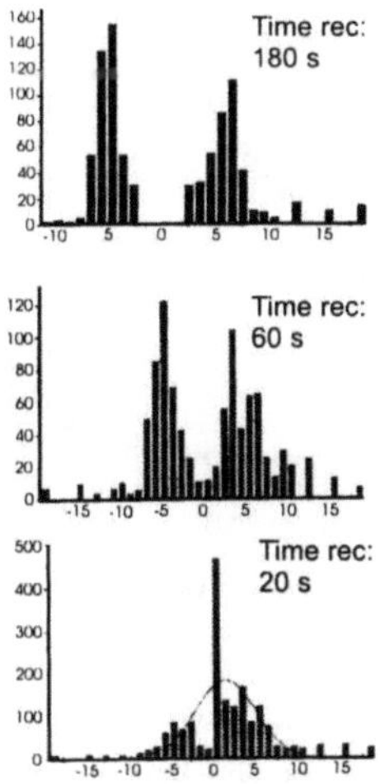

Figure 2. A rough measure of the large-scale velocity in the cell. The data are obtained by correlating the temperature signals from two neighboring probes, allowing for a time delay from one of them so as to maximize the correlation. The correlation is obtained by averaging the two signals over a certain amount of time. These time scales are different for the three cases shown here (180s, 60s, and 20s for (a), (b) and (c) respectively). By construction, velocities corresponding to time scales smaller than the averaging time cannot be discerned from these measurements. The distinctly bi-modal nature seen in (a) becomes less clear as the averaging time becomes smaller. The conclusion appears to be that the so-called mean wind is a manifestation of the large scale when small-scales are suitably averaged out.

shearing motion established by the somewhat random large-scales is perhaps adequate.

3. Prandtl number variation

In general, the Nusselt number depends not only on the Rayleigh number (the dynamical parameter), but also on the Prandtl number (which is a fluid property) and the aspect ratio (a geometric property). In our measurements, the Prandtl number (Pr) was constant up to an Ra of 10^{13} . Beyond this, Pr eventually increased to about 30 (Fig. 3), staying less than or of order unity for Ra up to 10^{15} and increasing and increasing to a maximum value of about 30 at the highest Rayleigh number. Using the first seven decades of the Rayleigh number available in the regime of strictly constant Prandtl number, we obtained

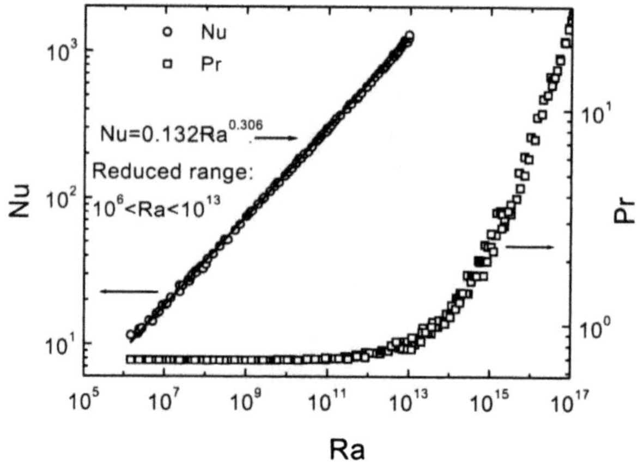

Figure 3. The Prandtl number corresponding to different Rayleigh numbers measured. The Nusselt number data for the first seven decades of Rayleigh number, over which the Prandtl number is strictly constant, are fitted by a power law in this figure. This power-law is given by $Nu = 0.132Ra^{0.306}$

the Rayleigh number scaling at fixed Prandtl number;this power-law fit is given by $Nu = 0.132Ra^{0.306}$. This does not differ significantly from the fit obtained earlier (Fig. 1) for the entire range of Ra. This means that the Prandtl number effects are relatively weak for Pr in the range considered here. These small effects can be estimated, assuming that no other transition occurs, by plotting the variation of the ratio $Nu/0.132Ra^{0.306}$ against Pr. This is done in Fig. 4. The Nusselt number ratio decreases weakly for increasing Prandtl numbers. If this decrease is fitted by a power law, even if less than convincingly, we obtain the Prandtl number effect to $Pr^{-0.09}$. Taking these results in conjunction with an earlier study of Verzicco et al. (1998) at low Prandtl numbers, we summarize the Prandtl number variation as follows:

$$\begin{array}{lll} Nu \sim Pr^{0.14} & \text{for} & Pr < 0.1 \\ Nu \sim Pr & \text{for} & 0.1 < Pr < 5 \\ Nu \sim Pr^{-0.09} & \text{for} & Pr > 5. \end{array}$$

This last formula is not inconsistent with the theory of Shraiman & Siggia (1990).

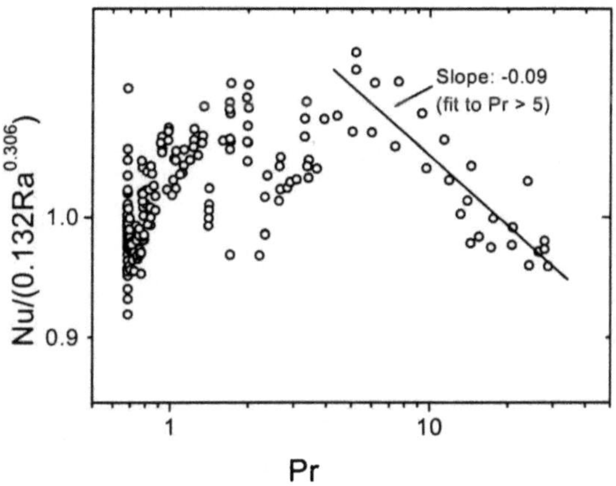

Figure 4. An estimate of the effects of variable Prandtl number. The ordinate is the measured Nusselt number divided by $0.132Ra^{0.306}$, the latter being the fit to the data in the region of constant Prandtl number. If the variability seen in the data is attributed to Prandtl number changes, we obtain, roughly, something like $Pr^{-0.09}$

4. Aspect ratio variation

The present experiments pertain to a fixed aspect ratio of 1/2. In order to get a sense of the effects of the aspect ratio, we collected various data on how the Nusselt number varies with aspect ratio, keeping both Rayleigh and Prandtl numbers fixed. This issue has been discussed by others (e.g., Castaing *et al.* 1989) before, but we carry it one modest step further. Figure 5 shows the results. To the lowest approximation, the Nusselt number decreases with increasing aspect ratio; the aspect ratio ceases to be important probably when it is as high as 5 to 10. Perhaps the increase for smaller aspect ratio is related to the increased importance of side-wall boundary layer. In detail, however, the Nusselt number dependence appears to be non-monotonic (see the dashed line in Fig. 5). That particular behavior is not understood, but is perhaps related to the accommodation of the most efficient heat transfer modes (largest scales) by the finite geometry. If so, it must vanish in the limit of very large aspect ratio, as it indeed seems to do.

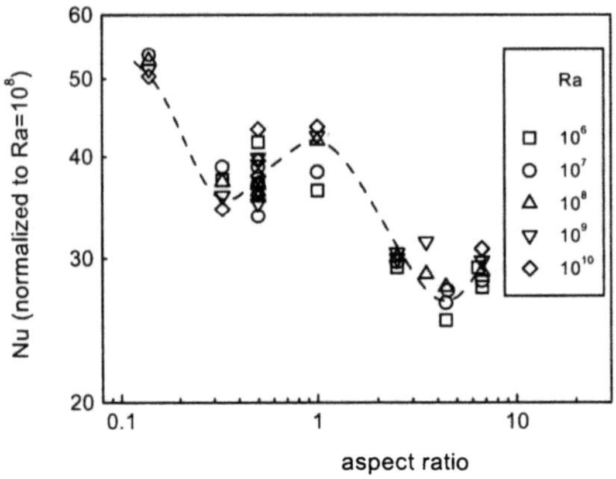

Figure 5. Effects of aspect ratio on the Nusselt number by fixing the Rayleigh number and the Prandtl number. (The latter is not strictly constant but does not vary over a large range, and the Pr variation over that range is small.) The data are normalized to those at Ra = 108, and are extracted from the following references: Rossby (1969), Garon & Goldstein (1973), Chu & Goldstein (1973), Threlfall (1975), Tanaka & Miyata (1980), Wu (1991), Chavanne (199?), and Niemela et al. (1999).

5. The dissipation rate

A few comments can be made on the dissipation rate in the convection cell. By integrating the equations of motion satisfying the Boussinesq approximation,it is possible to derive an exact equation for the turbulent energy dissipation averaged over the entire cell. The relevant expression for the non-dimensional dissipation rate is $(Nu-1)Ra$ (Howard 1963). This quantity is plotted in Fig. 6. There seems to be a unique power law for all Rayleigh numbers, as could have been guessed from Fig. 1. This power-law exponent is measurably distinct from 3/2 the latter being the expectation from Kolmogorov-type dimensional arguments.

Grossmann & Lohse (1999) have split the energy dissipation into a bulk contribution and the boundary layer contribution. Another parameter to contend with is the ratio of the thermal boundary layer to that of the momentum boundary layer. Depending on which effect dominates, these authors propose different power laws for the Nusselt/Rayleigh number relation; they note that, in a phase plane of $Ra-Pr$, different areas can be expected to have different scaling exponents. This is depicted in Fig. 7. The present experimental data are overlaid on

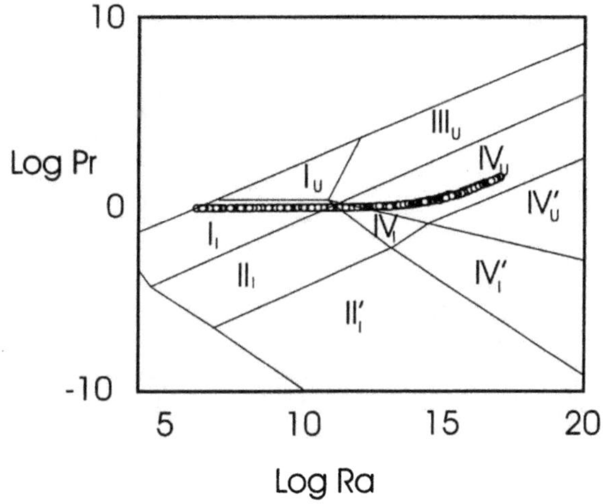

Figure 6. The product (Nu-1)Ra, which is the average energy dissipation in the whole convection cell, as a function of Ra. It is fitted essentially by a single power law, this being different from the 3/2 power expected by dimensional argument of the type employed by Kolmogorov in his 1941 theory.

that diagram. Although linear combinations of different power laws can mimic a single power law relationship over many decades (as noted by Grossman & Lohse 1999), no statistical advantage is gained over fitting the data with just one value of the power-law exponent, and the latter's simplicity makes its use somewhat more compelling. One possible explanation for why a single power law distinct from 3/2 can correctly describe the data is that the energy dissipation contained in the boundary layer never becomes unimportant (as would have to be the case beyond some Rayleigh number if the Kolmogorov scaling were to be valid). If, on the other hand, the boundary layer contribution remains the same fraction of the overall dissipation at all Rayleigh numbers, the power-law cannot be estimated by dimensional arguments. This may well be the case because, although the boundary layer becomes progressively thinner and occupies smaller volume at increasingly large Ra, the relevant velocity gradients within the boundary layer become correspondingly larger. It is thus conceivable that the boundary layer contribution to the dissipation never ceases to be important.

In order to test this idea, it is necessary to measure the boundary layer in detail. This has not yet been done. For a partial explanation, one can turn to the case of turbulent boundary layer flows, and inquire if the so-called wall region carries increasingly smaller or larger fraction as the Reynolds number is increased. Relegating details to another place,

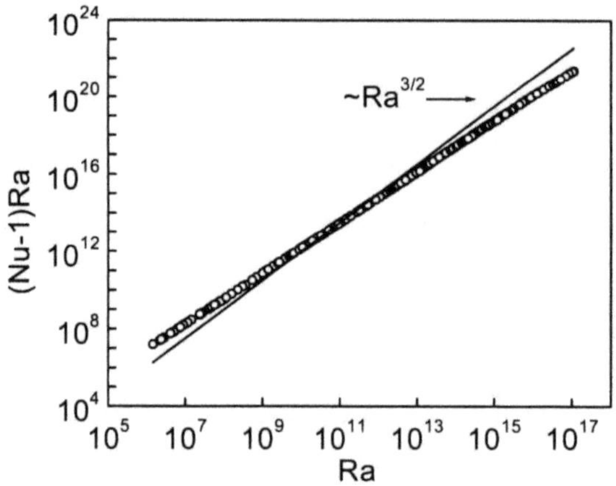

Figure 7. The different scaling regimes proposed by Grossmann & Lohse (1999) in the Pr-Ra plane. The darker regions are prohibited. The theory assigns a separate scaling exponent to each of the other regions. The interpretation is that the power law observed in an experiment is a superposition of more than one of those basic power laws. The present data are superimposed on this phase plane. They show that the data span more than one of the scaling regions of Grossmann & Lohse (1999), implying that the present power law is possibly a superposition of more than one of the basic power-laws. This exercise yields a reasonable fit to the data, but does not necessarily verify the basic tenets of the theory.

we simply note here that the basic idea presented here appears to be borne out roughly.

6. Concluding remarks

We have considered some aspects of high-Rayleigh-number convection. These aspects include the scaling of heat transport, the so-called mean wind, the effects of variable Prandtl number and aspect ratio, and the scaling of the energy dissipation rate. The problem of convection is extremely rich, and we have merely added to the existing knowledge on a few of its facets. Not all data acquired in our apparatus have been analyzed at this date. In particular, temperature fluctuation data from multiple probes have not yet been analyzed. We hope to be able to do this soon.

References

Castaing, B., Gunaratne, G., Heslot, F., Kadanoff, L.P., Libchaber, S., Thomme, S., Wu, X.-Z, Zaleski, S and Zanetti, G. 1989 *J. Fluid Mech.* **204**, 1

Chavanne, X. 1997 Etude du regime turbulent en convection de Rayleigh-Benard dans l'helium liquide ou gazeux autour de 5K. Ph.D. thesis, University of Grenoble

Chu, T.Y. & Goldstein, R.J. 1973 J. Fluid Mech., 60, 141

Constantin, P. and Doering, C.R. 1999 *J. Stat. Phys.* **94**, 159

Doering, C.R. and Constantin, P. 1996 *Phys. Rev. E*, **53**, 5957

Garon, A.M. & Goldstein, R.J. 1973 Phys. Fluids 16, 1818

Grossmann, S. and Lohse, D. 1999 *J. Fluid Mech.* (in print)

Howard, L.N. 1963 *J.Fluid Mech.* **17**, 405

Kraichnan, R.H. 1962 *Phys. Fluids*, **5**, 1374

Malkus, M.V.R. 1954 *Proc. Roy. Soc. Lond.* **A225**, 196

Niemela, J.J., Skrbek, L., Sreenivasan, K.R. and Donnelly, R.J. 2000 *Nature* **404**, 837.

Priestley, C.H.B. 1959 *Turbulent Transfer in the Lower Atmosphere* (U. Chicago Press)

Rossby, H.T. 1969 J. Fluid Mech. 36, 309

Shraiman, B.I. and Siggia, E.D. 1990 *Phys. Rev. E*, **42**, 3650

Tanaka, H. & Miyata, H. 1980 Int. J. Heat Mass Tr. 23, 1273

Threlfall, D.C. 1974 Natural convection in helium gas. Ph.D. thesis, Cambridge University

Threlfall, D.C. 1975 J. Fluid Mech., 67, 17

J. Toomre, D. O. Gough and E. A. Spiegel, 1977 Numerical solution of single-mode convection equation. *J. Fluid. Mech* 79, 1-31.

Verzicco, R., Camussi, R, 1998 in *Advances in Turbulence II* (ed. U. Frisch) (Kluwer Academic, Dordrecht), pp. 399-402

Wu, X.-Z. 1991 Along a road to developed turbulence: Free thermal convection in low temperature helium gas. Ph.D. thesis, University of Chicago

THE DYNAMICS OF STRUCTURES OF T-VORTICITY IN 2D FREE CONVECTION TURBULENCE

S. TOH AND T. MATSUMOTO
Division of Physics and Astronomy, Graduate School of Science
Kyoto University, Kyoto 606-8502, Japan

Abstract. Two dimensional free convection turbulence is examined by DNS. Coherent structures (CS) similar to worms in 3D Navier-Stokes turbulence are observed. These CSs are defined by the regions with high T-vorticity, $(\partial T/\partial y, -\partial T/\partial x)$. We obtained the Burgers T-vortex layer solutions, which are shown to be a good approximation of CSs. The intermittency of entropy $(T^2/2)$ dissipation rate is also examined in relation with the most singular structure.

1. Introduction

In 3D Navier-Stokes (3DNS) turbulence, there exist coherent fine structures, so-called worm, which are believed to be universal independent of the generation mechanism of turbulence in large scales. These structures are sometimes modeled by the Burgers vortex tube, but their role on the statistics is still unclear [1].

Here, we will report that coherent structures, which are quite similar to the worms observed in 3DNS turbulence, exist in two-dimensional free convection (2DFC) system. 2DFC system is a model of the central region of hard turbulence which is proposed to examine the power spectrum of temperature fluctuations, $P(\omega) \sim \omega^{-1.4}$. This power spectrum is explained by the Bolgiano-Obukhov(BO) scaling based on the entropy $(T^2/2)$ cascade [2]. The governing equations for the 2DFC system are based on the Boussinesq approximation:

$$\frac{\mathrm{D}\boldsymbol{u}}{\mathrm{D}t} = -\nabla p + \alpha g T \boldsymbol{e}_y + \nu \Delta \boldsymbol{u}, \tag{1}$$

T. Kambe et al. (eds.), IUTAM Symposium on Geometry and Statistics of Turbulence, 279–284.
© 2001 *Kluwer Academic Publishers.*

$$\frac{\mathrm{D}T}{\mathrm{D}t} = \kappa \Delta T, \tag{2}$$

where $\nabla \cdot \boldsymbol{u} = 0$, $D/Dt \equiv \partial/\partial t + (\boldsymbol{u} \cdot \nabla)$, and ν, κ, α and g are the kinematic viscosity, the heat diffusivity, the volume expansion coefficient, and the gravitational acceleration, respectively. The buoyancy acts along the y axis and $\boldsymbol{e}_y$ is the unit vector of this direction. This model assumes neutrally stable stratification (no mean temperature gradient) and homogeneity which is realized in the central region of hard turbulence [2], so that we should force the system to sustain turbulence. The 2DFC system has the conserved quantities, entropy $S \equiv \int |T|^2/2dV$ and total energy $E \equiv \int |\boldsymbol{u}|^2/2 + (\alpha g)yTdV$, in the inviscid case.

In the inertial range, the BO spectra for entropy and kinetic energy $S(k)$ and $E(k)$, are obtained based on entropy cascade:

$$S(k) \sim \epsilon_\theta^{4/5}(\alpha g)^{-2/5}k^{-7/5}, \tag{3}$$

$$E(k) \sim \epsilon_\theta^{2/5}(\alpha g)^{4/5}k^{-11/5}, \tag{4}$$

where ϵ_θ is the average dissipation rate of the entropy. The characteristic time corresponding to the eddy turnover time is also defined as follows:

$$\tau_l \sim \epsilon_\theta^{-1/5}(\alpha g)^{-2/5}l^{2/5}. \tag{5}$$

2. T-vorticity and Burgers T-Vortex layer

We introduce a new vector field called T-vorticity: $\boldsymbol{\chi} \equiv (\partial T/\partial y, -\partial T/\partial x)$. The evolution equation for T-vorticity reads

$$\mathrm{D}\boldsymbol{\chi}/\mathrm{D}t = \boldsymbol{\chi} \cdot \nabla \boldsymbol{u} + \kappa \Delta \boldsymbol{\chi}. \tag{6}$$

Since the square norm of T-vorticity, i.e., $|\boldsymbol{\chi}|^2$ is proportional to the entropy dissipation rate, $\kappa \nabla T \cdot \nabla T$, T-vortex layers themselves are also the structures of the latter. When T-vorticity is subjected to the stagnation flow, $\boldsymbol{U} = (-Ax, Ay)$, we obtain a linear steady solution similar to the Burgers vortex layer as follows. We call this solution the Burgers T-vortex layer. We assume temperature and vorticity depend only on x: $\boldsymbol{u} = \boldsymbol{U} + (0, v(x))$ and $T = T(x)$. Then, T-vorticity and vorticity are $(0, dT/dx) \equiv (0, \chi(x))$ and $(0, 0, \omega(x)) = (0, 0, dv/dx)$. Here we show the steady solution only for $Pr = 1$:

$$\chi(x) = \Theta_0 \sqrt{\frac{A}{2\pi\kappa}} \exp(-\frac{A}{2\kappa}x^2) \tag{7}$$

$$\omega(x) = -\frac{\alpha g \cos\theta}{A}\chi(x) \tag{8}$$

where θ is the angle between y-axis and the gravitational acceleration and $\Theta_0 = T(+\infty) - T(-\infty)$. Note that χ is independent of the direction of the gravitational acceleration. On the other hand, ω depends on the direction of the gravity, and is proportional to χ. This means that vorticity is induced by the buoyancy and localized on the T-vortex layer. T-vortex layer is a sharp interface, which separates two constant temperature regions. As known well, the entropy dissipation rate par unit length for Burgers T-vortex layer is proportional to $\sqrt{\kappa A}$, so that, if strain rate A is fixed, ϵ_θ tends to zero as κ goes to zero. However, we have found that the characteristic strength of the strain rate, such as the mean value and standard deviation, increases as κ decreases. Thus it is suggested that the Burgers T-vortex layer plays an important role on entropy transfer even in the invisid limit, although time dependence, that is, dynamics of such a coherent structure may be crucial in that case.

3. Direct numerical simulations

Direct numerical simulation is carried out with the eqs. (1) and (2) using the 4th-order Runge-Kutta method for time marching, and the pseudo-spectral scheme in a doubly periodical box, $[0, 2\pi] \times [0, 2\pi]$. Aliasing terms are removed in terms of the 1/2-shifted grids. Without loss of generality we can set $\alpha g = 1$. For simplicity, we set $Pr = 1$. The forcing term $F = 0.8\cos(2x)\cos(2y)$ is introduced to eq. (2) to drive temperature field. The drag term $D = 0.2(1 - \theta(|\boldsymbol{k} - 3|))\Delta^{-1}\boldsymbol{u}$, where $\theta(x)$ is Heviside's step function, is also introduced to eq. (1) to keep the system statistically stationary.

The numerical and flow parameters are listed in the table 1, where N^2 and Δx are resolution and mesh size; the characteristic lengths, η_θ, l_{BO}, λ_θ are Kolmogorov length, Bolgiano-Obukhov length, Taylor length; Ra_λ denotes Rayleigh number. These are defined as follows: $l_{BO} \equiv \epsilon^{5/4}/(\epsilon_\theta(\alpha g)^2)^{3/4}$, $\eta_\theta \equiv (\kappa^5/\epsilon_\theta(\alpha g)^2)^{1/8}$, $\lambda_\theta \equiv (\lambda_x + \lambda_y)/2$, $\lambda_{x_i} \equiv \{\langle T^2\rangle/\langle(\partial T/\partial x_i)^2\rangle\}^{1/2}$, $Ra_\lambda \equiv (Ra_{\lambda_x} + Ra_{\lambda_y})/2$, $Ra_{\lambda_{x_i}} \equiv (\alpha g\langle T^2\rangle^{1/2}\lambda_{x_i}^3)/\kappa\nu$ where ϵ is energy dissipation rate.

By comparing with DNS, we show that worms observed in 2DFC turbulence are described approximately by the T-vortex layer solution. Figure 1(a) shows the norm of T-vorticity obtained by the DNS with resolution 1024^2. Dark and slender regions are conspicuous and indicate high T-vorticity regions. We call these high T-vorticity regions coherent structures. Note that the coherent structures in 2DFC are layers or curves. The length scale of typical T-vortex layers is about integral scale. Their widths are about less than ten times Kolmogorov length η_θ. The characteristic sizes of the coherent structures are similar to those of the worms in 3DNS.

TABLE 1. Numerical and flow parameters.

N	κ	Δx	η_θ	l_{BO}	λ	Ra_λ
512	1×10^{-3}	$\pi/256$	$1.38\Delta x$	$1.3\eta_\theta$	$6.5\eta_\theta$	526
1024	2×10^{-4}	$\pi/512$	$1.14\Delta x$	$2.0\eta_\theta$	$8.0\eta_\theta$	1186
2048	1×10^{-4}	$\pi/1024$	$1.5\Delta x$	$1.9\eta_\theta$	$10\eta_\theta$	1860

Although roll-ups and spirals, which form plumes, are clearly visible, we only stress the fact that not only T-vorticity but vorticity and even strain rate are concentrated on T-vortex layers (see Fig. 2(c)).
In 3DNS, as shown by Kida and Miyauchi, the definition of coherent vortex structure is not trivial. On the other hand, in our case, it is easy to define coherent structure, because relatively high T-vorticity region is always elongated and concentrated on its ridge or skeleton. In Fig. 1(b), some of the skeletons of T-vortex layers observed in Fig. 1(a) are plotted.

In Fig.2(a), the skeleton of a typical T-vortex layer is plotted. Figure 2(b) shows the normalized ratio of χ to ω estimated on the skeleton. The ratio indicates the closeness to the Burgers T-vortex layer solution. On the relatively straight segment including the point p, the ratio is almost 1. This means that the Burgers T-vortex layer is a good approximation of the coherent structures as shown in Fig.2(c). However, in the parts with relatively large curvature, these correspondences are no longer satisfied, though vorticity yet concentrates on the coherent structures and T-vorticity still takes approximately a Gaussian form. This suggests that though vorticity is induced by buoyancy, the relaxation time seems to be so long that vorticity can not easily follow the motion of the coherent structure.

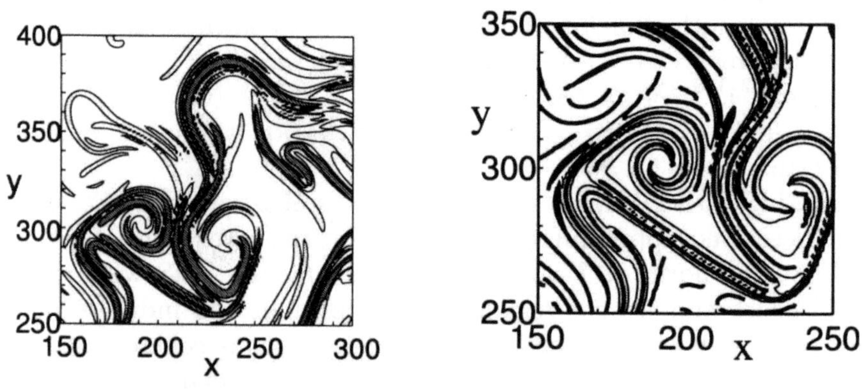

Figure 1 (a): Contours of $|\chi|$. The unit of length is mesh size Δx. (b): Tracked skeletons. Weak and short layers are also tracked. These skeletons are not necessarily parallel to T-vortex lines.

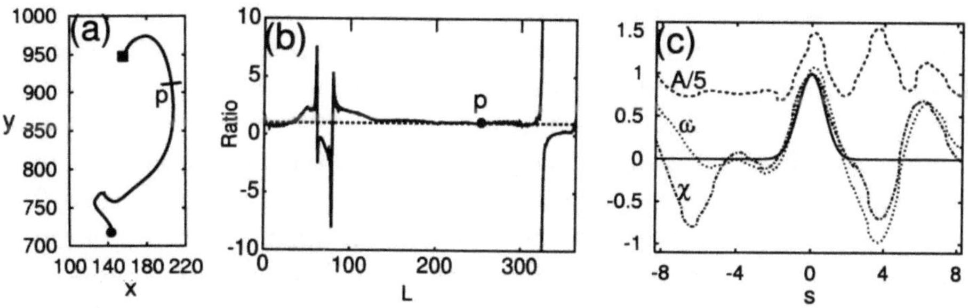

Figure 2 (a): The skeleton of a typical coherent structure in the case with $N = 1024$. The unit of length for coordinates is mesh size Δx. Black circle; head, black square; tail. Total length of the skeleton is 0.71π, which is of the order of the integral scale. (b): Normalized ratio of χ to ω on the skeleton. For the Burgers T-vortex layer, this ratio is 1 (dashed line). The abscissa is the distance from the head. (c): The sectional forms of χ (dash-dotted line) and ω (dotted line) at the section p in Figs.(a),(b). They are normalized as the Burgers T-vortex layer solutions (8) and (9) become $\exp(-x^2)$ (solid line). Dashed line shows $A/5$. The abscissa is the normalized distance from the skeleton: In this unit, η_θ is 0.96.

4. Intermittency and singular structure

Our goal is to understand the relation between coherent structures and statistics of turbulence. If some coherent structures actually play an important role in turbulence, there may exist singular structures, which should contribute to entropy dissipation even in the inviscid limit as mentioned in sec. 2. In this section, we consider such singular structures in relation with the intermittency of entropy dissipation rate [3].

We have estimated the intermittency exponent τ_p in terms of extended self similarity (ESS) scheme by use of the data obtained by DNS with resolution 1024^2: $< (\epsilon_\theta(l))^q > \sim < |\delta T(l_x)^2 \delta u_x(l_x)| >^{\tau_q}$ where $\delta T(l_x)$ and $\delta u_x(l_x)$ are the differences of T and u_x between the points separated by l_x. In Fig. 3, τ_q is plotted. We compare this result with She-Lévêque model, which is based on vortex filament – the most singular structure[4]. This model involves the two parameters: $\tau_q = -aq + C_0[1 - (1 - a/C_0)^q]$ where C_0 is the codimension of the structure, and a is the exponent of time scaling. They used $C_0 = 2$ and the exponent, 2/3, of the turnover time $t_l \sim l^{2/3}$. In 2DFC, we can assume $C_0 = 1$ because

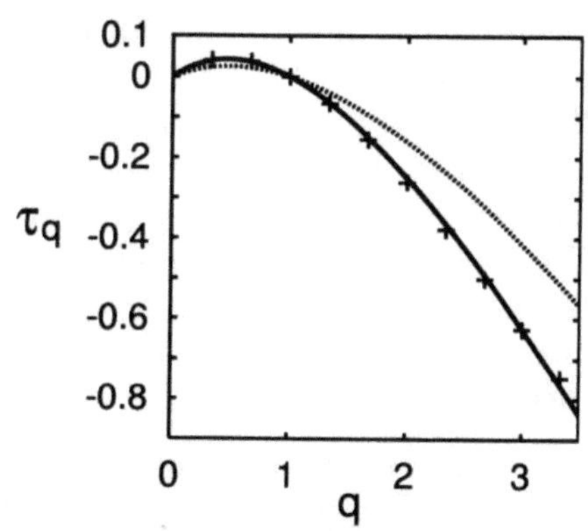

Figure 3 Comparison of τ_q estimated for DNS (+) with SL model. Solid line: SL model with $a = 1/2$. Dashed line: SL model with $a = 2/5$.

the most singular structure may also be layer. The BO time scaling (5) corresponds to the turnover time, so that we use $a = 2/5$. With this value SL model does not yield a good approximation. In fact, the best fit is obtained with $a = 1/2$ as shown in Fig. 3.

We have obtained a new time scaling [5]. Next we will show that this time scaling is related to a finite time singularity in the inviscid 2DFC system, although the existence of it is still an open question. Pumir and Siggia showed that maximum of the norm of $|\nabla T|$ blows up in proportional to inverse t squared approximately [6]. Here, we assume that their result holds just before the viscous effects become effective even in the viscous case. This scaling, $|\nabla T| \sim \Delta T_0/l(t) \sim (t - t_*)^{-2}$, is rewritten as follows: $|t - t_*| \equiv t_l \sim l^{1/2}$ where t_* is the time of blow-up and $l(t)$ is the width of the singular structure. Since temperature is Lagrangian conserved quantity, we can assume the order of T, ΔT_0 is nearly constant. From the evolution of the singular structure shown in their paper [5], it is supposed that the blow-up is the shrinkage of the width of T-vortex layer. This process is self-stretched and seems to correspond to the coherent cascade proposed by Jiménez and Wray [7].

5. Concluding remarks

We have confirmed the existence of T-vortical structure which are subjected to strain and dissipation. Some parts of T-vortical structure are well approximated by the Burgers T-vortex layer.

Intermittency of entropy dissipation rate has been examined. The obtained exponent τ_q was compared with SL model. We have shown that the exponent of the time scaling for the most singular structure is not 2/5 of the turn over time but 1/2. Although the existence of a finite-time singularity in 2DFC is still an open question, it is suggested that the singularity or at least the abrupt stretching of T-vortex layer is related to intermittency or the conceptual singular structure of S-L model.

We believe that quasi-equilibrium structures such as the Burgers T-vortex layer are too simple to describe turbulence in detail. We therefore require a new picture of structures maybe in a dynamical sense.

References

1. The papers on this proceedings by Kida et al., and Miyauchi and Tanahashi.
2. Toh, S. and Suzuki, E. (1994), *Phys. Rev. Lett.*, **73**, pp. 1501-1505
3. Suzuki, E. and Toh, T. (1995), *Phys. Rev. E* , **51**, pp. 5628-5635
4. She, Z. S. and Lévêque, E. (1994), *Phys. Rev. Lett.*, **72**, pp. 336-339
5. Pumir, A. and Siggia, E.D. (1992), *Phys. Fluids*, **A4**, pp. 1472-1491
6. Iima, M. and Toh, S. (1998), *J. Phys. Soc. Jpn.*, **67**, pp. 373-376 ;(Toh. S, and Iima, M. ,Dynamical Aspect of Entropy Transfer in Free Convection Turbulence , *to appear in Phys. Rev. E*)
7. Jiménez, J. and Wray, A. A. (1998), *J. Fluid Mech.*, **373**, pp. 255-285

STRUCTURE OF HOMOGENEOUS ROTATING TURBULENCE UNDER STABLE DENSITY STRATIFICATION

S. TSUJIMURA and O. IIDA
Department of Mechanical Engineering
Nagoya Institute of Technology
Gokiso-cho, Showa-ku, Nagoya 466-8555, JAPAN

Y. NAGANO
Department of Environmental Technology
Graduate School of Engineering
Nagoya Institute of Technology
Gokiso-cho, Showa-ku, Nagoya 466-8555, JAPAN

1. Introduction

Fluid motions subjected to Coriolis and buoyancy forces are important in many geophysical and engineering applications, such as in the Earth's atmosphere and ocean, and internal flows in turbomachinery, e.g., gas turbine. Recently, increasing attention has been paid to combined effects of these body forces and to new hydrodynamic phenomena generated by them[1].

The mixed effects of rotation and stable stratification are known to generate coherent vortices with a large vertical vorticity. One of the interesting phenomena in rotating turbulence is the preference in the sense of the rotation associated with coherent vertical vortices. For example, mesoscale eddies observed in the Arctic Ocean are mostly anticyclonic[2], while experimentally, cyclonic eddies are often observed in a rotating tank[3]. Previous numerical studies on rotating stratified turbulence have been mainly based on the quasi-geostrophic (QG) approximation[4, 5, 6], which clearly predicts the generation of coherent vertical vortices[5]. However, the QG form of the equations is symmetric with respect to vertical vorticity, and hence not appropriate to investigate the asymmetry of the coherent structures.

The main objectives of this study are to investigate the generation mechanism of the coherent vertical vortices and their asymmetry in the sense of rotation. We will also discuss the two-dimensionalization of turbulence under rotation. Our

T. Kambe et al. (eds.), IUTAM Symposium on Geometry and Statistics of Turbulence, 285–290.
© 2001 *Kluwer Academic Publishers.*

objective flow field is a homogeneous flow without any solid boundary. Hence, the dynamic effects of rotation and stratification must affect the vortices without the blocking effects of the wall.

2. Numerical procedure

The flow field is a homogeneous turbulence under system rotation around the x_3 axis. Both the mean temperature gradient and the gravitational acceleration are simultaneously imposed in the x_3 direction to include the effects of both buoyancy and rotation. The Boussinesq-approximated Navier-Stokes equation and the energy equation are numerically solved. The parameter S defined by the ratio of the Brunt-Väisälä frequency N to the Coriolis frequency of rotation 2Ω, i.e., $S = N/(2\Omega)$, is systematically varied from 0.1 to 5.0. The parameter S is also rewritten as Ro/Fr, where $Ro = \varepsilon/(2\Omega k)$ and $Fr = \varepsilon/(Nk)$ are the Rossby and Froude numbers, respectively. The initial Rossby and Froude numbers are 0.267, while the initial Reynolds number $Re = k^2/(\nu\varepsilon)$ is 201.4. In all cases, these three parameters decrease as the evolution of the time. Also, the Prandtl number Pr is varied from 0.025 to 5.0. The numerical procedure is based on the Fourier pseudo-spectral method with grid points of 128^3.

3. Results and discussion

3.1. EFFECTS OF PRANDTL NUMBER ON ASYMMETRY OF COHERENT VORTICES

In this study, two kinds of anisotropic flow fields are used as initial conditions, the two-component state (case 2C) where the vertical component of velocity is exactly zero and the quasi one-component state (case 1C) where the vertical velocity surpasses the horizontal ones. In both cases, the results of DNS are compared with those predicted by RDT to elucidate the effects of the nonlinear term. Moreover, the effects of the Prandtl number on the vertical vortices are also studied as well as the initial anisotropy of Reynolds stresses.

Figures 1(a) and (b) show the time evolution of the skewness factor $\overline{\tilde{\omega}_3^3}$ of the vertical vorticity ω_3 in the cases of $S = 1$. In what follows, a tilde denotes the normalization by the respective rms value. In case 2C, the skewness factors tend toward the negative side almost over the entire period of calculation in comparison to case 1C. In particular, the negative value of the skewness factor becomes more marked with a low Prandtl number, indicating the dominance of the anticyclones. On the other hand, in case 1C, they tend toward a positive value. It should be carefully noted that the skewness factors in the cases of $Pr = 0.71$ and 5.0 take definite positive values. The p.d.f. $P(\tilde{\omega}_3)$ of the vertical vorticity in Cases 2CL ($Pr = 0.025$) and 1CH ($Pr = 5.0$) are shown in Figs. 2(a) and (b), respectively.

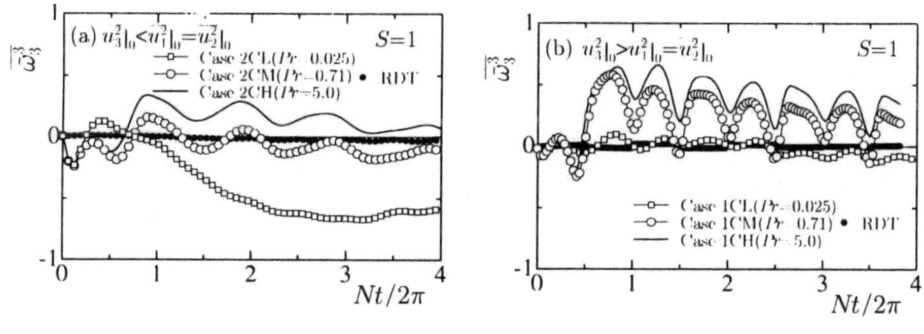

Figure 1. Time evolution of the skewness factor $\overline{\tilde{\omega}_3^3}$ of vertical vorticity ω_3 in the case of $S = 1$; (a) case 2C : $\overline{u_3^2}\big|_0 < \overline{u_1^2}\big|_0 = \overline{u_2^2}\big|_0$, (b) case 1C : $\overline{u_3^2}\big|_0 > \overline{u_1^2}\big|_0 = \overline{u_2^2}\big|_0$.

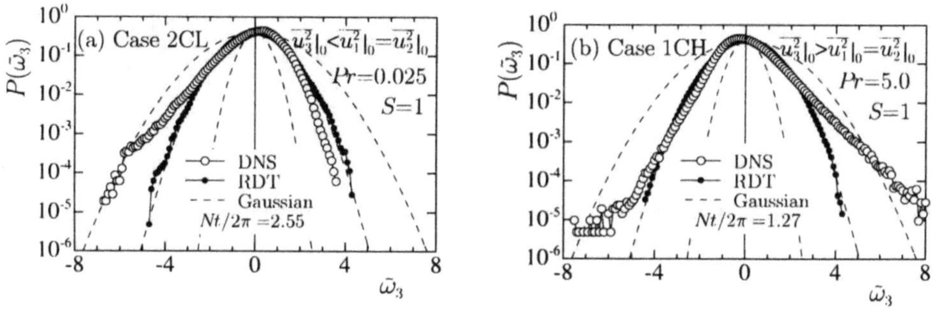

Figure 2. P.d.f.s of vertical vorticity ω_3; (a) Case 2CL, (b) Case 1CH.

The p.d.f.s of these figures are at the time when the skewness factor takes the marked non-zero value. Especially, in Fig. 2(b), the p.d.f. is represented at the time of the maximum $\tilde{\omega}_3^3$. The asymmetry of the vertical vorticity is found to occur at the intensive region of ω_3, indicating that the coherent vertical vortices are asymmetric. On the other hand, in the RDT, its p.d.f. is symmetric in both cases. Because the asymmetry of the vertical vorticity is not observed in RDT, the nonlinear term must affect their asymmetry.

Next, the mechanisms of the asymmetry of vertical vortices are discussed in terms of the transport equation of the vertical vorticity. Here, association between ω_3 and $\partial u_3/\partial x_3$ is investigated by using their joint weighted p.d.f.,

$$W_{\tilde{\omega}_3\tilde{\omega}_3\tilde{u}_{3,3}}(\tilde{\omega}_3, \tilde{u}_{3,3}) = \tilde{\omega}_3\tilde{\omega}_3\tilde{u}_{3,3}P(\tilde{\omega}_3, \tilde{u}_{3,3}), \tag{1}$$

which shows the contribution of ω_3 and $\partial u_3/\partial x_3$ to the production term $\overline{\omega_3\omega_3 u_{3,3}}$, i.e., the nonlinear vortex-stretching term in the transport equation for $\overline{\omega_3^2}$. The distributions of $W_{\tilde{\omega}_3\tilde{\omega}_3\tilde{u}_{3,3}}$ in Cases 2CL and 1CH are shown in Figs. 3(a) and (b). From the Case 2CL result, one can see that the large negative vertical vorticity is well correlated with positive strain rate $\partial u_3/\partial x_3$, indicating that the intensive anticyclonic vortices must be generated when the nonlinear vortex-stretching term becomes positive. In Case 1CH, the large positive vertical vorticity is again well

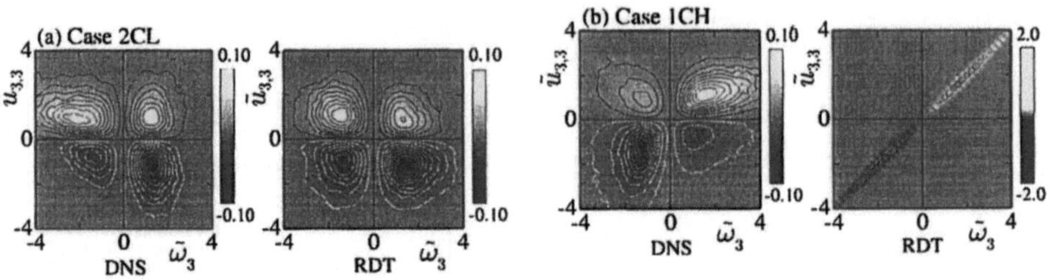

Figure 3. Weighted p.d.f. of $\tilde{\omega}_3\tilde{\omega}_3\tilde{u}_{3,3}$, white and black color maps respectively represent positive and negative values; (a) Case 2CL at $Nt/2\pi = 2.55$, (b) Case 1CH at $Nt/2\pi = 1.15$.

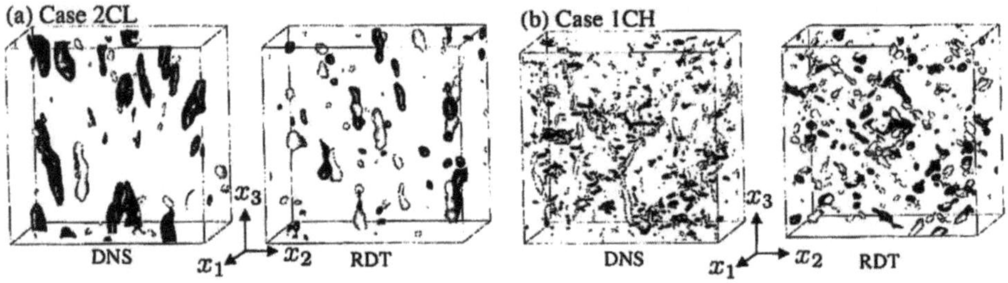

Figure 4. Iso-surfaces of the second invariant of deformation tensor, white and black iso-surfaces represent cyclonic and anticyclonic vortices; (a) Case 2CL at $Nt/2\pi = 2.55$, (b) Case 1CH at $Nt/2\pi = 1.27$.

correlated with the positive strain rate, indicating that the intensive cyclonic vortices are enhanced by the stretching in the vertical direction. In both cases, the joint p.d.f.s of RDT are symmetric without any preference in the sense of rotation. Hence, the vertical vorticity associated with the positive $\partial u_3/\partial x_3$ is enhanced in both cases of the initial conditions at any Prandtl number. It should also be noted that because the imposed mean rotation is cyclonic, both cyclones with $\partial u_3/\partial x_3 > 0$ and anticyclones with $\partial u_3/\partial x_3 < 0$ are enhanced. On the other hand, cyclones and anticyclones are deteriorated when they are correlated with $\partial u_3/\partial x_3 < 0$ and $\partial u_3/\partial x_3 > 0$, respectively. Thus, we can also say that when the vertical vorticity is enhanced (deteriorated) by the imposed rotation, cyclones (anticyclones) become dominant in comparison to anticyclones (cyclones).

Next, the instantaneous distributions of the vertical vortices are discussed. Figures 4(a) and (b) show the vortical structures represented by the second invariant of deformation tensor in Cases 2CL at $Nt/2\pi = 2.55$ and 1CH at $Nt/2\pi = 1.27$. In Case 2CL, the anticyclonic vortices are enhanced and vertically elongated in comparison to the cyclone. This tendency is more marked at a lower Prandtl number. The dominance of the anticyclonic vortices was also assured in the experiment on stably stratified rotating turbulence[7, 8], although there has been no related DNS study on homogeneous decaying turbulence. On the other hand, the cyclonic vortices are stretched in the vertical direction and become tube-like structures in Case 1CH. The elongation of the vertical vortices is more marked in the case of the high Prandtl number. In RDT, no difference is observed between the cyclonic and anticyclonic vortices. This is because the nonlinear vortex-stretching term intensifies and elongates those anticyclonic or cyclonic vortices in the vertical direction.

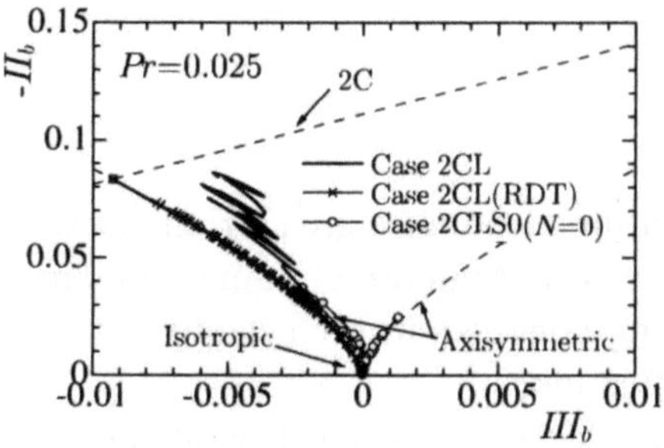

Figure 5. Anisotropy invariant map on Reynolds stresses in Cases 2CL, 2CLS0. Case 2CLS0 represents the neutrally stratified rotating flow of 2C initial condition.

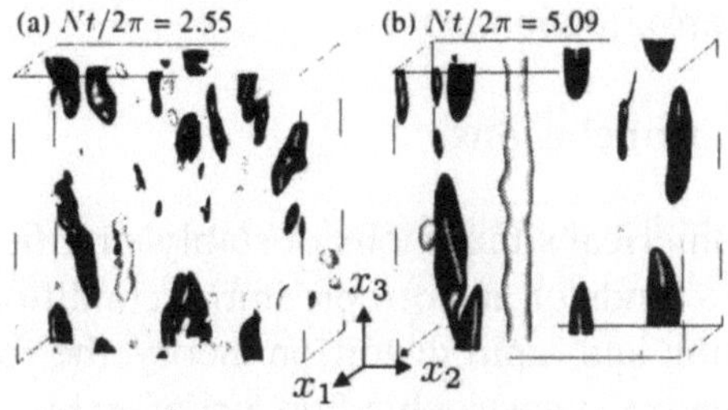

Figure 6. Time development of iso-surfaces of the second invariant of deformation tensor, white and black iso-surfaces represent cyclonic and anticyclonic vortices in Case 2CL; (a) $Nt/2\pi = 2.55$, (b) $Nt/2\pi = 5.09$.

3.2. MECHANISM OF TWO-DIMENSIONALIZATION AT LOW PRANDTL NUMBERS

At a low Prandtl number, turbulence becomes geostrophic and hydrostatic assumptions must be validated. The anisotropic invariant map on the Reynolds stresses is shown in Fig. 5. The second and the third invariant of the Reynolds stresses, i.e., II_b, III_b, are defined as follows:

$$b_{ij} = \frac{\overline{u_i u_j}}{2k} - \frac{1}{3}\delta_{ij}, \; II_b = -\frac{1}{2}b_{ij}b_{ji}, \; III_b = \frac{1}{3}b_{ij}b_{jk}b_{ki}. \quad (2)$$

It should be noted that when the vertical component of velocity becomes zero under the effect of the stable stratification, $(-II_b, III_b)$ must correspond to the upper side of the triangle. On the other hand, when the turbulence is isotropic, $(-II_b, III_b)$ must be in the origin. Interestingly, the two-dimensionalization of the Reynolds stresses is clearly observed at the low Prandtl numbers. It is also found that without stratification (Case 2CLS0), turbulence becomes almost isotropic and two-dimensionalization is not observed at all.

Figure 6 shows the time development of the vertical vortices represented by the iso-surfaces of the second invariant of the deformation tensor. The sense in the rotation of the vertical vortices is classified by the colors on the iso-surfaces. In the first period, the anticyclonic vortices (black iso-surfaces) are correlated with the strain rate $\partial u_3/\partial x_3$, and hence elongated in the vertical direction. The second important phenomenon is definitely observed in the cyclonic vortices (white iso-surfaces) after the enhanced anticyclones are generated. Some cyclonic vortices become intertwined to form a single, larger vortex with the vertical alignment.

Finally, the vortex population is reduced, and larger, vertically grouped vortices result. The mechanism of the vortex merging is similar to those observed in the numerical simulation of QG turbulence[5]. However, it should be noted that in the present study, the numerical simulations are carried out without QG approximation.

4. Conclusions

Numerical simulations of stably stratified rotating turbulence are carried out with two kinds of anisotropic initial conditions. By using both direct numerical simulation and rapid distortion theory, the following conclusions are established with regard to geostrophic flow under stable stratification.

(1) The effects of the anisotropy of Reynolds stresses and Prandtl number on the asymmetry of the vertical vortices are discussed. The asymmetry of the vertical vortices is associated with vortex stretching. Vertical vortices associated with $\partial u_3/\partial x_3 > 0$ are enhanced in comparison to the other vortices with $\partial u_3/\partial x_3 < 0$. In the case of two-component and low Prandtl number, the anticyclones are well correlated with stretching term $\partial u_3/\partial x_3 > 0$. On the other hand, in the case of one-component and high Prandtl number, the cyclones are correlated with it.

(2) At low Prandtl numbers, turbulence is two-dimensionalized and the merging of the cyclonic vortices results in the elongated Taylor columns observed in the numerical simulation with the quasi-geostrophic approximation.

This research was partially supported by a Grant-in-Aid for Scientific Research (B) from the Ministry of Education, Science, Sports and Culture of Japan (No. 10450085).

References

1. Iida, O. and Nagano, Y. (1999) Coherent structure and heat transfer in geostrophic flow under density stratification, *Phys. Fluids*, **11** (2), pp. 368–377.
2. Newton, J. L., Aagaard, K. and Coachman, L. K. (1974) Baroclinic eddies in the Arctic Ocean, *Deep-Sea Res.*, **21**, pp. 707–719.
3. Hopfinger, E. J., Browand, F. K.. and Gagne, Y. (1982) Turbulence and waves in a rotating tank, *J. Fluid Mech.*, **125**, pp. 505–534.
4. McWilliams, J. C. (1989) Statistical properties of decaying geostrophic turbulence, *J. Fluid Mech.*, **198**, pp. 199–230.
5. McWilliams, J. C., Weiss, J. B. and Yavneh, I. (1994) Anisotropy and coherent vortex structures in planetary turbulence, *Science*, **264**, pp. 410–413.
6. Lesieur, M. (1997) *Turbulence in Fluids*, 3rd ed., Kluwer Academic Publishers, The Netherlands.
7. Hopfinger, E. J. (1992) *Rotating fluids in geophysical and industrial applications*, Ed., Hopfinger, E. J., pp. 359–369, Springer-Verlag, Wien-New York.
8. Linden, P. F., Boubnov, B. M. and Dalziel, S. B. (1995) Source-sink turbulence in a rotating stratified fluid, *J. Fluid Mech.*, **337**, pp. 303–332.

LINEAR PROCESSES IN UNSTEADY STRATIFIED SHEARED TURBULENCE

H. HANAZAKI
Institute of Fluid Science
Tohoku University
2-1-1 Katahira, Aoba-ku, Sendai 980-8577
JAPAN

and

J.C.R. HUNT
Department of Applied Mathematics and Theoretical Physics
University of Cambridge
Silver Street, Cambridge, CB3 9EW
UK

1. Introduction

The transport of mass and heat in the atmosphere and ocean depends largely on how the turbulence is affected by the combined effects of the density gradient and the shear in the velocity. When there is no shear, the results of RDT (Hanazaki & Hunt, 1996) have shown how the initial conditions determine the subsequent time development of the partition of energy between the potential energy and the kinetic energy. In addition, the change of sign in the vertical density flux with time, which leads to the so called 'counter-gradient flux', could be explained by the simple linear oscillations due to buoyancy effects rather than by any new kind of nonlinear mixing processes.

In this study we extend the method to the flow with mean shear. There are two time scales $N^{-1}$and α^{-1} in this system, defined respectively by the Brunt-Väisälä frequency $N=\sqrt{-(g/\rho_0)(d\bar{\rho}/dx_3)}$ and the mean vertical shear $\alpha = dU/dx_3$. The Richardson number Ri is the square of the ratio of these two time scales, which has been usually used to determine the stability of one wave number component in the traditional linear stability theory. We will show how the linear processes exerted by stratification works in the

T. Kambe et al. (eds.), IUTAM Symposium on Geometry and Statistics of Turbulence, 291–296.
© 2001 *Kluwer Academic Publishers.*

shear flow turbulence. The RDT equations have been solved analytically and the time development of the spectra and the fluxes has been obtained by the short-time analytical approximations and also by the numerical integrations to investigate the moderate to long time developments. The results agree well with the DNS when the turbulent Froude number is small enough so that the nonlinear effects are negligible. For example, they explain the 'persistent' counter-gradient flux for high Richardson number ($Ri \geq 0.5$) flows observed in DNS as one of the most typical effects of the shear.

2. General solutions of the RDT equations

The RDT equations for stratified shear flow (Townsend 1976) with mean flow in the x_1 direction and stratification in the x_3 (vertical) direction are described as

$$\frac{d\hat{u}_i}{dt} = \alpha \hat{u}_3\left(\frac{2k_i k_1}{k^2} - \delta_{i1}\right) + \hat{\rho}\left(\frac{k_i k_3}{k^2} - \delta_{i3}\right) \quad (i = 1,2,3), \tag{1}$$

and

$$\frac{d\hat{\rho}}{dt} = N^2 \hat{u}_3, \tag{2}$$

where the Fourier components in the moving frame, i.e., $\hat{u}_i(t)$ and $\hat{\rho}(t)$ are defined by

$$u_i(\vec{x},t) = \sum_{\vec{k}(t)} \hat{u}_i(\vec{k}(t),t) e^{i\vec{k}(t)\cdot\vec{x}}, \quad \frac{g}{\rho_0}\rho(\vec{x},t) = \sum_{\vec{k}(t)} \hat{\rho}(\vec{k}(t),t) e^{i\vec{k}(t)\cdot\vec{x}} \tag{3}$$

The wave number vector $\vec{k}(t)$ develops with time as $\vec{k}(t) = (k_1, k_2, k_3)$ $= (k_{10}, k_{20}, k_{30} - \alpha t k_{10})$ where $\vec{k}_0 = (k_{10}, k_{20}, k_{30})$ is the initial wave number vector.

The most general solutions of the RDT equations give the three-dimensional spectra as

$$\begin{aligned}\Phi_{\rho\rho}(\vec{k}_0,t) = &\frac{N^4}{\alpha^2 \sin^4\theta \cos^2\phi}\left|P_\nu(z)Q_\nu(z_0) - P_\nu(z_0)Q_\nu(z)\right|^2 \Phi_{33}(\vec{k}_0, t=0) \\ &+ \frac{1}{\sin^4\theta}\left|P_\nu(z)Q_\nu'(z_0) - P_\nu'(z_0)Q_\nu(z)\right|^2 \Phi_{\rho\rho}(\vec{k}_0, t=0),\end{aligned} \tag{4}$$

$$\begin{aligned}\Phi_{33}(\vec{k}_0,t) = &\frac{1}{\sin^4\theta}\left|P_\nu'(z)Q_\nu(z_0) - P_\nu(z_0)Q_\nu'(z)\right|^2 \Phi_{33}(\vec{k}_0, t=0) \\ &+ \frac{\alpha^2\cos^2\phi}{N^4\sin^4\theta}\left|P_\nu'(z)Q_\nu'(z_0) - P_\nu'(z_0)Q_\nu'(z)\right|^2 \Phi_{\rho\rho}(\vec{k}_0, t=0),\end{aligned} \tag{5}$$

$$\Phi_{\rho 3}(\vec{k}_0,t) = \mathrm{Re}\Bigg[\frac{iN^2}{\alpha\sin^4\theta\cos\phi}\left(P_\nu(z)Q_\nu(z_0) - P_\nu(z_0)Q_\nu(z)\right)$$
$$\times\left(P_\nu'(z)^*Q_\nu(z_0)^* - P_\nu(z_0)^*Q_\nu'(z)^*\right)\Phi_{33}(\vec{k}_0,t=0)$$
$$+\frac{i\alpha\cos\phi}{N^2\sin^4\theta}\left(P_\nu(z)Q_\nu'(z_0) - P_\nu'(z_0)Q_\nu(z)\right) \tag{6}$$
$$\times\left(P_\nu'(z)^*Q_\nu'(z_0)^* - P_\nu'(z_0)^*Q_\nu'(z)^*\right)\Phi_{\rho\rho}(\vec{k}_0,t=0)\Bigg].$$

Here, θ and ϕ denote the polar angles of the initial wave number $\vec{k}_0$ which are given by

$$\sin\theta = \frac{\sqrt{k_{10}^2 + k_{20}^2}}{k_0}, \qquad \cos\phi = \frac{k_{10}}{\sqrt{k_{10}^2 + k_{20}^2}}. \tag{7}$$

The subscript $\nu = \frac{1}{2}\left(-1+\sqrt{1-4Ri/\cos^2\phi}\right)$ denotes the complex power of the Legendre functions of the first and second kind $P_\nu(z)$ and $Q_\nu(z)$, z and z_0 are defined respectively by $z = i(\cot\theta - \alpha\cos\phi)$ and $z_0 = i\cot\theta$, the primes denote the differentiation by z and the asterisks denote the complex conjugates.

In this study we assume all the initial density fluxes to be zero. If we assume that the turbulence is initially isotropic, the initial three-dimensional spectra are given by

$$\Phi_{ij}(\vec{k}_0,t=0) = \frac{E(k)}{4\pi k^2}\left(\delta_{ij} - \frac{k_{10}k_{20}}{k_0^2}\right), \qquad \Phi_{\rho\rho}(\vec{k}_0,t=0) = \frac{S(k)}{4\pi k^2}2N^2, \tag{8}$$

where $E(k)$ and $S(k)$ are the initial isotropic kinetic and potential energy spectra.

Then we can calculate the variances and the covariances for an arbitrary time. For example, the potential energy is calculated by

$$PE(t) = \frac{1}{2N^2}\overline{\rho^2(t)} = \int \Phi_{\rho\rho}(\vec{k})k_0{}^2 dk_0 \sin\theta d\theta d\phi$$
$$= \frac{Ri}{8\pi}KE_0\int\frac{1}{\sin\theta\cos^2\phi}\left|P_\nu(z)Q_\nu(z_0) - P_\nu(z_0)Q_\nu(z)\right|^2 d\theta d\phi$$
$$+\frac{1}{4\pi}PE_0\int\frac{1}{\sin^2\theta}\left|P_\nu(z)Q_\nu'(z_0) - P_\nu'(z_0)Q_\nu(z)\right|^2 d\theta d\phi. \tag{9}$$

We note that the time development of the variances and the covariances, such as the potential and kinetic energies and the vertical density flux, are solely determined by the initial total kinetic(KE_0) and potential(PE_0) energies and not by their specific spectral forms as far as the fluid is inviscid. This is the same as in the stratified turbulence with no shear (Hanazaki & Hunt, 1996).

3. Time development of the energy and the fluxes

The short-time approximations for the variances and the covariances are described by the initial potential and kinetic energy and also by the Richardson number *Ri*. For example, the potential energy is expressed by

$$PE(t) = PE_0 + \frac{1}{3}Ri(\alpha t)^2\left(KE_0 - 2PE_0\right) - \frac{2}{315}Ri(\alpha t)^4\left(3KE_0 - 2PE_0\right) \\ - \frac{4}{45}Ri^2(\alpha t)^4\left(KE_0 - 2PE_0\right). \tag{10}$$

Since $Ri\alpha^2 = N^2$, the leading-order correction to PE_0 does not depend on the shear represented by α. Equivalently the very initial time development of the potential energy is dominated only by stratification represented by N. The sign of $KE_0 - 2PE_0$ determines the initial decay or increase of the potential energy, in agreement with the previous DNS by Holt, Koseff & Ferziger (1992). It is of interest to note that the effect of shear appears only in the form coupled with the stratification, and it does not make separate contribution to the potential energy.

To investigate the long-time behaviour we integrated the spectra numerically (figure 1a). The time development of the potential energy shows good agreement with the previous DNS by Gerz, Schumann & Elghobashi (1989) shown in figure 1b, particularly when the stratification is strong ($Ri \geq 0.5$). In DNS the Richardson number is increased by increasing the stratification N, keeping the strength of shear constant. Then at high *Ri*, the turbulent Froude number Fr_t, defined by the Brunt-Vaisala frequency N, the integral scale and the initial rms velocity, is small. Then the buoyancy term becomes large compared to the nonlinear term and the linear processes become dominant at moderate Reynolds numbers.

The time development of the vertical density flux is shown in figure 2. The results of RDT show the 'persistent' counter gradient flux at *Ri*=1, which has been first observed in DNS by Gerz & Schumann (1991) and later by Holt et al. (1992). The RDT results show again good agreement with DNS in particular for large Richardson numbers ($Ri \geq 0.5$). The DNS results by Gerz, Schumann & Elghobashi (1989) and Gerz & Schumann (1991) generally give better agreement with RDT compared to Holt et al.'s results since both the stratification N and the shear α are stronger for the same *Ri* and the linear assumption holds better in Gerz et al., leading to the stronger counter-gradient flux.

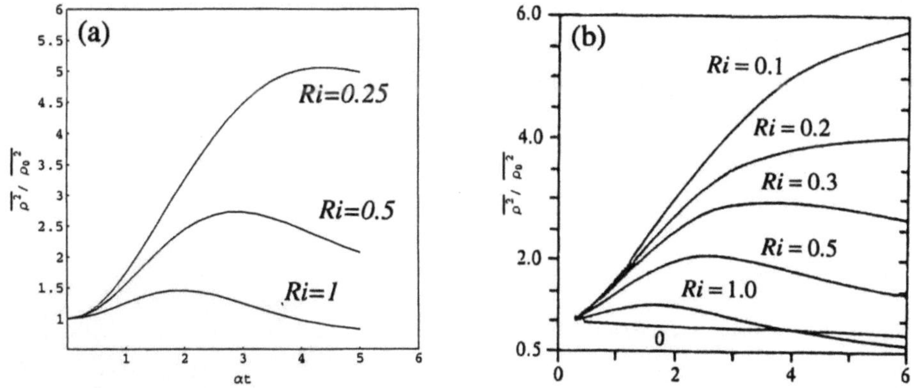

Figure 1. Time development of the potential energy $PE(t)$ for the initial condition of $PE_0/KE_0 = Ri/3$. (a)RDT, (b)DNS by Gerz, Schumann & Elghobashi (1989).

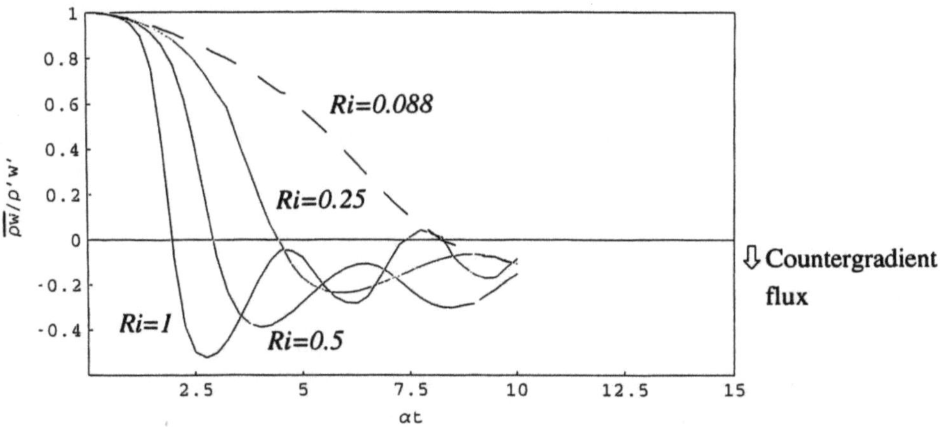

Figure 2. Time development of the vertical density flux obtained by RDT for initial condition of $PE_0/KE_0 = 0$. This should be compared with figure 15 of Holt, Koseff & Ferziger (1992).

The results for the vertical kinetic energy spectra $E_{33}(\chi)$ ($\chi = k(t)$) at large times show that even a weak stratification ($Ri < 0.1$) significantly reduces the vertical energy components at large scales in a long time, and the low wave number spectra asymptotically reach the same value irrespective of *Ri*. This is because all the (θ,ϕ)components diminish with time, except for those at $\phi = \pi/2$, for which direction the wave number vector is unchanged with time and the effects of shear are negligible. On the other hand, the energy is generally enhanced at small scales by stratification. This is shown in figure 3(a) for the case of $PE_0 = 0$ and the corresponding spectral components at a low wave number ($\chi = 0.1$) is shown in figure 3(b).

With the effects of viscosity and diffusion, the variances and the covariances will generally decay with time, while the normalized fluxes such as $\overline{\rho w}/\rho' w'$ may not be altered significantly from the inviscid value if the Prandtl number is equal to 1 and $E(k)$ and $S(k)$ have a same spectral form.

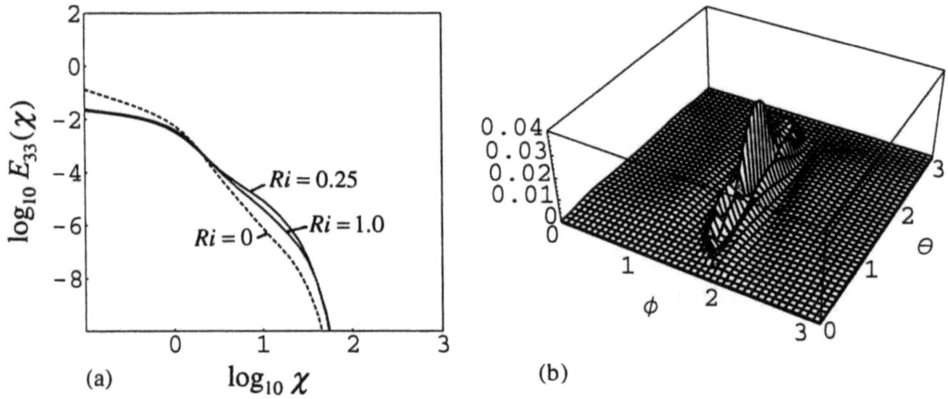

Figure 3. (a) Vertical kinetic energy spectra $E_{33}(\chi)$ at a large time ($\alpha t = 20$) for $PE_0 = 0$ and (b) the (θ, ϕ) components at large scales ($\chi = 0.1$) with weak stratification Ri=0.088 at $\alpha t = 20$.

4. Conclusions

In this study, we have investigated the stratified turbulence with mean shear by the rapid distortion theory (RDT). The results showed good agreement with DNS for low turbulent Froude numbers, in which the nonlinear effects are expected to be small. The initial total turbulent kinetic and potential energy are important in determining the initial direction of the vertical density flux. The persistent counter-gradient heat flux, identified as one of the main characteristics of the stratified sheared turbulence, was reproduced also in the linear processes described by RDT.

References

[1] H. Hanazaki and J.C.R. Hunt, Linear processes in unsteady stably stratified turbulence. J. Fluid Mech. 318, 303 (1996).

[2] A.A.Townsend, *The structure of turbulent shear flow*, Cambridge University Press (1976).

[3] S.E. Holt, J.R. Koseff and J.H. Ferziger, A numerical study of the evolution and structure of homogeneous stably stratified turbulence. J. Fluid Mech. 237, 499 (1992).

[4] T.Gerz, U.Schumann and S.E.Elghobashi, Direct numerical simulation of stratified homogeneous turbulent shear flows. J. Fluid Mech. 200, 563 (1989).

[5] T.Gerz and U.Schumann, Direct simulation of homogeneous turbulence and gravity waves in sheared and unsheared stratified flows, in *Turbulent Shear Flows* 7, Springer-Verlag, Berlin Heidelberg, 27-45 (1991).

H

Transition Mechanisms

DIRECT NUMERICAL SIMULATION OF TRANSITION IN PLANE POISEUILLE FLOW

KIYOSHI YAMAMOTO
National Aerospace Laboratory, Chofu, Tokyo 182-8522, Japan

NAOYA TAKAHASHI
University of Electro-Communications,
Chofu, Tokyo 182-8585, Japan

TSUTOMU KAMBE
Department of Physics, University of Tokyo,
Hongo, Bunkyou-ku, Tokyo 113-0033, Japan

1. Introduction

It is traditionally understood that laminar-turbulent transitions of wall shear flows are induced through two steps of instability, linear and nonlinear ones. The linear instability occurs at an early stage of the transition process and amplifies proper unstable waves selected from small disturbances existing in the flow. This instability has been investigated by many researchers with the Orr-Sommerfeld equation in the linear stability equations[1]. As a result, it has been found in plane Poiseuille flow that unstable Tollmien-Schlichting (hereafter, TS in short) waves are amplified exponentially with proper growth rates calculated as the eigenvalues of the Orr-Sommerfeld equation. Furthermore, the critical Reynolds number for the TS wave is derived as 5772[2]. On the other hand, the nonlinear instability is induced by the linearly amplified TS waves and amplifies various three-dimensional disturbances existing in the flow. Thus, this instability occurs usually at the final stage of the transition process and completes the transition process to turbulence.

Another type of the linear instability without eigenvalues, however, has been recognized recently to be important in the transition of the plane Poiseuille flow[3, 4]. This instability induces transient growth of streamwise vortices (hereafter, SV in short), which is described by the vorticity equation in the linear stability equations. The instability becomes to be

T. Kambe et al. (eds.), IUTAM Symposium on Geometry and Statistics of Turbulence, 299–304.
© 2001 *Kluwer Academic Publishers.*

dominant in the flow when its initial disturbances are considerably large. In this case, the strength of the amplified streamwise vortices exceed the threshold value which can trigger the nonlinear instability, then, the transition of the flow. In this paper, the direct numerical simulations (DNS) of transition in the plane Poiseuille flow with various initial disturbances are carried out to clarify how these instabilities become to be dominant.

2. Basic equations

The coordinate system $\{x, y, z\}$ in the flow is taken for x in the streamwise, y in the spanwise and z in the direction normal to wall. The computational domain is confined in the region $x \in [0, 4\pi]$, $y \in [0, 4\pi]$, and $z \in [-1, 1]$. The flow is described as $\{U + u, v, w\}$, where $U(z)(= 1 - z^2)$ is the basic flow and $\mathbf{u}\{u, v, w\}$ the disturbance velocity.

The Navier-Stokes equation for $\mathbf{u}$ is described by

$$\frac{\partial \mathbf{u}}{\partial t} = -U\frac{\partial \mathbf{u}}{\partial x} - w\frac{\partial U}{\partial z} - \omega \times \mathbf{u} - \nabla P + \frac{1}{R}\nabla^2 \mathbf{u}, \tag{1}$$

where $\omega = \nabla \times \mathbf{u}$, $P = p + |\mathbf{u}|^2/2$, and R is Reynolds number defined by the maximum value of U and a half width between walls, $R = U(0)/\nu$.

The incompressible condition is $\nabla \cdot \mathbf{u} = 0$.

3. Direct numerical simulation of the transition

The disturbance velocity $\mathbf{u}$ is expanded by the Fourier series for x, y directions, but for the z direction, it is calculated on the Chebyshev collocation points z_j as

$$\mathbf{u}(x, y, z_j, t) = \sum_k \mathbf{u}(k_x, k_y, z_j, t) \exp(ik_x x + ik_y y) \tag{2}$$

where $k^2 = k_x^2 + k_y^2$. Then, the Navier-Stokes equation (1) is calculated by the Fourier-Chebyshev spectral method[5]. The number of Fourier modes taken in the (x, y) directions is (42×42) and that of collocation points taken in the z direction 65.

Energy norm of Fourier mode (k_x, k_y) per unit mass is defined by

$$E(k_x, k_y) = \sum_j |\mathbf{u}(k_x, k_y, z_j, t)|^2 \tag{3}$$

The initial disturbance is given by the Fourier modes as following

$$\mathbf{u}(\mathbf{k}, z_j, 0) = \varepsilon \mathbf{q}(\mathbf{k}, z_j) \tag{4}$$

where ε is a constant amplitude parameter assigned as

$$\begin{aligned} \varepsilon &= const.\ for\ |k_x|, |k_y| \le 5.0, \\ &= 0 \qquad\qquad otherwise \end{aligned} \tag{5}$$

and $\mathbf{q}$ is a random function which satisfies the following constraint conditions [6].

$$i\left[k_x q_x(\mathbf{k}, z_n) + k_y q_y(\mathbf{k}, z_n)\right] + D_{nm} q_z(\mathbf{k}, z_m) = 0, \tag{6}$$

$$\sum_n |\mathbf{q}(\mathbf{k}, z_n)|^2 = 1, \tag{7}$$

where D_{nm} is the difference matrix on the collocation points.

DNS's of the transition have been carried out for various amplitude parameters ε and Reynolds numbers R. As a result, time evolution of the dominant Fourier modes ($0 \le k_x, k_y \le 3$) in the TS type transition for $R = 10000$ is shown in Fig. 1. Solid curves denote the two-dimensional modes and dashed curves the three-dimensional modes. Initial energies of Fourier modes are given to $10^{-9}(\varepsilon = 10^{-9/2})$. The curve denoted (1,0) shows a TS mode whose growth rate agrees well with the corresponding eigenvalue of the Orr-Sommerfeld equation. At $t \simeq 1700$, all modes behave abrupt growthes, which are induced by the nonlinear instability. After then, the flow breaks down to turbulence. This figure show a typical numerical result of the TS type transition for the super-critical Reynolds number. Even in this figure, however, the transient growth is observed in some modes whose energies reach to their maximum values at $t \simeq 300$ and after then, decay very slowly. As a matter of fact, all of these modes are streamwise components, $E(0, k_y)$.

Time evolution of Fourier modes in the SV type transition for the same Reynolds number is shown in Fig. 2. In this case, initial energies of modes are given to $10^{-8}(\varepsilon = 10^{-4})$, which is ten times larger than that of Fig. 1. The result shows that the transition process in this case completes within very short period in comparison with the TS type transition (see Fig. 1). This reflects the large growth rate of disturbance in the transient growth of streamwise vortices.

Similar simulations have been conducted to obtain the threshold value of initial disturbances which triggers the SV type transition. Figure 3 shows the numerical result obtained by the simulations. $|u(\mathbf{x})|_{\max}$ denotes the maximum value in the absolute streamwise velocities of the initial disturbance on the physical collocation points. Open circles show the minimum values in the set of $|u(\mathbf{x})|_{\max}$ which have induced the SV type transition for the same Reynolds number. On the other hand, closed circles show the maximum value in the set of $|u(\mathbf{x})|_{\max}$ which have never induced the SV type transition for the same Reynolds number. The solid line shows the inclination of $R^{-7/4}$ as a function of Reynolds number[7].

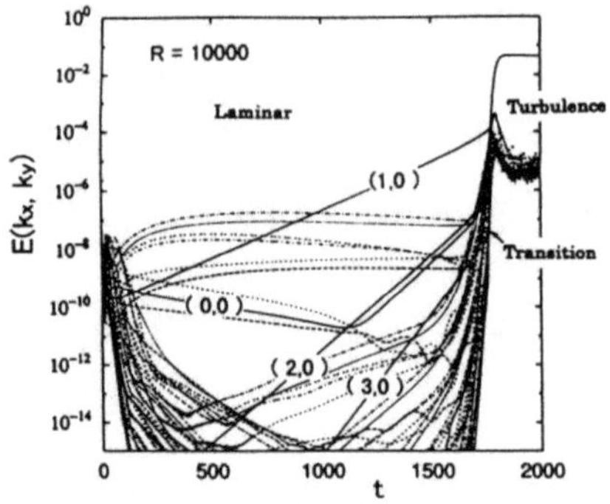

Figure 1. Time evolution of dominant Fourier modes in the TS type transition for $R = 10000$. Initial mode energies are given to 10^{-9}. Solid curves denote two-dimensional modes and dashed curves three-dimennsional modes.

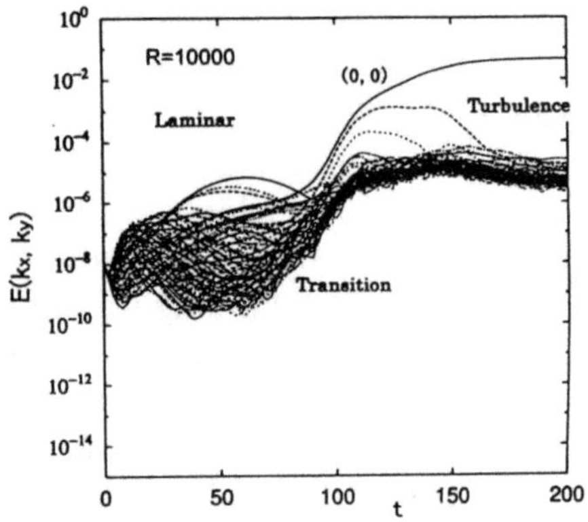

Figure 2. Time evolution of dominant Fourier modes in the SV type transition for $R = 10000$. Initial mode energies are given to 10^{-9}. Solid curves denote two-dimensional modes and dashed curves three-dimennsional modes.

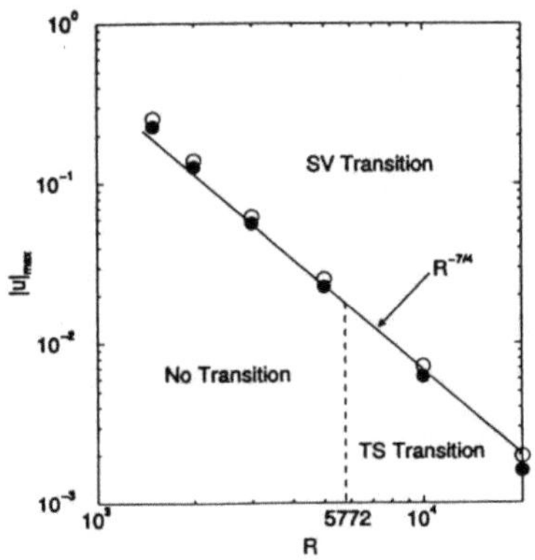

Figure 3. Threshold amplitudes of the streamwise velocity in the initial disturbances for the SV type transition versus Reynolds number. O denotes the amplitude with the transition and ● the amplitude without the transition.

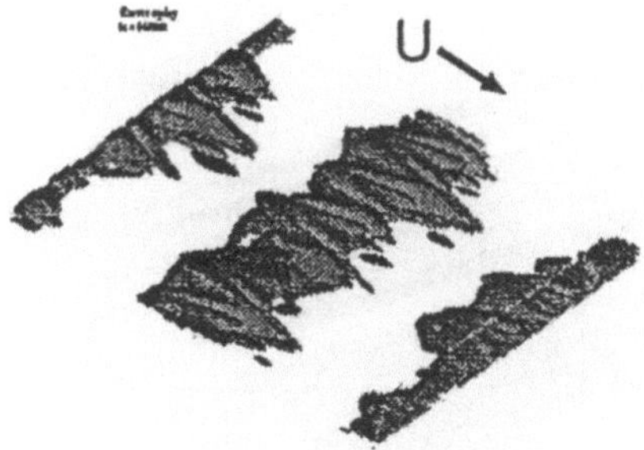

Figure 4. Λ vortices appearing in the TS type transition at $t = 1600$ for $R = 10000$.

An initial disturbance with $|u(\mathbf{x})|_{\max}$ smaller than the closed circles can not induce the SV type transition. Even in this case, however, if the Reynolds number is greater than 5772, then the TS type transition will occur. In other case with Reynolds number smaller than 5772, probably the transition does not occur. It should be noted that the SV type transition explains the subcritical transition for TS modes observed in the experiments[8].

4. Vortical structure in the transition region

The transition is completed by the nonlinear instability, whose mechanism depend on the Reynolds number. To understand the mechanism, it is instructive to observe the spatial vortical structure.

Figure 4 shows vortices which appear on the negative crest of the TS wave in the TS type transition in Fig. 1 at $t = 1600$. The vortices have been observed as the peak-valley structure of the TS waves in the experiments. The Λ vortices lift up low speed fluid into an upper high speed position and make local high shear regions in the flow, which enhance the nonlinear interaction more and more.

On the other hand, Fig. 5 shows the streamwise vortices which appear in the SV type transition Fig. 2 at $t = 80$. In this transition, the streamwise vortices are amplified at first. When initial disturbances in the flow is large, these vortices are amplified until enough to trigger the transition (See Fig. 3). As seen in Fig. 1, the transient growth is always observed at an early stage even in the TS type transition, the TS type transition occurs when the transient growth of SV does not trigger the transition.

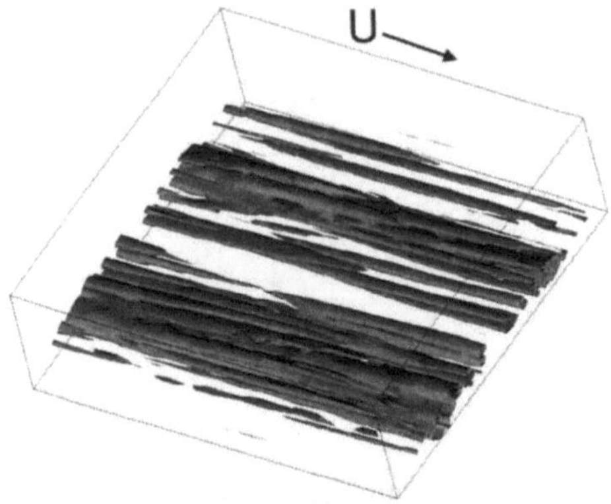

Figure 5. Streamwise vortices appearing in the SV type transition at $t = 80$ for $R = 10000$.

5. Conclusions

Laminar-turbulent transition process of plane Poiseuille flow is studied by mean of DNS. It is found that the transition is induced by TS wave instability or transient growth of SV, depending on the initial disturbance level existing in the flow. The threshold amplitudes of the disturbance, above which the SV type transition occur, are numerically estimated as a function of Reynolds number, which are proportional to $R^{-7/4}$.

References

1. Drazin, P. G. and W. H. Reid: 1981, *Hydrodynamic Stability.* Cambridge Univ. Press.
2. Orszag, S. A.: 1971, 'Accurate solution of the Orr—Sommerfeld stability equation'. *J. Fluid Mech.* **50**, 693–703.
3. Trefethen, L. N., A. E. Trefethen, S. C. Reddy, and T. A. Driscoll: 1993, 'Hydrodynamic stability without eigenvalues'. *Science* **261**, 578–584.
4. Reddy, S. C., P. J. Schmid, J. S. Baggett, and D. S. Henningson: 1998, 'On stability of Streamwise Streaks and Transition Thresholds in Plane Channel Flows'. *J. Fluid Mech.* **365**, 269–303.
5. Canuto, C., M. Y. Hussaini, A. Quarteroni, and T. A. Zang: 1988, *Spectral Methods in Fluid Dynamics.* Berlin: Springer-Verlag.
6. Hosokawa, I. and K. Yamamoto: 1986, 'On Direct Numerical Simulation of Incompressible Shear-Flow turbulences by the Fourier-SPectral method'. *J. Phys. Soc. Japan* **55**(3), 1030–1031.
7. Lundbladh, A., D. S. Henningson, and S. C. Reddy: 1994, 'Threshold Amplitudes for Transition in Channel Flow'. In: T. B. M.Y. Hussaini and T. Jackson. (eds.): *Transition, turbulence and combustion*, Vol. 1. pp. 309–318.
8. Patel, V. C. and M. R. Head: 1969, 'Some observation on skin friction and velocity profiles in fully developed pipe and channel flows'. *J. Fluid Mech.* **38**, 181–201.

ON THE REGENERATION MECHANISM OF TURBULENCE IN THE CHANNEL FLOW

~ *Role of the Traveling-wave Solution*

S. TOH AND T. ITANO
Division of Physics and Astronomy, Graduate School of Science
Kyoto University, Kyoto 606-8502, Japan

Abstract. Direct numerical simulations (DNS) for the incompressible Navier Stokes equations are employed to study three-dimensional plane Poiseuille turbulent flow (Re=3000). A traveling wave solution (TWS) is found. A scenario of turbulent production is proposed in terms of the TWS and its manifolds. It is concluded that a low-dimensional dynamics confined to near-wall region exists in the scenario and plays an active role in turbulent production and maintenance.

1. Introduction

One of the most important characteristics of the wall turbulence is that the turbulence production occurs locally both in space and time close to the wall, which is called burst [1]. However, the mechanism for maintaining turbulence including the burst is still unclear.

Hamilton *et al.* [4] proposed the idea of self-sustaining process (SSP) to explain this mechanism through the studies of Couette flow turbulence at low Reynolds number. In this model, streaks develop and become unstable with respect to the three dimensional, i.e., streamwisely-dependent disturbance. Through this instability, longitudinal vortices emerge and then are destroyed by nonlinear interactions producing turbulent fluctuation. After these processes, the system settles into another calm quasi two-dimensional, i.e., streamwisely-independent state and the next cycle sets in. It is accepted that SSP is the generic mechanism for the regeneration of turbulence in wall turbulence.

In this report, by means of DNS we show that SSP really works in near-wall region of Poiseuille flow turbulence in more global sense. The key

T. Kambe et al. (eds.), IUTAM Symposium on Geometry and Statistics of Turbulence, 305–310.
© 2001 *Kluwer Academic Publishers.*

points of our study are the existence of TWS and the low-dimensionality of the dynamics governing coherent structures in the near-wall region.

2. Traveling wave solution and its manifolds

The numerical scheme employed here is basically the same as those of Kim *et al.*[2]. For all simulations reported, we use periodic boundary conditions in x and z with period lengths $L_x = \pi \sim 420^+$ and $L_z = 0.4\pi \sim 170^+$, and apply no-slip boundary conditions at $y = \pm 1$: x, y and z represent the streamwise, normal to the walls and spanwise directions, respectively. The value of Reynolds number is fixed to 3000. Though somewhat large, this system belongs to the minimal flow unit [3].

First we summarize the method used to obtain the TWS. An idea underlying this method is the linear instability of streaks mentioned in SSP. To define the streak formally, we separate the flow into quasi two-dimensional (Q2D) and three-dimensional (3D) components as follows:

$$\boldsymbol{u}^{Q2D}(y,z,t) = \frac{1}{L_x}\int_{L_x} \boldsymbol{u}(x,y,z,t)dx, \qquad \boldsymbol{u}^{3D} = \boldsymbol{u} - \boldsymbol{u}^{Q2D}.$$

An initial condition for this shooting method is constructed by adding 3D (linearly small) random disturbance to the Q2D component of a snapshot (SNAP1 hereafter). SNAP1 is a special state where its Q2D component takes the maximum value over our simulations, i.e., an extremely developed low-speed streak is formed in the lower wall region. By adjusting the norm of the initial 3D disturbance, we can make the trajectory approach a TWS as close as possible.

Figure 1(a) shows typical two trajectories obtained by this method. The trajectories are drawn in the phase space spanned by $(E_y^{Q2D}(t), E_y^{3D}(t))$, where

$$E_y^{Q2D}(t) = \int_V |u_y^{Q2D}(y,z,t)|^2 dxdydz, \; E_y^{3D}(t) = \int_V |u_y^{3D}(x,y,z,t)|^2 dxdydz.$$

The TWS corresponds to the saddle point in the center of Fig.1(a) and has both stable and unstable manifolds. Along these trajectories, the flow field shows the instability of the streak, stays around the TWS for a while and finally develops into either laminar or turbulent state. The initial stage of the process of escape from the the saddle is due to the linear instability of the TWS, so that E_y^{Q2D} and E_y^{3D} evolve abruptly.

On the attractor of actual turbulence, the trajectory projected to this 2D phase space does not pass so close to the saddle and manifolds. We can, however, observe similar abrupt evolutions in actual bursting processes in the near-wall region.

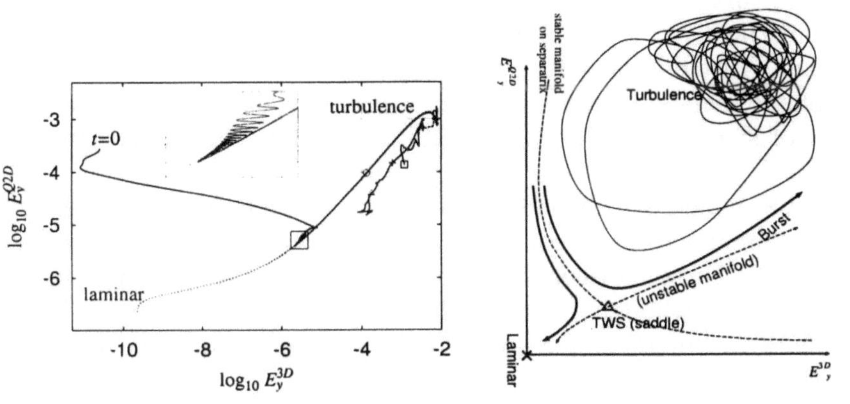

Figure 1(a) (left). Log-log plots of evolutions of $(E_y^{Q2D}(t), E_y^{3D}(t))$ in 2D phase space, where a bursting part of the trajectory of actual turbulence (from $\triangle$ to via $+$) is appended for comparison. Solid line: The value of the norm of 3D random disturbance (F_{ac}) is $9.116010224 \times 10^{-8}$. Dashed line: $F_{ac} = 9.1160104 \times 10^{-8}$. Dotted line: $F_{ac} = 9.11601 \times 10^{-8}$. Dashed and solid lines are almost collapsed.
Figure 1(b) (right). Schematic view of the proposed scenario.

3. Scenario of turbulent production

Based on the aspects mentioned above and SSP picture, we propose a scenario of turbulent production and maintenance in this section (*cf.* Fig.1(b)).

The dynamics of near-wall region in wall turbulence is quasi-cyclic. The following two main phases consist in each of quasi-cycles.

1. The formation phase of coherent structures is defined as the process of approach toward the TWS along the stable manifold. In this phase, almost two-dimensional and streaky structures in near-wall region e-merge from turbulent state via linear instability of Q2D streaks.
2. The burst in wall turbulence is defined as the process of escape out of the TWS along the unstable manifold. We assume bursts observed in actual turbulence are prescribed by the linear instability of the TWS and the following nonlinear evolution in this phase.

In the 2D phase space, the stable manifold is extended close to E_y^{Q2D} axis, so that the attractor of actual turbulence itself is assumed to be close to this axis locally. This means that streaky coherent states can be observed easily even in actual turbulence.

Note that the instability of streaks lead not directly to fully-developed turbulence but to a saturated state, i.e., TWS in our scenario. However, such saturation is not observed in actual wall turbulence. This is because the trajectory of turbulence does not necessarily approach the stable manifold so close that the TWS is not observed. Thus the SSP picture seems to work well even in Poiseuille flow turbulence. Even in this situation, we expect that the phenomena are understood more clearly by dividing the conceptual instability of streaks mentioned in SSP into the two elementary processes listed above.

The TWS, which is localized to the lower wall, consists of a streak and a pair of coherent vortices with positive and negative streamwise vorticity over the streak. These vortices correspond to so-called "longitudinal vortex" (LV1 hearafter), although they are flatter than those supposed. This is because LV1s are emerged on a kind of critical layer of the linearly unstable mode of the streak.

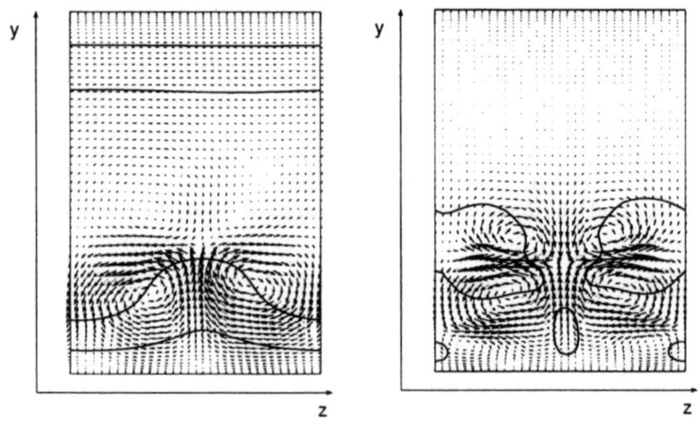

Figure 2. Q2D vector field (u_z^{Q2D}, u_y^{Q2D}) is shown. (left) TWS ,(right) most unstable mode. A part of some longitudinal vortices are projected in this y-z transverse plane. Solid line is contour of $u_x^{Q2D} \sim 0$, hill of which corresponds to low-speed streak. In particular, we note that the new streak and LV2s exist in unstable mode which are absent from TWS.

In the burst phase, it is observed that the longitudinal vortices are developed into the well-known tube-like structures owing to the linear instability of TWS: Through this instability LV1s are stretched roughly in the streamwise direction. In addition, the linearly unstable mode of TWS includes new another streak and a pair of longitudinal vortices (LV2) which are located closer to the wall (see Fig.2). In the succeeding nonlinear stage, the lower parts of LV1 are combined with LV2 in the near-wall region and forms horseshoe vortices. Shortly after, smaller vortices or turbulent fluctuations are created abruptly by these horseshoe vortices and others mentioned already. These turbulent fluctuations are finally ejected by the upper part of LV1.

4. Low-dimensional dynamics in near-wall region

In order to ascertain the scenario to be realized in actual turbulence, we need compare the burst due to this scenario with bursts observed in actual turbulence. For this purpose, we compare the actual burst which follows SNAP1 in DNS with the burst which evolves along the unstable manifold. It should be noted that the latter starts from the Q2D component of SNAP1 with small random 3D disturbance. In this sense, we examine whether 3D component extended in the whole channel affects the nature of bursts or not. Although the flow in this actual bursting process is fully-developed, streaks and longitudinal vortices are observed in near-wall region. Figure 3 shows clearly that in the near-wall region of lower side, the structures in this burst are almost as same as that on the unstable manifold. This suggests that the low-dimensional dynamics governing near-wall structures is confined to near-wall region and not significantly affected by turbulent fluctuations in the outer region. In fact, Jiménez and Moin showed that intermittency cycle acts independently at each wall in a single channel [3]. We infer that this is the reason why coherent structures such as streaks and longitudinal vortices are observed clearly even in the experiments of fully-developed wall turbulence.

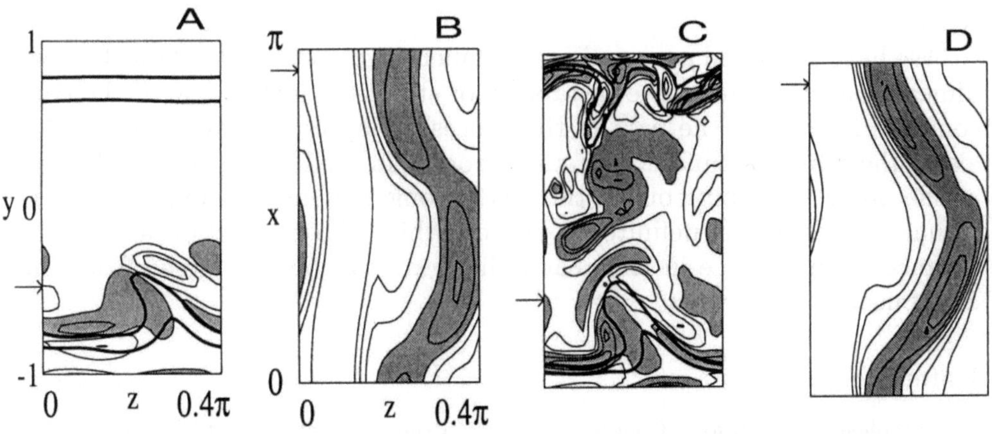

Figure 3. Comparison between (A,B) the burst model and (C,D) a instantaneous field of real turbulence obtained from DNS. (A,C) Contours of $\omega_x(x, y)$ at $z^+ = 39.8$. (B,D) Contours of $u_x(x, z)$ at $y^+ = 71$. Shade indicates $u_x < 0.65$ or $\omega_x < -0.1$. Thick solid lines are for $u_x = 0.4$ and 0.6. The phase of (c) and (d) is shifted for ease of comparison.

5. Concluding remarks

We have found a traveling wave solution (TWS) which corresponds to the fixed point in the phase space spanned by E_y^{Q2D} and E_y^{3D} (Figure 1-(a)). This indicates the low-dimensionality of the dynamics governing at least the flow in the near-wall region. Comparing the ideal bursting process defined on the manifolds of the TWS with the corresponding actual burst, we infer that this dynamics is confined to near-wall region and not significantly affected by the turbulent fluctuations exsisting in the outer region. From the point of view of dynamical system, this fact suggests the followings: The functional subspaces in which this low-dimensional dynamics and the huge-dimensional dynamics of the outer region are embedded respectively, are almost orthogonal to each other, and these dynamicses are weakly coupled by intermittent transport of tubulent energy through the buffer region from the near-wall region.

The bursting processes observed in actual turbulence have statistical property in common, although each of bursts evolves in different way. Actually, coherent structures have been successfully educed in bursting processes by means of conditional sampling methods, such as VITA [5]. We conclude that these common statistical natures of bursts derive from the low-dimensional dynamics governing coherent structures in the near-wall region.

We have proposed a scenario of turbulent production based on the low-dimensional dynamics. The instability of streaks, which is one of the core concepts in SSP, is the part of the process of approach toward the TWS along the stable manifold. Bursting process in wall turbulence is defined as the process escaping out of the TWS along the unstable manifold. This scenario gives us the definite ground on which the production and maintenance of turbulence in the near-wall region can be considered from the dynamical point of view.

References

1. Stephen K. Robinson, 'Coherent motions in the turbulent boundary layer', *Annu. Rev. Fluid Mech.* vol 23, pp.601-639, 1991.
2. John Kim, Parviz Moin and Rovert Moser, 'Turbulence statics in fully developed channel flow at low Reynolds number' *J.Fluid Mech.*, vol 177, pp.133-166 1987.
3. Javier Jiménez and Parviz Moin, 'The minimal flow unit in near-wall turbulence', *J.Fluid Mech.*, vol 225, pp.213-240, 1991.
4. James M. Hamilton, John Kim and Fabian Waleffe, 'Regeneration mechanisms of near-wall turbulence structures', *J.Fluid Mech.*, vol 287, pp.317-348, 1995.
5. Jinhee Jeong, Fazle Hussain,Wade Schoppa and John Kim, 'Coherent structure near the wall in a turbulent channel flow', *J.Fluid Mech*, vol 332, pp.185-214, 1997.
6. Sadayoshi Toh and Tomoaki Itano, 'Low-dimensional dynamics embedded in a plane Poiseuille flow turbulence' *online preprint server:arXiv.org e-Print archive*, http://xxx.lanl.gov/abs/physics/9905012, 1999.

I

Modulation of Turbulence

TURBULENCE IN POLYMER SOLUTIONS

MICHAEL CHERTKOV
Theoretical Division,
Los Alamos National Laboratory,
Los Alamos, NM 87545, USA

Abstract. The constitutive (rheological) equation of state, relating local stress to the rate-of-strain tensor, is derived consistently within the kinetic theory approximation for a dilute solution of elastic polymers in a strong flow. Microscopic polymer parameters (i.e. nonlinear elastic potential and the equilibrium polymer length) enter explicitly this coarse-grained description. This theory describes recent experiments by Groisman and Steinberg [1] on pure elastic turbulence (i.e. a steady state, realized at very low *Re* number, which is characterized by a non-smooth small-scale velocity distribution). The measured temporal spectra of velocity fluctuations are theoretically explained. A coarse-grained hydrodynamic description of pure elastic turbulence is given by a nonlinear diffusion equation for the velocity gradient tensor. Nonlinear elastic dissipation (originating from the anharmonic polymer elasticity) plays a key role in the emergence of the small scale turbulence.

The motion of a polymer solution fluid element is described by the non-Newtonian generalization of the Navier-Stokes equation,

$$\partial_t u^\alpha + (\mathbf{u}\nabla)\, u^\alpha = -\nabla^\alpha p + \nu \nabla^\beta \sigma^{\alpha\beta} - \nabla^\beta \tau^{\alpha\beta} + f^\alpha. \tag{1}$$

Here $\sigma^{\alpha\beta} \equiv \nabla^\beta u^\alpha$ and $\mathbf{f}$ is the divergenceless external driving force. $\hat{\tau}$ is the elastic part of the stress tensor. The incompressibility of $\mathbf{u}$, $\nabla^\alpha u^\alpha = 0$, unambiguously fixes the pressure, p. An additional relation between $\hat{\tau}$ and $\hat{\sigma}$ is provided by the so-called constitutive equation (also called the rheological equation of state), e.g. [2, 3]. The derivation of the constitutive equation obtained by averaging over the microscopic polymer related scales is required for any analysis of (1). This paper begins with the microscopic kinetic theory calculations based on the assumption that the polymer solution is dilute, i.e. the polymers do not contact each other directly and are influenced only through their collective contributions in the rate-of-strain

T. Kambe et al. (eds.), IUTAM Symposium on Geometry and Statistics of Turbulence, 313–318.
© 2001 *Kluwer Academic Publishers.*

tensor. (Hydrodynamic interactions which may also contribute the polymer induced stress will not be discussed here.) Further, we apply the derived constitutive relation to the case of pure elastic flows, in which the inertial terms (the second term on the rhs of (1) and the corresponding part of the pressure) are neglected. The elastic turbulence regime is developed (with a gradual increase of the external force) in two steps: first, the elastic instability takes place [4, 5, 6, 7], and, second, dissipation due to elasticity dominates the viscous dissipation. Those two subsequent steps can be separated or combined in one (the later is the case of [1]). A simple scaling theory for the fully developed pure elastic turbulence (elasticity dominated region above the threshold) is presented which explains the multi-scaling nature of the small scale velocity fluctuations. We conclude this letter with an analysis of the high Re inertia-elastic turbulence.

A theory of passive polymer stretching in a statistically fixed large-scale flow (based on deterministic and linear elasticity case studied by Lumley [8], see also [9], and also based on the extensive study of the so-called Batchelor model of passive scalar turbulence [10]) was recently presented [11, 12]. This theory cannot be directly applied to the problem of self advection of the polymer solution. However, the previously mentioned scale separation does allow one to extend (and in a sense even to simplify) the nonlinear theory of polymer stretching [12] to the active case. Averaging over the polymer separations, which are assumed to be much larger than the stretched polymer length, one may introduce averages, and moments of the polymer length, conditioned on the yet unknown statistics of $\hat{\sigma}$. The elastic part of the stress tensor (elastic tension force per unit area) is given by

$$\tau^{\alpha\beta} = -\frac{n}{N^3}\left\langle \int_0^N ds \rho^\alpha(s)\, F_{el}^\beta(s) \right\rangle_N \tag{2}$$

Here n is the polymer solution concentration; N is the equilibrium polymer length; $\rho(s)$ and $\mathbf{F}_{el}(s)$ are the vector of polymer displacement (measured from an end point of the polymer) and the elastic tension force, respectively, taken at a polymer position parametrized by s ($\int_0^N ds = N$). In (2), we have also assumed that the turbulent stretching is much stronger than that caused by thermo-diffusion. The averaging over the distribution of the polymer length is introduced in (2). (If the solution is mono-disperse, the average can be dropped. Later we will be primarily discussing the mono-disperse case.) The relation between ρ and $\hat{\sigma}$ is fixed by the balance of force acting on a polymer [12]

$$\frac{d}{dt}\rho^\alpha(s) - \sigma^{\alpha\beta}\rho^\beta(s) = -F_{el}^\alpha(s) = -\partial^\alpha U(\partial_s \rho(s)). \tag{3}$$

Here the polymer's friction against the flow (lhs of (3) with the kinetic coefficient set to 1) is equilibrated by the elastic (and generally thermal forces which are dropped in (3)); $U(s)$ is the elastic potential of the polymer (string). The free (zero-derivative) conditions at the polymer end points, $\partial_s\rho(0) = \partial_s\rho(N) = 0$, complete the description of the polymer dynamics. All of the fields in (2,3) can be subscripted by the label $\mathbf{r}$, standing for the spatial position of the region subject to the coarse grained description. The kinetic theory approximation consists in neglecting the spatial derivatives of $\sigma^{\alpha\beta}$ in the microscopic (polymer length scales) calculations.

The experiment [1] showed that chaotic flows, driven by the external shear occur at a very low Re-number (i.e. the kinetic inertia is negligible). The relatively weak bare flow (pure shear flow in [7]) is stable. An elastic instability is observed if the (pure shear) Deborah number (the ratio of the rate-of-strain to the inverse relaxation time, $De \equiv ||\hat{\sigma}||/U''(0)$, also called the Weissenberg number) is sufficiently large. The onset of the pure elastic instability was studied both experimentally [4, 5, 7] and theoretically [6, 7]. The theoretical approach was based on the so-called Oldroyd-B model [3] corresponding to the Hook (linear) elasticity of polymers. There, the constitutive equation is

$$\frac{d}{dt}\hat{\tau} - \hat{\sigma}\hat{\tau} - \hat{\tau}\hat{\sigma}^{+} = -U''(0)\hat{\tau} - \eta_p\left(\hat{\sigma} + \hat{\sigma}^{+}\right), \tag{4}$$

(This equation directly follows from the dumb-bell, i.e. $N = 2$, versions of (3,2) with the thermo-diffusion noise added to (3). η_p is the viscosity of the solvent, which is related to the rate-of-strain and the thermo-diffusion coefficient through the Einstein relation. The constitutive equation for a long linear polymer was discussed in [13].) The threshold of the elastic instability occurs at $De\sqrt{a\eta_p/\eta} \sim 1$ [7], where a is a geometrical (the gap ratio) factor: the chaotic component of the rate-of-strain has a well pronounced jump. However, as shown in [8, 11, 12], a polymer stretching by a random flow with $De_{chaot} > 1$ cannot be described by the linear version of (3): there is an abrupt jump (i.e. a first order phase transition) of the elastic contribution into the stress tensor at $De_{chaot} = 1$. The experiments [7, 1] show a jump of the chaotic component to $De_{chaot} > 1$ immediately at the elastic instability threshold, i.e. the two transitions occurs at once. The transitions can probably be separated (then, the ratio of the polymer contribution into the stress tensor to the viscous stress is less than unity at the instability threshold) at a lower polymer solution concentration.

In the following nonlinear (and therefore chaotic $\hat{\sigma}$) study let us consider, first, the case of the non-degenerate largest eigenvalue of $\hat{\sigma}$, denoted by λ. From the point of view of the longest (macroscopic) times, the polymer adjusts itself to the rate-of-strain and rapidly reaches its steady state (in a period of time less than $1/\lambda$). Therefore, the temporal derivative on the lhs of (3) can also be neglected in order to find the optimal stretched polymer configuration in the given (frozen) field $\hat{\sigma}$. The optimal configuration of the polymer is a stretched one elongating $\mathbf{k}_\lambda$, the unit vector parallel to the eigenvector of $\hat{\sigma}$ corresponding to λ. It is given by the steady solution of (3) (see [12] for more details),

$$\left[s + (N-2s)\theta\left(s - \frac{N}{2}\right)\right] = \int_0^z \frac{U''(x)dx}{\sqrt{2\lambda \int_x^{z_*} yU''(y)dy}}, \tag{5}$$

$$N\sqrt{\lambda} = \sqrt{2}\int_0^{z_*} \frac{U''(x)dx}{\sqrt{\int_x^{z_*} yU''(y)dy}}. \tag{6}$$

Here, $z \equiv \partial_s \rho$, with the maximal tension, τ_*, reached at the middle of the string, $s = N/2$. (5) defines the dependence of τ on s, τ_*, λ and N, while (6) gives the implicit dependence of τ_* on λ and N. For the mono-disperse polymer solution one has

$$\tau^{\alpha\beta} = \frac{n\mu^{\alpha\beta}}{2\lambda N^3}\int_0^{z_*} \frac{dzU'(z)U''(z)}{\sqrt{\int_z^{z_*} yU''(y)dy}} \int_0^z \frac{U''(x)xdx}{\sqrt{\int_x^{z_*} yU''(y)dy}}. \tag{7}$$

This is the required constitutive equation (actually, it is an explicit relation between $\hat{\tau}$ and $\hat{\sigma}$). Here in the nondegenerate case, $\mu^{\alpha\beta} \equiv k_\lambda^\alpha k_\lambda^\beta$. The degenerate case, when the two eigenvalues of $\hat{\sigma}$ have the same positive real part, λ, can also be considered. There is no steady solution of the projected equation in the case, because of the rotation caused by the asymmetric (vorticity) part of $\hat{\sigma}$ projected on the plane. However, once averaging with respect to the fast rotation and over the initial polymer orientations is performed, the expression for the stress tensor is given by (5,6) with $\mu^{\alpha\beta} = \left[\delta^{\alpha\beta} - k_{-2\lambda}^\alpha k_{-2\lambda}^\beta\right]/2$, where $\mathbf{k}_{-2\lambda}$ is the unitvector parallel to the eigenvector of the contracting direction. Consequently,

$$\mu^{\alpha\beta} \equiv \begin{cases} k_\lambda^\alpha k_\lambda^\beta, & \text{nondegenerate;} \\ \left[\delta^{\alpha\beta} - k_{-2\lambda}^\alpha k_{-2\lambda}^\beta\right]/2, & \text{degenerate.} \end{cases} \tag{8}$$

Based on the nonlinear constitutive relation derived above, we will show how the experimental phenomenon of pure (zero Re) elastic turbulence [1] can be described. The experiment shows that above the $De_{chaot} = 1$ threshold the elastic contribution into the stress tensor dominates the viscous one. In the elasticity-dominated regime all the viscous (stress and pressure) terms in (1) can be dropped and we are left with the equation of what we call (due to the presence of fine scale structures, see below) the pure elastic dissipative turbulence equation

$$\partial_t \sigma^{\alpha\gamma} = -\nabla^\gamma \nabla^\beta \tau^{\alpha\beta} + \nabla^\gamma \nabla^\alpha \Delta^{-1} \nabla^\mu \nabla^\nu \tau^{\mu\nu} + \nabla^\gamma f^\alpha . \tag{9}$$

Because this system is in the chaotic regime $\nabla^\gamma f^\alpha$ is renormalized, i.e. it can be considered to be a large scale (system size) random noise. (9), supplied by the constitutive relation (7,8,6), is nothing but a non-linear diffusion equation forced at large scales. Consider the case of an algebraic anharmonic potential

$$U_{al}(x) = q x^{2+\phi}, \quad \phi > 0. \tag{10}$$

The special case of interest, $\phi = 2$, accounts for the first regular in the x^2-expansion anharmonic correction. Substituting (10) in (7) gives the following explicit result:

$$\tau^{\alpha\beta} = -\mu^{\alpha\beta} \frac{\lambda^{1+2/\phi} N^{1+4/\phi} n}{q^{2/\phi}} \frac{2+\phi}{[2\pi(1+\phi)]^{2+2/\phi}} \tag{11}$$
$$\times \left(\frac{2+\phi}{2} + \frac{(1+\phi)\sqrt{\pi}\,\Gamma\left(\frac{2+2\phi}{2+\phi}\right)}{\Gamma\left(\frac{6+5\phi}{4+2\phi}\right)} \right) \left(\frac{\Gamma\left(\frac{4+3\phi}{4+2\phi}\right)}{\Gamma\left(\frac{1+\phi}{2+\phi}\right)} \right)^{4+4/\phi}$$

The onset value of the strain rate, λ_{os}, marking the equilibration of the bare (viscous) and elastic stresses, is estimated by

$$\lambda_{os} = \frac{q}{(n b_\phi)^{\phi/2} N^{2+\phi/2}}, \tag{12}$$

where b_ϕ is the ϕ-dependent coefficient of $\tau^{\alpha\beta}$ on the rhs of (11). The elasticity dominated regime of (9) is $\lambda \gg \lambda_{os}$.

We present here a simple dimensional analysis of (9) postponing a more accurate theoretical and numerical study of (9) to a future publication. The structure of (9,11), under the assumption that the pumping $\nabla^\gamma f^\alpha$ is large-scale delta-correlated in time (the later is not crucial), suggests the following scaling relations

$$\delta\sigma_r \sim \delta\lambda_r \sim \sqrt{Pt} \sim \left[\frac{q P^{\phi/2} r^\phi}{N^{2+\phi/2} n^{\phi/2}} \right]^{1/(1+\phi)} \tag{13}$$

Here P is the pumping amplitude of σ (i.e. P stand for the temporal integral of the ∇f different time pair correlation function); $\delta\sigma_r$ stands for the strain rate fluctuation coarse-grained at the observation scale, r. According to (13)

$\delta\lambda_r$ is nonsmooth in r at any finite ϕ. The nonlinear stochastic problem can be analyzed using Wyld diagrammatic technique [14]). Making the scaling ansatz (13), one finds (at least for $\phi = 2$) that diagrams of all the orders are convergent both infrared and ultraviolet length- and time-scales. This does not prove that the scaling estimate (13) holds for all order correlation functions of $\delta\sigma_r$. It is however plausible to conjecture that (13) holds, at least, for some low-order (particularly, second) moments of $\delta\sigma_r$.

In the experiment of Groisman and Steinberg [1], the spectra of the temporal (in the frequency domain) velocity fluctuations, $K_\omega \equiv \int dt \exp[i\omega t] \langle \delta u_t^2 \rangle$, was measured in the elastic turbulence regime at many different shear rates . The measurements can be approximated by a power law, $K_\omega^{(\text{exper})} \sim \omega^{-3.5}$. The theoretical scaling law (13) gives (the obvious $t \sim 1/\omega$ should be added)

$$K_\omega^{(\text{theory})} \sim \frac{\sigma_r^2 r^2}{\omega} \sim \frac{N^{1+4/\phi} P^{1+1/\phi} n}{q^{2/\phi}} \omega^{-3-1/\phi}. \tag{14}$$

For $\phi = 2$, (14) explains the experimental results of [1]. The scaling of (13,14) suggests the existence of multi-scale structures as it shows a nonsmooth spatio-temporal behavior of the rate-of-strain fluctuations.

Very fruitful discussions with A. Groisman are gratefully acknowledged. This work was supported by a J. R. Oppenheimer fellowship.

References

1. A. Groisman, V. Steinberg, to appear in "Nature".
2. M. Doi, S. Edwards, *The theory of polymer dynamics*, (Clarendon Press, Oxford 1986).
3. R. B. Bird, C. F. Curtiss, R. C. Armstrong, O. Hassager, *Dynamics of polymeric liquids*, John Wiley, NY, 1987.
4. J.J. Magda, R.G. Larson, J.Non-Newt. Fluid Mech. **30**, 1 (1988).
5. S.J. Muller, R.G. Larson, E.S.G. Shaqfeh, Rheol. Acta **28**, 499 (1989).
6. R.G. Larson, E.S.G. Shaqfeh, S.J. Muller, J. Fluid Mech **218**, 573 (1990).
7. A. Groisman,V. Steinberg, Phys.Rev.Lett**77**, 1480; **78**, 1460 (1997);Phys.Fluids **10**, 2451 (1998); Europhys.Lett **43**, 165 (1998).
8. J. L. Lumley, Symp. Math. 9, 315 (1972).
9. E. J. Hinch, Phys. of Fluids, **20** S22 (1977).
10. G. K. Batchelor, J. Fluid. Mech. **5**, 113 (1959); R. H. Kraichnan, Phys. Fluids **11**, 945 (1968); B.I. Shraiman, E.D. Siggia, Phys.Rev.E**49**, 2912 (1994); M. Chertkov, G. Falkovich, I. Kolokolov, V. Lebedev, Phys. Rev. E **51**, 5609 (1995); M. Chertkov, A. Gamba, I. Kolokolov, Phys. Lett. A. **192**, 435 (1994);E. Balkovsky, M. Chertkov, I. Kolokolov and V. Lebedev, JETP Lett. **61**, 1049 (1995); B.I. Shraiman and E.D. Siggia, CRAS **321**, 279 (1995); D. Bernard, K. Gawedzki, and A. Kupiainen, J. Stat. Phys. **90**, 519 (1998); M. Chertkov, G. Falkovich, I. Kolokolov, Phys.Rev.Lett. **80**, 2121 (1998); E. Balkovsky, A. Fouxon, Phys. Rev. E**60**, 4164(1999).
11. E. Balkovsky, A. Fouxon,V. Lebedev, chao-dyn/9911014.
12. M. Chertkov, chao-dyn/9911011.
13. Y. Rabin, S.Q. Wang, K.F. Freed, Macromolecules **22**, 2420 (1989).
14. H. Wyld, Ann. of Phys **14**, 143 (1961); P.C. Martin, E.D. Siggia, H.A. Rose, Phys.Rev.A **8**, 423 (1973).

TURBULENCE MODULATION DUE TO SOLID PARTICLES MOVING WITH VORTEX SHEDDING

T. KAJISHIMA, S. TAKIGUCHI AND Y. MIYAKE
Department of Mechanical Engineering, Osaka University
2-1 Yamadaoka, Suita, Osaka, 565-0871 Japan

Abstract. Turbulence modulation in particle-laden fluid flow, especially the influence of vortex shedding, was investigated by means of the direct numerical simulation. To this end, we developed a finite-difference scheme to resolve the flow around each particle moving in turbulence. Then, we simulated an upward flow in a vertical channel including solid particles. The velocity and vorticity fluctuations as well as Reynolds shear stress were strongly affected by wakes from particles. The shed vortices were elongated in the mainstream direction by the velocity gradient and resulted in the hairpin vortices.

1. Introduction

To understand the mechanism of turbulence modulation due to dispersed particles, bubbles or droplets, direct numerical simulations (DNS) of two-phase turbulence have been conducted. Recently, complete DNS without any point-source models have been realized. Pan and Banerjee(1997) used a body-force model at the grid points inside particle in couple with the spectral method for fluid turbulence. Particle Reynolds numbers in their DNS were 133 for stationary particles and 15 for moving particles. The present authors proposed another approach by the finite-difference method (Kajishima et al., 1998, 1999; Takiguchi et al., 1999). We directly integrated the fluid stress on the particle surface and successfully simulated the vortex shedding in the flow past a sphere up to $Re = 600$. Though the relatively low resolution for each particle, these DNS efficiently reproduced the flow including a number of particles.

In the present paper, our numerical method and the result of accuracy test will be briefly summarized in Sec.2. The modulation in fully developed

T. Kambe et al. (eds.), IUTAM Symposium on Geometry and Statistics of Turbulence, 319–324.
© 2001 *Kluwer Academic Publishers.*

turbulence in a vertical channel including large particles will be discussed in Sec.3.

2. Numerical Method

The incompressible flow of Newtonian fluid including rigid particles is considered. The fluid turbulence is solved by the finite-difference method using the fixed Cartesian coordinate system.

The basic equations are

$$\nabla \cdot \boldsymbol{u} = 0\,, \quad \frac{\partial \boldsymbol{u}}{\partial t} + \boldsymbol{u} \cdot \nabla \boldsymbol{u} = -\frac{\nabla p}{\rho} + \nu \nabla^2 \boldsymbol{u} + \boldsymbol{g} + \boldsymbol{f}_p\,, \tag{1}$$

where $\boldsymbol{g}$ is body force such as gravity. $\boldsymbol{u}$ is a representative velocity at the grid defined by $\boldsymbol{u} = (1-\alpha)\boldsymbol{u}_f + \alpha\boldsymbol{u}_p$, where α denotes the volumetric fraction of particle in the grid, $\boldsymbol{u}_f$ the fluid velocity and $\boldsymbol{u}_p$ the velocity inside the particle defined by $\boldsymbol{u}_p = \boldsymbol{v}_p + \boldsymbol{r} \times \boldsymbol{\omega}_p$. $\boldsymbol{v}_p$ is the velocity at the particle center, $\boldsymbol{\omega}_p$ the angular velocity and $\boldsymbol{r}$ the relative position from the momentum center of particle. The last term in the equation of motion is represented as

$$\boldsymbol{f}_p = \alpha\,(\boldsymbol{u}_p - \boldsymbol{u})\,/\Delta t\,. \tag{2}$$

where Δt is the time-increment. It is interpreted as a particle force on the fluid at the interface ($0 < \alpha < 1$) grid (Kajishima et al., 1999).

Equations for momentum and angular momentum of each particle are

$$\frac{d(m_p \boldsymbol{v}_p)}{dt} = -\int_{V_p} \boldsymbol{f}_p dV + \boldsymbol{G}_p\,, \quad \frac{d(\boldsymbol{I}_p \cdot \boldsymbol{\omega}_p)}{dt} = -\int_{V_p} \boldsymbol{r} \times \boldsymbol{f}_p dV + \boldsymbol{T}_p \tag{3}$$

where m_p denotes the mass, $\boldsymbol{I}_p$ the inertia tensor and V_p the volume including the particle. $\boldsymbol{G}_p$ and $\boldsymbol{T}_p$ represent the external force and the external moment, respectively. Particles are tracked in Lagrangian frame of reference without any models for drag force, lift force, memory effect, etc.

The spatial derivative is approximated by the central finite-difference method of the forth-order accuracy. The time-marching procedure is the fractional-step method based on the 2nd-order Adams-Bashforth method. Details are given in Kajishima, et al.(1999).

To examine the accuracy of our method, three-dimensional flow around a sphere fixed in a uniform stream was simulated (Takiguchi et al., 1999). The computational grid was cubic of the width Δ. Three steps of grid resolution, D_p/Δ =11, 8 and 5 were examined where D_p denotes the particle diameter. Finer cases D_p/Δ =11 and 8 reproduced wavy wake for $Re > 130$ as well as vortex shedding for $Re > 300$. Coarser case D_p/Δ =5, however, could not reproduce the unsteady flow in high Reynolds number region. Our numerical result agreed acceptably with experimental data up to the Reynolds number 600. Thus the desirable resolution of grid is $D_p/\Delta \geq 8$ for our analysis.

Prior to the simulation including solid particles, we obtained a fully developed single-phase flow in a channel. The Reynolds number based on the channel width H and mean friction velocity on the wall u_τ was fixed as $Re_\tau (= Hu_\tau/\nu) = 300$. The numbers of grid points were $N_x = 512$ in the streamwise (upward, x), $N_y = 128$ in the wall-normal (y) and $N_z = 256$ in the spanwise (z) directions. The computational grid was made almost cubic at the channel center and getting finer in y direction in the near-wall region. The grid spacings in wall-coordinate were $\Delta_x^+ = 4$, $\Delta_y^+ = 0.7 \sim 3.8$ and $\Delta_z^+ = 4$. The time increment was $\Delta_t^+ = 0.03$. The Reynolds number based on H and the centerline velocity resulted in $Re_c = 5400$. In the following figures, results by the spectral method at the corresponding Reynolds number (Kasagi et al., 1992) will be shown for comparison with ours. The agreement between them was good even for higher order statistics, but we will not mention one by one in this paper. We think our fourth order finite-difference method is reliable for DNS of the turbulent flow.

3. Results and Discussion

Then solid particles were supplied randomly in an instantaneous flow field. The particle density ρ_p was supposed to be larger than the fluid density ρ_f. Thus an additional pressure gradient was necessary to maintain the flow. The external force was $2 + (\rho_p/\rho_f - 1)\Phi/Fr^2$ (non-dimensionalized by $\rho_f u_\tau^2/H$), where the gravity force due to fluid is omitted. Φ is the mean volumetric fraction of particles, $Fr(= u_\tau/\sqrt{gH})$ the Froude number. We decided these parameters considering the aim of this research, which is to observe the influence of vortex shedding. The number of particles was $N_p = 36$ and the diameter was $D_p^+ = 32$, giving $\Phi = 0.001$. The density ratio was determined as $\rho_p/\rho_f = 10$ to make the relative Reynolds number to be $Re_p \simeq 400$. Fr was adjusted so that the wall friction velocity was almost equal to that in single-phase flow. The rotation of particles was ignored in this simulation for simplicity.

After the statistically steady state was reached, we obtained the time-averaged flow field as shown in Fig.1. Particles moved into the near-wall region. In the particle-laden flow, the fluid velocity is decreased in the high concentration region near the wall but the logarithmic law profile is recovered in the rare region. Every component of velocity fluctuation is increased. Spanwise ω_z and wall-normal ω_y components of vorticity fluctuation are intensified around the high concentration region, which is corresponding to

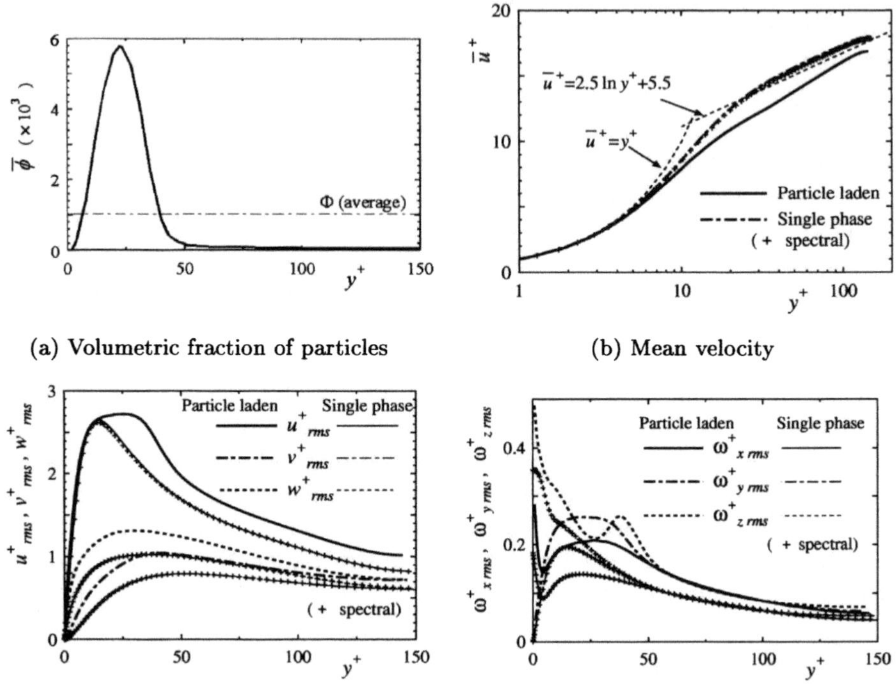

(a) Volumetric fraction of particles

(b) Mean velocity

(c) Intensity of velocity fluctutation

(d) Intensity of vorticity fluctutation

Figure 1. Time-averaged flow field of particle-laden turbulence in a vertical channel

Figure 2. Time evolution of shedded vortices from a sphere visualized by $\nabla^2 p$ isosurfaces

vortex rings shed from particles. In addition, we must pay attention to increase in the streamwise vorticity ω_x since it may be related to the production of Reynolds shear stress.

To observe the vortex structure in the wake of particles, time evolution of flow near a typical particle is visualized by the iso-surfaces of $\nabla^2 p$ in Fig.2. Vortex rings shed intermittently from a sphere is stretched due to the high velocity gradient in near wall region. They grow into hairpin like vortices and move downstream as well as away from the wall. Legs of these

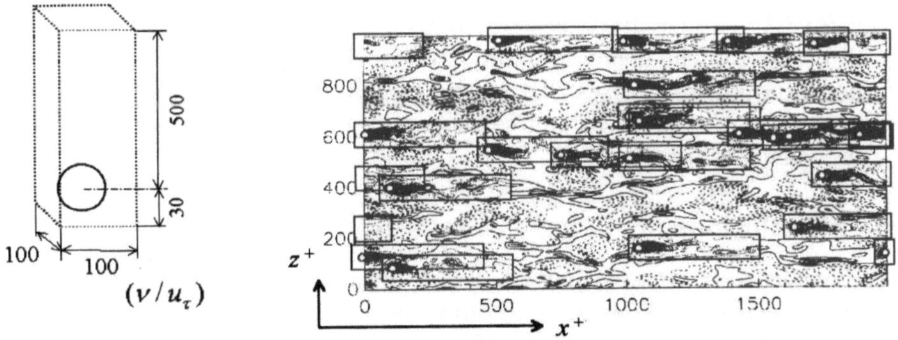

(a) Definition (b) x-z cross section at $y^+ = 22.5$

Figure 3. Wake regions: flow field is visulalized by contours of $-u'|u'|$

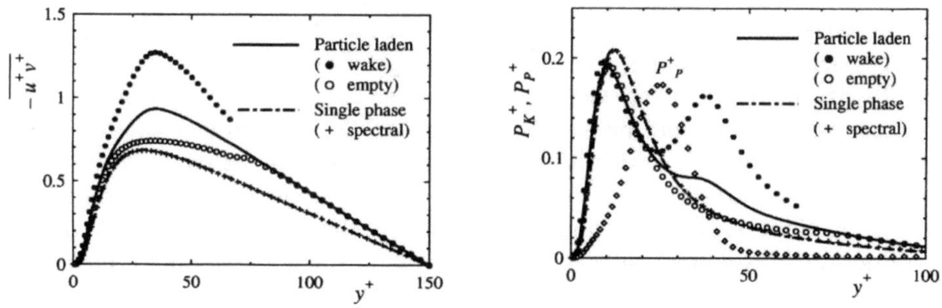

(a) Reynolds shear stress (b) Energu production rate

Figure 4. Profiles of shear stress and energy production rate

vortices are supposed to be the origin of increase in ω_x intensity.

Hereafter, the flow field is decomposed into "wake region" and "empty region." The size of wake region is assumed to be about 500 wall units in downstream and 100 wall units in lateral directions. Fig.3 indicates that this decomposition is appropriate considering the size of particle wakes.

Note again that most particles accumulate in $y^+ < 50$ as shown in Fig.1(a). Fig.4(a) indicates that the turbulence shear stress is produced especially in the particle wake in this region. The hairpin like vortex structure shown in Fig.2 might cause it and also increase in shear stress gradient in the particle-rare region to be balance with additional driving force.

Figure 4(b) shows the profile of energy production rate. In this flow field, the particle-induced force in addition to the shear produces the turbulence energy. The former, $P_P = \overline{u_i' f_{pi}'}$, is due to the momentum exchange between phases and it could be expressed by point force models. The latter is $P_K =$

$\overline{u_i'u_j'}\partial\overline{u}_i/\partial x_j$. In the particle-laden flow, P_K decreases in the vicinity of the wall because of the decrease in $\partial\overline{u}_i/\partial x_j$. In the region $y^+ = 30 \sim 60$, on the other hand, P_K increases due to the contribution of vortices in the wake region. This is the extra source of energy production in addition to the direct effect of particles, P_P.

4. Conclusions

We carried out DNS of particle-laden upward flow in a two-dimensional vertical channel taking into account the unsteady vortex shedding from particles. The vortex shedding affected the profiles of velocity fluctuation, vorticity fluctuation and Reynolds shear stress.

Since the particle density is larger than fluid density in our simulation, additional pressure gradient was needed to maintain the flow in comparison with the single-phase flow. The particle distribution was not uniform because particles were transported upward almost along the wall. Therefore, the particle gravity and additional pressure gradient was not in equilibrium locally. In the particle-rare region, the increase in shear stress gradient to be balance with additional pressure gradient should be supplemented by the turbulence modulation. In the near-wall region, on the other hand, vortex rings shed by particles grew into the hairpin like structure by the high velocity gradient. Our result suggested that this structure contributed to the production of Reynolds shears stress as well as turbulence kinetic energy.

References

Kajishima,T., Takiguchi,S. and Miyake,Y. (1998) Direct Numerical Simulation of Interaction between Turbulence and Particles, *Proc. ASME Fluids Engineering Division Summer Meeting, Washington,DC*, No.FEDSM98-5021 (CD-ROM)

Kajishima,T., Takiguchi,S. and Miyake,Y. (1999) Modulation and Subgrid Scale Modeling of Gas-Particle Turbulent Flow, to appear in *Proc. 2nd AFOSR Int. Conf. on Direct Numerical Simulation and Large Eddy Simulation, Rutgers*

Kasagi,N., Tomita,Y. and Kuroda,A., (1992) Direct numerical simulation of passive scalar field in a turbulent channel flow, *Trans. ASME, J. Heat Transf.*, **114**, pp.598-606

Pan,Y. and Banerjee,S. (1997) Numerical Investigation of the Effects of Large Particles on Wall-Turbulence, *Phys Fluids.*, **9**-2, pp.3786-3807

Takiguchi,S., Kajishima,T. and Miyake,Y. (1999) Numerical Scheme to Resolve the Interaction between Solid Particles and Fluid Turbulence, *JSME Int. J., Ser.B*, **42**-3, pp.411-418

SPECTRAL FEATURES OF AXIAL VELOCITY AND CONCENTRATION IN A HIGH SCHMIDT NUMBER TURBULENT JET

Y. SAKAI, Y. OKADA and I. NAKAMURA
Department of Mechano-Informatics & Systems,
Nagoya University
Furo-cho, Chikusa-ku, Nagoya 464-8603, Japan

N. KOBAYASHI
HONDA R&D CO., LTD.
Chuo 1-4-1, Wako city 351-0113, Japan

1. Introduction

The important information related to the structure of turbulent diffusion field will be obtained from the simultaneous measurements of velocity and a passive scalar in turbulent flows. Although there are some reports concerned with scalar flux or energy budgets (e.g. [1, 2, 3, 5, 6, 8]), there are still few reports including discussions of velocity-scalar cross spectrum or the structure of velocity-scalar correlations.

Here, we present the experimental results of the simultaneous one-point measurements of the axial velocity and concentration in a high Schmidt number axisymmetric turbulent jet. We investigate the features of each distribution of the velocity and concentration spectrum. Then we discuss the interaction of the velocity and concentration in the frequency space by examining the cross-spectrum obtained by the simultaneous measurements.

2. Experimental apparatus

The diffusing fluid is a water solution of the commercial dye (C.I. Direct Blue) and it is issued into the quiescent water from the nozzle of the exit diameter $D = 1.02$ mm. The issuing Reynolds number, $U_J D/\nu$, is 6300 and the Schmidt number is about 3800. The initial concentration C_J and velocity U_J are 3.0 kg/m^3 and 6.3 m/s, respectively. The turbulent Reynlods number, $\lambda u'/\nu$, is about 100 which is approximately constant on the jet centerline. The simultaneous measurements were performed by combining a single hot film probe (its sensing element has a

T. Kambe et al. (eds.), IUTAM Symposium on Geometry and Statistics of Turbulence, 325–330.
© 2001 *Kluwer Academic Publishers.*

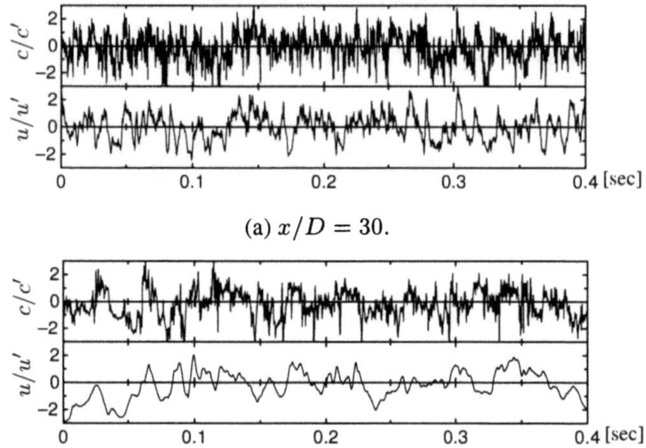

(a) $x/D = 30$.

(b) $x/D = 90$.

Figure 1. Fluctuation signals of concentration c and axial velocity u on the jet centerline. They are normalized by their standard deviations c', u'.

length of $l_{HF} = 1.0$mm, and a diameter of $d_{HF} = 51\mu$m) for the axial velocity and a fiber optic probe (length $l_{FO} = 0.475$mm, diameter $d_{FO} = 10\mu$m) for the concentration of the dye. The most crucial scale of sampling volume for the simultanesou measurement of velocity and concentration by the combined two probes seems to be l_{HF}, which is about ten times of the Kolmogorov microscale in the present experiments. See Sakai et al.[7] on the details of the experimental apparatus.

3. Results and discussions

3.1. FLUCTUATION SIGNALS

The normalized fluctuation signals of concentration and axial velocity are shown in figure 1. We can see that the concentration signals have higher frequency components than the velocity signals and that the concentration and velocity signal have similar large scale fluctuations.

3.2. AXIAL VELOCITY AND CONCENTRATION SPECTRA

The normalized frequency spectra of the axial velocity and the concentration on the jet centerline are shown in figure 2. The frequency is normalized by the integral time scale of the axial velocity, t_{uI}, which is calculated by integrating the autocorrelation coefficient curve from the time lag $\tau = 0$ to the zero-cross point. The spectra of velocity and concentration are drawn up to the frequency corresponding to l_{HF} and l_{FO}, respectively. We find that the axial velocity spectra show the $-5/3$ scaling range, but the concentration spectra have the slope less steep than $-5/3$ since the turbulent Reynolds number is small ($Re_\lambda \simeq 100$)[9, 10]. It is notable to find the $-5/3$ scaling law in the lower frequency range than $t_{uI} = 1.0$, where the greater part of kinetic turbulence energy is contained.

Figure 3 shows the spectral slopes in log-log coordinates, i.e. the logarithmic derivative of the spectral curves in figure 2. Miller and Dimotakis[4] suggested that the concentration spectra in the high Schmidt number turbulent jet can be approximated by the log-normal functions in the higher frequency range than the power-law range. If the spectra exhibit the log-normal behavior, their logarithmic derivatives must show the straight lines with negative slopes. The spectral slopes of the concentration in figure 3 show the straight lines in the higher frequency range than the power-law range. And also the spectral slopes of the axial velocity seem to show the similar tendencies.

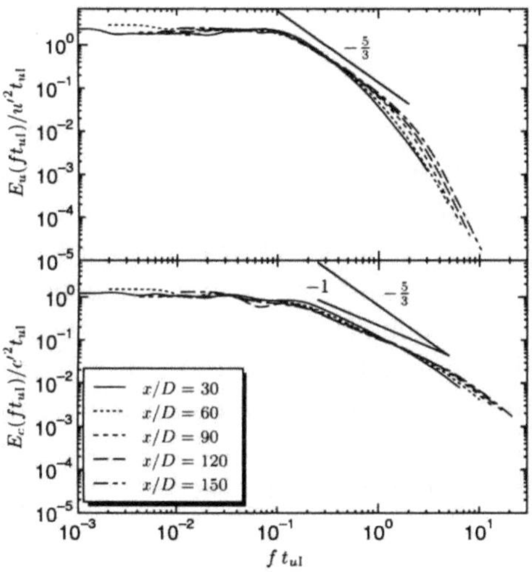

Figure 2. Axial velocity (upper) and concentration (lower) spectra on the jet centerline.

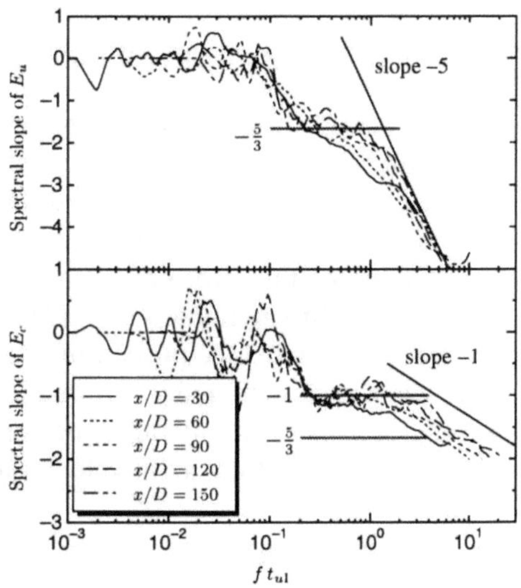

Figure 3. Spectral slopes in log-log coordinates, i.e. logarithmic derivatives, on the jet centerline.

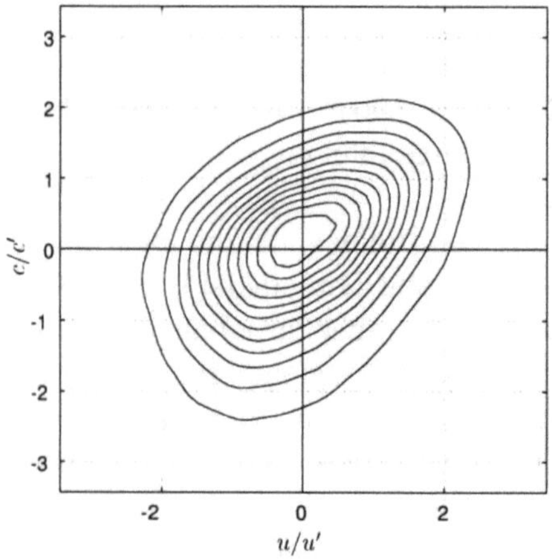

Figure 4. Equiprobability contours of the joint p.d.f. of axial velocity and concentration on the jet centerline ($x/D = 90$).

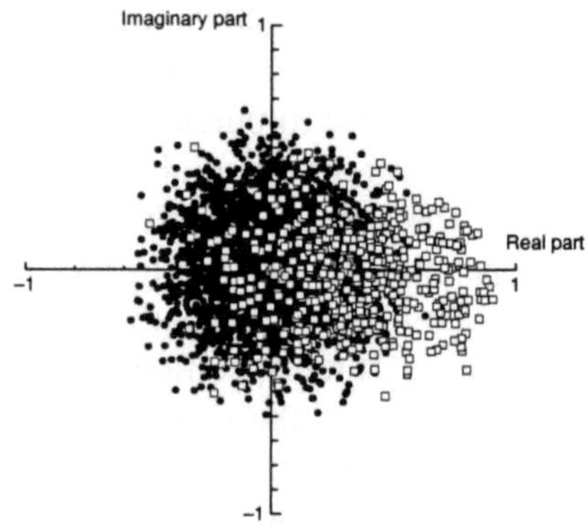

Figure 5. Normalized cross-spectrum, $S_{cu}(f)/\sqrt{E_c(f)E_u(f)}$, plotted in a complex plane ($x/D = 90$).

3.3. JOINT PROBABILITY DENSITY FUNCTION

Figure 4 shows the equiprobability contours of the joint probability density function (joint pdf) of the axial velocity and concentration on the jet centerline. We can find that the contours are slanted to the upper right and so that the axial velocity and concentration have a positive correlation. The contours shown in figure 4 and the contours of the joint pdf of temperature and axial-velocity in a turbulent jet obtained by Venkataramani et al.[11] are very much similar. However, it should be here noted that since probability density functions have no information on the time, we cannot obtain any informations on the structure of the correlation, such as which scales of the fluctuations contribute to the correlations. So we examine the cross-spectra of the axial velocity and concentration in the next section.

3.4. CROSS SPECTRA

The cross-spectrum is the complex number which contains the information of the correlation and the phase angle of two fluctuating quantities for each frequency component. In figure 5, the values of the normalized cross-spectrum for each frequency component are plotted in a complex plane. The squares represent the values of cross-spectrum for lower frequency than the frequency corresponding to the integral time scale of the velocity field, and the black circles represent those for higher frequency. It is notable that the large scale (i.e. the low frequency) behavior is obviously different from the small scale (i.e. the high frequency) behavior.

To examine the difference between their behavior, "coherence" (defined as the magnitude of the normalized cross-spectrum) and "phase" (defined as the phase angle of the cross-spectrum) are calculated. The coherence and phase mean the

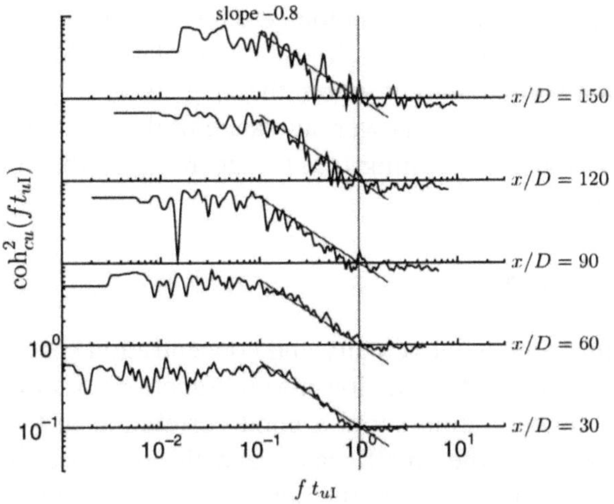

Figure 6. Coherence of axial velocity and concentration on the jet centerline.

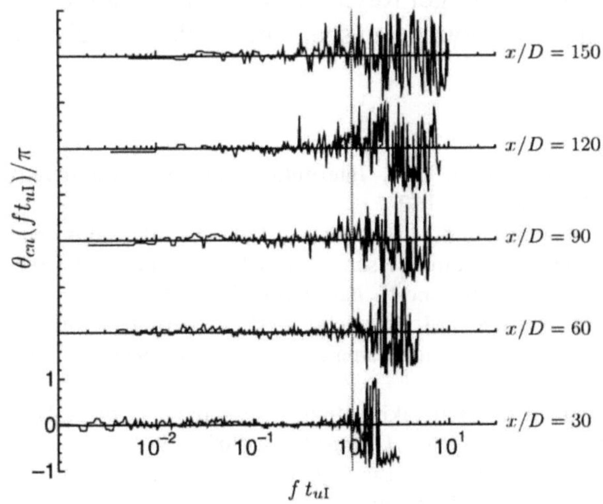

Figure 7. Phase of axial velocity and concentration on the jet centerline.

degree of the correlation and the phase angle between two Fourier components, respectively. Figure 6 and figure 7 show the coherence coh^2_{cu} and phase θ_{cu} of the axial velocity and concentration on the jet centerline. The largest frequency of coh^2_{uc} and θ_{cu} drawn in the figures is one corresponding to l_{HF}. In the range of $f t_{uI} \lesssim 10^{-1}$, the coherence shows the large value and approximately constant, and the phase remains zero at any downstream locations. Thus the large scale fluctuations of the velocity and concentration have large correlations with no time delay. On the other hand, in the range of $10^0 \lesssim f t_{uI}$, the coherence becomes smaller and the phase fluctuates randomly. Thus the small scale fluctuations of the velocity and concentration have almost no correlation. Moreover, the decrease of their correlation seems to occur within the range of $10^{-1} \lesssim ft_{uI} \lesssim 10^{0}$, which is almost the same range as the power law range of the axial velocity spectra (See figure 3). And it is also interesting that the decrease of the coherence seems to obey a power law.

4. Concluding remarks

We have found that both axial velocity and concentration spectra have the ranges which can be approximated by the log-normal function. Also we have found that the velocity and concentration show a good correlation in the large scales and almost no correlation in the small scales, and the decrease of their correlation seems to occur in the same range as the power law range of the velocity spectrum. However, our experiments have some limitations of small Reynolds number and not so small sampling volume (about ten times of the Kolmogorov microscale). So the experiments with larger Reynolds number and smaller sampling volume in various flows with or without mean shear are needed to make sure our conclusions.

References

1. Chevray, R. and N. K. Tutu: 1978, 'Intermittency and preferential transport of heat in a round jet'. *J. Fluid Mech.* **88**, 133–160.
2. Lemoine, F., M. Wolff, and M. Lebouche: 1996, 'Simultaneous concentration and velocity measurements using combined laser-induced fluorescence and laser Doppler velocimetry: Application to turbulent transport'. *Exp. Fluids* **20**, 319–327.
3. Lemoine, F., M. Wolff, and M. Lebouche: 1997, 'Experimental investigation of mass transfer in a grid-generated turbulent flow using combined optical methods'. *Int. J. Heat Mass Transfer* **40**, 3255–3266.
4. Miller, P. L. and P. E. Dimotakis: 1996, 'Measurements of scalar power spectra in high Schmidt number turbulent jets'. *J. Fluid Mech.* **308**, 129–146.
5. Panchapakesan, N. R. and J. L. Lumley: 1993, 'Turbulence measurements in axisymmetric jets of air and helium. Part2. Helium jet'. *J. Fluid Mech.* **246**, 225–247.
6. Papanicolaou, P. N. and E. J. List: 1988, 'Investigations of round vertical turbulent buoyant jets'. *J. Fluid Mech.* **195**, 341–391.
7. Sakai, Y., Y. Okada, N. Kobayashi, I. Nakamura, and M. Miwa: 1999, 'Measurements of velocity-scalar joint statistics in a high Schmidt number axisymmetric turbulent jet.'. In: *Proc. of the 3rd ASME/JSME Joint Fluids Eng. Conf.* San Francisco, U.S.A.
8. Simoens, S. and M. Ayrault: 1994, 'Concentration flux measurements of a scalar quantity in turbulent flows'. *Exp. Fluids* **16**, 273–281.
9. Sreenivasan, K. R.: 1991, 'On local isotropy of passive scalars in turbulent shear flows'. *Proc. R. Soc. Lond. Ser. A* **434**, 165–182.
10. Sreenivasan, K. R.: 1996, 'The passive scalar spectrum and the Obukhov-Corrsin constant'. *Phys. Fluids* **8**, 189–196.
11. Venkataramani, K. S., N. K. Tutu, and R. Chevray: 1975, 'Probability distributions in a round heated jet'. *Phys. Fluids* **18**, 1413–1420.

J

Pipe-Flow and Channel-Flow Turbulence

STRUCTURES AND INTERMITTENCY IN ROTATING PIPES

P. ORLANDI
Dipartimento di Meccanica e Aeronautica
Universita' di Roma "La Sapienza" via Eudossiana n° 18, 00184 Roma, Italy

R.A. ANTONIA & Q. ZHAO
Department of Mechanical Engineering
The University of Newcastle NSW 230

1. Introduction

The study of turbulence through a pipe which rotates about its axis is of fundamental interest, partly because of the practical applications but also because the rotation introduces a variety of new flow structures. Since the intermittency phenomenon is likely to be related to these structures, the effect of rotation on velocity structure functions of different orders may allow some insight into the relationship between moments and structures. Orlandi & Fatica [1] , using a direct numerical simulation at low Reynolds number and at relatively high rotation rate, showed that the structures everywhere across the pipe differ significantly from those in the stationary pipe. Clearly, the major interest is on the changes to the near-wall structures because of their interrelationship to the wall friction, turbulent energy production and energy dissipation rate. In a subsequent study [2], the simulations were extended to higher rotation rates with the aim of providing budgets for each of the Reynolds stresses which would in turn be used for improving/streamlining RANS turbulence models of rotating flows. The budgets were found to change at all locations across the pipe although the major changes occurred near the wall.

As is well known, the complexity of turbulence is linked to the interaction between structures of different scales, and significant effort has recently been devoted to studying the role of very intense vortical structures in isotropic or locally isotropic flows. To a significant degree, the intermittent character of turbulence reflects the formation of these intense structures, and their role in forced or decaying isotropic turbulence has

T. Kambe et al. (eds.), IUTAM Symposium on Geometry and Statistics of Turbulence, 333–338.
© 2001 *Kluwer Academic Publishers.*

been investigated in detail, e.g. by Jimenez *et al.*[3]. The deviation of high-order exponents $\zeta(n)$ of longitudinal structure functions, from "$n/3$", the asymptotic high Reynolds number limit of Kolmogorov (1941), reflects, in part, the presence of vortical structures. The evaluation of these exponents can be somewhat ambiguous [4], and the larger n is, the greater the number of samples required to achieve convergence; this requirement is more easily met by experiments than simulations. Taylor's hypothesis, applied to a suitably long time series of the signal at one point, permits evaluation of structure functions at relatively large values of the separation ρ. In numerical simulations the number of samples is related to the amount of storage available in the computer; the limited storage capacity in our case allows only low order moment structure functions to be correctly evaluated. Furthermore, direct numerical simulations are limited to low Re, where there is no inertial range.

In order to partially validate the present simulations, a comparison has been made with experiments in stationary pipe flow and with the numerical simulation, also for stationary pipe flow [5]. The latter simulation was carried out with a different numerical method and better resolution in both streamwise and radial directions near the central region of the pipe. The validation of the results in the rotating case was not possible because our database is the only one currently available; however, since the flow structures in the rotating pipe have larger length scales than in the stationary case, adequate resolution is expected with rotation. This claim is supported by the smoothly decaying nature of the spectra in the axial and tangential directions at high wave numbers.

2. Results

The evaluation of the structure functions of any order n can be obtained by evaluating the probability density function $p(x_{ij})$ where $x_{ij} = \delta u_i(\rho_j)/\langle\delta u_i(\rho_j)^2\rangle^{1/2}$. The cases $i = j$ and $i \neq j$ correspond to longitudinal and transverse structure functions respectively. The structure function of order n is

$$F_{ij}^n = \langle\delta u_i(\rho_j)^n\rangle = \langle\delta u_i(\rho_j)^2\rangle^{n/2} \int x_{ij}^n p(x_{ij}) d(x_{ij}) \ . \tag{1}$$

F_{ij}^2 can also be inferred from the one-dimensional energy spectrum, or more directly, as reported by Ref. [6], from the two-point correlation. This second method has been used to evaluate the second-order structure functions from Loulou's data [5]. The measurements were made in fully developed pipe at a distance of about $100D$ from the entrance to the pipe. The Reynolds number $Re = U_b D/\nu$ was 5500, somewhat larger than for the simulation ($Re = 4900$). A X-wire probe (2.5 μm dia. hot wires) was operated with a constant temperature anemometer (overheat ratio of 1.5)

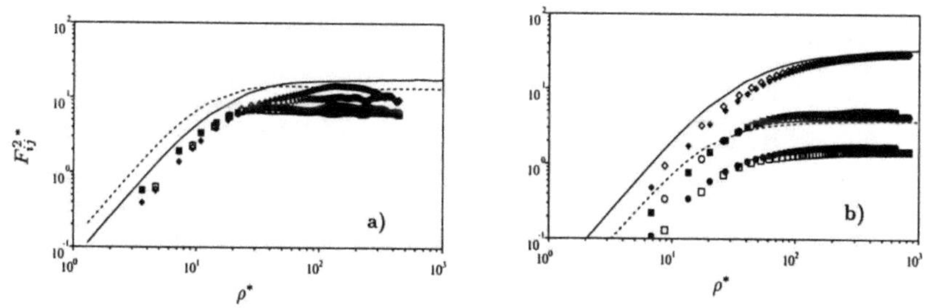

Figure 1. Second order longitudinal and transverse structure functions in the streamwise directions in Kolmogorov units; a) center of the pipe, b) at $y^+ = 20$. Lines experiments, open symbols present, closed symbols Loulou (1996). (——— , ◇ longitudinal; ---- □ $v'_\theta(\Delta z)$; ○ $v'_r(\Delta z)$.

to measure velocity fluctuations in the axial (z) and radial (r) directions. Taylor's hypothesis was used to convert temporal increments into spatial increments. Initially, the measured values of F_{ij}^2 were Kolmogorov-normalized using the isotropic estimate of $\langle\epsilon\rangle$, viz. $\langle\epsilon\rangle_{iso} = 15\nu\langle(\partial u/\partial z)^2\rangle$. This is only likely to be reasonable near the axis, the ratio $\langle\epsilon\rangle/\langle\epsilon\rangle_{iso}$ increasing as the wall is approached. The present DNS value of this ratio ($\simeq 2.8$ at $y^+ = 20$) was therefore used to rescale the measured structure functions shown in Figure 1. The present simulation was performed in a periodic pipe of length $L_z = 15$ with $192\times128\times96$ intervals respectively in the z, θ and r directions. Several values of $N = \Omega U_b/D$, where Ω is the angular velocity of the pipe, were used.

The comparison shows that, in the central region, there is poor agreement between the present simulations and that of Loulou, reflecting the unsatisfactory resolution of the present simulation. On the other hand, there is satisfactory agreement at $y^+ = 20$ (Figure 1b), between the simulations, so that the present resolution should allow changes to the near-wall structure functions caused by the rotation to be estimated satisfactorily.

As is well known, numerical simulations allow visualizations of velocity and vorticity fields, generally too difficult a task experimentally. In Figure 2 visualizations of the fluctuating azimuthal and axial velocity components, scaled by the local rms values, in the $z-\theta$ plane at $y^+ = 20$, show that the flow structures change with rotation. The first observation is that the rotation makes the low and high speed streaks wider and longer. In addition, whereas for $N = 0$, the distribution of v'_θ is concentrated in scales much smaller than for v'_z, the rotation inclines and lengthens the contours. The inclination seems to be independent of the rotation rate but the amplitude increases with N. This single flow realization can of course only qualitatively depict the effect of N on the flow structures. More quantitative information may however be derived from two-point velocity correlations in the streamwise and azimuthal directions. Although not shown here, the

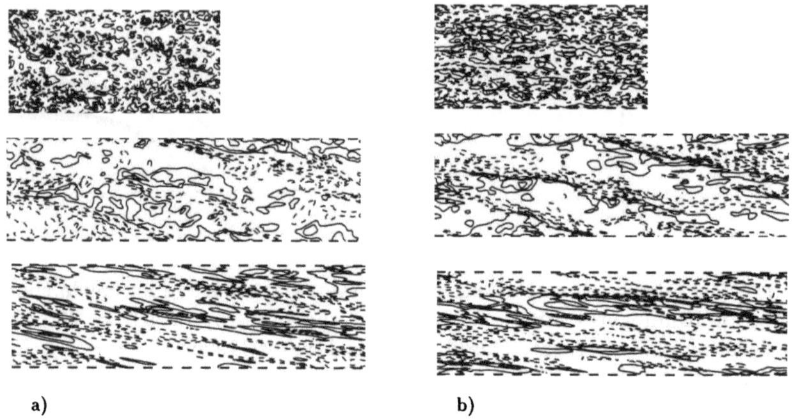

Figure 2. Contour plots of fluctuating velocity components scaled by the local rms value at a distance $y^+ = 20$ from the wall; a) axial, b) azimuthal components. From top to bottom $N = 0$, $N = 2$ and $N = 10$. Contours intervals $\Delta v = 0.75$.

radial velocity correlation, drops rapidly to zero with and without rotation; in contrast, the azimuthal velocity correlation reaches a minimum and does not return to zero. This behavior suggests that the structures are helically shaped over the full length of the pipe.

The previous physical information is also, in principle, contained in the structure functions, which as shown in Eq. (1), can be evaluated from the pdf of increments of the velocity fluctuations displayed in Figure 2. To evaluate the pdf, approximately 70 fields have been used and, since 192×128 grid points are used in the homogeneous directions, the total number of samples is approximately 1.5×10^6. We expect that the vortical structures are responsible for the departure of the pdfs from Gaussianity. In the context of the wall region of a channel flow, Orlandi *et al.*[7] claimed that the more intermittent character of the longitudinal velocity differences is due to the elongated structures which make quiescent periods more dominant than regions in which rapid changes occur. This condition is typical of intermittent phenomena. For $N = 0$, the pipe should exhibit a similar behavior to channel flow; indeed, the pdf of x_{13} at very large separations, e.g. $85\Delta z$, approaches a Gaussian distribution. This agrees with the qualitative picture implied by the flow visualizations (Figure 2) where concentrations for v_z' are more elongated than those for v_θ'. Figure 3 shows that, by reducing the separation, the pdf deviates from the Gaussian distribution. In fact, at $15\Delta z$, even if the intensity of the negative events dominates over that of the positive ones, the distribution is more Gaussian (Figures 3b) than for a separation equal to Δz (Figure 3a). At the smallest separation (Figure 3a), the pdf is almost exponential, characteristic of very

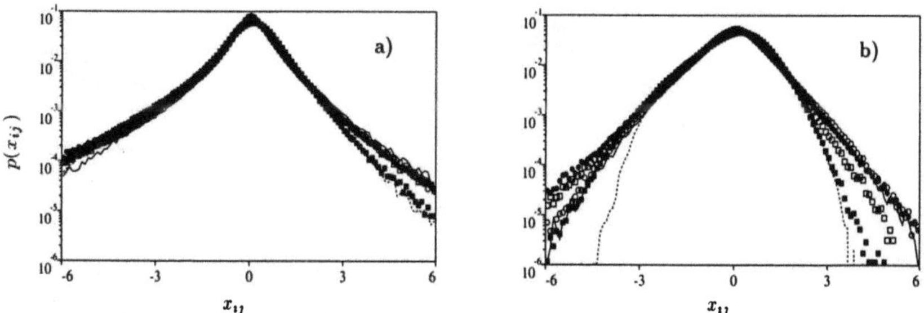

Figure 3. Pdf of $x_{iJ}(\rho)$ at $y^+ = 20$: (a) $\rho = \Delta z$, (b) $\rho = 15\Delta z$; x_3 (longitudinal —— $N = 0$, ∘ $N = 2$, • $N = 10$), x_1 (transverse ---- $N = 0$, □ $N = 2$, ▪ $N = 10$)

intermittent phenomena. In addition, Figures 3a show that, at the smallest separation, the longitudinal and transverse distributions are very similar to each other. In contrast, at larger separations (Figure 3b), the changes are more pronounced, consistent with the visualizations for v'_θ and v'_z. In summary, the rotation increases the intermittent character of the flow as a result of the formation of wider and longer vortical structures.

Second-order longitudinal and transverse structure functions have been calculated at several distances from the wall and are presented at three locations: close to the wall ($y^+ = 5$), where the maximum production occurs ($y^+ = 20$) and on the pipe axis. At any r, the velocity increment for small separations is proportional to ρ. This implies a disentanglement from the energy containing eddies and a tendency towards local isotropy. The natural consequence is that the two-point correlations near $\rho = 0$ have a parabolic shape. Near the centreline (Figure 4a), as the separation increases, the present small Reynolds number simulation shows that, for $N = 0$, there is near-saturation of all velocity components implying that the large scales are not far from isotropy. This aspect was investigated in previous low R_λ studies [7, 8] where statistics of the velocity and vorticity fields, at or near the centre of the channel, were compared with isotropy. Figure 4a, on the other hand, shows that when the pipe rotates, the saturation disappears, implying that the structures become very elongated. The helical shape of these structures follows from the observation that the longitudinal and transverse structure functions are almost coincident.

At $y^+ = 20$, the anisotropy of the flow in the stationary pipe is highlighted by the significantly different asymptotic limits indicated by the three curves in Figure 4b. The saturation at smaller separation of the transverse structure functions implies that the scales of the axial velocity are more elongated. The rotation leads to a tendency of small scales towards isotropy and an elongation of the length scales associated with axial and azimuthal fluctuations.

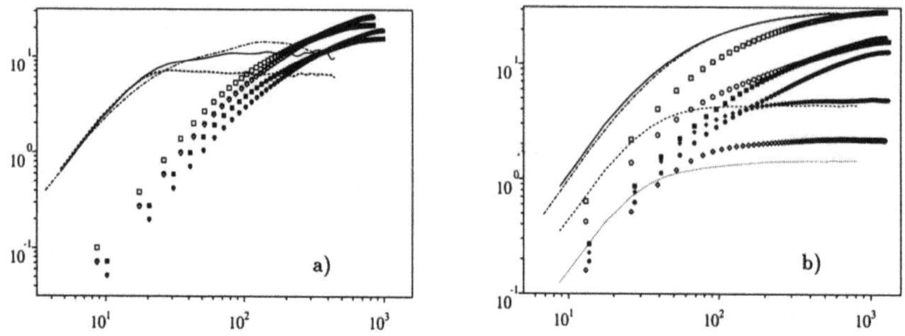

Figure 4. Longitudinal and transverse structure functions at different distances from the wall; a) centreline b) $y^+ = 20$. Lines $N = 0$, open symbols $N = 2$ and closed symbols $N = 10$. (x_{33} ———— , □) ; (x_{13} ---- , ○) ; (x_{23} ······· , ◊).

3. Conclusions

In this paper, a numerical database for turbulent pipe flow with relatively high rotation rates has been used to examine how near-wall vortical structures influence different flow quantities. The qualitative picture implied by the flow visualization has been quantitatively confirmed by the second-order structure functions. In previous studies, e.g. [7], the deviation of the exponents for the high-order structure functions from "$n/3$" was used to quantify the intermittent character of the intermediate scales in locally isotropic flows. In this low Reynolds number study, the intermittency of the large scales has been artificially increased by imposing an external rotation. The structures become longer and wider and these changes are reflected in changes in the longitudinal and transverse structure functions.

References

1. Orlandi P. & Fatica, M. "Direct simulations of a turbulent pipe rotating along the axis" *J.Fluid Mech* ,**343**, 43–72 (1997) .
2. Orlandi, P. and Ebstein, "Turbulent budgets in rotating pipes by DNS", *Proc Turbulence and Shear Flow Phenomena-1*, 468 (1999).
3. Jimenez, J., Wray, A. A., Saffman, P. G. and Rogallo, R. S. 1993. "The structure of intense vorticity in homogeneous isotropic turbulence," (1993)*J. Fluid Mech.*, **255**, 65-90.
4. Antonia, R. A. and Pearson, B. R. 1997. "Velocity structure functions in a turbulent plane jet,"*Proc. Eleventh Turbulent Shear Flow Conference*, Vol. 3, 3-117 to 3-121.
5. Loulou, P. "Direct numerical simulation of incompressible pipe flow using a B-spline spectral method".(1996) Thesis Department of Aeronautics and Astronautics, SUDAAR 683, Stanford University.
6. Monin, A. S. and Yaglom, A. M. 1972. *Statistical Fluid Mechanics*, MIT Press.
7. Orlandi, P. ,Antonia, R. A. and Esposito, P. G. "Structure functions in homogeneous and non-homogeneous turbulent flows" Fundamental Problematic Issues in Turbulence Gyr, Kinzelbach, Tsinober Editors Birkhauser Verlag (1999).
8. Kim, J. and Antonia, R. A. 1993. "Isotropy of the small-scales of turbulence at small Reynolds number," *J. Fluid Mech.*, **251**, 219-238.

STATISTICS OF OPEN-CHANNEL TURBULENCE BY USING SIMULTANEOUS MEASUREMENTS OF ALL THREE COMPONENTS OF TURBULENT FLUCTUATIONS WITH TWO-SETS OF LDAs

Iehisa NEZU and Kouki ONITSUKA
Dept. of Civil & Global Environment Eng., Kyoto University
Kyoto 606-8501, Japan

1. Introduction

In the case of 2-D uniform open-channel flows, the bursting phenomenon is analyzed by making use of two-components of the turbulent fluctuations in the streamwise and vertical directions. However, the flow structure is three dimensional near the side-wall in rectangular open-channel flow. In such a situation, the bursting phenomenon may be affected by both the side-wall and bed-wall.

In this study, three-dimensional(3-D) measurements of an open-channel flow were conducted by making use of two-sets of fiber-optic laser Doppler anemometers. The contribution rates to the Reynolds stress fluctuations considering the spanwise turbulent component were analyzed. Further more, the distribution rates of the contributions were indicated in the case that a hole value was set to be zero. It was found that the ejection occurs when the spanwise fluctuation is positive and also that the sweep occurs when the spanwise fluctuation is negative near the side-wall.

2. Theoretical Considerations

Nakagawa & Nezu(1977) proposed the "u-v quadrant threshold method", which is one of conditional sampling functions, in the case of two-dimensional(2-D) flows. The 2-D conditional function is defined as follows:

$$\langle uv \rangle_{i,H} = \lim_{T\to\infty} \frac{1}{T}\int_0^T u(t)v(t)\cdot I_{i,H}(u,v)dt \tag{1}$$

T. Kambe et al. (eds.), IUTAM Symposium on Geometry and Statistics of Turbulence, 339–344.
© 2001 *Kluwer Academic Publishers.*

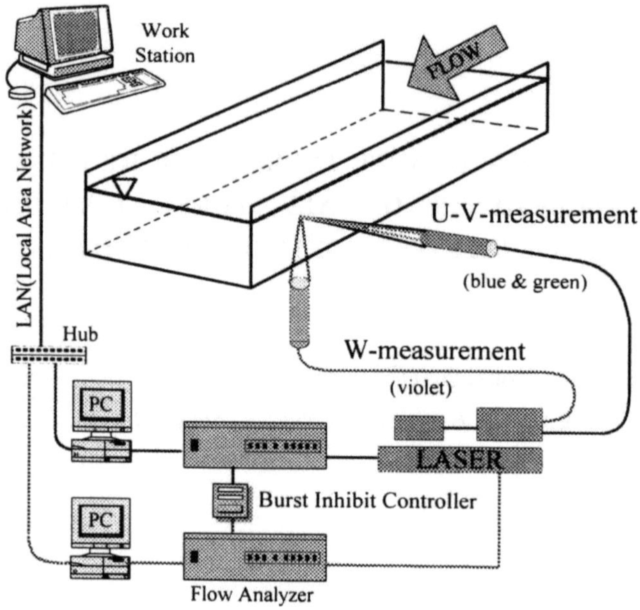

Fig.1 Experimental Setup

$$I_{i,H}(u,v)=\begin{cases}1, \text{if } (u,v) \text{is in the quadrant } i \text{ and if } |uv| \geq H|\overline{uv}| \\ 0, \text{otherwise}\end{cases} \tag{2}$$

$$RS_i(H)=\frac{\langle uv\rangle_{i,H}}{\overline{uv}} \tag{3}$$

in which, u, v are the streamwise and vertical components, respectively. $-\overline{uv}$ is the Reynolds stress, H is a hole value and $RS_i(H)$ is a contribution rate of each event. In contrast, the structure is three dimensional near the side-wall. Therefore, a new conditional sampling function considering the spanwise turbulent component is proposed here as follows:

$$I_{i,p,H}(u,v,w)=\begin{cases}1, \text{if } (u,v,w) \text{is in the quadrant } i,\ w>0 \text{ and if } |uv| \geq H|\overline{uv}| \\ 0, \text{otherwise}\end{cases} \tag{4.a}$$

$$I_{i,m,H}(u,v,w)=\begin{cases}1, \text{if } (u,v,w) \text{is in the quadrant } i,\ w<0 \text{ and if } |uv| \geq H|\overline{uv}| \\ 0, \text{otherwise}\end{cases} \tag{4.b}$$

By using Eqs.(1), (3) and (4), eight types of the contribution rates are defined to be as RS_{1p}, RS_{2p}, RS_{3p}, RS_{4p}, RS_{1m}, RS_{2m}, RS_{3m} and RS_{4m}. The suffices of 1, 2, 3 and 4 mean the quadrant and the suffices of m and p mean the sign of minus and plus of spanwise fluctuation component, $w(t)$, respectively.

3. Experimental Set-up and Experimental Conditions

The open-channel flume was 10m long, 20cm wide, and 50cm deep tilting innovative one, which is an automatically controlled discharge system as shown in Fig.1. Three-dimensional measurements of an open-channel flow were conducted by making use of two-sets of fiber-optic laser Doppler anemometers (LDA, DANTEC-made). One LDA fiber probe was located at side of the channel, which has two-color system (blue and green) and this probe can measure the instantaneous streamwise velocity, $\tilde{u}(t) = U + u(t)$ and vertical velocity, $\tilde{v}(t) = V + v(t)$. The other probe was located under the channel bottom, which has one-color system(violet) and this probe can measure the instantaneous spanwise velocity, $\tilde{w}(t) = W + w(t)$. The clocks of these two LDAs are synchronized by a burst inhibit controller. Experimental condition is as follows: the water depth h is 10.0 (cm), the channel width B is 20 (cm), the Froude number is 0.3 and the Reynolds number is 60,000.

4. Experimental Results

4.1 Contribution Rates Considering the Spanwise Turbulent Component

The original contribution rate to the Reynolds stress $-\overline{uv}$ can be calculated by Eqs.(1), (2) and (3) which were suggested by Nakagawa & Nezu(1977). They indicated that the contribution rates of the ejection event, $RS_2(H)$ and sweep event $RS_4(H)$ are dominant ones as

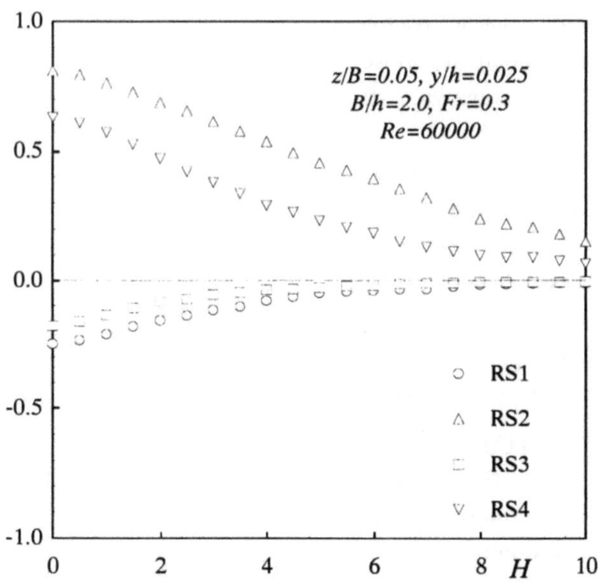

Fig.2 (a) Contribution Rate of Each Event

compared with those of the outward interaction, $RS_1(H)$ and inward interaction, $RS_3(H)$, e.g. see Nezu & Nakagawa(1993). Fig.2(a) shows the original contribution rates at the near bed and side-wall. The magnitudes of the $RS_2(H)$ and $RS_4(H)$ are larger than $RS_1(H)$ and $RS_3(H)$. This shows the ejection and sweep events have a large contribution to the Reynolds stress $-\overline{uv}$. These results correspond to those of Nakagawa & Nezu(1977).

Fig.2(b) shows the contribution rates to the Reynolds stress $-\overline{uv}$ calculated by Eqs.(1), (3) and (4) in the condition of $w(t)>0$ in Eq.(4). Fig.2(c) shows the contribution rates in the condition of $w(t)<0$. The magnitude of the contribution rate of

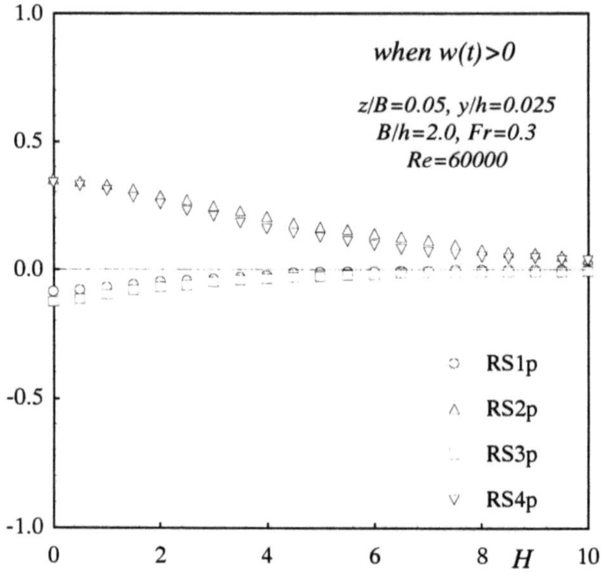

Fig.2 (b) Contribution Rate in the Condition $w(t)>0$

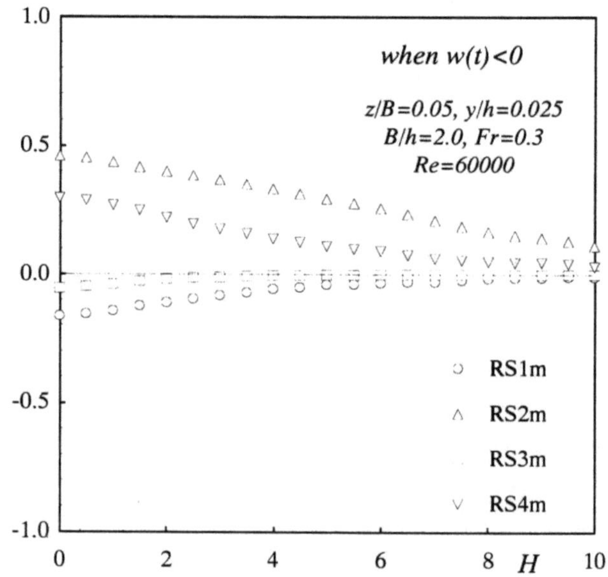

Fig.2 (c) Contribution Rate in the Condition $w(t)<0$

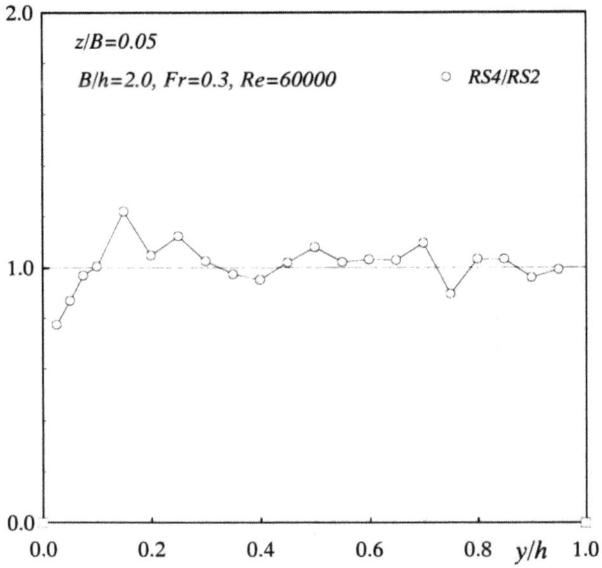

Fig3 (a) Ratio of Contribution Rates of sweep and ejection

the ejection is almost same as that of sweep in the condition of $w(t)$ >0. In contrast, the contribution rate of the ejection is larger than that of sweep in the condition of $w(t)$ <0. This suggests that the bursting phenomenon near the side-wall is affected not only by the bed-wall but also by the side-wall. This means that the bursting events are not symmetric.

4.2 Distributions of the Ratio of Contribution Rate Considering the Spanwise Turbulent Component

Nakagawa & Nezu(1977) have measured the 2-D uniform open-channel flows by making use of hot-film anemometers and indicated that the ratio between the contribution rates of sweep and ejection, i.e., RS_4 / RS_2, decreases against the vertical direction in the near-wall region when the hole value is zero. Fig.3(a) shows the vertical distribution of the ratio RS_4 / RS_2 at z / B =0.05. The ratio changes against the vertical direction in a complicated manner. This tendency does not correspond to that of Nakagawa & Nezu(1977), because the present flow is affected not only by the bed-wall but also by the side-wall.

Fig.3(b) shows the vertical distribution of the ratio between the fractional contribution to the Reynolds stress $-\overline{uv}$ in the condition of $w(t)$ >0 and that in the condition of $w(t)$ <0 near the side-wall when the hole value H is zero. The ratio of RS_{2p} / RS_{2m} is greater than the unit value (=1.0) except near the bed. In contrast, the ratio of RS_{4p} / RS_{4m} is smaller than the unit value except near the bed. This fact shows that the ejection occurs when the spanwise fluctuation is positive ($w(t)$ >0)

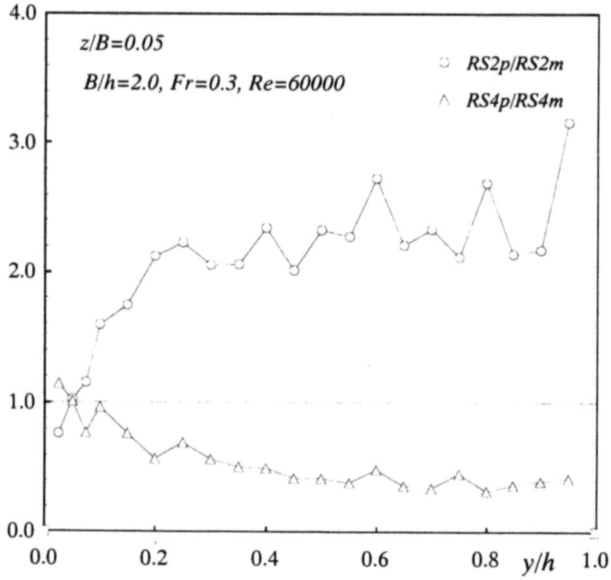

Fig3 (b) Ratio between the fractional contribution in the condition of $w(t)>0$ and that in the condition of $w(t)<0$

and also that the sweep occurs when the spanwise fluctuation is negative ($w(t)$ <0) near the side-wall. In general, the fluid including the negative spanwise fluctuation ($w(t)$ <0) has a high momentum ($u(t)$ >0). This is because the streamwise velocity increases toward the central region of the channel. Of course, the sweep occurs when the streamwise fluctuation has a negative value ($u(t)$ <0). Therefore, the ejection occurs when the spanwise fluctuation is positive ($w(t)$ >0) and also that the sweep occurs when the spanwise fluctuation is negative ($w(t)$ <0) near the side-wall.

5. Conclusion

All three components of velocities in an open-channel flow were measured firstly by the use of two sets of laser Doppler anemometers(LDAs). The contribution rates to the Reynolds stress $-\overline{uv}$ considering the spanwise turbulent fluctuation $w(t)$ were analyzed. It was found that the bursting phenomenon near the side-wall is affected not only by the bed-wall but also by the side-wall. Near the side-wall, the ejection occurs when the spanwise fluctuation is positive and also that the sweep occurs when the spanwise fluctuation is negative. .

References

Nakagawa, H. and Nezu, I. (1977) Prediction of the contributions to the Reynolds stress from the bursting events in open-channel flows, *J. Fluid Mech.* 80, 99-128.

GENERATION MECHANISM OF TURBULENCE-DRIVEN SECONDARY CURRENTS IN OPEN-CHANNEL FLOWS

Kouki ONITSUKA and Iehisa NEZU
Dept. of Civil & Global Environment Eng., Kyoto University
Kyoto 606-8501, Japan

1. Introduction

In general, there are two kinds of secondary currents. One is called the 'secondary currents of Prandtl's first kind'. The other is called the 'secondary currents of Prandtl's second kind'. The vorticity equation is obtained from the equations of motion in the vertical(y) and spanwise(z) directions, as follows:

$$V\frac{\partial \Omega}{\partial y} + W\frac{\partial \Omega}{\partial z} = \frac{\partial^2}{\partial y \partial z}\left(\overline{v^2} - \overline{w^2}\right) + \left(\frac{\partial^2}{\partial z^2} - \frac{\partial^2}{\partial y^2}\right)\overline{vw} + \nu\nabla^2\Omega \qquad (1)$$

$$\Omega = \frac{\partial W}{\partial y} - \frac{\partial V}{\partial z} \qquad (2)$$

in which, V and W are the time averaged vertical and spanwise velocities, respectively. v and w are the turbulence fluctuations of the vertical and spanwise velocities, respectively. The first term and second term on the left-hand side are the advection terms. The first term and second term on the right-hand side are the generation term and Reynolds stress term, respectively. Nezu & Nakagawa(1984) have implied experimentally and Demuren & Rodi(1984) implied numerically that the generation term and the Reynolds-stress term are the predominant ones. However, they could not measure the Reynolds stress term of $\overline{vw}$. Therefore, the Reynolds-stress term was evaluated as (advection term)-(generation term)-(viscous term).

In this study, the cross-sectional flow fields in open-channel flow were measured by both an innovative Particle Image Velocimetry(PIV) with dual pulse Nd : YAG laser system and laser Doppler anemometer(LDA).

T. Kambe et al. (eds.), IUTAM Symposium on Geometry and Statistics of Turbulence, 345–350.
© 2001 *Kluwer Academic Publishers.*

Table-1 Hydraulic Conditions

Case	h (cm)	B (cm)	B/h	Fr	Re ($\times 10^3$)	
AR2-P	6.0	12.0	2.0	0.10	9.3	by PIV
AR6-P	2.0	12.0	6.0	0.28	7.6	by PIV
AR2-L	10.0	20.0	2.0	0.30	59.0	by LDA

2. Experimental Set-up and Hydraulic Conditions

The experiments were conducted in a 10-m-long and 50-cm-deep tilting flume. The experiments were conducted by both the PIV and LDA.

In the case of PIV measurements, the thin light sheets, which were generated by the dual pulse Nd : YAG laser(13mJ/pulse), were scattered in the cross-sectional plane. A couple of images were taken by a high performance progressive scan interline CCD camera, which has 1008×1018 pixels. Both images were transferred to the FlowMap PIV2000 (DANTEC made) through a digital connector. In this system, the minimum time between these two frames is 1μsec. Two components of instantaneous velocities, i.e., the vertical velocity $\tilde{v}(t)$ and spanwise velocity $\tilde{w}(t)$, were calculated with both images by the FFT method. The sampling frequency was set to be 3(Hz).

In the case of LDA measurements, three components of instantaneous velocities, i.e., the streamwise velocity $\tilde{u}(t)$, the vertical velocity $\tilde{v}(t)$ and spanwise velocity $\tilde{w}(t)$, were measured with a six-beam LDA system. All of data of the LDA were recorded in a HDD of a personal computer with a sampling frequency with more than 200(Hz) and sampling time of 120s.

The hydraulic conditions are summarized in Table 1. The channel width B was adjusted by using a steel plate. In which, $Fr = U_m / \sqrt{gh}$ is the Froude number, $Re = 4RU_m / \nu$ is the Reynolds number, U_m is the bulk mean velocity, R is the hydraulic radius, g is the gravitational acceleration, h is the flow depth and ν is the kinematic viscosity.

3. Experimental Results

3.1 Instantaneous Flow Fields

Figure 1 shows the instantaneous flow fields in the cross section of case AR2-P which was measured with the present PIV. At t=0.0(s), the vectors point to each corner of channel from near the free surface. In this moment, large scale vortex can not be recognized. In contrast, at t=0.33(s), a large scale vortex can be seen clearly near the left side wall. At t=0.67(s), a new vortex is generated near the right side wall. These two vortices move toward the center of the channel from the both side walls. It can be seen that the scale of these vortices is controlled by both the flow depth and the channel width. It can be said that the instantaneous structure of the secondary currents is not symmetric in spite that the geometry of channel is symmetric at the center.

Figure2 shows the instanta- neous flow fields in the cross section of case AR6-P which was measured with the present PIV. The vertical left-side axis in this figure is at the center of the channel. Several small vortices can be seen at each time. The scale of these vortices is also controlled by the flow depth. The pattern of these instantaneous vortices does not coincide with that of time averaged secondary currents (see Fig.3). It is well known that the time averaged cellular secondary currents might exist in wide open-channels on the basis of the surprising fact that the concentration of suspended sediment varied cyclically in the spanwise direction. The picture of Fig.2 is similar to that of secondary currents in natural wide open-channel flows (see Nezu & Nakagawa(1993)). It was found that the pattern of the instantaneous vortices is quite deferent from that of time averaged secondary currents.

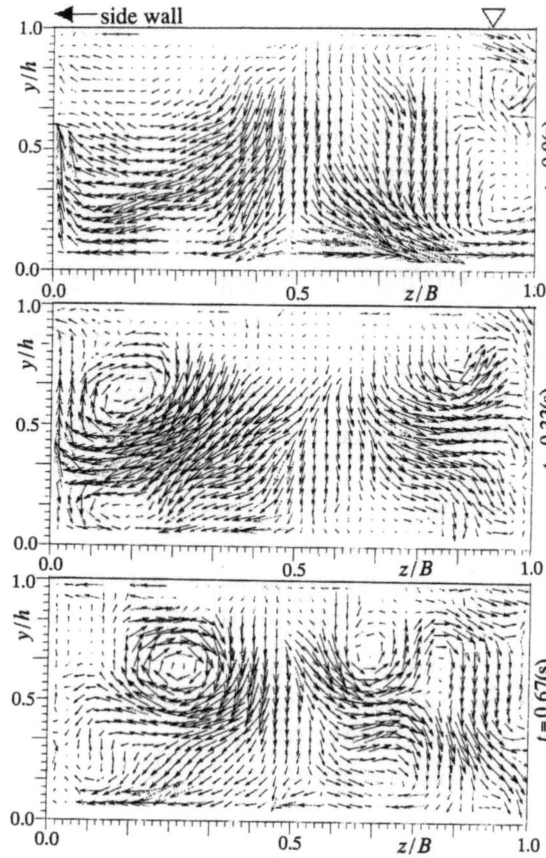

Fig.1 Instantaneous Flow Fields (case AR2-P)

3.2 Time Averaged Flow Fields

Figure 3 shows the secondary currents which were directly measured by the six-beams LDA. Two sets of vortices can be seen. The upper vortex is called the "surface vortex" and the lower one is called the "bed vortex". The scale of the surface vortex is lager than that of the bed vortex. This feature is original character- ristics in open-channel flow. This is caused by the boundary condition of the free surface.

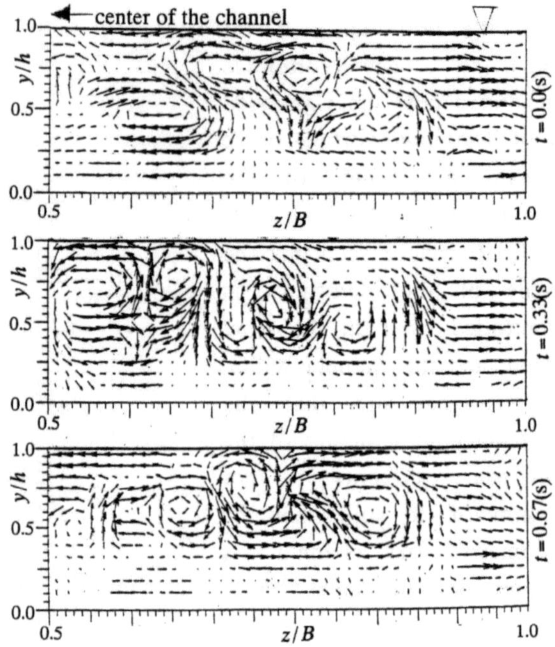

Fig.2 Instantaneous Flow Fields (case AR6-P)

Figure 4 shows the primary isovel lines of the streamwise velocity U normalized by the maximum velocity U_{max}. The isovel lines do not parallel to the bottom and side walls. The isovel lines of 0.95 and 0.90 bulge toward the side wall at the near-half depth. The pattern of these isovel lines in open-channel flow is quite different from those in closed-duct flows.

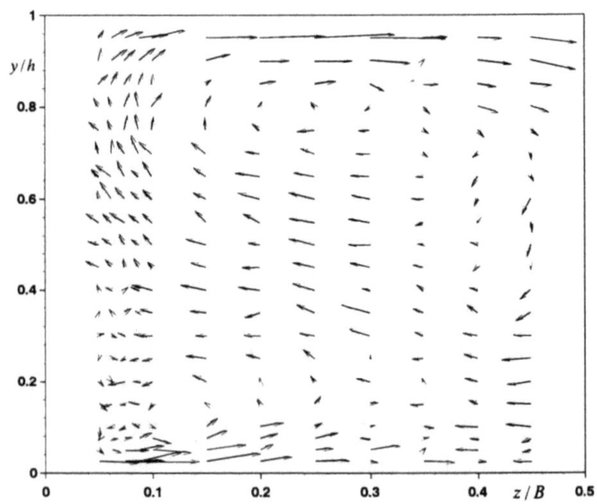

Fig.3 Secondary Currents (case AR2-L)

3.3 Evaluation of the Vorticity Equation

Figure5 shows the advection term in the vorticity equation normalized by the flow depth and friction velocity $\left(h/\overline{U}_*\right)^2$. The advection term takes a positive value near the free surface and also takes a negative value near the channel bed. These regions correspond to the surface vortex region and the bottom vortex region, respectively. The maximum absolute value near the free surface is larger than that near the channel bed, because the strength of the free surface vortex is greater than that of the bed vortex.

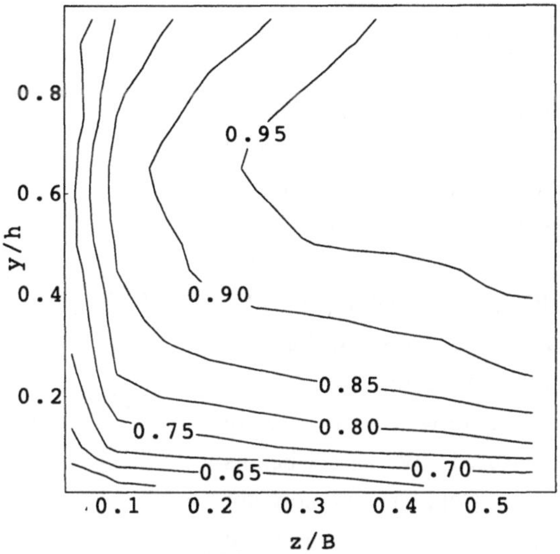

Fig.4 Isovel Lines (case AR2-L)

Figure6 shows the generation term normalized by $\left(h/\overline{U}_*\right)^2$. Figure7 shows the Reynolds stress term normalized by $\left(h/\overline{U}_*\right)^2$. The generation term takes a negative value at a region where is located upper than a boundary line that links (y/h , z/B)=(1, 0) and (y/h , z/B)=(0.5, 0.5), and it takes a positive value at the lower region than the boundary line except for a small area around z/B =0.2 very near the bed. In contrast, the Reynolds-stress term takes a positive value at a region upper than that boundary and takes a negative value at a lower region than that boundary. Furthermore, the order of the values of the generation term and the Reynolds stress term are almost same. In contrast, the value of the convection term is smaller than that of generation and Reynolds stress terms. Therefore, it can be said that the generation term and the Reynolds-stress term are the predominant ones and that the secondary currents in a straight channel are generated by the anisotropy of the Reynolds normal stresses.

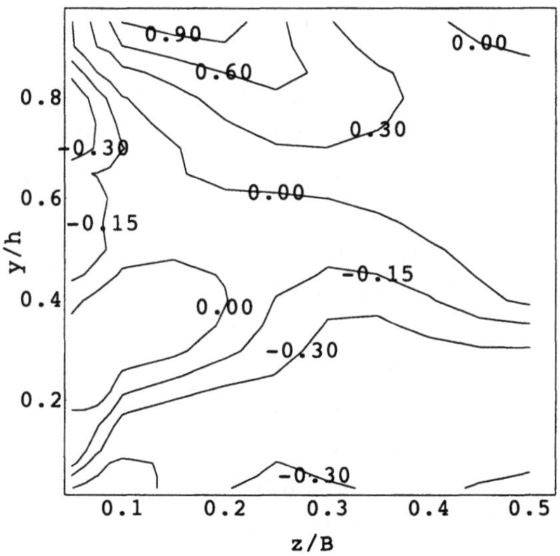

Fig.5 Advection Term (case AR2-L)

4. Conclusion

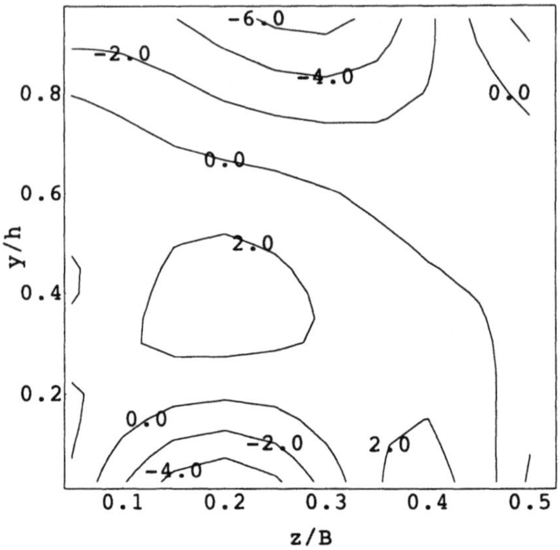

Fig.6 Generation Lines (case AR2-L)

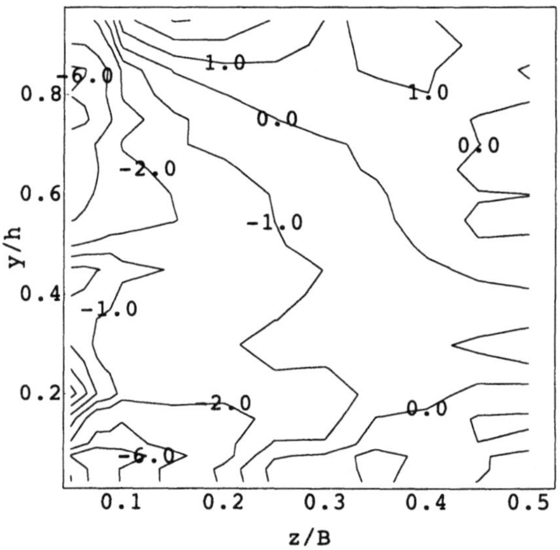

Fig.7 Reynolds Stress Term (case AR2-L)

Instantaneous cross-sectional flow fields in open-channel flow were measured by making use of innovative Particle Image Velocimetry (PIV) with dual pulse Nd : YAG laser system. Furthermore, all three components of the flow velocities were measured by a six-beams laser Doppler anemometer (LDA) with high accuracy. It was found that the instantaneous structure of the secondary currents is not symmetric in spite that the geometry of channel is symmetric in rectangular o-en-channel flow and also that the small vortices are generated instantaneously in wide smooth open-channel flow. The Reynolds-stress term in the vorticity equation is firstly measured directly and investigated quantitatively. It was verified that the Reynolds-stress term and the generation term have a tendency of balance with each other.

References

Nezu, I. and Nakagawa, H.(1984) *J. Hydraulic Eng.*, ASCE **110,** 173.

Demuen, A.O. and Rodi, W.(1984) *J. Fluid Mech.* **140,** 189.

Nezu, I. and Nakagawa, H.(1993) *Turbulence in Open-channel Flows*, IAHR Monograph, Balkema.

TURBULENCE MEASUREMENTS AND CONDITIONAL SAMPLING ANALYSIS ON COHERENT VORTICES IN UNSTEADY OPEN-CHANNEL FLOWS OVER DUNE BED

AKIHIRO KADOTA[1)], IEHISA NEZU[2)] AND KOICHI SUZUKI[1)]
1) *Dept. of Civil and Environmental Eng., Ehime University, Bunkyo-cho 3, Matsuyama 790-8577, Japan*
2) *Dept. of Global Environment Eng., Kyoto University Sakyo-ku, Kyoto 606-8501, Japan*

1. Introduction

Large-scale coherent vortices such as separated vortex and kolk-boil vortex are often observed and generated behind the crest of sand dunes in fluvial flooded rivers. The kolk vortex is a strong upward-tilting streamwise vortex and it seems like a tornado motion in the atmosphere. The kolk vortex develops up to the free surface and then becomes a boil. On the contrary, the separated vortex is generated from the crest of dunes then reattaches down to the dune bed surface. The upward motion of boil is mainly generated by the reattachment to the dunes. The behavior of these coherent structures is quite different between the rising and falling stages of the flood period. In open-channel flows such as rivers and estuaries, various coherent vortices are often observed and greatly affect sediment transport. The coherent structures often occur especially in flooded fluvial rivers (Coleman 1969, Kinoshita 1984). It is very important to investigate unsteadiness effects on the unsteady open-channel flows so that the growing process of the dunes and the behavior of the suspended sediment can be explained dynamically. In this study, two velocities at different points were accurately measured by means of the simultaneous use of two sets of laser Doppler anemometers, to analyze conditional space-time correlation and coherent structures in unsteady open-channel flows over dune bed. Also, measurements of concentration by using dye-injection technique and fiber-optic concentration instruments were conducted. The differences of correlation coefficient and dye concentration between the rising and falling stages were discussed.

2. Experimental Setup and Conditions

Experiments were conducted in a 10m long, 40cm wide and 50cm deep tilting flume. Figure 1 illustrates the measuring system for space-time correlation analyses. Two sets of LDA (DANTEC), a water-wave gauge and an automatic traversing system were connected to two set of IBM personal computers which enabled us to control parameters such as bias, Bragg cell and the position of the movable probe. The Argon-Ion fiber-optic LDA with a 2W high-power water-cooling laser light and 4

T. Kambe et al. (eds.), IUTAM Symposium on Geometry and Statistics of Turbulence, 351–356.
© 2001 *Kluwer Academic Publishers.*

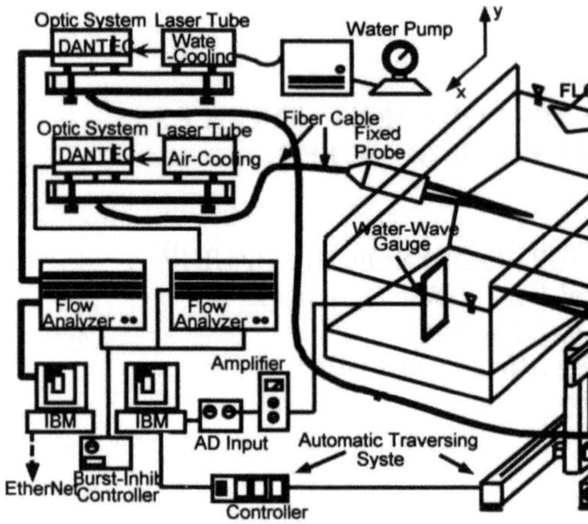

Figure 1. Measuring system for simultaneous measurements of two instantaneous velocities.

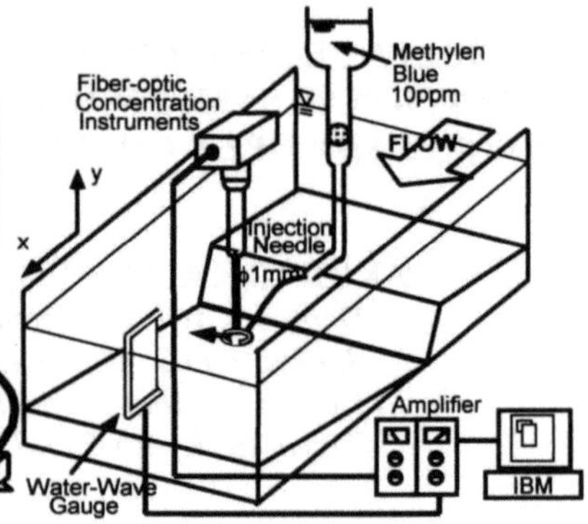

Figure 2. Measuring system for dye-concentration.

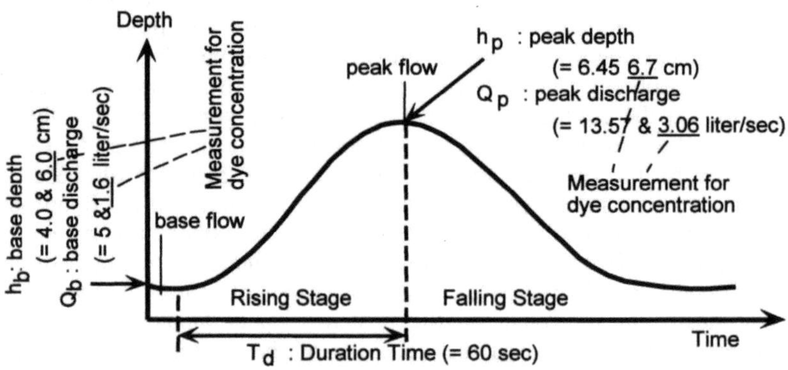

Figure 3. Experimental conditions.

beams in backscattering mode was adopted for the movable probe. The other LDA for the fixed probe has the same specification as the water-cooled LDA except for 100mW power and the air-cooled laser light. The automatic traversing system has an error of ± 0.1mm. The fixed-probe measuring points adopted here are both the separation point (crest of the dune) and the reattachment point. The latter is defined as the point at which the streamwise mean velocity becomes just zero on the channel bed. On the other hand, the fiber-optic concentration instrument with the above water-wave gauge was adopted to measure the dye concentration as shown in figure 2. The experimental condition and hydrograph are shown in figure 3. In this study, the time t is normalized by the flood period T_d, i.e., $T=t/T_d$. Therefore, the period from T=0 to 1 corresponds to the rising stage, whereas the period from T=1 to 2 corresponds to the falling stage. The most difficult and important aspect to investigate turbulent structures in unsteady flows is how to determine the mean velocity component because the long-time measurements are necessary for one-wave as seen in figure 3. Nezu and Nakagawa (1991) suggested that the Fourier component method was best comparing with the ensemble and moving averaging methods. In the present study, the Fourier component method was adapted.

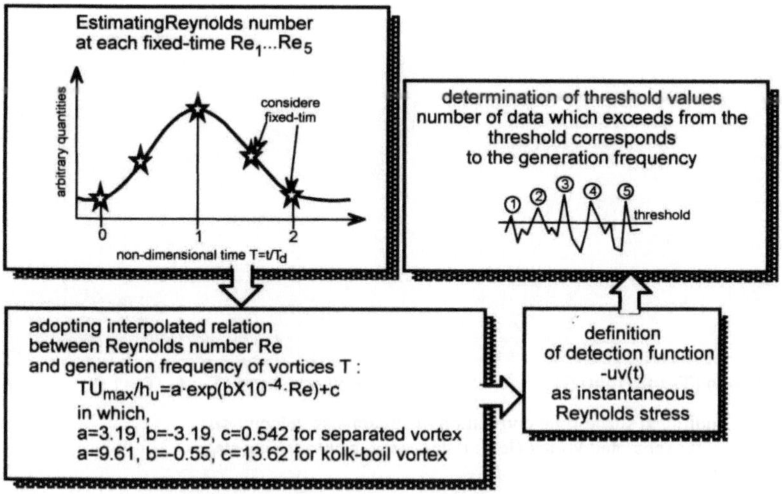

Figure 4 Definition procedure of threshold values.

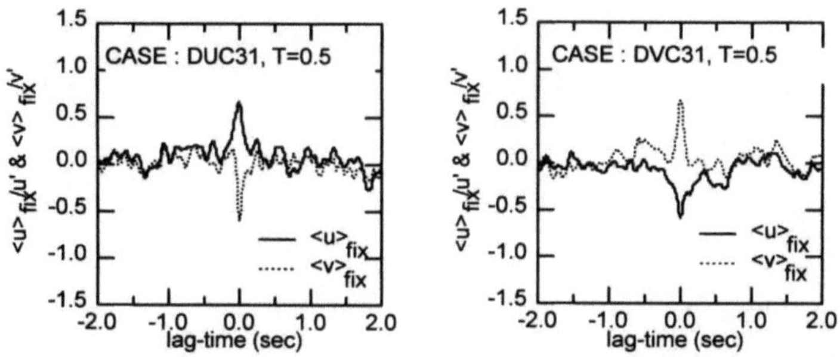

Figure 5 Conditional auto correlation (left : separated vortex, right : kolk-boil vortex).

3. Detection Function and Threshold Value in Conditional Sampling Analysis

The conditional space-time correlation coefficient is generally defined as :

$$\langle q(x_0, y_0, \Delta x; \Delta y; t_{fix} + t, \tau)\rangle = \frac{\int_T q(x_1, y_1, t_{fix} + t + \tau)\cdot I(x_0, y_0, t_{fix} + t)dt}{\int_T I(x_0, y_0, t_{fix} + t)dt} \tag{1}$$

where, q denotes an arbitrary sampling signal such as instantaneous velocity, the parenthesis < > indicates the averaged value after conditional sampling. The subscripts 0 and 1 are fixed and movable points, respectively. τ is lag-time and (Δx, Δy) are distances from the fixed point (0) to the movable one (1).

Furthermore, $I(x,y,z,\tau)$ is the detection function of coherent vortices. For the purpose of strict conditional sampling analysis, the detection function and threshold value should be reasonably defined. Figure 4 explains the procedure of the definition. First, the fixed-time points t_{fix} were defined as shown in the figure so that the coherent vortices at the arbitrary phase can be examined. Then, the Reynolds numbers Re at each t_{fix} were evaluated in order to determine the generation period T of the separated and

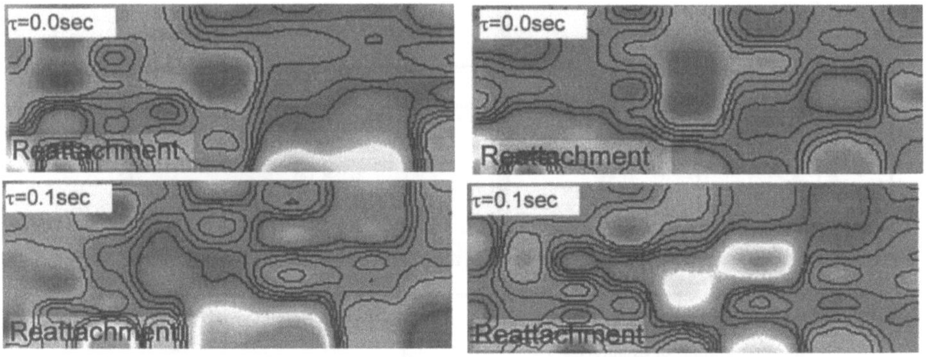

Rising Stage ($T_{fix}=t_{fix}/T_d=0.5$) **Falling Stage ($T_{fix}=t_{fix}/T_d=1.0$)**

Figure 6 Conditional space-time correlation of streamwise velocity <*u*> for separated vortex (left : rising stage, right : falling stage).

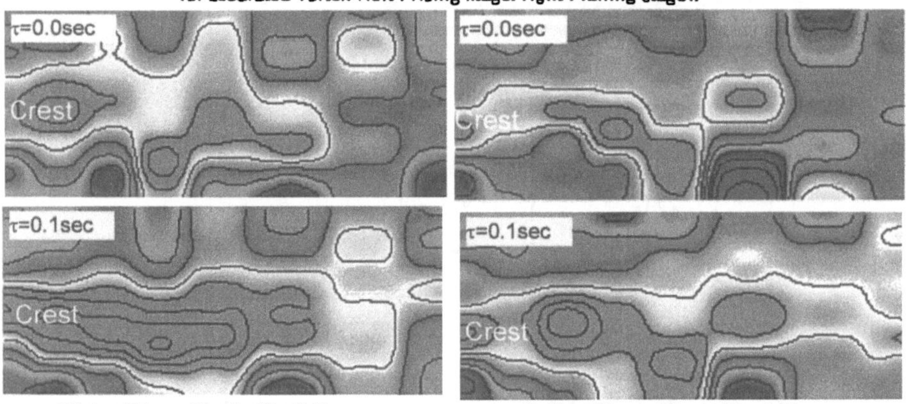

Rising Stage ($T_{fix}=t_{fix}/T_d=0.5$) **Falling Stage ($T_{fix}=t_{fix}/T_d=1.0$)**

Figure 7 Conditional space-time correlation of vertical velocity <*v*> for kolk-boil vortex (left : rising stage, right : falling stage).

kolk-boil vortices by using the approximated relations between *Re* and *T* (Nezu et al., 1996). These relations consist of exponential functions on the basis of the assumption that the value of *T* may tend to a constant value as *Re* becomes larger. In the present study, instantaneous Reynolds stress $-uv(t)$ was adopted as detection the function *I*(*t*). The detailed expressions of detection functions for separated vortex I_s and kolk-boil vortex I_k are expressed in the followings :

$$I_s(t) \equiv \begin{cases} 1 : u > 0, v < 0 \ \& \ |uv/u'v'| \geq H_s \\ 0 : otherwise \end{cases}, \quad I_k(t) \equiv \begin{cases} 1 : u < 0, v > 0 \ \& \ |uv/u'v'| \geq H_k \\ 0 : otherwise \end{cases} \tag{2}$$

in which, *H* is the threshold value. Subscripts *s* and *k* denote the separated and kolk-boil vortices, respectively. I_s means the condition where the high-speed separated vortex generated from the dune crest descends to the reattachment point, whereas the I_k indicates that the low-speed fluid ascends suddenly from the reattachment point. The threshold values H_s and H_k for these detection functions are determined so that the sampling number which exceeds the threshold value coincides with the generation frequency *T* of coherent vortices. The values of H_s and H_k are almost constant during the flood period and they are 0.1 - 0.15 and 1.5 - 1.6 for separated and kolk-boil vortices,

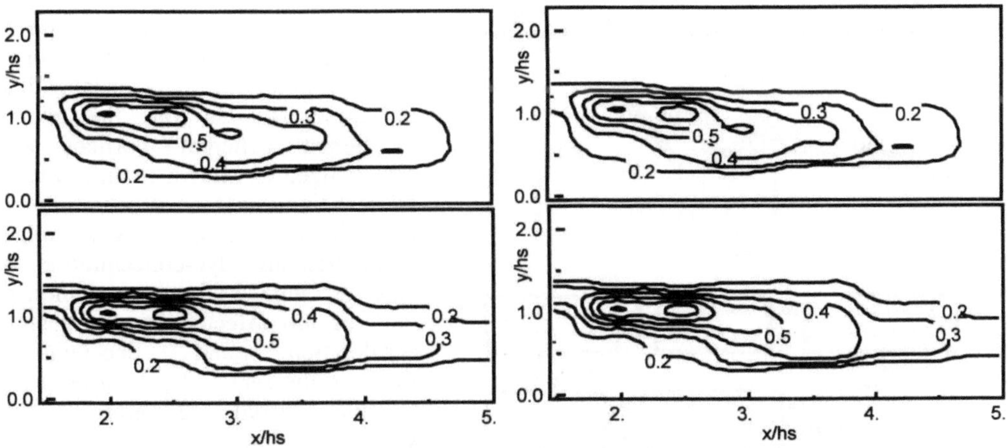

Figure 8 Convection of dye-concentration (left : separated vortex, right : kolk-boil vortex).

respectively. Figure 5 shows the conditional auto-correlations of $\langle u\rangle(t)$ and $\langle v\rangle(t)$ for separated and kolk-boil vortices ($\Delta x=\Delta y=0$). The separated vortex shows a positive value of $\langle u\rangle$ and negative $\langle v\rangle$ with lag-time of 0sec, whereas the kolk-boil vortex shows the opposite tendency. It is seen that the coherent vortices can be reasonably sampled in the lag-time of 0sec.

4. Results and Discussion

Figure 6 shows the distributions of conditionally sampled streamwise velocity $\langle u\rangle(t)$ for separated vortex. The sequence of figures indicates the evolution of separated vortex generated at lag-time t=0 sec from the location of fixed probe. The high positive parcel of vortex is transported to the streamwise direction. Especially in the falling stage, the high parcel drops down to the reattachment point (x/h_s=4, h_s : height of dune crest) and streamwise convection velocity becomes larger as compared with the rising stage. On the contrary, the high parcel in the rising stage is enlarged to vertical direction from dune bed to free surface and it is restrained in streamwise direction. The conditional space-time correlation of vertical velocity $\langle v\rangle(t)$ for the evolution of kolk-boil vortex is shown in figure 7. The high-correlation region of $\langle v\rangle$ indicates the strong upward flow from the reattachment point in the falling stage. The high correlation existing near the free surface in the rising stage is relatively larger than in the falling stage and the same tendency as field observations can be recognized. The kolk-boil vortex with weak energy is reasonably sampled in the outer region.

Figure 8 shows the convection process of dye concentration for separated and kolk-boil vortices in which the concentration are expressed in ppm. With respect to 0.4ppm line of separated vortex, the separation point that corresponds to the reattachment point, is shifted far as the Reynolds number increases. It is considered that the profiles dye corresponds to the locus of separated vortex. Also, the head of line is shorter in rising stage so that the reattachment point becomes shorter. On the other hand, the dye of kolk-boil vortex, diffusion is larger in the falling stage because effect of spatial acceleration is larger in streamwise direction. Also the head of 0.06ppm line is shorter. From the results of coherent structure generated over dune bed by simultaneous

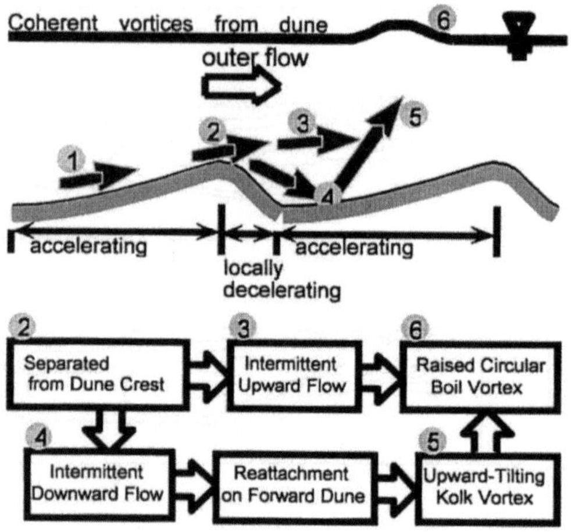

Figure 9 Generation mechanism of separated and kolk-boil vortices in vertical section.

measurements of laser Doppler anemometers, water-wave gauge, dye-concentration, visualization technique and so on, the above mentioned coherent vortices are schematized as shown in figure 9 to show their convection structure. From the above results of conditional correlation and dye-concentration, it was recognized that the coherent vortices are affected by the spatial acceleration and deceleration. Because the rising stage corresponds to the spatial deceleration, the vortices are not so transported to the streamwise direction comparing with the vertical whereas the streamwise convection of vortices is dominant for streamwise convection..

5. Conclusion

In the present study, turbulence and dye-concentration measurements by using two sets of LDA and fiber-optic concentration instruments were conducted in unsteady open-channel flows over dune bed. Convection of coherent vortices was discussed by conditional space-time correlation coefficient for the differences of coherent vortices between the rising and falling stages. As the results, coherent vortices are affected by the spatial acceleration and deceleration. The streamwise convection of the correlation and dye-concentration are dominant for the falling stage which corresponds to the spatial acceleration whereas the vertical convection is large in the rising stage.

6. References

Coleman, J. M.(1969) Brahmaputra River, *Channel Process and Sedimentation. Sediment. Geol.*, vol.3, pp.129-239.

Kinoshita, R.(1984) Present Status and Future Prospects of River Flow Analysis by Aerial Photograph, *J. Hydraulic, Coastal and Environmental Eng.*, JSCE, pp.1-19, (in Japanese)

Nezu, I., Kadota, A. and Kurata, M.(1996) Free-Surface Flow Structures of Space-Time Correlation of Coherent Vortices Generated behind Dune Bed, *Proc. of the 6th Int. Symp. on Flow Modeling and Turbulence Measurements*, Tallahasee, pp.695-702.

Nezu, I. and Nakagawa, H. (1991) Turbulent Structures over Dunes and its Role on Suspended Sediments in Steady and Unsteady Open-channel Flows, *Proc. of Int. Symp. on Transport of Suspended Sediments and its Mathematical Modeling*, IAHR, Firenze, pp.165-189.

DYE VISUALIZATION AND P.I.V. IN THE CROSS-STREAM PLANE OF A TURBULENT CHANNEL FLOW

Evidence for Oblique Structure

J.C. WELLS, Y. YAMAMOTO, Y. YAMANE, S. EGASHIRA AND H. NAKAGAWA
Dept. of Civil & Environmental Engineering,
Ritsumeikan University
Noji Higashi 1-1-1, Kusatsu, Shiga 525-8577 Japan

Abstract. Illuminating in a quasi-cross-stream plane of an open channel flow of water at low *Re* and injecting fluorescene dye from a bed slit, mushroom patterns reliably inferred to result from counter-rotating streamwise vortices are found to be typically oriented at some 30° from the cross-stream vertical. PIV in the cross-stream plane shows that the averaged field, and two-point correlation, of the quantity $Q_x = \left(\omega_x^2 - S_x^2\right)/2$, ($S_x = \frac{\partial v}{\partial z} + \frac{\partial w}{\partial y}$) exhibit preferred directions in the cross-stream plane at about 45 degrees to the vertical. Spatio-temporal measurements of streamwise vorticity provide direct experimental evidence, possibly the first, for the staggered arrangement of streamwise vortices previously educed from DNS data.

1. Dye Visualization

Head & Bandyopadhyay [1] injected smoke from the wall of a wind tunnel and observed mushroom patterns, interpreted to mark the counter-rotating legs of hairpin vortices, in a laser sheet tilted 45 degrees above the upstream direction. Though not mentioned by those workers, their film gives the impression that mushrooms tend be oriented at a distinct angle from the wall-normal streamwise plane. We confirmed this impression for patterns of wall-injected dye in an open channel flow of water [2]; results are summarized here for workers outside Japan.

Fluorescene dye was injected from a 2-*mm* wide slit in the bottom of a 50 *cm* wide, 8 *m* long horizontal smooth-bed water channel, at 2.5 *m* from the downstream end. Flow depth d was 10.0 *cm* . Three flow rates gave bulk Reynolds numbers Ud/ν of 3860, 5020, and 6090 (with friction velocity u_* estimated from flow rate *via* the Blasius friction law, this yields

T. Kambe et al. (eds.), IUTAM Symposium on Geometry and Statistics of Turbulence, 357–364.
© 2001 *Kluwer Academic Publishers.*

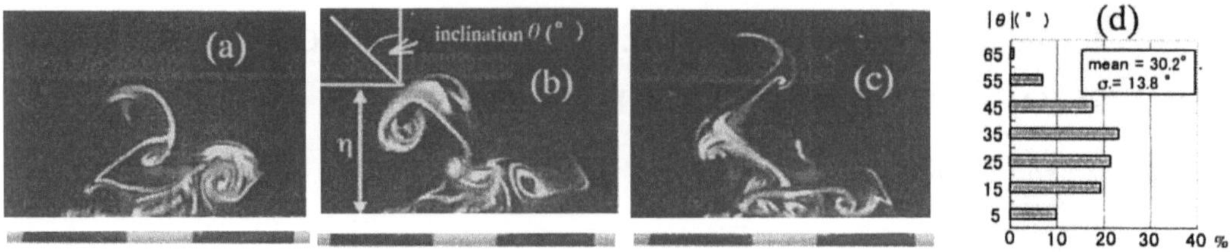

Figure 1. (a)-(c) Fluorescene dye visualized with a light sheet tilted up from the upstream horizontal at 60 degrees, so that the light strikes the bottom wall 30 *cm* downstream of the dye injection slit, at instants new mushroom patterns first appear. Time proceeds from left to right, the second frame coming 7/30 *s* ($= 1.46\ \nu/u_*^2$), and the third frame 17/30 *s* ($= 3.54\ \nu/u_*^2$), after the first frame. Width of white markers is 2 *cm* ; flow depth was 10.0 *cm* . (d) Histogram of the absolute value of the angle θ of the "stalks" of mushroom-shaped patterns at the moment they first appear in a light sheet tilted up at 60 degrees from the upstream horizontal plane. The 130 data points were distributed fairly uniformly in the range of y^+ from 25 to 120. The average value of $|\theta|$ over all the data was 30.2 degrees, with averages over three subranges of y^+ indicating a slight increase of $|\theta|$ with y^+ [2].

$Re_* = u_* d/\nu$ of 228, 291, and 327). A projector light sheet, tilted up at 60 degrees from the upstream horizontal, intersected the bed at a distance of 30 *cm* or 60 *cm* downstream of the dye slit.

NTSC video recordings revealed mushroom-shaped patterns of dye which are reliably inferred to result from counter-rotating streamwise vortices (*cf.* Figure 1a). One finds that succeeding streamwise vortices ("SV") usually appear one by one, tending to pair with one of the SV from an old mushroom to produce a new mushroom pattern (*cf.* Figs. 1b and 1c), and in such a way that the orientation of the newly-appearing mushrooms alternates between oblique left and oblique right. The experimental distribution of inclinations of such "new" mushrooms (fig. 1(d)) indicates a clear tendency to lie at a distinct angle from the vertical plane. Observation of video sequences, exemplified by fig. 1 (a-c), suggest that straight portions of rising dye (*e.g.* the "stalks" of the mushrooms) mark the boundaries between counter-rotating SV. By this interpretation, the result that newly-appearing mushrooms lie at a distinct angle from the vertical suggests that successive SV appear to pair obliquely at their "contact point" when viewed in cross-stream. It may be argued [2] that such oblique instantaneous pairing, and the observed tendency to oscillate from oblique left to oblique right, is consistent with the staggered arrangement of SV proposed by Stretch.

2. Particle Image Velocimetry

We have performed particle image velocimetry ("PIV") in the same flow studied by dye injection, with Re_* now restricted to a single value of 300 ($Ud/\nu = 5250$). A 5 mm thick double-pulsed laser sheet[1] illuminated the flow in a cross-stream plane 5.5 m from the channel entrance. 1K*1K video images were recorded at 30 f.p.s.[2], looking upstream through a bottom-mounted mirror positioned 50 cm downstream of the light sheet, and through a glass window at the water surface. Light pulses from one laser were synchronised with a 250 μs exposure of the camera; pulses from the second laser were triggered 5 ms later, in the next camera exposure. 10 μm diameter hollow glass spheres of a density very slightly greater than water[3] served as flow tracers. The field of view was 4.96 cm wide * 4.69 cm high, so that only the lower half of the channel was observed. Velocity components perpendicular to rays from the lens center were calculated by maximising the cross- correlation of 49*49 pixel templates between image pairs. Tests with a rotating clear plastic plate simulating the particle- laden fluid showed that streamwise vorticity could be estimated with acceptable accuracy by first smoothing velocity vectors with the 9 surrounding values.

Averaged Flow Quantities. We now present averaged and r.m.s. flow quantities from a run of 350 velocity fields measured at 1 s intervals. The decorrelation time of streamwise vorticity near the bed was found to be about 3 s, so the samples are, *a priori*, only weakly correlated. The averaged velocity field, presented in figure 2 (b), shows mean vectors flowing outward near the edges, since we are measuring components of velocity normal to rays from the lens center. Our analysis here concentrates on vorticity and associated quantities; contours of mean vorticity, overplotted in figure 2 (b), show very nearly zero average vorticity near the left and right edges, except very near the bottom, so pollution of measured ("streamwise") vorticity by spanwise vorticity does not appear to be a problem. The profile of r.m.s. vorticity (fig. figure 2 (a)) agrees closely with data from DNS [5].

Consider the quantity Q_x defined, with $S_x = \frac{\partial v}{\partial z} + \frac{\partial w}{\partial y}$, by

$$Q_x = \left(\omega_x^2 - S_x^2\right)/2 \ , \tag{1}$$

which is analogous to the second invariant of the deformation rate tensor in 3D flow, and expresses the relative strength of rotational motion and straining motion. The average value of Q_x at each point in the field of view

[1]double Nd-YAG laser(Quantel S.A., France) run at 15 Hz; 100 mJ/pulse nominal
[2]Kodak ES1.0 progressive-scan, 8 bit resolution, with Nikkor 105 mm microlens
[3]Sphericel powder, Potters Industries, USA

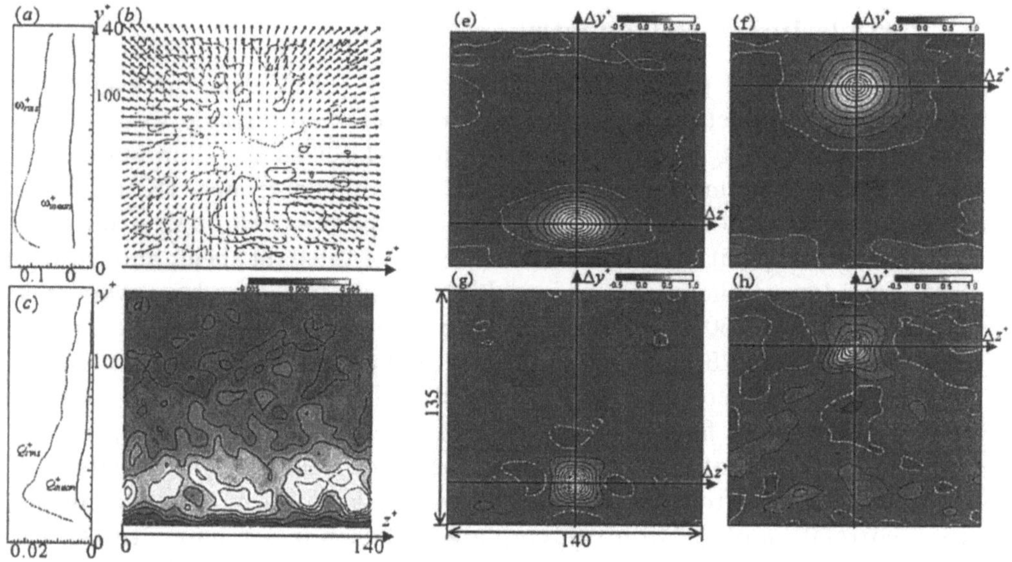

Figure 2. Flow quantities over a sample of 350 cross-stream velocity fields taken at 1 s intervals ($\Delta t^+ = 10$). (a) Profiles of average and r.m.s. streamwise vorticity ω_x^+. (b) Averaged cross-stream velocity field. Magnitude of velocity at left, right, and top edges is about 0.4 wall units; arrows are magnified 2.8 times more than in instantaneous samples of figure 3 . Contours of $\omega_x^+ = -0.0125, 0, 0.0125$ are superposed; the zero contour is a dotted curve. (c) Profiles of average and r.m.s. Q_x, defined in equation (1), as expressed in wall units. (d) Sample average of Q_x^+. Contour intervals of 0.001 wall units, zero contour, in white, is seen at bottom edge. (e-h) Two-point spatial correlations $C_\omega(y, \Delta y, \Delta z)$ of streamwise vorticity (e,f) and Q_x (g,h), (*cf.* equation (2)), with the base point at values of $y^+ = 31$ and 104. Contour intervals of 0.1, zero contour in white.

is shown in figure 2 (d); one observes a remarkable tendency for contours to be oriented at about 30 to 45 degrees from vertical. In the range $y^+ < 40$, contours are tilted both ways. In the range $40 < y^+ < 80$, there is a tendency to be oriented from lower right to upper left, while farther from the bottom, the contour lines swing generally the other way. The ribs of this structure seem to be ordered rather regularly in space, and furthermore one has the overall impression of a hierarchical structure. The data reported in figure 2 were recorded over a period of about 170 large eddy turnover times based on mean velocity and flow depth, and a very similar field was found by averaging three runs at 15 Hz, so it would appear that inclined ribs of high Q_x are a quasi-permanent feature of the flow.

Two-point Correlations. Figure 2 (e)-(h) show two-point correlations defined with $f = \omega_x, Q_x$ by

$$C_f(y, \Delta y, \Delta z) = \left(\sum \int f(y,z) f(y+\Delta y, z+\Delta z) dz\right) / < f^2(y) > \quad (2)$$

where the sum, and the average indicated by brackets, is over all 350 samples. The correlations of ω_x, shown in (e) and (g), show some evidence of a diamond pattern, but this tendency is not very stable as y^+ is varied, nor between experiments, and in any case is weak. By contrast, C_Q (f,h) shows a clear tendency for preferred directions at about $\pm 45^o$ to the vertical that is consistent over the whole range of y^+ between 30 and 140.

Instantaneous Flow Fields. To help interpret the differences between the spatial correlations of ω_x and Q_x, figure 3 shows two instantaneous cross-stream velocity fields, separated by 1.2 s (= 12 ν/u_*^2), that exemplify two common patterns of vorticity. The first snapshot (a) is characterised by narrow "jets" rising at about 45 degrees from horizontal, switching from oblique left to oblique right, and snaking between vorticity concentrations of opposite sign. The vorticity field at this instant (c) has a structure of orthogonal ribs, joined together at T-junctions, that is reflected in the oblique rectangular aspect of the instantaneous two-point vorticity correlation C'_ω whose origin is scanned over the entire instantaneous field;

$$C'_\omega(t, \Delta y, \Delta z) = \int \omega_x(t,y,z)\omega_x(t, y+\Delta y, z+\Delta z)\, dy\, dz / < \omega_x^2 > \quad (3)$$

(In fact, such oblique vorticity correlation is more commonly observed to show banding in a single direction instead of two near-orthogonal directions.) Similar remarks apply to the field of Q_x, shown in graph (g), and corresponding correlation C'_Q (graph h) at this instant. Observations of instantaneous vorticity fields frequently reveal such an oblique pattern. Our impression is that such patterns occur as arrangements of somewhat weak vorticity patches. When stronger structures appear, their orientations and relative positioning seem to be much more varied.

Another common pattern of instantaneous vorticity correlation C'_ω, that of horizontal banding, is seen in figure 3 (f). The correlation of Q_x at this instant, shown in (j), retains the oblique checkerboard character observed in (h). This contrast in the behavior of C'_Q and C'_ω reflects the very different forms of the two-point correlations C_ω and C_Q discussed above.

We believe that the obliquely oriented "preferred directions" observed in these, and many other, instantaneous velocity and vorticity fields, have a strong connection with our observations of the flow by dye injection. Upwelling jet structures carry fluid from the near-wall region, so these might

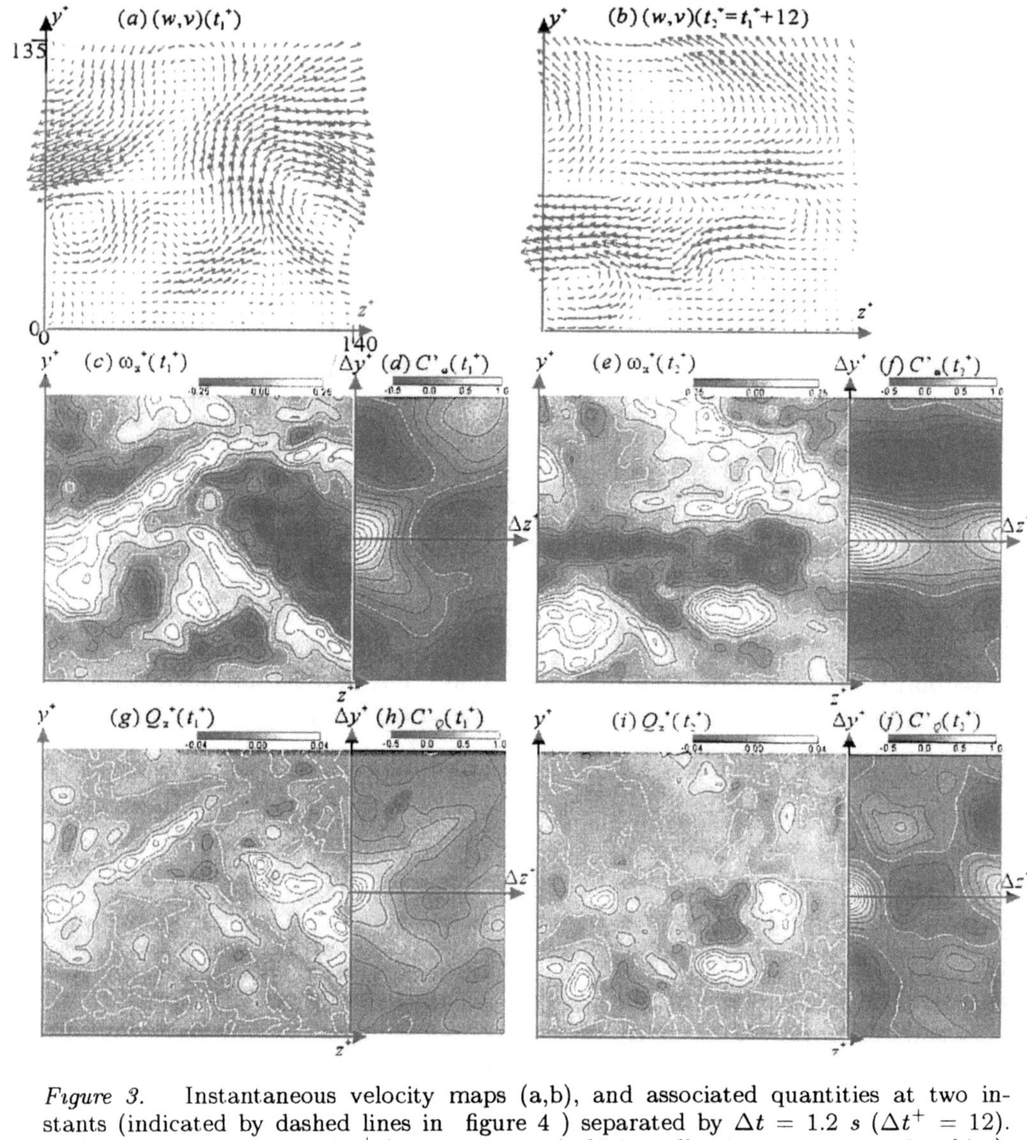

Figure 3. Instantaneous velocity maps (a,b), and associated quantities at two instants (indicated by dashed lines in figure 4) separated by $\Delta t = 1.2\ s$ ($\Delta t^+ = 12$). (c,e):instantaneous maps of ω_x^+ (contour interval of 0.05 wall units, zero contour in white); (g,i): instantaneous maps of Q_x^+ (contour interval of 0.008 wall units, zero contour in white); (d,f): instantaneous two-point correlations C'_ω of ω_x over each sample (*cf.* equation (3))(contour interval 0.1, zero contour in white); (h,j): corresponding instantaneous correlations of Q_x (contour interval 0.1, zero contour in white)

well correspond with the "stems" of the mushroom patterns, which consist of dye being pumped away from the wall. Our impression that these jets are often oriented obliquely concurs with the data presented in figure 1 on the orientation of the mushroom stems. Furthermore, the tendency of these

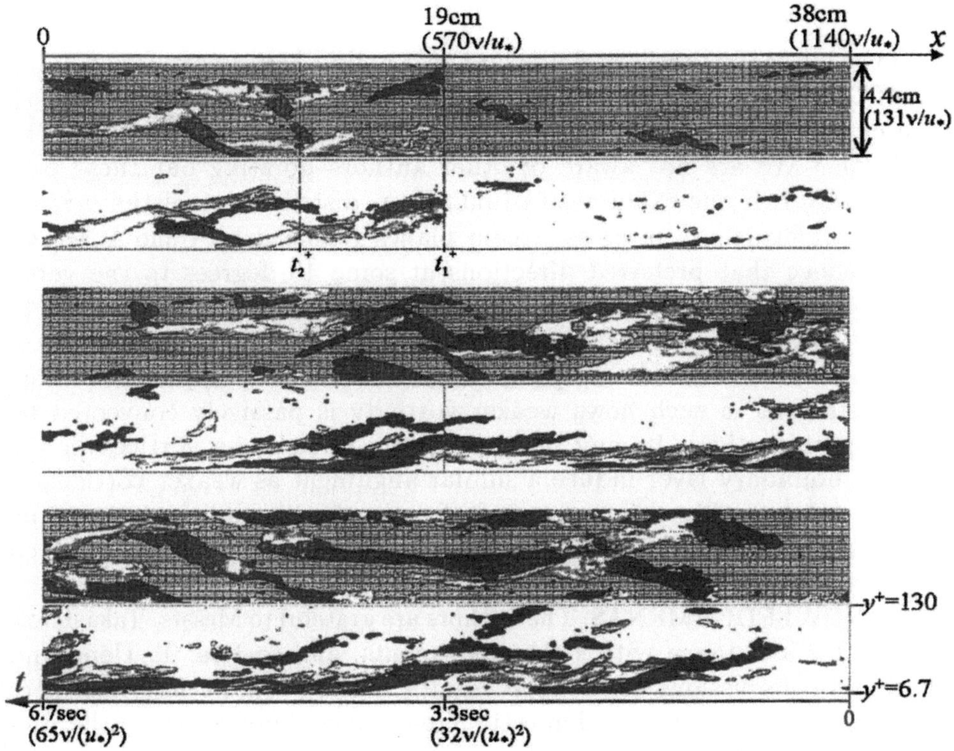

Figure 4. Spatiotemporal structure of ω_x in a record at 15 velocity fields/s expressed by isosurfaces $\omega_x^+ = 0.25$ (white) and $\omega_x^+ = -0.25$ (black) seen in top view (first, third, fifth strips from top) and side view. Recording started at right bottom. Time t converted to streamwise distance x by multiplication with the mean flow velocity; structures can be thought of as flowing from left to right, and top to bottom.

jets to alternate direction as they snake upwards may have a connection with the oscillation of mushroom patterns referred to above.

Spatio-temporal Relationship of Streamwise Vortices. Continuously measured cross-stream slices of streamwise vorticity are displayed as a three-dimensional plot in figure 4, with time converted to space according to Taylor's hypothesis. The arrangement of near-wall SV thus visualized generally agrees with the staggered arrangement proposed by Stretch [3]. In plan view, such staggered vortices are often oriented at about 25 degrees from the mean stream direction, while other vortices seem to be oriented almost exactly streamwise. This impression appears to contradict the rather weak average SV orientation of $\pm 4^o$ proposed by Jeong *et al.* [4], while it agrees roughly with the preferred directions revealed in the experimental data of *wall-normal* vorticity in reference [6].

3. Conclusion

Both instantaneous vorticity fields and two-point spatial correlation functions of Q_x suggest that there is an oblique, sometimes lattice-like pattern in the cross-stream plane at medium to large scale (*i.e.* 1/4 to 1/2 of the flow depth.) We are not aware of other authors pointing out these patterns. However, if one examines instantaneous distributions of the *vertical* component of vorticity in cross-stream planes reported by Ueno & Utami [7], we believe that preferred directions at some 45 degrees to the vertical are apparent. We remark that the oblique ribs of vorticity and dye to which we draw attention seem to qualitatively resemble the filaments of weak vorticity that form X-like patterns in 2D numerical turbulence. Farge [8] suggests that in such flows weaker vorticity is passively convected by the strongest vortices. It may be that strong streamwise vortices in the turbulent boundary layer induce a similar alignment as weaker vorticity is stretched out into thin sheets oriented close to the streamlines induced by the strong structures. Why such vortical sheets should tend to be oriented in a particular direction is not clear to us.

ACKNOWLEDGEMENTS. The authors are grateful to Messrs. Takashima and Ukai for assistance with PIV experiments, and to Drs. T. Ueno and T. Utami for discussions about their results. The PIV work described here expands on that reported in domestic Japanese conference papers [9] and [10]. This work has been supported in part by a Monbusho Grant-in-aid for Scientific Research (lead researcher; J. Wells) and the Monbusho Gakujutsu Frontier program.

References

1. Head, M.R. and Bandyopadhyay, P. (1981) *J. Fluid Mech.* **Vol. 107**, pp. 297–338
2. Yamane, Y., Wells, J.C., Egashira, S., Nakagawa, H. (1998) *Proc. of the 30th Fluid Dynamics Conference, Aerospace Society of Japan* pp. 113–116 (In Japanese)
3. Stretch, D. (1990) *Center for Turbulence Research Annual Research Briefs* pp. 145–157
4. Jeong, J., Hussain, F., Schoppa, W., and Kim, J. (1997) *J. Fluid Mech.* **Vol. 332** pp. 185– 214
5. Nagaosa, R. (1999) *Physics Fluids* **Vol. 11 no. 6** pp. 1581–1595
6. Ueno, T. and Utami, T. (1991) *Proceedings, Eighth Symposium on Turbulent Shear Flows*
7. Ueno, T. and Utami, T. (1993) *Proceedings, Ninth Symposium on Turbulent Shear Flows* paper P205
8. Farge, M. (1992) *Ann. Rev. Fluid Mech.* **Vol. 24** pp. 395–458
9. Yamamoto, Y., Yamane, Y., Wells, J.C., Egashira, S., Nakagawa, H. (2000) *Proc. 49th Japanese Conference on Theoretical and Applied Mechanics* (in Japanese)
10. Wells, J.C., Yamane, Y., Yamamoto, Y., Egashira, S., Nakagawa, H. (2000) *Annual J. Hydr. Eng., JSCE* **Vol. 44** pp. 491-496

K

Boundary Layers and Near-Wall Turbulence

VORTICITY STRUCTURES AND INTERMITTENCY IN NEAR WALL TURBULENCE

R. PIVA*, C. M. CASCIOLA*, G. AMATI+, P. GUALTIERI*
* *Dip. Meccanica e Aeronautica, Università di Roma, La Sapienza via Eudossiana 18, 00184 Roma, Italy*
+ *CASPUR, P. A. Moro 5, 00185 Roma, Italy*

1. Introduction

Statistical properties of wall bounded flows manifest an increased intermittency towards the wall together with a larger population of coherent structures as confirmed in recent years by many investigations, either experimental or numerical.

A quantitative description is achieved by considering, within the inertial range, the moments of the longitudinal increments of streamwise velocity, $\delta V = V(x+r) - V(x)$. In the context of homogeneous isotropic turbulence, the Kolmogorov-Obukhov refined similarity (RKSH) provides a scaling law, in terms of the separation r,

$$< \delta V^p > \quad \propto \quad < \epsilon_r^{p/3} > r^{p/3}, \tag{1}$$

where the moment of order $p/3$ of the energy dissipation rate ϵ_{loc} spatially averaged over a volume of characteristic dimension r, accounts for the intermittent behavior in the inertial range. The anomalous correction to the purely dimensional K41-exponent, $p/3$, directly follows from the statistical properties of the dissipation field, $< \epsilon_r^q > \propto \; r^{\tau(q)}$. As a consequence, a generic structure function of order p can be expressed as a power law of the separation with exponent $\zeta_p \; = \; \tau(p/3) \, + \, p/3$.

As well known, the scaling laws are displayed only for sufficiently large values of the Reynolds number, Re. For moderate values, instead, an inertial scaling law in terms of the separation can hardly be detected. However, when assuming as independent variable the third order structure function instead of separation [1], in the spirit of extended self similarity (ESS), scal-

T. Kambe et al. (eds.), IUTAM Symposium on Geometry and Statistics of Turbulence, 367–372.
© 2001 *Kluwer Academic Publishers.*

ing laws emerge also at the relatively small values of the Reynolds number that can be achieved by DNS.

The RKSH in its extended form,

$$< \delta V^p > \quad \propto \quad \frac{< \epsilon_r^{p/3} >}{< \epsilon >^{p/3}} < \delta V^3 >^{p/3}, \tag{2}$$

has been extensively checked against most of the available data. Clearly, for large Re, eq. (2) is an immediate consequence of the classical form (1), when use is made of the forth-fifths law, relating $< \delta V^3 >$ to the separation, r. At moderate Re the new form allows for an extension of the range of scales where similarity laws are observed and for the extraction of scaling exponents which match those of large Reynolds number turbulence.

The kind of phenomenology behind eq. (2) is the same of the classical RKSH, and is related to the effect that the fluctuations of energy dissipation induce on the energy transfer through the inertia-dominated scales.

2. Scaling law for the wall region

As discussed in [2], the scaling behavior observed in terms of ESS in the near wall region is not consistent with the refined Kolmogorov similarity hypothesis, as expressed by eq. (2). Actually, here the relevant physical processes which are quite different from the energy cascade towards small scales implied by (2), may induce the observed failure of the RKSH. In particular, close to the wall the leading process is the momentum transfer occurring in the wall normal direction, due to the large local mean shear S, which corresponds to production of turbulent kinetic energy via Reynolds stresses, $S < uv >$.

Based on this analysis, a new form of scaling law has been proposed [2] for the wall region,

$$< \delta V^p > \quad \propto \quad \frac{< \epsilon_r^{p/2} >}{< \epsilon >^{p/2}} < \delta V^2 >^{p/2}, \tag{3}$$

in terms of the structure function of order two. The new similarity law has been shown to be consistent with the available data from numerical simulations in a channel, fig. 1, and gives reason for the large intermittency [2] without invoking dramatic changes in the field of energy dissipation which are not observed in the DNS.

These findings are consistent with the presence of a scale, related to the shear S and to the averaged energy dissipation ϵ, defined as $L_s = \sqrt{\epsilon/S^3}$ [3] which becomes very small in the wall region. We expect the new form of RKSH (3) to hold for values of the separation r significantly larger than

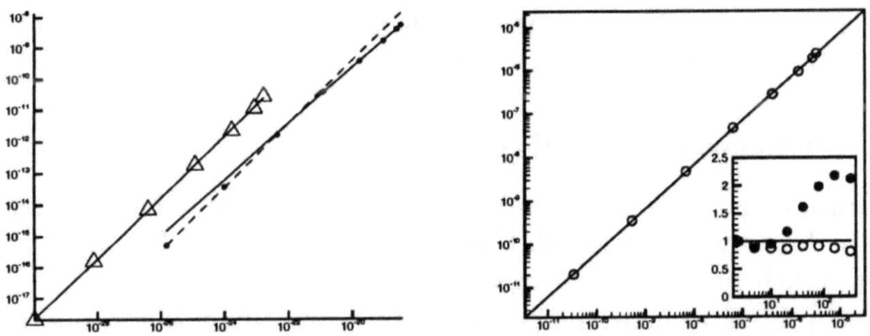

Figure 1. Left part: $< \delta V^6 >$ vs. $< \epsilon^2 >< \delta V^3 >^2$ in turbulent channel flow for two different wall normal distances. Bulk region ($y^+ = 151$): data (triangles) and their fit in the region $r^+ \in [20, 320]$ (solid line with slope 0.99). Wall region ($y^+ = 31$) data (circles) and their fits in two regions $r^+ \in [1, 20]$ and $r^+ \in [20, 320]$, solid line with slope 0.99 and dotted line with slope 0.88, respectively. Right part: check of consistency for eq. (3) at $y^+ = 31$: $< \delta V^4 >$ vs. $< \epsilon^2 >< \delta V^2 >^2$. The solid line (slope 1.01) gives the fit in the whole range. In the inset compensated plot for eq. (2) vs. r^+ (filled circles); corresponding plot for eq. (3), (circles).

L_s, a condition which is verified only near the wall. On the other hand the original form of RKSH (2) remains perfectly valid where $r << L_s$, i.e. in the bulk at the center of the channel, fig. 2. Presently, we have no clear evidence of the simultaneous presence of both similarity laws at a single location in the intermediate region, where the separation is of the same order of the shear scale L_s. However the numerical data for the channel suggest the presence of a transition which is, however, not very sharply detected since the available range of scales is not sufficiently large.

Unfortunately, the continuous variation in the intensity of the mean shear with the distance from the wall, the influence of the boundary condition suppressing the wall-normal fluctuations [4], as well as the non-uniform momentum flux across adjacent fluid layers make the data analysis for the channel flow particularly complex.

For the above reasons we consider here a flow with a constant mean shear rate in the absence of solid walls. In fact, the homogeneous shear flow presents two main advantages: it isolates the effect of the shear from other concurrent effects and it improves the statistical analysis by exploiting homogeneity in all the spatial directions. Clearly the possibility to reach a stationary state [5] is a necessary prerequisite. Under this respect, an unbounded homogeneous shear flows does not achieve a steady state since its characteristic length keep increasing with time [4]. In fact [5], the confinement of the flow to a finite box determines an upper limit for the integral scale and allows an equilibrium state to be eventually achieved. By this approach we are able to construct a shear-dominated system, certainly not easy to be reproduced in experiments, which is very suitable to DNS.

3. Homogeneous shear flow

We consider a periodic box with an imposed mean velocity gradient, as already done by Lee *et al.* [4], Kida and Tanaka [6] and Pumir [5] following the approach proposed by Rogallo [7].

The proper dimensionless parameter controlling the strength of the shear is given by $S^* = Sq^2/\epsilon$ with q the rms value of velocity [4]. In particular, we have analyzed the scaling law of the structure functions for $S^* = 7$. As in the channel flow we check the consistency for the two forms of similarity laws, equations (2) and (3). We can observe, also here, a quite clear failure, although more contained with respect to the channel flow, of the classical RKSH form (2) for the shear flow while, see figure 3, it holds for homogeneous and isotropic turbulence.

It's worth noticing that the parameter S^* may be interpreted as a ratio between an inertial characteristics scale $l_d = q^3/\epsilon$ and the shear scale L_s, $S^* = (l_d/L_s)^{2/3}$. To get a deeper insight into the physical meaning of L_s let us make the following consideration in terms of velocity fluctuations. By denoting by $\delta U = Sr$ the velocity difference at scale r originated by the mean shear and denoting by δu the turbulent velocity fluctuation, we may find the scale at which the shear becomes relevant by the condition $\delta U >> \delta u$.

By using the expression $\delta u \propto \epsilon^{1/3} r^{1/3}$ the dynamics induced by the shear is predicted to be dominant at the scales $r/L_s >> 1$. This dimensional argument shows that L_s is the characteristic length that separates the range

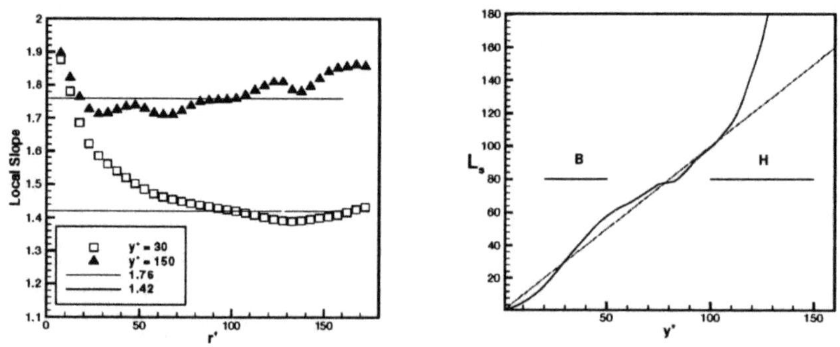

Figure 2. Left part: local slope of the structure function of order six for the turbulent channel flow. Right part: shear scale in the channel flow.

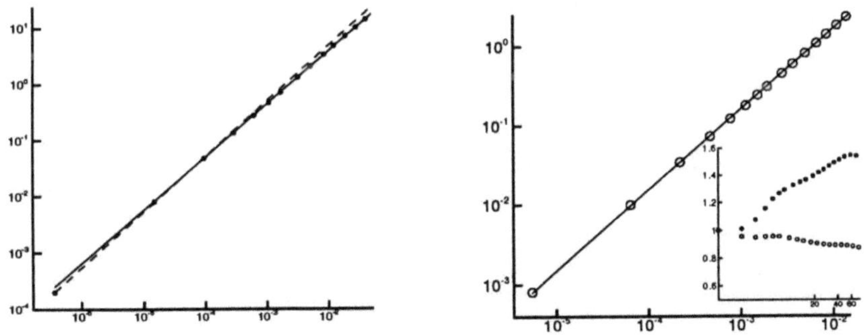

Figure 3. Left part: check of the RKSH in the form of equation (2): $<\delta V^6>$ vs. $<\epsilon_r^2><\delta V^3>^2$ for the shear flow (filled circles). The solid line has a slope of 0.94 while the dotted line has a slope of 0.99, instead of 1. as a single slope as for the homogeneous isotropic turbulence. Right part: check of equation (3) for the shear flow in the form of $<\delta V^4>$ vs. $<\epsilon_r^2><\delta V^2>^2$. The solid line has a slope of 1.01. In the inset compensated plot of $<\epsilon_r^2><\delta V^2>^2 / <\delta V^4>$ (circles) and $<\epsilon_r^2><\delta V^3>^2 / <\delta V^6>$ (filled circles) as a function of $r^+ = r/r_*$ where $r_* = (\nu/S)^{1/2}$.

where the energy cascade is driven by the non linear terms from that where the shear acts substantially. Clearly, L_s is always expected to be greater than the Kolmogorov scale, η, since, for $r = \eta$, we would obtain $\nu S^2 >> \epsilon$. Hence, in principle, a range of scales may exist, $\eta << r << L_s$, where the dynamics is essentially inertia-dominated and unaffected by the mean shear. The behavior of these fine scales of the flow could be, thus, very similar to that of homogeneous and isotropic turbulence.

The present considerations lead to the conjecture that the two similarity laws may coexist, namely the classical form in the range $\eta << r << L_s$, the new form for $r >> L_s$. In order to verify this conjecture a large separation of scales is required. Unfortunately the present capabilities do not allow to reach these conditions by DNS and suitable experiments should be devised to this purpose. The coexistence of the two scaling laws would require, in any case, the cascade process to become increasingly important for $r << L_s$.

To attempt a scale by scale analysis, we have applied a sharp low-pass filter with cut-off wave number k_f to the DNS data and we have measured the tensors $b_{ij} =< u_i u_j > / < u^2 > -1/3\delta_{ij}$ and $V_{ij} =< \zeta_i \zeta_j > / < \zeta^2 > -1/3\delta_{ij}$, typically used as indicators of isotropy. Once evaluated for the low-pass filtered velocity field, these quantities, considered as functions of k_f, manifest a clear saturation at $k_f \simeq k_s = 2\pi/L_s$. In other words, for $k_f > k_s$, any further increase of the cut-off wavenumber k_f seems to produce no appreciable changes in the values of the indicators, see fig. 4.

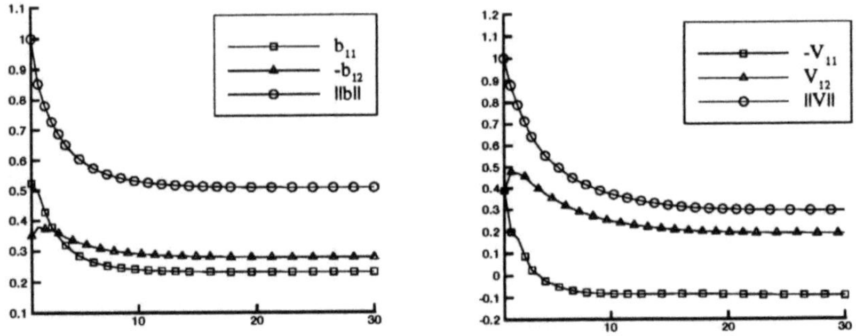

Figure 4. Different components and norm of b_{ij} (left) and V_{ij} (right), as functions of k_f (definitions in text), normalized to have maximum of the norm equal to one. $k_s \simeq 9$.

As it will be discussed in a separate paper, this kind of behavior may be shown to be consistent with the idea of a dominating energy cascade in the range where saturation is observed. Clearly, these preliminary impressions about the coexistence of the two scaling laws should be substantiated by further work, e.g. along the lines suggested in [9].

4. Final remarks and open problems

In conclusion, some evidence has been provided that the new scaling law, originally proposed for the wall region of the channel flow [2], still holds in the confined homogeneous shear flow, that we have considered to single out the effect of shear. The failure of the classical RKSH is here less dramatic than in the channel flow, presumably due to the lower value of the shear parameter S^*. The new scaling law is observed as for the channel flow, despite the substantial difference in the coherent structures geometry [8].

References

1. Benzi, Biferale, Ciliberto, Struglia, Tripicccione, Phy. D., **96**, 162-181, 1996.
2. Benzi, Amati, Casciola, Toschi, Piva, Phys. Fluids, **11**(6), 1-3, 1999.
3. Toschi, Amati, Succi, Benzi, Piva, Phys. Rev. Lett., **82**(25), 5044-5047, 1999.
4. Lee, M.J., Kim, J., Moin, P., J. Fluid Mech., **216**, 561-583, 1990.
5. Pumir, A., Phys. Fluids, **8**(11), 3112-3127, 1996.
6. Kida, S., Tanaka, M., J. Fluid Mech., **274**, 43, 1994.
7. Rogallo, R.S, NASZ T.M., **81315**, 1981.
8. Casciola, C.M., and Amati, G., 8th Int. Symp. on Flow Vis., September 1998.
9. Arad, Biferale, Mazzitelli, Procaccia, Phys. Rev. Lett., **82** 50-40, 1999.

THE USE OF PIV IN TURBULENT BOUNDARY LAYER FLOWS

KRISTIAN ANGELE and BARBRO MUHAMMAD-KLINGMANN
Department of Mechanics, Royal Institute of Technology (KTH), S-100 44 Stockholm, Sweden.

1. Introduction

Due to the recent technical development, digital PIV (Particle Image Velocimetry) has become a widespread tool for overall flowfield measurements. PIV is often used in industrial applications as a simultaneous measurement and visualization technique, while basic research is still focussed on investigating the capabilities and limitations of the technique as such, see [1]. Although the digital technique can not compete with photographic film when it comes to spatial resolution, the present study shows that accurate near wall measurements can also be obtained with digital PIV, provided the image size is small enough and sufficient care is taken to eliminate optical disturbances. This is also the first time that PIV-measurements have been made in a boundary layer at high Reynolds numbers with a sufficiently large data set to form a statistically safe average.

2. Experimental setup

Two different experiments where performed, one in a turbulent boundary layer with zero pressure gradient (ZPG) and the other in an adverse pressure gradient (APG), both developing on a flat plate. In both cases the images cover a region which includes the entire logarithmic region.

PIV-measurements were made in the xy-plane where x is the streamwise and y is the wall-normal. A large number of images were captured and processed statistically, and the results were compared to those obtained with conventional measurement techniques such as hotwire anemometry and LDV.

2.1. PIV EQUIPMENT

The PIV equipment consists of a 400 mJ double pulsed Nd:Yag laser and a digital Kodak ES1.0 CCD camera, containing 1018 x 1008 pixels. The measurement process is synchronized and controlled by a hard- and software from Dantec. Figure 1 shows the setup for measurements in the xy-plane in the ZPG case. The flow was seeded by means of a smoke generator using either paraffin oil or glycol. Different

T. Kambe et al. (eds.), IUTAM Symposium on Geometry and Statistics of Turbulence, 373–378.
© 2001 *Kluwer Academic Publishers.*

techniques were tried for injecting the particles into the flow, and it was found that the best method was to introduce smoke into the entire wind tunnel circuit by placing the smoke generator inside the stagnation chamber.

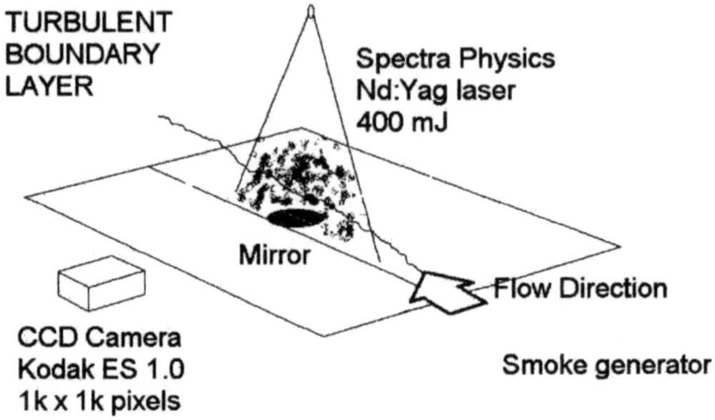

Figure 1. *Experimental setup in the ZPG experiment.*

Large ensembles of image pairs were collected at each position and submitted to validation and statistical processing. The interrogation area size was 32 x 32 pixels with an overlap of 50%. Each interrogation area is weighted with a Gaussian window function, in order to reduce signal noise caused by so-called loss of pairs. The final velocity vectors were validated using two criteria : a lower limit of 1.2 for the signal-to-noise ratio, taken as the ratio between the highest and the second highest peak in the correlation plane, and a second criterion for a maximum allowable displacement of 1/4 of the interrogation area size. Vectors not fulfilling these criteria were removed without replacing them.

The presented data are ensemble averages of 2700 and 1300 image pairs for the ZPG and APG cases respectively. Within each image, velocity profiles in the y-direction were averaged in the x-direction over 12 and 17 interrogation areas respectively.

The laser frequency was 15 Hz, ensuing that the image pairs were statistically independent.

2.2. THE ZPG SETUP

The ZPG experiment was carried out in the MTL wind tunnel at KTH. The tunnel has a 7 meter long test section with cross section area 0.8 m x 1.2 m, and an exceptionally good flow quality. Measurements were done on a flat aluminum plate at a distance x=5.5 m from the leading edge and a free stream velocity U_∞=13.2 m/s. Re_x, the Reynolds number based on x-position, was 4.8 millions and Re_θ, based on momentum loss thickness, was 8300.

A mirror was used to reflect the laser sheet back into itself, in order to avoid reflexions of the light sheet from the plate surface which might corrupt the data in the near wall region. The mirror was flush-mounted to the wall and painted with a thin layer of black paint, except for a narrow slit along the incidence of the laser sheet. The camera was equipped with a 300 mm lens.

The reference measurements, made with CTA hot wire anemometry were made in the same setup [5].

2.3. THE APG SETUP

The APG experiment was carried out in a small low speed wind tunnel with a 0.4 m x 0.5 m test section on a 1.15 m long flat plate. A diverging flow was obtained by adjusting the test section ceiling, which consisted of a flexible perforated sheet. By applying suction through the holes, the boundary layer was removed from the ceiling and an APG boundary layer developed along the plate. To ensure a fully developed turbulent flow a trip was placed a few centimeters downstream of the leading edge.

The boundary layer development is characterized by its shape factor, H, which is the ratio between the displacement thickness and the momentumloss thickness. The value of H varies from 1.3 when the pressure gradient is zero to a value of 3-4 at separation. In the present case, the flow was non-uniformly decelerated, giving a increasing value of H in the downstream direction. Further details of the boundary layer development are described in [3].

The present measurements were made at a position 0.925 m downstream of the leading edge, where the local value of H was 1.7. The free stream velocity was U_{∞}=28.5 m/s, giving a Reynolds number Re_x of 1.76 millions. The Reynolds number based on momentum loss thickness, Re_θ, was about 6000.

The reference measurements presented in here were made with a 1-component LDV equipment (Dantec FlowLite).

3. Results

3.1. COMPARISON BETWEEN PIV AND CONVENTIONAL TECHNIQUES

3.1.1. *ZPG boundary layer*

Figure 2 shows the boundary layer profile of the the streamwise velocity component (U) and its fluctuations (u_{rms}). Here, PIV measurements are compared to hot wire

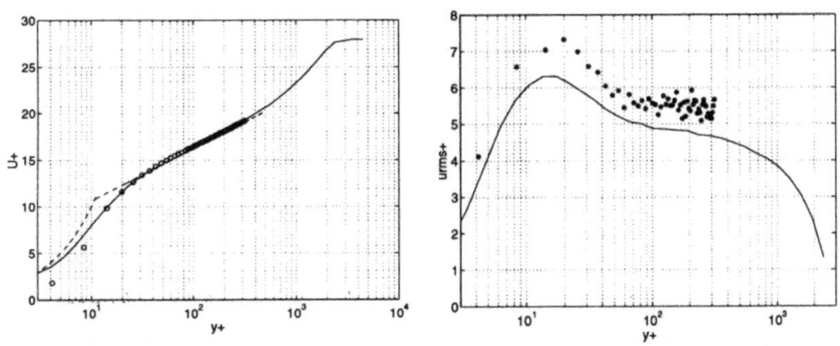

Figure 2. ZPG boundary layer profile in inner variables, comparison between PIV (symbols) and hot wire anemometry (line). Left: mean, right: rms of the streamwise velocity.

data obtained earlier in the same experimental setup [5]. The data are plotted in wall units, defined using a wall shear stress determined by means of oil film interferometry, see [6], giving a viscous length scale 31 μm. The dashed line in the figure shows the logarithmic law and a linear scaling $u^+=y^+$ for $y^+\leq 10$. The total boundary layer thickness was 70 mm and the PIV image size was 11.2 mm x 11.1 mm giving an interrogation area size of 0.35 mm, i.e 11.4 viscous units.

The agreement between PIV and hot wire measurements is fair all the way down to the measurement point at $y^+=8$. Also the profile of u_{rms} shows the correct behavior. The inner peak at $y^+=17$ is well captured, however, the overall level of the u_{rms} values is higher than those measured with the hotwire.

With use of the mirror the wall distance can accurately be determined and optical disturbances are efficiently eliminated. It is possible to measure all the way down to the innermost point at $y^+=4.5$ but due to the large velocity gradient close to the wall the velocity is underestimated in the averaging over a rather large area.

The hot wire used in the ZPG case has a much smaller extent normal to the wall than can be achieved with PIV, however, it is affected by heat convection into the wall. As a result, the two methods are comparable in terms of actual near wall resolution.

3.1.2. *APG boundary layer*

Figure 3 shows the boundary layer velocity profile measured with PIV and LDV respectively. The data are plotted in wall units, defined using a wall shear stress determined from a Clauser plot fitted to the LDV data. Earlier hot wire experiments [3] showed that APG boundary layers have a well defined logarithmic region up to H-values well above the present value of 1.7. This is also clearly visible in Figure 3. The standard constants $\kappa=0.41$ and B=5 were used. This gave a viscous length scale of approximately 18μm (the same scaling is used for plotting both the LDV and PIV data).

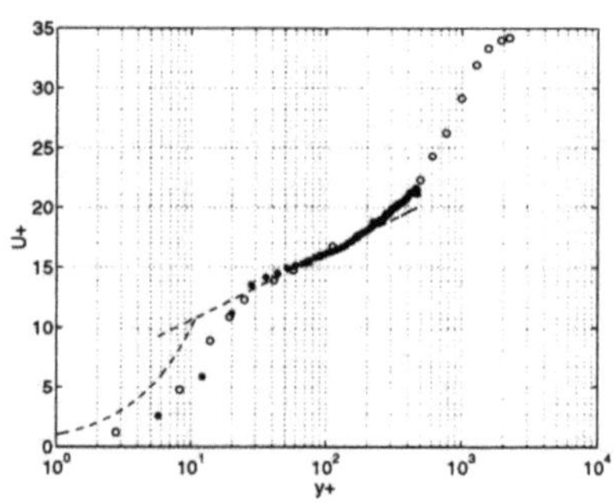

Figure 3. Streamwise velocity component in inner variables, comparison between PIV (full symbols) and LDV (open symbols).

The total boundary layer thickness was 35 mm, and the size of the images captured was 9.1 mm x 9.2 mm, which covers the viscous, buffer and logarithmic region. Each PIV measurement volume (i.e. interrogation area) covered about 16 viscous units, however, due to the overlap, the distance between measurement points is 8 viscous unit. The "center" of the first interrogation area is at a distance of only 4 viscous units from the wall sice it partly contains the wall.

With LDV, the measurement volume was about 5 viscous units, and its position relative to the wall could be determined with an accuracy of about 1 viscous unit. No mirror was used in the APG case so the wall distance could not be determined accurately.

The agreement between the two measurements is good throughout the logarithmic region, whereas the linear region is not well represented with any of the measurement techniques. This may be an effect of averaging over a rather large measurement volume in a region where the velocity gradient is large causing a bias towards lower velocity.

4. Discussion of the results

All PIV results suffer from the well-known "peaklocking" problem which is inherent to the discrete nature of CCD imaging, see [4][9]. The present evaluation algorithm uses a Gaussian peak fit for the sub-pixel interpolation, as recommended by [4]. This does not affect the mean velocity, however, if the probability density function is not symmetric (as is the case in the near wall region), one may expect that peak locking affects the u_{rms} values. In the present study, u_{rms} is overestimated by about 8 percent, while the data reported by Westerweel [8] in turbulent pipe flow show an underestimate of the velocity fluctuations.

Insufficient averaging is another error source, which directly affects the measured mean velocity field. In order to obtain statistically safe data it is necessary to average over image pairs acquired at a time interval larger than the integral time scale. Averaging within an image may give a smoother distribution but does not increase the amount of statistically independent data.

In the present study, the Reynolds number is a factor 10 larger than that of Westerweel [8], hence the extent of the boundary layer, measured in viscous units, is correspondingly larger. Willert et al. [7] describe a PIV-study of a boundary layer at similar Reynolds number and viscous length scale as in the present case, where 100 image pairs were found to be insufficient. In the present studies, 2700 and 1300 image pairs were used, which is still small compared to the number of samples collected with CTA or LDV. The validation rate in the present study was about 50-60 percent. This may be due to several factors, e.g. the steep velocity gradient close to the wall and the low seeding density, which in turn is related to the small image size.

5. Summary and future work

In summary, the present experiments show that PIV can give accurate results for the mean velocity in the near wall region of a turbulent boundary layer, provided the image is focused in this region, reflexions are eliminated, and sufficiently large data sets are used. Accurate measurements are made down to $y^+=8$ in a high Reynolds number turbulent boundary layer. This is comparable to what can be achieved with hotwire.

The present study in APG is of great practical importance to decelerating and separating flows in diffusers and on turbine blades, where it is of interest to accurately determine the point of separation, see [2].

So far, only the streamwise velocity has been obtained with sufficient accuracy. The future aim is to evaluate two velocity components in order to derive correlations such as u′v′ and u′w′, which are needed for the validation of turbulence models in decelerating flows.

References

1. I. Grant, "Particle image velocimetry: a review", IMechE C09494/97, (1997)
2. R. Holm and J. Gustavsson, "A PIV study of separated flow around a 2D airfoil at high angles of attack in a low speed wind tunnel", FFA TN 1999-52
3. B. G. B. Muhammad-Klingmann and J. P. R. Gustavsson, "Experiments on turbulent flow separation", IMechE C557/153/99, (1999)
4. J. Westerweel, "Digital Particle Image Velocimetry: theory and application", Doctoral Thesis, (1993)
5. Österlund, J. M., "Experimental Studies of Zero Pressure-Gradient Turbulent Boundary-Layer Flow", (1999)
6. Österlund, J. M. and Johansson, A. V. and Nagib, H. M. and Hites, M. H., "Wall Shear Stress Measurements in High Reynolds Number Boundary Layers from Two Facilities", AIAA paper 99-3814 (1999)
7. C. Willert, M. Raffel, J. Kompenhans, B. Stasicki, C. Khler, "Recent applications of particle image velocimetry in aerodynamic research", Flow Meas. Instrum. Vol. 7, No. 3/4, pp. 247-256, (1996)
8. J. Westerweel, A.A. Draad, J.G. Th. van der Hoeven, J. van Oord, "Measurements of fully-developed turbulent pipe flow with digital particle image velocimetry", Experiments in Fluids 20 (1996) 165-177
9. J. Westerweel,"Fundamentals of digital particle image velocimetry", Meas. Sci. Technol. 8 (1997) 1379-1392

BOUNDARY LAYER INTERMITTENCY MODEL

Boundary layer model based on Intermittency concept and consequent Length Scales, Skewness, and Kurtosis distribution laws on the ground of experimental data

P. GUALTIERI, G. PULCI DORIA
Università degli Studi di Napoli Federico II - Dipartimento di Ingegneria Idraulica ed Ambientale Via Claudio 21 80125 Napoli – Italy
Phone 0039 81 7683460 Fax 0039 81 5938936 E-mail pulci@unina.it

1. Introduction

At the present time, the impact of free stream turbulence on fully turbulent boundary layer flow has been examined in numerous experimental, analytical and computational studies. Many of these studies [1]..[12] deal with the effects of the free stream turbulence on the main statistical turbulence quantities (for example mean velocities, turbulent shear stress, turbulent kinetic energy): in particular some workers have studied Length Scales, Skewness and Kurtosis, which will be used in this paper.

Previous workers [13] [14] have also studied the effects of the free stream turbulence on the irregularity of the boundary between the boundary layer fluid and the external stream fluid. In fact, as is well known, an indented interface arises, continuously variable, separating the whirling turbulent boundary layer fluid from the external stream fluid. Quantitatively this is best expressed in terms of the behaviour of the Intermittency Factor (I), defined as the fraction of total time for which the boundary layer is present at the considered point. The distribution of I (called here the Intermittency Function I(Y)) is described in literature by the Erf function [15]. Consequently, in the whole flow, there is a band where, at each point, the boundary layer is alternatively present or absent.

2. Contribution of the present paper

The present study refers to boundary layer beneath free stream turbulence with approximately the same turbulence level.

The aims are: a) to measure the Intermittency in new way, starting from distributions of longitudinal integral Length Scales (L), and also from Skewness (S) and Kurtosis (K) of velocity fluctuations; b) to lightly modify the Intermittency Function in order to attain a more physically based expression. c) as a corollary, distributions of Length Scales, Skewness, and Kurtosis independent of free stream turbulence will also be obtained.

Consequently, the authors performed a set of measurements of instantaneous velocities values at various points in a boundary layer under the earlier defined free stream turbulence. Using these measurements the authors obtained experimental distributions of

T. Kambe et al. (eds.), IUTAM Symposium on Geometry and Statistics of Turbulence, 379–384.
© 2001 *Kluwer Academic Publishers.*

the quantities recalled at point a) [16]..[18], and a rough first attempt model of Intermittency (three band model) [19].

In this paper the research about this topic finds a first point of arrival. In particular:

1) The external flow has different turbulence levels, always approximately of the same order of the boundary layer turbulence level.

2) The Intermittency Function is expressed through a modified Erf function, with a logarithmic function introduced, which allows that: a) the Function starts from the origin with a 0 derivative (the Erf function starts with non 0 derivative, so that it is generally supposed not to be valid near the origin); b) the Function is no more symmetrical (whereas the Erf is symmetrical) around the point where I=0,5 (it is not possible that I Function be symmetrical, as the boundary conditions are not). The I(Y) Function has two free parameters which have to be defined through experimental fit.

3) The experimentally obtained values of Length Scales, Skewness, and Kurtosis, allow a definition of the Intermittency Function free parameters.

4) The so-called associated distribution functions are found for Length Scales, Skewness, and Kurtosis, which are independent of the boundary layer section and of the free stream turbulence level.

3. The model

3.1. THE INTERMITTENCY FUNCTION I(Y)

As recalled earlier, the Intermittency Factor I at a point is defined as the fraction of total time for which boundary layer is present at that point. In this model the following Intermittency Function I(Y) is chosen:

$$I(Y) = \frac{1}{\sqrt{2\pi}} \int_{-\infty}^{\frac{\ln Y}{H}} e^{-\frac{\eta^2}{2}} d\eta$$

In this expression Y is the non dimensional distance y/δ of the point from the plate, where y is the true distance and δ is the physically based boundary layer thickness. In particular, δ is defined as the distance from the plate where I=0,5 [20].

In this model, in each working condition (different free stream turbulence levels) and in each test section, the expression (1) is assumed to be valid, and the ratio $R=\delta/\delta_c$ (where δ_c is the classical boundary layer thickness following Coles) is assumed to be constant.

It is worth emphasising that:

1) The derivative of I(Y) Function is 0 when Y = 0;

2) The I(Y) Function is not symmetrical with respect to the point where Y=1 and I=0,5.

Both features are inconsistent with the classical relation of the Intermittency Factor, but are consistent with the use of a logarithmic function into I(Y) Function.

3.2. THE TURBULENCE STATISTICAL QUANTITIES DISTRIBUTIONS

If we consider any turbulence statistical quantity Z at a point in the boundary layer, its local mean value can be expressed as a linear combination between the values it assumes when the boundary layer is present or absent, giving as weights I and (1-I). We can define Z_t the first value and Z_∞ the second value, so that:

$$Z = Z_t * I + Z_\infty * (1\text{-}I) \tag{2}$$

This expression can be assumed only if boundary layer and free stream turbulence have approximately the same turbulence level.
Yet the model proposed supposes a number of things about Z_t and Z_∞.
At each point, the Z_∞ value depends on the working conditions and on the test section, but does not depend on the Y distance of the point from the plate. On the contrary, the Z_t value is independent either of the working conditions, or of the test section (equilibrium boundary layer). Consequently the result in each test section and at each point is:

$$Z(Y) = Z_t(Y) * I(Y) + Z_\infty * [1\text{-}I(Y)] \tag{3}$$

In each test section the true distribution of Z against Y depends specifically on the Z_∞ value of external flow characteristic of that test section and of that working condition.
As a consequence of statements made earlier it is possible to define a so called associated distribution $Z_a(Y)$ of the same quantity Z, which does not depend on Z_∞ value. Indeed we can write in turn, and equivalently:

$$Z_a(Y) = Z_t(Y) * I(Y) \tag{4}$$
$$Z_a(Y) = Z(Y) - Z_\infty * [1\text{-}I(Y)] \tag{5}$$

It is possible to state that the associated distribution so defined is universal with respect to the considered boundary layer types (equilibrium boundary layer, zero longitudinal pressure gradient, smooth flat plate, different free stream turbulence levels).
Furthermore, if the model proves satisfactory, it will be possible to choose suitable values for the two free parameters (δ and consequently R_t and H) such that the associated distribution of any statistical quantity Z, defined through (4) and (5), is independent of the external flow turbulence and consequently of Z_∞ value.

4. Validation/verification of the model

The model has been tested using three particular statistical functions Z (namely the L, S, and K, functions). The Length Scales, in particular, have been made non-dimensional in respect of the distance from the boundary layer's attack, following [16].
All the experimental data for testing the model were obtained on a boundary layer developing along a smooth flat bottom of a rectangular channel coming from a sluice gate. In this plant the external flow was naturally turbulent: different levels were attained by use of grids in the sluice gate. Instantaneous velocity measurements were performed in four test sections of the boundary layer, at distances 0.15 m apart. In each test section more than 20 points along the vertical have been sampled up to and including points whose y values were well out of the boundary layer. At each point the instantaneous longitudinal component of velocity was measured through an LDA device, and more

than 200.000 sample data were collected at time intervals of 0,005 s apart. A more detailed description of these tests is to be found in [21].

Consequently, more than 20 experimental Z_{exp} values have been obtained in each test section and in each working condition, based on the instantaneous velocity samples. These values defined experimental distributions of Z along y in each test section and in each working condition. Furthermore, in each test section and in each working condition, the Z_{exp} values relative to points well out of the boundary layer have been considered representative, in their mean, of the Z_{∞} value.

At this point pairs of R and H values are assumed to be valid. In relation to each pair the following data processing was performed.

1) The non-dimensional free variable values $y/\delta_c = Y_c$ have been substituted by the non-dimensional values $Y=y/\delta=Y_c/R$.

2) Experimental values of associated functions $Z_{a\text{-}exp}$ have been obtained starting from the Z_{exp} values through equation (5).

3) Experimental values of turbulent functions $Z_{t\text{-}exp}$ have been obtained starting from the $Z_{a\text{-}exp}$ values through equation (4).

4) All experimental (Y, $Z_{t\text{-}exp}$) points have been collected in the same plot.

5) On the basis of the plots obtained in 4), simple interpolating turbulent functions $Z_{t\text{-}int}$ (Y) have been assumed (linear for Length Scales and Skewness, and fractional polynomial for Kurtosis). In this elaboration it is worth noting that the experimental values obtained for $Z_{t\text{-}exp}$ show a large scatter, either because the original experimental Z_{exp} values were very spread, or because the presence of the I(Y) Function in the denominator increases further the spreading, especially where I is very small. Consequently, as the standard deviation of the experimental points is large and not alike for different points, a moving mean technique has been adopted for the experimental values arising from equation (4), using a differing number of points in the different zones of the boundary layer, in order that each mean value could be characterised by the same value of its standard deviation.

6) Interpolating associated functions $Z_{a\text{-}int}$ (Y) have been obtained starting from the $Z_{t\text{-}int}$ (Y) through equation (4).

7) The $Z_{a\text{-}int}(Y)$ behaviour has been compared with the $Z_{a\text{-}exp}$ values. Depending on the way $Z_{a\text{-}int}(Y)$ function has been drawn, the $Z_{a\text{-}int}(Y)$ values necessarily tend to 0 when $Y\rightarrow\infty$; and, on the other hand, the $Z_{a\text{-}int}(Y)$ values necessarily well interpolate the experimental data when $Y\rightarrow 0$ (if interpolations in 5) fit well). On the contrary, the $Z_{a\text{-}int}(Y)$ values when $Y\approx 1$ can interpolate well the experimental points $Z_{a\text{-}exp}$ only if the model is robust. This is the main characteristic of the model which will be used in order to validate it.

8) Many pairs of R and H values have been verified, in order to choose that pair which gave the best fit in the comparison performed in 7). The results of these tests gave R=1,4 and H=0,29.

5. Test results

In fig. 1 the I(Y) function with H=0.29 is shown.

In relation to the results of interpolations in 5) of previous paragraph, the $L_{t\text{-}int}(Y)$, $S_{t\text{-}int}(Y)$, and $K_{t\text{-}int}(Y)$ functions show the following behaviours:

$$L_{t\text{-}int}(Y) = 0.522\,Y + 0.160 \qquad (6)$$
$$S_{t\text{-}int}(Y) = -2.100\,Y + 0.449 \qquad (7)$$
$$K_{t\text{-}int}(Y) = (15.17\,Y^2 - 5.68\,Y + 0.45) / (Y + 0.18) \qquad (8)$$

In figs. 2, 3, 4, the $Z_{a\text{-}int}(Y)$ functions are shown compared with the experimental $Z_{a\text{-}exp}$ points. A light moving mean technique has been applied to these points in order to better verify the fit.
As result the interpolating curves follow with a great accuracy the behaviour of experimental points, in particular in the zone near the value Y=1.

6. Conclusions

The Intermittency Function in a boundary layer beneath free stream turbulence with approximately the same turbulence level, can be obtained in an innovative way, starting from measurements of other statistical quantities. In particular Length Scales, Skewness and Kurtosis have been here employed.
The Intermittency Function can be approximated through a modified Erf function, with the presence of the logarithmic function, to better reproduce the real behaviour of the Intermittency phenomenon.
Lastly and in particular, so-called associated distribution functions can be obtained for Length Scales, Skewness, Kurtosis which are valid in boundary layers along a smooth flat plate, with zero longitudinal pressure gradient, and with free stream turbulence.

7. References

1 Kline, S.J., Lisin, A.V., Waitman, B.A.(1960) Preliminary experimental investigation of the effect of free-stream turbulence on turbulent boundary layer growth *N.A.C.A. T.N. D 368.*

2 Charnay, G., Comte-Bellot, G., Mathieu, J. (1971) Development of a turbulent boundary layer on a flat plate in an external turbulent flow" *AGARD CPP 93-71.*

3 Huffmann, G.D., Zimmermann, D.R., Bennett, W.A. (1972) The effect of free stream turbulence level on turbulent boundary layer behaviour *Paper I-5 AGARDograph n.164*

4 Robertson, J.M., Holt, C. (1972) Stream turbulence effects on turbulent boundary layer *ASCE*

5 Evans, R.L., Horlock, J.H. (1974) Calculation of the development of turbulent boundary layer with a turbulent free stream *JFE.*

6 McDonald, H., Krecovsky, J.P. (1974) Effect of free stream turbulence on the turbulent boundary layer *JHMT.*

7 Meier H.U., Kreplin H.P. (1980) Influence of free stream turbulence on boundary layer development *AIAA*

8 Hancock P.E., Bradshaw P. (1983) The effect of free stream turbulence on turbulent boundary layers *JFE*

9 Castro, I.P. (1984) Effects of free stream turbulence on low Reynolds number boundary layer *JFE.*

10 Evans, R.L. (1985) Free stream turbulence effects on turbulent boundary layers in an adverse pressure gradient *AIAA.*

11 Hoffmann J. A. and Mohammadi K. (1991) Velocity profiles for turbulent boundary layers under free stream turbulence *JFE.*

12 Bandyophadyai, P.R. (1992) Reynolds number dependence of the free stream turbulence effects on turbulent boundary layers *AIAA.*

13 Charnay G., Comte-Bellot G., Mathieu J.(1976) Response of a turbulent boundary layer to random fluctuations in the external stream *The Physics of Fluids*.
14 Hancock P. E. and Bradshow P. (1989) Turbulence structure of a boundary layer beneath a turbulent free stream *JFM*.
15 Hinze, J.O. (1987) *Turbulence* McGraw-Hill
16 Gualtieri P., Pulci Doria G. (1996) Researches about boundary layer: a way of making non dimensional longitudinal integral length scales and fixing the thickness of the intermittency layer" (in italian) *IV A.I.VE.LA.* Ancona, Italy.
17 Gualtieri P., Pulci Doria G. (1996) Distributions of skewness and kurtosis in a boundary layer either with external undisturbed flow or with external turbulent flow"(in italian) *XXV Conv. Idr. Costr. Idr.*, Torino, Italy
18 Gualtieri P. and Pulci Doria G.(1997) Non dimensional distribution of longitudinal integral length scales in a turbulent boundary layer *Fifth Int. Symp. FLUCOME*, Hayama Japan
19 Gualtieri P. and Pulci Doria G. (1998) Skewness, Kurtosis, and Length Scales experimental distribution laws based on a three bands model in a turbulent boundary layer *Excerpta of italian contribution to the field of hydraulic engineering,*
20 Gualtieri P., Pulci Doria G. (1999) A proposal of a physically based thickness definition and a new mean velocities distribution law in a turbulent boundary layer on the ground of LDA measurements *XIII Australasian Fluid Mechanics Conference* Monash University, Melbourne, Australia.
21 Gualtieri P. (1995) Data acquisition by LDA and data processing techniques in order to fix the distribution of longitudinal integral length scales in a turbulent boundary layer" (in italian) *III A.I.VE.LA.* Ancona, Italy

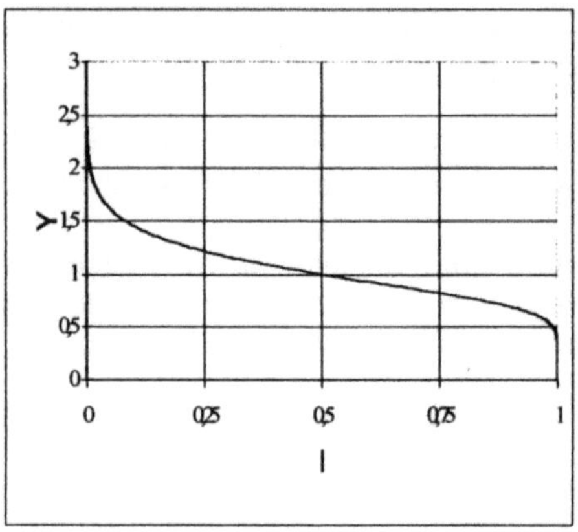

Fig. 1

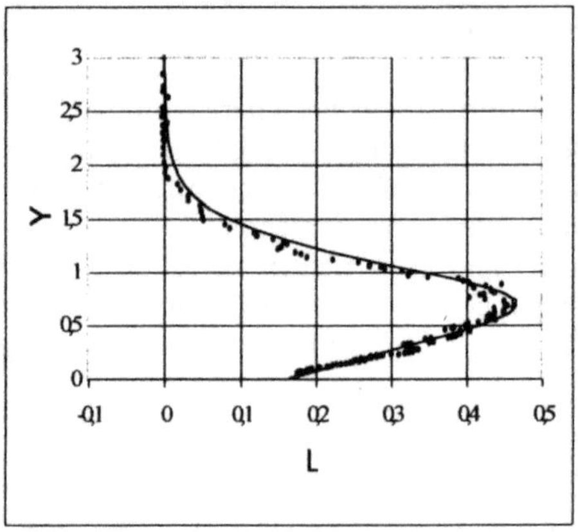

Fig. 2

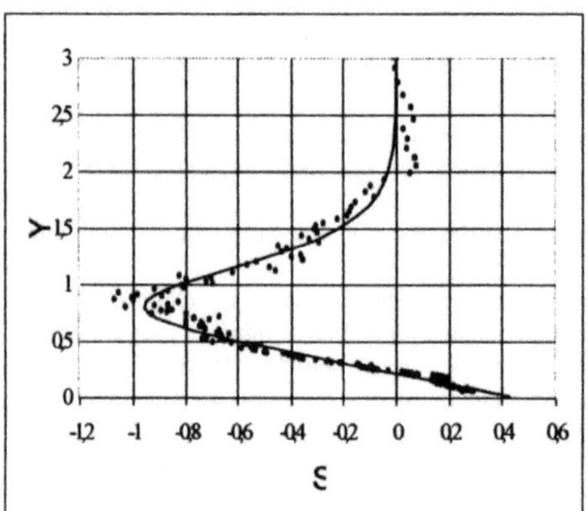

Fig. 3

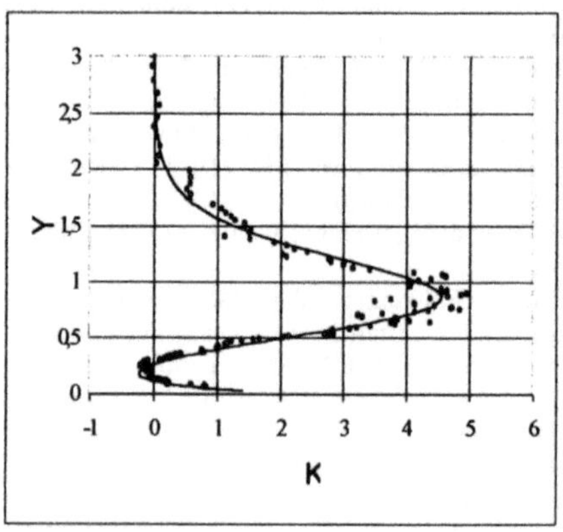

Fig. 4

UNIVERSAL PROPERTY OF AUTONOMUS LAYER IN NEAR-WALL TURBULENCE

K. TSUJIMOTO AND Y. MIYAKE
Department of Mechanophysics , Osaka University
2-1 Yamada-oka, Suita, Osaka, 565-0871 Japan

1. Introduction

In turbulent channel flow, the flow receives energy from outside to sustain itself and the most part is lost by viscous dissipation due to the strain rate of mean flow. The remainder is supplied to the turbulence energy and is dissipated by the viscosity. As the consequence, the energy balance holds. In the wall turbulence, it is understood that the most of the turbulence activity occurs in the near-wall region and turbulence is mainly supplied from there to the far layer by turbulent diffusion. That is, the eddies of various scales exchange momentum between the layers somehow, and this contributes to sustain the turbulence in the whole layer. Therefore, if the turbulence energy production is suppressed by modification of interaction, turbulence control such as drag reduction is realized. However, in case of the smooth wall, near-wall layer and far layer seem to be distinctly separated, as exemplified by the fact that the former is scaled with wall unit which depends on the wall friction velocity and kinematic viscosity and the latter, scaled by global scale of flow passage or channel width. The authors confirmed in the previous work that the turbulence in the far layer depends only on the magnitude of the total shear stress irrespective of its origin, friction drag or profile drag. These facts suggest that each near-wall and far layer has the turbulence sustenance mechanism of each own. Recently Jiménez *et al.*(1999) reported an autonomous cycle of turbulence in the near-wall layer derived from numerical experiments, demonstrating that necessary and sufficient conditions of the autonomous behavior are the existence of streaks. Here, the autonomous cycle means that the near-wall layer maintains the turbulent flow without assistance of the turbulence from the upper layer. However, their results do not give full explanation for the interaction between the autonomous near-wall layer and above it.

T. Kambe et al. (eds.), IUTAM Symposium on Geometry and Statistics of Turbulence, 385–390.
© 2001 *Kluwer Academic Publishers.*

Table 1. Calculation condition of data bases

No.	Volume(L_x, L_y, L_z)	Grid(N_x, N_y, N_z)	Re_τ
1	$4\pi H \times 2H \times \pi H$	$128 \times 129 \times 64$	150
2	$2\pi H \times 2H \times \pi H$	$128 \times 257 \times 128$	395

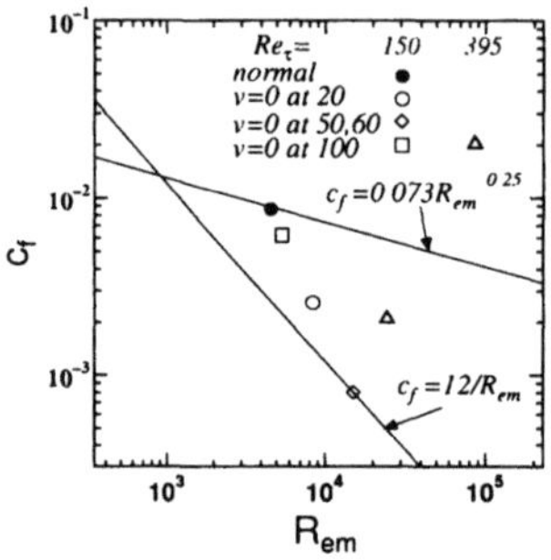

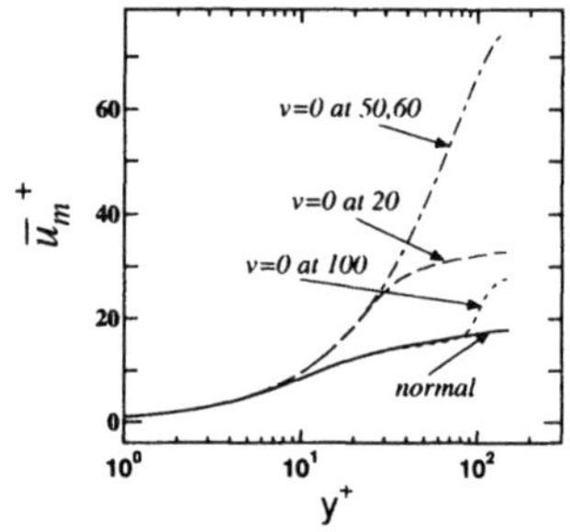

Figure 1. Friction coefficient and Velocity distribution for the case of a channel flow having a interceptive plane, $(Re_\tau = 150)$

Thus, it is intended in this paper to examine the property of autonomous layer focusing on the interaction between the layers, more in detail.

2. Outline of the numerical procedure

A direct numerical simulation is carried out for a turbulent channel flow. Numerical procedures are given in Miyake *et al.*(1995) in which periodic boundary conditions are assumed both in streamwise and spanwsie directions and no-slip condition is applied on the walls. The spatial discretization is a Fourier series expansion in streamwise and spanwise direction, and a Chebyschev polynomial expansion in wall-normal direction. The mean pressure gradient is kept constant throughout the computation. A Cartesian coordinate system x, y, z which are in streamwise, wall-normal, and spanwise directions, respectively, is employed and the velocity components in the respective directions are denoted by u, v, w. The computational domain is a rectangular volume with a wall separation $L_y = 2H$, a streamwise length L_x, and a spanwise width L_z. The Reynolds number is defined as $Re_\tau = H\bar{u}_\tau/\nu$ where $\bar{u}_\tau$ is mean friction velocity and ν is kinematic viscosity. The grid number and the Reynolds number are shown in Table 1.

3. Result

3.1. THE PROPERTY OF INTERACTION BETWEEN THE LAYERS

In the numerical experiments of Jiménez *et al.* (1999), the autonomous behavior of the near-wall layer is examined by removing all the turbulence fluctuation artificially over some height. In this work, another way of investigating the interaction between the layers is adopted. That is, wall-normal velocity in a specified plane parallel to the wall at some height y_s is forced to vanish. In order to realize this control, the external force defined in the following equation is added to the momentum equation.

$$\frac{\mathrm{D}u_i}{\mathrm{D}t} = -\frac{1}{\rho}\frac{\partial p}{\partial x_i} + \nu\nabla^2 u_i + f_i \tag{1}$$

$$f_i(\boldsymbol{x}_s, t) = \alpha \int_0^t u_i(\boldsymbol{x}_s, t')dt' + \beta u_i(\boldsymbol{x}_s, t) \tag{2}$$

where $\boldsymbol{x}_s$ is the point in the interceptive plane, α, β are negative constants and are optionally determined so that the wall-normal velocity vanishes in the plane. This numerical treatment of external force is based on the method proposed by Goldstein *et al.* (1993).

Figure 1(a) shows the change of wall friction coefficient for the Reynolds number. In the figure the straight line $C_f = 0.073Re_m^{-0.25}$ is for the normal turbulent channel flow and $C_f = 12/Re_m$, for the laminar flow. $\circ$, $\diamond$, $\square$ are for the case of $y_s^+ = 20, 50, 100$, $\bullet$ is for usual channel flow where $y_s^+ = y_s\bar{u}_\tau/\nu$. In case of $y_s^+ = 20$, the friction coefficient is in the mid of turbulent and laminar flow and the layer below $y^+ \leq 20$ is found to be laminarized as observed by the mean velocity profiles shown in figure 1(b). While, in case of $y_s^+ = 50, 60$, C_f takes the value of laminar flow and the flow is actually proven to be perfectly laminar in the whole region as demonstrated by the mean velocity profile in Fig.1(b). The reason for the top of the interceptive layer to be turbulent in the cases of $y_s^+ = 50, 60$ is because Re_τ is too small. These results show that the near-wall layer thinner than some specific height can not keep turbulent regime by itself, if energy supply from the upper layer is removed. The minimum thickness which can sustain turbulent regime by itself is around $y^+ = 60$ which confirms the finding of Jiménez *et al.*(1999) .

Figures 2 (a)(b) show the distribution of turbulence intensity and vorticity, respectively in the case of $y_s^+ = 100$. Noticeable is that mean velocity distribution in Fig.1(b) for $y_s^+ = 100$ shows identical curve with normal channel flow below $y^+ \leq 100$ suggesting that when the turbulent flow is maintained, the near-wall layer is not affected by upper layer. That is, turbulence in this layer is autonomous in the sense that it is self-sustenance without momentum exchange with outside. Since the wall-normal velocity vanishes at the interceptive plane, the slip wall effect manifests itself there and causes the turbulence production on the wall side of interceptive plane. The similar effect also appears on the vorticity there. Figures 2(c)(d) show the balance of the turbulent kinetic energy and the mean flow energy. In the figure, the solid line is for the normal turbulent channel, and the symbols, for the case interception at $y_s^+ = 100$. In the layer $y^+ \leq 60$, the energy balance is not much different from the normal turbulent channel flow. Thus, it

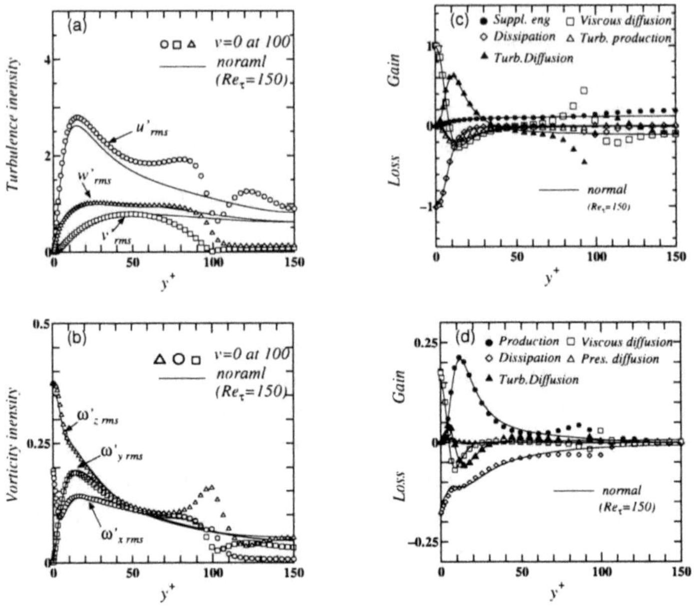

Figure 2. Mean turbulence properties in a channel having with momentum interception at $y_h^+ = 100$ at $Re_\tau = 150$, (a)turbulent intensity, (b) vorticity intensity, (c)mean energy budget and (d) turbulent energy budget

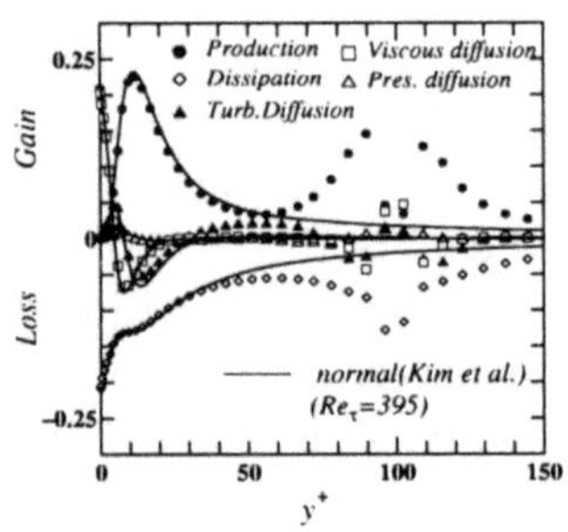

Figure 3. Budget of kinetic energy of fluctuating velocity for the case of $y_h^+ = 100$ at $Re_\tau = 395$

is concluded that the autonomous layer has the universal structure concerning turbulence production and dissipation. Figures 4 shows the distribution of the turbulence energy balance in case of $Re_\tau = 395$ suggesting that the above-mentioned conclusion does not depend on the Reynolds number.

3.2. THE ENERGY EXCHANGE BETWEEN THE LAYERS

By modifying artificially the mode of energy supply due to mean streamwise pressure gradient ($d\bar{p}/dx$), the property of energy supply to the autonomous layer can be examined. For this artificial energy supply, it is assumed that $d\bar{p}/dx$ is allowed only in the specific layer of some thickness, elsewhere being removed. Fig.4(a) shows the distribution of the shear stress in the case of

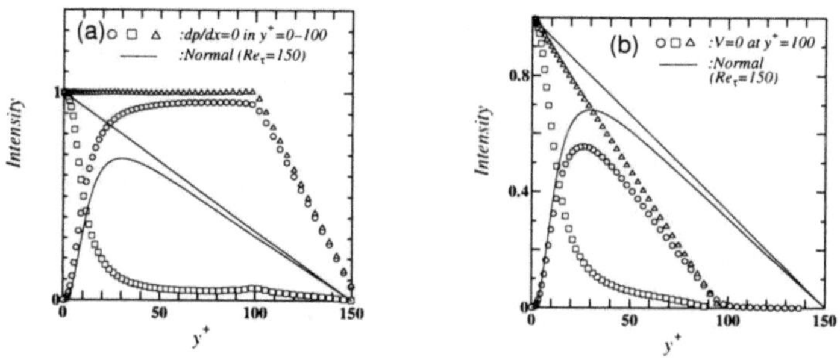

Figure 4. The distribution of Reynolds stress in case of $d\bar{p}/dx = 0$ in $y^+ = 0 \sim 100$

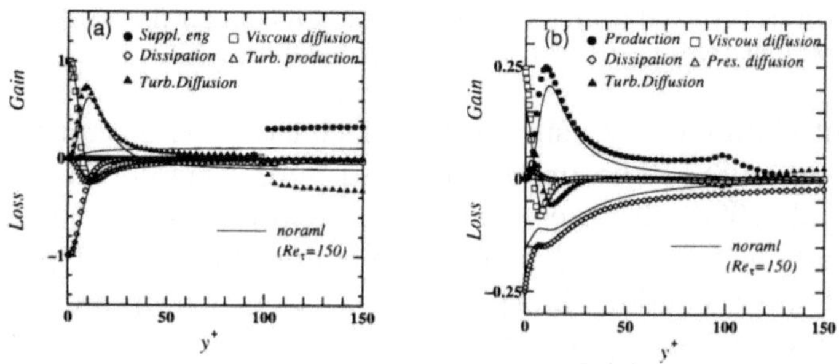

Figure 5. The balance of (a) mean flow energy and (b)turbulent energy in case of $dp/dx = 0$ in $y^+ = 0 \sim 100$

work done by pressure gradient only in the layer $y^+ > 100$. In Fig.4, □, ○, △ mean viscous, turbulent and total shear stress, respectively. In the near-wall layer, constant-stress layer is established by the mechanism similar to that of Couette flow and turbulent regime is maintained in the whole region. The energy balance of both mean and fluctuating part of the flow there is also nearly identical with universal one appearing in normal channel flow, as shown in fig.5. These property is also reproduced in the case when pressure gradient is allowed only in $y^+ > 50$. In the meantime, in the inverse case where work done by the pressure gradient is allowed only in inner side, the turbulent regime is maintained only in $y^+ < 100$, as shown in Fig.4(b). However, in the case of $y^+ < 50$, the flow field is perfectly laminarized. So, it is shown that the energy supply from the upper layer is indispensable to make the layer thinner than $y^+ < 50$ form the wall turbulence. In this sense, the layer is not autonomous, though when turbulent regime is established, its property becomes universal.

3.3. THE SIMULATION OF THE GROWTH OF A VORTEX PAIR

The growth of a single vortex pair is simulated to examine the energy transfer of the layer closest to the wall. For this purpose, a small vortex pair is introduced into a laminar flow field having the mean velocity of the turbulent channel flow. The initial vortex pair is extracted from the flow field

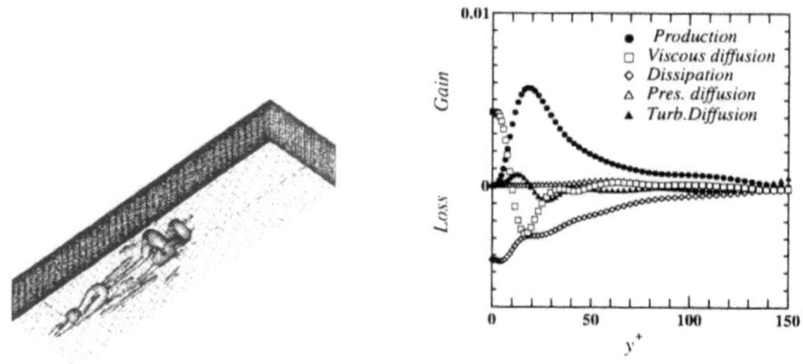

(a) (b)

Figure 6. Vortices evolved from a hairpin vortex in a laminar flow, (a) second invariance of velocity gradient tensor (b)turbulence energy balance.

of turbulent channel flow using linear stochastic estimation (LSE) method (Adrian *et al.* 1988). The procedure used here is based on the method by Zhou it et al.(1996). Figures 6 show the contour of second invariance of velocity gradient tensor and turbulence energy balance . Up to this stage, the head of this secondary hairpin vortex has grown up and the number of the secondary quasi-streamwise vortices has increased as reported by Zhou *et al.*(1996). In Fig.6(b), the pattern of energy balance is almost similar to that of fully-developed turbulent channel flow. Therefore, the group of the quasi-streamwise vortices which are existing near the initial hairpin vortex, generated from it, constitute basic structure for the transport of turbulence energy in near-wall turbulence.

4. Conclusions

A layer closest to the wall can be autonomously turbulent even if no momentum interaction with outside is allowed, if the layer is thicker than around 50~ 60 in wall unit. This turbulent layer has universal property. However, this layer has to receive energy from outside in order to sustain itself as turbulent regime. The energy balance of the universal layer is governed by coherent vortices of group of quasi-streamwise vortices.

References

Adrian,R.J. and Moin, P. (1988) Stochastic estimation of coherent structure : homogeneous shear flows *J. Fluid Mech.*, **190**, 531.

Goldstein, D., Handler, R. and Sirovich, L. (1993) Modeling a no-slip flow boundary with an external force field *J. Fluid Mech.*, **105**, 354.

Jiménez,J. and Pinelli, A. (1999) The autonomous cycle of near-wall turbulence. *J.Fluid Mech.***389**, 335.

Miyake, Y. and Tsujimoto, K. (1995) Direct numerical simulation of turbulent channel flow with periodic pressure gradient. *Trans JSME, ser. B* **61**-587, 2401. (*in Japanese*)

Zhou, J., Adrian, R. J. and Balachandar, S. (1996) Autogeneration of near-wall vortical structures in channel flow. Phys. Fluids A 8, 288.

LIST OF PARTICIPANTS

ANGERE Kristian
KTH Mekanik
Institutionen for Mekanik
Stockholm 10044
SWEDEN
kristian@mech.kth.se

ANTONIA Robert
Dept of Mechanical Engineering
University of Newcastle
University Drive Callaghan
NSW 2308 AUSTRALIA
meraa@cc.newcastle.edu.au

ATOBE Takashi
National Aerospace Laboratory
7-44-2 Jindaiji-Higashimachi Chofu
Tokyo 182-8522
JAPAN
atobe@nal.go.jp

CHAVANIS Pierre-Henri
Laboratoire de Physique
Quantique Irsamc
Universite Paul Sabatier
118, route de Narbonne
Toulouse 31062 FRANCE
chavanis@irsamc2.ups-tlse.fr

CHERTKOV Michael
Los Alamos National Laboratory
Los Alamos
NM 87545
USA
chertkov@princeton.edu

CONSTANTIN Peter
Department of Mathematics
The University of Chicago
5734 University Avenue
Il 60637 USA
const@cs.uchicago.edu

DOERING Charles
Department of Mathematics
University of Michigan
Ann Arbor
Michigan 48109-1109 USA
doering@umich.edu

FRISCH Hélène
Observatoire de la Cote d'Azur
BP 4229
06304 Nice Cedex 4
FRANCE
helene@obs-nice.fr

FRISCH Uriel
Observatoire de la Cote d'Azur
BP 4229
06304 Nice Cedex 4
FRANCE
uriel@obs-nice.fr

FUKAYAMA Daigen
Department of Physics
Chuo University
1-13-27 Kasuga Bunkyo-ku
Tokyo 112-8551 JAPAN
daigen@phys.chuo-u.ac.jp

FUKUMOTO Yasuhide
Graduate School of Mathematics
Kyushu University
33, Fukuoka 812-8581
JAPAN
yasuhide@math.kyushu-u.ac.jp

FUKUNISHI Yu
Dept of Mach Intell & Sys Engng
Tohoku University
Aramaki-Aoba 01 Aoba-ku
Sendai 980-8579 JAPAN
fushi@fluid.mech.tohoku.ac.jp

FUKUZATO Katsuhiko
SEIKA Corporation
1-5-3 Koraku, Bunkyo-ku
Tokyo 112-0004
JAPAN
kfukuzato@jp.seika.com

GOTO Susumu
Theory and Comp Simul Center
National Inst for Fusion Science
Oroshi-cho 322-6
Toki 509-5292 JAPAN
goto@toki.theory.nifs.ac.jp

GOTOH Toshiyuki
Department of Systems Engng
Nagoya Institute of Technology
Showa-ku
Nagoya 466-8555 JAPAN
gotoh@system.nitech.ac.jp

HAMBA Fujihiro
Institute of Industrial Science
University of Tokyo
7-22-1 Roppongi Minato-ku
Tokyo 106-8558 JAPAN
hamba@iis.u-tokyo.ac.jp

HANAZAKI Hideshi
Institute of Fluid Science
Tohoku University
2-1-1 Katahira Aoba-Ku
Sendai 980-8577 JAPAN
hanazaki@ifs.tohoku.ac.jp

HASEGAWA Tatsuya
Dept Envir Tech & Urb Planning
Nagoya Institute of Technology
Gokiso-cho Showa-ku
Nagoya 466-8555 JAPAN
hasegawa@mech.nitech.ac.jp

HATAKEYAMA Nozomu
Institute of Fluid Sceience
Tohoku University
2-1-1 Katahira Aoba-ku
Sendai 980-8577 JAPAN
hatake@miro.ifs.tohoku.ac.jp

HATTORI Yuji
Kyushu Institute of Technology
1-1 Sensui-cho Tobata-ku
Kitakyushu 804-8550
JAPAN
hattori@keith.mns.kyutech.ac.jp

HINO Mikio
4-20-6 Kataseyama
Fujisawa,
Kanagawa 251-0033
JAPAN
fax: +81-4-6626-6272

HORIUTI Kiyosi
Dept. of Mechano-Aerospace Engng.
Tokyo Institute of Technology
2-12-1 O-okayama Meguro-ku
Tokyo 152-8552 JAPAN
khoriuti@mes.titech.ac.jp

HUSSAIN Fazle
Dept of Mechanical Engineering
University of Houston
4800 Calhoun Road, Houston
Texas 77204-4792 USA
fhussain@uh.edu

IMAI Isao
Member of The Japan Academy
1-11-5 Suido Bunkyo-ku
Tokyo 112-0005
JAPAN
fax: +81-3-3816-0513

ISHIHARA Takashi
Dept Comput Science and Engng
Nagoya University
Furo-cho Chikusa-ku
Nagoya 464-8603 JAPAN
ishihara@cse.nagoya-u.ac.jp

ISHII Katsuya
Dept of Comput. Science and Engng.
Nagoya University
Furo-cho Chikusa-ku
Nagoya 464-8603 JAPAN
ishii@cse.nagoya-u.ac.jp

ISHIZAWA Akihiro
Dept. of Fusion Plasma Research
Naka Fusion Res. Establ. JAERI
801-1 Mukouyama Naka-machi
Naka-gun, Ibaraki 311-0193, JAPAN
ishizawa@supra.naka.jaeri.go.jp

ITANO Tomoaki
Department of Physics
Kyoto University
Kitashirakawa Oiwakecho Sakyo-ku
Kyoto 606-8502 JAPAN
itano@kyoryu.scphys.kyoto-u.ac.jp

IWASAKI Takao
Zexel Corporation
3-13-26 Yakyucho
Higashi-Matsuyama
Tokyo 355-8603, JAPAN
iwasaki448@zexel.co.jp

JIMÉNEZ Javier
School of Aeronautics
Universidad Politecnica
Pl. Cardenal Cisneros 3
Madrid 28040 SPAIN
jimenez@torroja.dmt.upm.es

KADOTA Akihiro
Dept Civil and Envir Engng
Ehime University
3 Bunkyo-cho Matsuyama
Ehime 790-8577 JAPAN
kadota@coe.ehime-u.ac.jp

KAJISHIMA Takeo
Dept of Mechanical Engineering
Osaka University
2-1 Yamadaoka Suita
Osaka 565-0871 JAPAN
kajisima@mech.eng.osaka-u.ac.jp,

KAMBE Tsutomu
Departmemt of Physics
University of Tokyo
7-3-1 Hongo Bunkyo-ku
Tokyo 113-0033 JAPAN
kambe@phys.s.u-tokyo.ac.jp

KANEDA Yukio
School of Engineering
Nagoya University
Furo-cho Chikusa-ku
Nagoya 464-8603 JAPAN
kaneda@fluid.nuap.nagoya-u.ac.jp

KATO Sei
Grad School of Math Sciences
University of Tokyo
3-8-1 Komaba Meguro-ku
Tokyo 153-8914 JAPAN
kato@gfd01.ms.u-tokyo.ac.jp

KAWAHARA Genta
Dept of Mechanical Engineering
Ehime University
3 Bunkyo-cho Matsuyama
Ehime 790-8577 JAPAN
kawahara@dpc.ehime-u.ac.jp

KAWAMURA Tetsuya
Grad. School of Humanities and Sci.
Ochanomizu Univ.,
2-1-1 Ootsuka Bunkyo-ku
Tokyo 112, Japan
kawamura@is.ocha.ac.jp

KIDA Shigeo
Theory and Comp. Simul. Center
National Inst. for Fusion Science
Oroshi-cho 322-6
Toki 509-5292 JAPAN
kida@toki.theory.nifs.ac.jp

KITANO Yasuhiro
Grad School of Math Sciences
University of Tokyo
3-8-1 Komaba Meguro-ku
Tokyo 153-8914 JAPAN
kitano@ms.u-tokyo.ac.jp

KUROKAWA Michihiro
Sanyo Electric Co., Ltd.
1-18-13 Hashiridahi Hirakata
Osaka 573-8534
JAPAN
kurokawa@mech.rd.sanyo.co.jp

LAVAL Jean-Philippe
CEA Saclay, DAPNIA/SAp,
L'Orme des Merisiers,
Gif sur Yvette 91191
FRANCE
laval@discovery.saclay.cea.fr

MAEKAWA Hiroshi
Dept Mech Engng & Intel Sys
Univ of Electro-Communication
1-5-1 Chofugaoka Chofu
Tokyo 182-8585 JAPAN
ctr@maekawa.mce.uec.ac.jp

MATSUMOTO Takeshi
Department of Physics
Kyoto University
Kitashirakawa Oiwakecho Sakyo-ku
Kyoto 606-8502 JAPAN
takeshi@kyoryu.scphys.kyoto-u.ac.jp

MISIOLEK Gerard
Department of Mathematics
University of Notre Dame
Indiana 46556
USA
misiolek.1@nd.edu

MIYAKE Yutaka
Dept of Mechanical Engineering
Osaka University
2-1 Yamada-oka Suita
Osaka 565-0871 JAPAN
miyake@mech.eng.osaka-u.ac.jp

MIYAUCHI Toshio
Tokyo Institute of Technology
2-12-1 Ohokayama Meguro-ku
Tokyo 152-8552
JAPAN
tmiyauch@mes.titech.ac.jp

MIYAZAKI Takeshi
Dept Mech Engng & Intel Sys
Univ of Electro-Communication
1-5-1 Chofugaoka, Chofu
Tokyo 182-8585 JAPAN
miyazaki@mce.uec.ac.jp

MOFFATT Keith
Isaac Newton Inst for Math Sci
University of Cambridge
20 Clarkson Road
Cambridge CB3 0EH UK
hkm2@newton.cam.ac.uk

NAGAYA Kougaku
HPC Marketing Department
Supercomp Marketing Promot Div
NEC Corporation
7-1 Shiba 5-chome Minato-ku
Tokyo 108-8001 JAPAN
nagaya@sxsmd.ho.nec.co.jp

NAKANO Tohru
Department of Physics
Chuo University
1-13-27 Kasuga Bunkyo-ku
Tokyo 112-8551 JAPAN
nakano@phys.chuo-u.ac.jp

NOMURA Keiko K.
Dept Mech and Aerospace Engng
Univ of California San Diego
9500 Gilman Dr., La Jolla
CA 92093-0411 USA
knomura@ucsd.edu

NOULLEZ Alain
CNRS Obs de la Cote d'Azur
Blvd. de l'Observatoire
B.P. 4229
06304 Nice Cedex 4, FRANCE
anz@obs-nice.fr

OHASHI Hideo
President, Kogakuin Univ
Nishi-Shinjyuku 1-24-2
Shinjyuku-ku, Tokyo 163-8677
JAPAN
ohashi@cc.kogakuin.ac.jp

OHKITANI Koji
Institute for Mathematical Sciences
Kyoto University
Kitashirakawa Oiwake-cho Sakyo-ku
Kyoto 606-8502 JAPAN
ohkitani@kurims.kyoto-u.ac.jp

OKADA Yoshihiro
Dept of Mechano-Infomatics & Sys
Nagoya University
Furo-cho Chikusa-ku
Nagoya 464-8603 JAPAN
okaday@sps.mech.nagoya-u.ac.jp

ONITSUKA Kouki
Dept of Global Envir Engng
Kyoto University
Yoshida-Honcho Sakyo-ku
Kyoto 606-8501 JAPAN
onitsuka@nezu.gee.kyoto-u.ac.jp

ORLANDI Paolo
Dipt. di Meccanica e Aeronautica
Universita' di Roma "La Sapienza"
Via Eudossiana 16
Roma 184 ITALY
orlandi@navier.ing.uniroma1.it

PASMANTER Ruben A.
KNMI
P.O.Box 201
De Bilt 3730 AE
THE NETHERLANDS
pasmante@knmi.nl

PEINKE Joachim
University of Oldenburg
PB 8-Physics, D 26111
Oldenburg 26111
GERMANY
peinke@uni-oldenburg.de

PIVA Renzo
Dept of Mechanics and Aeronautics
University of Rome "La Sapienza"
Via Eudossiana, 18 Rome I-00184
ITALY
piva@dma.ing.uniroma1.it

PROCACCIA Itamar
Department of Chemical Physics
Weizmann Institute of Science
Rehovot 76100
ISRAEL
Itamar.Procaccia@weizmann.ac.il

PULCI DORIA Guelfo
Univ di Napoli "Federico II"
Dipt di Ingegn Idraulica ed Ambient
Via Claudio 21 - 80125 - Napoli
Italy
pulci@unina.it

RUBINSTEIN Robert
ICASE
NASA Langley Res Center
MS 132C 3 West Reid Street
VA 23681 USA
bobr@icase.edu

SAKAI Yasuhiko
Dept of Mechano-Informatics & Sys.
Nagoya University
Furo-cho Chikusa-ku
Nagoya 464-8603 JAPAN
ysakai@everest.mech.nagoya-u.ac.jp

SANO Masaki
Departmemt of Physics
University of Tokyo
7-3-1 Hongo Bunkyo-ku
Tokyo 113-0033 JAPAN
sano@phys.s.u-tokyo.ac.jp

SANO Osamu
Tokyo Univ of Agri and Techn
3-5-8 Saiwai-cho Fuchu
Tokyo 183-0054
JAPAN
sano@cc.tuat.ac.jp

SCHORGHOFER Norbert
Department of Physics
Chinese Univ of Hong Kong, Shatin
New Territories
HONG KONG
norbert@sun1.phy.cuhk.edu.hk

SHRAIMAN Boris
Bell Laboratories
1-D236 700 Mountain Ave
Murray Hill
NJ 7974 USA
boris@physics.bell-labs.com

SREENIVASAN K. R.
Mason Laboratory
Yale University
P.O. Box 208286 New Haven
CT 6520 USA
krs@kolmogorov.ENG.YALE.EDU

SUZUKI Katsuhiro
Tokyo Univ of Agri and Techn
3-5-8 Saiwai-cho Fuchu
Tokyo 183-0054
JAPAN
suz@cc.tuat.ac.jp

SUZUKI Takeshi
Mechanical Engineering Laboratory
1-2 Namiki Tsukuba
Ibaraki 305-8564
JAPAN
tsuzuki@mel.go.jp

SUZUKI Yukihito
Fuji Research Institute Corporation
2-3 Takebashi Square
Kandanishiki-cho Chiyoda-ku
Tokyo 101-8443 JAPAN
y_suzuki@star.fuji-ric.co.jp

TAKAHASHI Naoya
Dept Mech Engng & Intell Sys
Univ of Electro-Communication
1-5-1 Chofugaoka, Chofu
Tokyo 182-8585 JAPAN
naoya@miyazaki.mce.uec.ac.jp

TAKAOKA Masanori
Div of Physics and Astronomy
Graduate School of Science
Kyoto University
Sakyo-ku, Kyoto 606-8502 JAPAN
takaoka@kyoryu.scphys.kyoto-u.ac.jp

TANAHASHI Mamoru
Dept of Mechano-Aerospace Engng
Tokyo Institute of Technology
2-12-1 Ookayama Meguro-ku
Tokyo 152-8552 JAPAN
mtanahas@mes.titech.ac.jp

TATSUMI Tomomasa
Professor Emeritus, Kyoto University
26-6 Chikuzendai-cho Fushimi-ku
Kyoto 612-8032
JAPAN
tatsumi@skyblue.ocn.ne.jp

TOH Sadayoshi
Department of Physics
Kyoto University
Kitashirakawa Oiwakecho Sakyo-ku
Kyoto 606-8502 JAPAN
toh@kyoryu.scphys.kyoto-u.ac.jp

TOKUGAWA Naoko
National Aerospace Laboratory
7-44-1 Jindaiji-Higashimachi Chofu
Tokyo 182-8522 JAPAN
nao@nal.go.jp

TSUJI Yoshiyuki
Dept of Energy Engng and Science
Nagoya University
Furo-cho
Nagoya 464-8603 JAPAN
c42406a@nucc.cc.nagoya-u.ac.jp

TSUJIMOTO Kouichi
Dept of Mechanical Engineering
Osaka University
2-1 Yamada-oka Suita
Osaka 565-0871 JAPAN
tsujimoto@mech.eng.osaka-u.ac.jp

TSUJIMURA Shinji
Dept of Mechanical Engineering
Nagoya Institute of Technology
Gokiso-cho Showa-ku
Nagoya 466-8555 JAPAN
tujimura@heat.mech.nitech.ac.jp

UMEKI Makoto
Department of Physics
University of Tokyo
7-3-1 Hongo Bunkyo-ku
Tokyo 113-0033 JAPAN
umeki@phys.s.u-tokyo.ac.jp

WELLS John Craig
Dept of Civil Engineering
Ritsumeikan University
1-1-1 Noji-Higashi Kusatsu
Shiga 525-8577 JAPAN
jwells@se.ritsumei.ac.jp

YAKHOT Victor
Dept of Aerospace and Mech Engng
Boston University
MA 02215, USA
vy@bu.edu

YAMADA Michio
Grad School of Math Sciences
University of Tokyo
3-8-1 Komaba Meguro-ku
Tokyo 153-8914 JAPAN
yamada@ms.u-tokyo.ac.jp

YAMAMOTO Kiyoshi
National Aerospace Laboratory
7-44-1 Jindaijihigashi-Machi
Chofu, Tokyo 182-8522, JAPAN
yamamo@nal.go.jp

YOSHIDA Kyo
Dept Comput Science and Engng
Nagoya University
Furou-cho Chikusa-ku
Nagoya 464-8603 JAPAN
kyo@cse.nagoya-u.ac.jp

YOSHIZAWA Akira
Institute of Industrial Science
University of Tokyo
7-22-1 Roppongi Minato-ku
Tokyo 106-8558 JAPAN
yosizawa@iis.u-tokyo.ac.jp

Mechanics

FLUID MECHANICS AND ITS APPLICATIONS

Series Editor: R. Moreau

24. R. Benzi (ed.): *Advances in Turbulence V.* 1995 ISBN 0-7923-3032-3
25. B.I. Rabinovich, V.G. Lebedev and A.I. Mytarev: *Vortex Processes and Solid Body Dynamics.* The Dynamic Problems of Spacecrafts and Magnetic Levitation Systems. 1994 ISBN 0-7923-3092-7
26. P.R. Voke, L. Kleiser and J.-P. Chollet (eds.): *Direct and Large-Eddy Simulation I.* Selected papers from the First ERCOFTAC Workshop on Direct and Large-Eddy Simulation. 1994 ISBN 0-7923-3106-0
27. J.A. Sparenberg: *Hydrodynamic Propulsion and its Optimization.* Analytic Theory. 1995 ISBN 0-7923-3201-6
28. J.F. Dijksman and G.D.C. Kuiken (eds.): *IUTAM Symposium on Numerical Simulation of Non-Isothermal Flow of Viscoelastic Liquids.* Proceedings of an IUTAM Symposium held in Kerkrade, The Netherlands. 1995 ISBN 0-7923-3262-8
29. B.M. Boubnov and G.S. Golitsyn: *Convection in Rotating Fluids.* 1995 ISBN 0-7923-3371-3
30. S.I. Green (ed.): *Fluid Vortices.* 1995 ISBN 0-7923-3376-4
31. S. Morioka and L. van Wijngaarden (eds.): *IUTAM Symposium on Waves in Liquid/Gas and Liquid/Vapour Two-Phase Systems.* 1995 ISBN 0-7923-3424-8
32. A. Gyr and H.-W. Bewersdorff: *Drag Reduction of Turbulent Flows by Additives.* 1995 ISBN 0-7923-3485-X
33. Y.P. Golovachov: *Numerical Simulation of Viscous Shock Layer Flows.* 1995 ISBN 0-7923-3626-7
34. J. Grue, B. Gjevik and J.E. Weber (eds.): *Waves and Nonlinear Processes in Hydrodynamics.* 1996 ISBN 0-7923-4031-0
35. P.W. Duck and P. Hall (eds.): *IUTAM Symposium on Nonlinear Instability and Transition in Three-Dimensional Boundary Layers.* 1996 ISBN 0-7923-4079-5
36. S. Gavrilakis, L. Machiels and P.A. Monkewitz (eds.): *Advances in Turbulence VI.* Proceedings of the 6th European Turbulence Conference. 1996 ISBN 0-7923-4132-5
37. K. Gersten (ed.): *IUTAM Symposium on Asymptotic Methods for Turbulent Shear Flows at High Reynolds Numbers.* Proceedings of the IUTAM Symposium held in Bochum, Germany. 1996 ISBN 0-7923-4138-4
38. J. Verhás: *Thermodynamics and Rheology.* 1997 ISBN 0-7923-4251-8
39. M. Champion and B. Deshaies (eds.): *IUTAM Symposium on Combustion in Supersonic Flows.* Proceedings of the IUTAM Symposium held in Poitiers, France. 1997 ISBN 0-7923-4313-1
40. M. Lesieur: *Turbulence in Fluids.* Third Revised and Enlarged Edition. 1997 ISBN 0-7923-4415-4; Pb: 0-7923-4416-2
41. L. Fulachier, J.L. Lumley and F. Anselmet (eds.): *IUTAM Symposium on Variable Density Low-Speed Turbulent Flows.* Proceedings of the IUTAM Symposium held in Marseille, France. 1997 ISBN 0-7923-4602-5
42. B.K. Shivamoggi: *Nonlinear Dynamics and Chaotic Phenomena.* An Introduction. 1997 ISBN 0-7923-4772-2
43. H. Ramkissoon, *IUTAM Symposium on Lubricated Transport of Viscous Materials.* Proceedings of the IUTAM Symposium held in Tobago, West Indies. 1998 ISBN 0-7923-4897-4
44. E. Krause and K. Gersten, *IUTAM Symposium on Dynamics of Slender Vortices.* Proceedings of the IUTAM Symposium held in Aachen, Germany. 1998 ISBN 0-7923-5041-3
45. A. Biesheuvel and G.J.F. van Heyst (eds.): *In Fascination of Fluid Dynamics.* A Symposium in honour of Leen van Wijngaarden. 1998 ISBN 0-7923-5078-2

Mechanics

FLUID MECHANICS AND ITS APPLICATIONS

Series Editor: R. Moreau

46. U. Frisch (ed.): *Advances in Turbulence VII*. Proceedings of the Seventh European Turbulence Conference, held in Saint-Jean Cap Ferrat, 30 June–3 July 1998. 1998 ISBN 0-7923-5115-0
47. E.F. Toro and J.F. Clarke: *Numerical Methods for Wave Propagation*. Selected Contributions from the Workshop held in Manchester, UK. 1998 ISBN 0-7923-5125-8
48. A. Yoshizawa: *Hydrodynamic and Magnetohydrodynamic Turbulent Flows*. Modelling and Statistical Theory. 1998 ISBN 0-7923-5225-4
49. T.L. Geers (ed.): *IUTAM Symposium on Computational Methods for Unbounded Domains*. 1998 ISBN 0-7923-5266-1
50. Z. Zapryanov and S. Tabakova: *Dynamics of Bubbles, Drops and Rigid Particles*. 1999 ISBN 0-7923-5347-1
51. A. Alemany, Ph. Marty and J.P. Thibault (eds.): *Transfer Phenomena in Magnetohydrodynamic and Electroconducting Flows*. 1999 ISBN 0-7923-5532-6
52. J.N. Sørensen, E.J. Hopfinger and N. Aubry (eds.): *IUTAM Symposium on Simulation and Identification of Organized Structures in Flows*. 1999 ISBN 0-7923-5603-9
53. G.E.A. Meier and P.R. Viswanath (eds.): *IUTAM Symposium on Mechanics of Passive and Active Flow Control*. 1999 ISBN 0-7923-5928-3
54. D. Knight and L. Sakell (eds.): *Recent Advances in DNS and LES*. 1999 ISBN 0-7923-6004-4
55. P. Orlandi: *Fluid Flow Phenomena*. A Numerical Toolkit. 2000 ISBN 0-7923-6095-8
56. M. Stanislas, J. Kompenhans and J. Westerveel (eds.): *Particle Image Velocimetry*. Progress towards Industrial Application. 2000 ISBN 0-7923-6160-1
57. H.-C. Chang (ed.): *IUTAM Symposium on Nonlinear Waves in Multi-Phase Flow*. 2000 ISBN 0-7923-6454-6
58. R.M. Kerr and Y. Kimura (eds.): *IUTAM Symposium on Developments in Geophysical Turbulence* held at the National Center for Atmospheric Research, (Boulder, CO, June 16–19, 1998) 2000 ISBN 0-7923-6673-5
59. T. Kambe, T. Nakano and T. Miyauchi (eds.): *IUTAM Symposium on Geometry and Statistics of Turbulence* held at the Shonan International Village Center, Hayama (Kanagawa-ken, Japan November 2–5, 1999). 2001 ISBN 0-7923-6711-1
60. V.V. Aristov: *Direct Methods for Solving the Boltzmann Equation and Study of Nonequilibrium Flows*. 2001 ISBN 0-7923-6831-2

Kluwer Academic Publishers – Dordrecht / Boston / London

MIX
Papier aus verantwortungsvollen Quellen
Paper from responsible sources
FSC® C105338

If you have any concerns about our products,
you can contact us on
ProductSafety@springernature.com

In case Publisher is established outside the EU,
the EU authorized representative is:
Springer Nature Customer Service Center GmbH
Europaplatz 3, 69115 Heidelberg, Germany

Printed by Libri Plureos GmbH
in Hamburg, Germany